T0319272

The Agronomy and Economy of Turmeric and Ginger

The Agronomy and Economy of Turmeric and Ginger
The Invaluable Medicinal Spice Crops

K.P. Prabhakaran Nair

Director, Rallis India Ltd (A Tata Group), India

ELSEVIER

AMSTERDAM • BOSTON • HEIDELBERG • LONDON • NEW YORK • OXFORD
PARIS • SAN DIEGO • SAN FRANCISCO • SINGAPORE • SYDNEY • TOKYO

Elsevier
32 Jamestown Road London NW1 7BY
225 Wyman Street, Waltham, MA 02451, USA

Notices
Knowledge and best practice in this field are constantly changing. As new research and
experience broaden our understanding, changes in research methods, professional practices, or
medical treatment may become necessary.

Practitioners and researchers must always rely on their own experience and knowledge in
evaluating and using any information, methods, compounds, or experiments described herein.
In using such information or methods they should be mindful of their own safety and the
safety of others, including parties for whom they have a professional responsibility.

To the fullest extent of the law, neither the Publisher nor the authors, contributors, or editors,
assume any liability for any injury and/or damage to persons or property as a matter of
products liability, negligence or otherwise, or from any use or operation of any methods,
products, instructions, or ideas contained in the material herein.

British Library Cataloguing-in-Publication Data
A catalogue record for this book is available from the British Library

Library of Congress Cataloging-in-Publication Data
A catalog record for this book is available from the Library of Congress

ISBN: 978-0-12-394801-4

For information on all Elsevier publications
visit our website at store.elsevier.com

This book has been manufactured using Print On Demand technology. Each copy is
produced to order and is limited to black ink. The online version of this book will show color
figures where appropriate.

This book, written under very trying circumstances, is dedicated to my wife, Pankajam, a Nematologist trained in Europe, but one who gave up her profession, and instead, chose to be a homemaker almost four decades ago, when we had our son and daughter. She is my all, and she sustains me in this difficult journey, that is, life.

Contents

Acknowledgments

I find it a very great pleasure to place on record, first, my sincere gratitude to Ms. Lisa Tickner, Director, Continuity Publishing, Elsevier, who invited me to compile this book, and to Dr. Erin Hill-Parks and Ms. Tracey Miller, in Oxford, both of whom worked with incessant enthusiasm and promptness to help me at every stage of the book production. It was, indeed, a very great joy to work with them.

1 Turmeric: Origin and History

When one leafs through ancient scripture, primarily Indian, the most important plant that one comes across is turmeric. Turmeric, also known as "Indian saffron" has been in use dating back to 4000 BC. It is mentioned in *Ayurveda*, the age-old Indian system of medicine, and one encounters its name and use recorded in Sanskrit, the ancient Indian language describing the ageless *Vedas* (ancient Indian scriptures), between 1700 and 800 BC during the period known as the *Vedic* age. In fact, the use of turmeric spans many purposes, as a dye, condiment, and medicine. In *Sanskrit*, it is referred to as "*Haridara*," a word which has two parts: "Hari" and "Dara," meaning Vishnu, also known as "Hari," the omnipotent and omnipresent Hindu deity; and "Dara" meaning what one wears, obviously referring to the fact that Vishnu used it on his body. In India, it is put to several uses, as a coloring material, flavoring agent with digestive properties, and in fact, no Indian preparation (vegetarian or nonvegetarian) is complete without turmeric as an ingredient. The bright yellow color of the now famous Indian *curry* is due to turmeric. Turmeric is much revered by the Hindus, and interestingly, is given as "*Prasad*" (a benedictory material) in powdered form, in some temples. Obviously, whoever originated this idea had two purposes in mind—to bless the recipient and to give him or her material that has great medicinal value. *Characa* and *Susruta*, the great ancient Indian physicians who systematized the *Ayurvedic* system of medicine, have cataloged the various uses of turmeric (Anon, 1950; Nadkarni, 1976). Also the Greek physician Dioscorides, in the Roman Army (AD 40–90) makes a mention of turmeric. In Malaysia, a paste of turmeric is spread on the mother's abdomen and on the umbilical cord after childbirth in the belief that it would ward off evil spirits, and also would provide some medicinal value, primarily antiseptic. Both the East and the West have held turmeric in high esteem for its medicinal properties. The Indus Valley Civilization dating back to 3300 BC in western India was involved in the spice trade, of which turmeric was an important constituent. The Greco-Roman, Egyptian, and Middle East regions were all familiar with turmeric (Raghavan, 2007). The crushed and powdered rhizome of turmeric was used extensively in Asian cookery, medicines, cosmetics, and fabric dying for more than 20,000 years (Ammon and Wahl, 1991). Early European explorers to the Asian continent introduced turmeric to the Western world in the fourteenth century (Aggarwal et al., 2007). About 40 species of the genus *Curcuma* are indigenous to India, which point to its Indian origin (Velayudhan et al., 1999). Apart from *Curcuma longa*, several species of economic importance are available, such as *Curcuma aromatica* Salisb., *Curcuma amada* Roxb., *Curcuma caesia* Roxb., *Curcuma aeruginosa* Roxb., and *Curcuma zanthorrizha* Roxb. About 70–110 species of the genus have been reported throughout tropical Asia. The species in India, Myanmar, and Thailand show the greatest diversity. Some species are seen as far away as China, Australia, and the South Pacific, while some other popular species are cultivated all over the tropics.

The Agronomy and Economy of Turmeric and Ginger. DOI: http://dx.doi.org/10.1016/B978-0-12-394801-4.00001-6

Turmeric, originating from India, reached the coast of China in AD 700 and reached East Africa 100 years later and West Africa 500 years later. Arab traders were instrumental in spreading the plant to the European continent in the thirteenth century. There is a parallel here between black pepper and turmeric. The first explorers who went out in search of both spices were Arabs, and in fact the sea route was a secret until the Europeans came on to the scene, as exemplified by the landing of Vasco da Gama in coastal Malabar in Kappad, in Kozhikode district in Kerala State, India. The exact location in India where turmeric originated is still in dispute, but all the available details point to its origin in western and southern India. Turmeric has been in use in India for more than 5000 years now. Marco Polo described it in AD 1280 in his travel memoirs about China. During his several legendary voyages to India via the "Silk Route," Marco Polo was so impressed by turmeric that he had mentioned it as a vegetable which possesses properties akin to saffron but is not actually saffron (Parry, 1969). Probably that is also the reason why it was then known as "Indian saffron."

Turmeric derives its name from the Latin word "*terra merita*," meaning meritorious earth, which refers to the color of ground turmeric, resembling a mineral pigment. The botanical name is *Curcuma domestica* Val. Syn.. *Curcuma longa* L. belongs to the family Zingiberaceae. The Latin name for turmeric is *Curcuma longa*, which has its origin in the Arabic name "*Kurkum*," for this plant (Willamson, 2002). In Sanskrit, it is called "*Haridara*" ("The Yellow One"), "*Gauri*" ("The one whose face is light and shining"), "*Kanchani*" ("Golden Goddess"), and "*Aushadi*" ("Herb"). "*Haridara*" also comes from the *Mundas*, a pre-Aryan population, who lived through much of their life in northern India (Frawley and Lad, 1993). The ancient Indian *Vedas* also refer to a set of people called *Nishadas*, literally translated as "Turmeric Eaters." Turmeric has also been used as a dye for mustards, canned chicken broth, and pickles. It has been coded as food additive "E 100" in canned beverages, baked products, dairy, ice cream, yogurts, yellow cakes, biscuits, popcorn, sweets, cake icing, cereal, sauces, gelatin, and also direct compression tablets.

Because of its unique color and history, turmeric has a special place in both Hindu and Buddhist religious ceremonies. Initially, it was cultivated as a dye because of its brilliant yellow color. With the passage of time, ancient populations came to know of its varied uses and they began introducing it into cosmetics. The plant's roots are used in one of the most popular Indian *Ayurvedic* preparations called "*Dashamularishta*," a concoction prepared from 10 different types of roots, which relieve fatigue, and have been in use since thousands of years. The plant's flowers are used as an antidote against worms in the stomach of humans and can also cure jaundice and venereal diseases, and have been known to have specific properties to combat mental disorders. Human breast tumors can be treated with turmeric leaf extracts.

Area and Production

About 80% of world turmeric production is from India. India is the largest producer, consumer, and exporter. The plant grows extensively in the country, but the southern states of Tamil Nadu and Andhra Pradesh, Maharashtra in central, and West Bengal in east India respectively grow it extensively (Spices Board, 2007). Overseas producers are Thailand, China, Taiwan, South America, and the Pacific islands. Major importers are Japan, the United States, the United Kingdom, Sri Lanka, North African countries and Ethiopia in East Africa, and Middle Eastern countries. Iran is the largest importer. China produces about 8%, followed by Myanmar (4%), while Nigeria and Bangladesh combined contribute 6% of world production. Recent statistical estimates indicate Indian production at 856,464 metric tons from a total acreage of 183,917 hectares (Spices Board). In 2006–07, India exported 51,500 metric tons valued at US$35.77 million. In 2007–2008, world export totaled 49,250 metric tons valued at US$33.87 million, and in the following year, the corresponding figures were 52,500 metric tons valued at US$35.77 million. From India's total export, 65% is exported to the United Arab Emirates (UAE), the United States, Japan, Sri Lanka, the United Kingdom, and Malaysia. The institutional sector in the West buys ground turmeric and oleoresins, while dry turmeric is preferred by the industrial sector. Table 1.1 gives the details about the turmeric scenario in India.

In India, turmeric is produced in 230 districts in 22 states (Table 1.1). Andhra Pradesh, Tamil Nadu, Odisha, Karnataka, and West Bengal are the major turmeric-producing states which contribute 90% of the production in the country. Turmeric is available in two seasons in India (February–May and August–October). The different varieties of turmeric traded in India are Alleppey Finger, from the State of Kerala; Erode Turmeric and Salem Turmeric, both from the State of Tamil Nadu; Rajapore Turmeric and Sangli Turmeric from the State of Maharashtra; and Nizamabad Bulb from the State of Andhra Pradesh. The major turmeric trading centers in India are Nizamabad and Dugirala in Andhra Pradesh; Sangli in Maharashtra; and Salem, Erode, Dharmapuri, and Coimbatore in Tamil Nadu.

Global Turmeric Scenario

The global turmeric production is around 1,100,000 tons per annum. India's position in global turmeric trade is formidable, with a total of 48% in volume and 44% in value. Table 1.2 gives a country-wise breakdown.

India is the global leader in turmeric export and its value-added products. The UAE is the major importer of turmeric from India, and it accounts for about 18% of the total export volume. The UAE is followed by the United States with 8%. The other leading importers are Bangladesh, Pakistan, Sri Lanka, Japan, Egypt, the United Kingdom, Malaysia, South Africa, the Netherlands, and Saudi Arabia. These countries together account for 75% of the total import volume. Asian countries are the main suppliers of turmeric with India leading the pack. The remaining 25% of the total global import volume is met by Europe, North America, and Central and Latin American countries. The United States imports 97% of its turmeric requirement from India and the remaining 3% from the islands of the Pacific and Thailand. Of the total global production, the UAE accounts for 18% of the imports, followed by the United States (11%), Japan (9%), Sri Lanka, the United Kingdom, and Malaysia put together 17%.

Table 1.1 Turmeric Area, Production, and Productivity in Indian States

State	Area (ha)	Production (t)	Productivity (t/ha)
Southern States			
Andhra Pradesh	64,500	400,920	6.22
Tamil Nadu	30,530	175,390	5.74
Karnataka	12,720	82,470	6.48
Kerala	3920	9980	2.54
Central States			
Maharashtra	6798	8508	1.25
Gujarat	1297	16,909	13.03
Rajasthan	140	620	4.43
Northern States			
Himachal Pradesh	187	99	0.53
Madhya Pradesh	670	585	0.87
Jammu and Kashmir	0	0	0
Chhattisgarh	879	743	0.85
Eastern States			
Odisha	24,730	59,360	2.40
West Bengal	13,660	30,070	2.20
Assam	11,740	8540	0.72
Bihar	3038	2981	0.98
Tripura	1150	3380	2.94
Uttar Pradesh	2000	6000	3.00
Meghalaya	1910	14,350	10.59
Nagaland	18	62	3.40
Sikkim	580	1920	3.31
Uttarakhand	630	6068	9.63
Arunachal Pradesh	600	2300	3.83
Manipur	400	280	0.70
Mizoram	1740	24,460	14.17
Union Territory			
Andaman and Nicobar	80	469	5.86
Total	183,917	856,464	4.66

Source: Spices Board, Kerala State, India.

Table 1.2 Export of Turmeric from India Around the World in US$ (million)

Destination	2006–07		2007–08	
	Quantity	Value	Quantity	Value
UAE	7823.8	4.818	5150.6	3.121
Japan	2631.9	2.572	2797.1	2.676
USA	2460.6	2.983	2648.6	2.609

(Continued)

Table 1.2 (Continued)

Destination	2006–07		2007–08	
	Quantity	Value	Quantity	Value
Iran	6094.7	3.151	3708.7	2.032
Malaysia	2263.5	1.647	2895.4	1.969
UK	2896.1	2.313	2460.6	1.852
Egypt (ARE)	2057.0	1.259	2438.8	1.529
Bangladesh	4039.2	2.245	2879.5	1.503
Pakistan	47.6	0.024	2756.2	1.480
Sri Lanka	3725.0	1.496	3453.0	1.382
South Africa	2195.2	1.563	1842.8	1.364
The Netherlands	1816.7	1.528	1700.3	1.325
Morocco	736.3	0.439	1772.0	1.105
Germany	1155.8	1.052	1255.2	1.009
Saudi Arabia	1406.1	1.070	1239.0	0.988
France	627.5	0.497	761.3	0.626
Canada	347.4	0.416	600.3	0.485
Singapore	622.5	0.471	868.1	0.588
Kuwait	320.1	0.281	519.4	0.417
Russia	567.3	0.378	635.3	0.409
Israel	632.7	0.363	621.7	0. 367
Others	7033.1	5.483	6246.3	5.120
Total	51500.0	35.74	49250.1	33.96

Source: Spices, Kerala State, India.

References

Aggarwal, B.B., Sundaram, C., Malani, N., Ichikawa, H., 2007. Curcumin: the Indian solid gold. Adv. Exp. Med. Biol. (595), 1–75.

Ammon, H.P.T, Wahl, M.A., 1991. Pharmacology of *Curcuma longa*. Planta Med. 57, 1–7.

Anon, 1950. Curcuma in "The Wealth of India", Raw Materials. Publications and Information Directorate, CSIR, New Delhi, 401, 11.

Frawley, D., Lad, Vasant, 1993. The Yoga of Herbs. Lotus Light Publications, New Delhi.

Nadkarni, K.M., 1976. In: Nadkarni, A.K. (Ed.), Indian Matera Medica, Popular Prakashan Publications, Bombay.

Parry, J.W., 1969. Spices. Vol I. The Story of Spices and Spices Described. Vol II. Morphology, Histology and Chemistry. Chemical Publishing, New York, NY.

Raghavan, S. (Ed.), 2007. Handbook of Spices, Seasonings and Flavourings. (ISBN O-8493–2842-X).

Spices Board, 2007. <http://www.indianspices.com/>.

Velayudhan, K.C., Muralidharan, V.K., Amalraj, V.A., Gautam, P.L., Mandal, S., Dinesh Kumar, 1999. Curcuma Genetic Resources. Scientific Monograph No. 4. National Bureau of Plant Genetic Resources, New Delhi, p. 149.

Willamson, E., 2002. Major Herbs of Ayurveda. Churchill Livingstone, UK.

2 The Botany of Turmeric

Curcuma longa L. belongs to the family Zingiberaceae which falls under the order Zingiberales of monocots and is an important genus in the family. The family is composed of 47 genera and 1400 species of perennial tropical herbs, found usually in the ground flora of lowland forests. It is a very popular family which includes other important spices, such as cardamom (*Elettaria cardamomum* Maton.), large cardamom (*Amomum subulatum*), and ginger (*Zingiber officinale*).

Origin and Distribution

The exact geographic origin of turmeric is unknown, but it is a safe bet that it could be Southeast Asia (Velayudhan et al., 1999). Watt (1972) reported that there is no conclusive evidence to show that *C. longa* is a native of India, though several species of *Curcuma* are found in India. The greatest diversity of turmeric species is found in India, Myanmar, and Thailand. Table 2.1 gives a geographic distribution worldwide.

Turmeric Taxonomy

Despite systematic investigation by taxonomists, starting from Linnaeus, Hooker, Rendle, Watt, Valeton, and Hutchinson (Hooker, 1894; Hutchinson, 1934; Valeton, 1918), the classification and nomenclature of *Curcuma* remained quite confusing. Hooker (1894) described *Curcuma* under the natural order Scitamineae and tribe Zingibereae. However, Rendle (1904) introduced the subfamily Zingiberoideae under Zingiberaceae and described *Curcuma* under the tribe Hedychieae, which was corroborated by Hutchinson (1934). Holtum's (1950) classification of the Zingiberaceae family is presumed to be the most authoritative to date, wherein he divided the family into two subfamilies, namely Zingiberoideae and Costoideae, and *Curcuma* was included in Zingiberoideae, under the tribe Hedychieae. The description of the *Curcuma* genus (Holtum, 1950) as referred by Ravindran et al., (2007) is presented below.

A fleshy complex rhizome, the base of each aerial stem consisting of an erect, ovoid, or ellipsoid structure (primary tuber), ringed with the bases of old-scale leaves, bearing several horizontal or curved rhizomes, when mature, which are again branched. Fleshy roots, many of them bearing ellipsoid tubers. Leafy shoots bearing a group of leaves surrounded by bladeless sheaths, the leaf sheaths forming a pseudostem; total height of leafy shoots ranging from 1 to 2 m. Leaf blades usually more or less erect, often with a purple-flushed strip on either side of the midrib; size and proportional width varying from the outermost to the innermost (uppermost) leaf. Petioles of outermost leaf short or none, of inner leaves fairly long, channeled. Ligule forms a narrow upgrowth across the

The Agronomy and Economy of Turmeric and Ginger. DOI: http://dx.doi.org/10.1016/B978-0-12-394801-4.00003-X

Table 2.1 *Curcuma* Distribution World-Wide

Country	Approximate Number of Species
Bangladesh	16–20
China	20–25
India	40–45
Cambodia, Vietnam, and Laos	20–25
Malaysia	20–30
Nepal	10–15
The Philippines	12–15
Thailand	30–40
Total	100–110

Source: Ravindran et al. (2007).

base of the petiole; its ends join to form thin edges of the sheath, the ends in most species simply decurrent, rarely raised as prominent auricles. Inflorescence either terminal on the leafy shoot, the scape covered by rather large bladeless sheaths. Bracts are large, very broad, each joined to those adjacent to it for about half of its length, the basal parts thus forming enclosed pockets, the free ends more or less spreading, the whole forming a cylindrical spike; uppermost bracts usually larger than the rest and differently colored; a few of them sterile (the group is called coma). Flowers in cincinni of two to seven, each cincinnus in the axil of a bract. Bracteoles thin, elliptic with the sides inflexed, each one at right angles to the last, quite enclosing the flower buds but not tubular at the base. Calyx short, unequally toothed, and split nearly halfway down one side. Corolla tube and stamina tube tubular at the base, the upper portion half cupped, the corolla lobes inserted on the edges of the cup, and the lip, staminodes, and stamen just above them. Corolla lobes thin, translucent white or pink to purplish, the dorsal one hooded and ending in a hollow hairy point. Staminodes elliptic-oblong, their inner edges folded under the hood of the dorsal petal. Labellum obovate, consisting of a thickened yellow middle band which points straight toward or somewhat reflexed, its tip slightly cleft, and thinner pale (white or pale yellow) side-lobes upcurved and overlapping the staminodes. Filament of stamen short and broad, constricted at the top, anther versatile, the filament joined to its back, the pollen sacs parallel, with usually a curved spur at the base of each; connective sometimes protruded at the apex into a small crest. Stylodes cylindrical, 4–8 mm long. Ovary trilocular; fruit ellipsoid, thin-walled, dehiscing and liberating the seeds in the mucilage of the bracht pouch; seeds ellipsoid with a lacerate aril of few segments which are free to the base. The above description was adapted by many of the later taxonomists and reviewers as a basis for describing or redescribing the genus *Curcuma* (Velayudhan et al., 1999).

Taxonomic Investigations in *Curcuma*

It was in 1753 that the genus *Curcuma* was established by Linnaeus in his *Species Plantarum* (Linnaeus, 1753). This was based on a plant observed by Hermann in what was then known as Ceylon (now Sri Lanka). The generic name might have originated from the Arabic word "*Kurcum*," meaning yellow color, and *Curcuma* is the Latinized version (Islam, 2004; Ravindran et al., 2007).

Curcuma was described early (1678–1693) by Van Rheede (1678) in *Hortus Indicus Malabaricus*. He recorded two species of *Curcuma* under the local names "*Kua*" and "*Manjella Kua*," which were later identified as *C. zedoaria* Rosc. and *C. longa* L., respectively (Burtt, 1977). "*Manjella Kua*" was selected as lectotype of *C. longa* by Burtt (1977).

Baker (1890, 1898) confirmed 27 species of *Curcuma* in British India (The Flora of British India) and subdivided the genus into three sections, namely *Exantha*, *Mesantha*, and *Hitcheniopsis*. The section *Exantha* comprises 14 species, including turmeric and other economically important species, such as *C. augustifolia* Roxb., *C. aromatica* Salisb., and *C. zedoaria* Rose (Velayudhan et al., 1999). Valeton (1918) classified the genus into two subgenera, namely *Paracurcuma* and *Eucurcuma*, based on the presence or absence of the anther spur. He included two species *C. ecalcarata* and *C. aurantiaca* in *Paracurcuma*, as they lack the anther spur or possess a very short spur. *Eucurcuma* was further divided into three sections, namely *tuberosa* (presence of sessile root tubers), *non tuberosa* (absence of sessile root tubers), and *stolonifera* (presence of stoloniferous tubers). He identified *C. longa* as *C. domestica*, which was later accepted as a synonym for *C. longa* L.

Fischer (1928) reported eight species of *Curcuma* from South India, and Kumar (1991) reported the occurrence of five species, namely *C. augustifolia* Roxb., *C. aromatica* Salisb., *C. caesia* Roxb., *C. longa* L., and *C. zedoaria* (Christm.) Rose in different altitudinal zones of Sikkim State in the Himalayas, India.

Taxonomic and polygenic studies on South Indian Zingiberaceae by Sabu (1991) revealed that 12 *Curcuma* species, including *C. coriaceae*, *C. ecalcarata*, *C. haritha*, *C. kudagensis*, *C. neilgherrensis*, *C. raktakanta*, *C. vamana*, and *C. oligantha* var. *lutea*, are endemic to South India. This investigation included the identification of eight new taxa from South India which included four new *Curcuma* species, namely *C. coriaceae* Mangaly and Sabu, *C. haritha* Mangaly and Sabu, *C. raktakanta* Mangaly and Sabu, and *C. vamana* Sabu and Mangaly.

The South Indian *Curcuma* was revised in 1993 (Mangaly and Sabu, 1993). They identified 17 species and included 16 of them in the subgenus *Eucurcuma* and a single species *C. ecalcarata* under subgenus *Hitcheniopsis* as it had unspurred anther. They prepared artificial keys for identification of the taxa, their descriptions, illustrations, and other relevant notes.

Forty *Curcuma* species were reported from India (Velayudhan et al., 1999), which were accommodated into two subgenera proposed by Valeton (1918), based on the presence or absence of anther spurs. These authors described new species, namely *C. malabarica*, *C. kudagensis*, and *C. thalaaveriensis* (Velayudhan et al., 1990). These 40 *Curcuma* species reported from India (Velayudhan et al., 1999) have already been detailed in Table 3.2. New species of *Curcuma*, namely *C. rubrobracteata* (Mizoram State, India), *C. codonantha* (Andaman Islands, Indian Union territory), *C. mutabilis* (South India), were reported by Skornickova (Sirirugsa, 1997; Skornickova et al., 2004). Skornickova and Sabu (2005) provided detailed description of *C. roscoeana*

Wall in India based on live specimens and historical nomenclatural details and also presented a detailed account of identity and description of *C. zanthorrhiza* (Skornickova and Sabu, 2005). The genus *Curcuma* was recircumscribed to include the monotypic genus *Paracautleya* and renamed *Paracautleya bhatii* as *C. bhatii* by the same authors (Skornickova and Sabu, 2005).

Skornickova et al. (2008) investigated the identity and nomenclatural history of *C. zedoaria* and *C. zerumbet* in India. This investigation revealed that the name *C. zedoaria* (Christm.) Roscoe is currently applied to several superficially similar taxa in different parts of India and Southeast Asia. They explained the identity of plant representing *C. zedoaria* in the sense lectotypified by Brutt (1977), with photographic evidence. The plant described and depicted by Roxborgh as *C. zerumbet* Roxb. is illegitimate and is named *C. picta*.

Skornickova et al. (2008) discussed the typification of *C. longa* while investigating the identity of turmeric in 2008. They concluded that the genus *Curcuma* is one of those in which the meanings of words, and often also the inadequate state of herbarium specimens, do not convey the necessary information for unambiguous application of specific names. Although the correct application of the Linnaean name *C. longa* was first questioned by Guibourt, his observations and the proposal of the new name *C. tinctoria* did not affect the situation, as his remarks have remained obscure. Yet, despite early warnings by Trimen (1887) of the existence of Hermann's specimen and its importance to Linnaeus, confusion about the identity of *C. longa* was perpetuated by various authors Valeton (1918) and Burtt and Smith (1972), blurring the situation more and leaving the turmeric without a type. Skornickova et al. (2008) based on their analysis and examination designated the Hermann's specimen, which is the basis of Linnaeus' *C. longa* as the lectotype of *C. longa*.

The taxonomical investigations reveal that sufficient attention has been given to identify the *Curcuma* species present in Asia, in general, and India, in particular. *Curcuma* species in South India have been thoroughly investigated by many researchers. However, confusion in the classification of *Curcuma* still persists, considering the interspecific and intraspecific variations. This has to be resolved by detailed investigations on *Curcuma* species from different parts of the world and simulation data from morphology, cytology, and molecular markers. Correct phylogeny of cultivated *C. longa* is yet to be established.

Use of biochemical and molecular markers to determine phylogeny and diversity of *Curcuma* in the current decade has shown that biochemical and molecular markers have been widely employed to elucidate taxonomical relationships, phylogeny, and genetic diversity in *Curcuma*.

Use of Isoenzymes

Monomorphism of malate enzyme and glutamate oxaloacetate transaminase was found in four species of *Curcuma* including *C. domestica*, *C. manga*, *C. zanthorrhiza*, and *C. zedoaria* and polymorphism of esterase, and peroxidase isoenzymes with 2–6 and 3–11 bands, respectively (Ibrahim, 1996). Oischi (1996) reported the monomorphism found at two loci for leucine aminopeptidase (LAP) and one locus each for glucose-6-phosphate isomerase (GPI) and phosphoglucomutase (PGM) in *C. alismatifolia* samples collected from Thailand which suggested that the material investigated were of clonally propagated sources. Apavatjrut et al. (1999) used isoenzymes to identify some early flowering *Curcuma* species, namely *C. zedoaria* Rosc., *C. zanthorrhiza* Roxb., *C. rubescens* Roxb., *C. elata* Roxb., *C. aeruginosa* Roxb., and two unidentified species. Of the 21 isoenzymes initially tested, 8 showed reliable polymorphism to distinguish between the taxa analyzed. The cluster analysis of data indicated that these taxa are not as closely related as one may assume from the overall morphology, and the similarity in their growth and reproductive habits.

Paisooksantivatana and Thepsen (2001) studied seven enzyme systems to reveal the genetic diversity among the natural populations of *C. alismatifolia* Gagnep., compared to cultivated populations from Thailand. Of the seven enzyme systems analyzed in this study, five enzymes (ADH, GDH-1, LAP-1, GPI-2, and PGM) which showed reproducible and consistent bands were used to determine the diversity. Mean genetic diversity over all loci across all populations was 0.444. Mean genetic identity between cultivated populations (IC), lowland populations (IL), highland populations (IH), and across all populations (IAP) were 0.950, 0.947, 0.944, and 0.922, respectively.

Molecular Markers

The phylogeny of the members of Zingiberaceae was investigated using morphological and molecular markers (Kress et al., 2002). For this DNA sequences of the nuclear ITS and plastid matK were employed and the authors suggested a new classification. Their studies suggest that at least some of the morphological traits based on which members of Zingiberaceae are classified are homoplasious and three of the tribes are paraphyletic. The African genus *Siphonochilus* and the Bornean genus *Tamijia* are basal clades. The former Alpineae and Hedychieae for the most part are monophyletic taxa with the Globbae and Zingibereae included within the latter. They proposed a new classification of the Zingiberaceae that recognizes subfamilies and tribes into four groups as detailed here: (i) subfamily Siphonochiloideae (tribe—Siphonocilieae), (ii) subfamily Tamijioideae (tribe—Tamijieae), (iii) subfamily Alpinioideae (tribe—Alpineae, Riedelieae), and (iv) subfamily Zingiberoideae (tribe–Zingibereae, Globbae). As per the above classification, the genus *Curcuma* was included in the tribe Zingibereae, instead of Hedychieae as in earlier classifications. To establish a rapid and simple molecular identification of six species of *Curcuma*, namely *C. longa*, *C. phaeocaulis*, *C. cichuanensis*, *C. chuanyujin*, *C. chuanhuangjiang*, and *C. chuanezhu* in Sichuan Province, the trnK nucleotide sequencing was used (Cao and Komatsu, 2003) and they stated that sequence data were potentially informative in the identification of these six species at the DNA level. Ngamriabsakul et al. (2004) performed a phylogenetic analysis of the tribe Zingibereae (Zingiberaceae) using nuclear ribosomal DNA (ITS1, 5.8S, and ITS2) and chloroplast DNA (trnL (UAA) 5'exon to trnF (GAA)). They stated that the tribe is monophyletic with two major clades, the *Curcuma* clade and the *Hedychium* clade. The genera *Boesenbergia* and *Curcuma* are apparently not monophyletic. Cao et al. (2001) analyzed medicinally used Chinese and Japanese *Curcuma* based on 18*S rRNA* gene and *trnK* gene sequences and reported that the molecular data can be used to confirm the *Curcuma* species and their derived drugs. Single-nucleotide polymorphism based on sequence of *trnK* gene to identify the plants and drugs derived from *C. longa*, *C. phaeocaulis*, *C. zedoaria*, and *C. aromatica* was used by Sasaki et al. (2004). Xia et al. (2005) used 5S rRNA spacer domain-specific primers to verify the component species of *Curcuma* used in the Chinese medicinal formulation *Rhizoma curcumae* (Ezhu). They found that apart from the three genuine ingredients *C. wenyujin*, *C. phaeocaulis*, and *C. kwangsiensis*, other species, namely *C. longa* and *C. chaniyujin*, are used as adulterants. The phylogenetic analysis by comparing the sequence data showed that *C. phaeocaulis*, *C. kwangsiensis*, and *C. wenyujin* formed a single group with closest homology between *C. phaeocaulis* and *C. wenyujin*, while *C. longa* and *C. chanyujin* showed only 50–55% DNA similarity between them. They also reported the taxonomic confusion regarding the position of *C. wenyujin* and *C. chanyujin*, the latter growing in the Sichuan province of China is also known as *C. sichuanensis*.

   RAPD analyses to estimate the level of genetic diversity within and between natural populations of *C. zedoaria* in Bangladesh was performed by Islam et al. (2005). They observed that hilly populations maintain rather higher genetic diversity than that of the plains and plateau land populations. Syamkumar and Sasikumar (2007) prepared molecular genetic fingerprints of 15 *Curcuma* species using Inter Simple Sequence Repeats (ISSR) and RAPD markers to elucidate the genetic diversity and relatedness among the species. Cluster analysis of data using UPGMA algorithm placed the 15 species into 7 groups in partial agreement with the morphological grouping proposed by the earlier investigators. The investigation also pointed out the limitations of the conventional taxonomic tools to resolve the taxonomic confusion prevailing in the genus and suggested the need to use molecular markers in conjunction with morpho-taxonomic and cytological studies while revising the genus. They observed the maximum molecular similarity between the two of the *Curcuma* species, namely *C. raktakanta* and *C. montana*, suggesting the important need to reexamine the separate status given to these two species. These investigators also suggested a reassessment of the status of the two species, namely *C. montana* and *C. pseudomontana*, based on the presence of sessile tubers.

   The SCAR DNA markers technique to identify important *Curcuma* species of Thailand and their hybrids was employed by Anuntalabhochai et al. (2007). The genetic variability of 12 starchy *Curcuma* species was investigated. All of the 12 species investigated were separated into 3 clusters using the UPGMA technique. *C. aromatica*, *C. leucorrhiza*, and *C. brog* formed a cluster within which *C. longa* and *C. zedoaria* formed a subgroup. *C. haritha* was genetically distinct from all the other *Curcuma* species. A set of 30 accessions of 4 *Curcuma* species, namely *C. latifolia*, *C. malabarica*, *C. manga*, and *C. raktakanta*, and 13 morphotypes of *C. longa* conserved *in vitro* was subjected to RAPD analysis. Mean genetic similarities based on Jaccard's similarity coefficient ranged from 0.18 to 0.86 in accessions of cultivated species and from 0.25 to 086 in wild species. They observed the primers

OPC-20, OPO-06, OPC-01, and OPL-03 to be highly informative in discriminating the germplasm of *Curcuma*. Ahmed et al., while investigating the genetic variation of chloroplast DNA in Zingiberaceae from Myanmar using PCR–RFLP polymorphism analysis, observed that the two *Curcuma* species, namely *C. zedoaria* and *C. zanthorrhiza*, appeared to be identical, supporting their recent classification as synonymous. It is evident from literature that isozyme analysis was attempted only by a limited number of authors for diversity analysis in *Curcuma* even though it is a reliable marker system. Molecular markers, such as RAPDs, ISSR, SCAR DNA markers, PCR–RFLP, and DNA sequence-based analysis were performed by different investigators. But more molecular data on different species of *Curcuma* has to be generated to have a clear understanding of phylogeny and diversity among the species.

Morphology of Turmeric

Holtum (1950) presented the morphological description of turmeric, which was subsequently cited by several other authors (Purseglove et al., 1981; Ravindran et al., 2007).

The following descriptions need to be noted.

Habit

Turmeric is an erect perennial herb, grown as an annual and in certain cases as a biennial as well. It grows to a height of around 120 cm but, significant variations exist in plant height, among varieties as well as in plants grown under different agroclimatic conditions (Rao et al., 2006).

Leaves

Leaves are borne in a tuft, alternate, obliquely erect or subsessile, with long-leaf stalks or sheaths forming a pseudostem or the aerial shoot. The leafy shoots rarely exceed 1 m in height and are erect. Usually, there will be 6–10 leaves in a leafy shoot. The thin petiole is rather abruptly broadened to the sheath. The ligule lobes are small and sheath near the ligules have ciliate edges. The lamina is lanceolate, acuminate, and thin, dark green above and pale green beneath with pellucid dots; it is usually upto 30 cm long and 7–8 cm wide and is rarely over 50 cm long (Purseglove et al., 1981). Foliar anatomy of different species of *Curcuma* including *C. longa* has been investigated by Das et al. (2004) and Jayasree and Sabu (2005). Scanning electron microscopy (SEM) investigations of the turmeric leaf showed dense, uneven, and waxy cuticle depositions, uniformly spread over epidermal boundaries (Das et al., 2004). The stomata are tetracytic type with long axis of the pore parallel to the veins. They are mostly seen on the abaxial epidermis. Stomata are very few on the adaxial epidermis (Jayasree and Sabu, 2005). The stomatal aperture is elliptic with a somewhat incomplete cuticle rim around it. Transection of turmeric leaf across the mid-vein showed the following structure (Das et al., 2004).

Epidermis

Uniseriate, thin-walled, barrel-shaped parenchymatous cells. Trichomes are less frequent at the adaxial surface. At the abaxial surface, small, unicellular, hook-like trichomes are present with a slightly bulbous base.

Hypodermis

Multiseriate, mostly one or two layered, composed of irregularly polygonal colorless cells, present interior to both upper and lower epidermis.

Mesophyll

Mesophyll is not differentiated into palisade and spongy tissue according to Das et al. (2004). But Jayasree and Sabu (2005) reported one-layered palisade tissue in all species of *Curcuma.*

Mesophyll tissue is traversed by a single layer of abaxial air canals alternating with vascular bundles, which are embedded in a distinct abaxial band of chlorenchyma. Air canals are traversed by thin-walled trabeculae, which form a loose mesh within.

Vascular Bundles

The vascular bundles are arranged in three layers, developing unequally at different levels. Main vascular bundles form a single conspicuous abaxial arc, alternating with air canals, and embedded in chlorenchyma. The abaxial conducting systems consists of an arc of vascular bundles of different sizes that are circular in outline. The adaxial conducting system consists of vascular bundles that are similar in appearance to the main vascular bundles, but are sclerenchymatous sheath above the xylem and below the phloem, extruded protoxylem, small mass of metaxylems and phloem tissue. Vascular bundles of accessory arcs have reduced vascular tissues and contracted protoxylem. Abaxial bundles are enveloped within almost a complete fibrous sheath.

Curcuma species differ in fine anatomical features (Das et al., 2004; Jayasree and Sabu, 2005). The latter authors presented a detailed account of such variations in 15 *Curcuma* species found in India with reference to dermal morphology, petiole, midrib, leaf margin, and venation pattern of brachts. Using the additional information generated, these authors prepared an anatomical key to identify these species.

The epidermal and stomatal structures of turmeric and *C. amada* were investigated by Raju and Shah (1975). They have reported that the upper epidermis consists of polygonal cells which are predominantly elongated at right angles to the long axis of leaf. Irregular polygonal cells are present on the lower epidermis, except at the vein region, where they are vertically elongated and thick-walled. The epidermal cells in the scale and sheath leaves (the first 2–5 leaves above ground without the leaf blade) are elongated parallel to the axis of the leaf. Oil cells are rectangular thick-walled and suberized and are frequent in the lower epidermis. They observed that the leaves are amphistomatic, with a distinct substomatal cavity and stomata may be diperigenous, tetraperigenous, or anisocytic. Often, two subsidiary cells align completely with guard cells. Stomatal development was also described in detail by the aforementioned authors.

Leaf Sheath

Das et al. (2004) investigated the transections of the sheathing petiole and found that the sheathing petiole is horseshoe shaped in outline, and the marginal parts are inflexed adaxially. Vascular bundles are arranged in three systems forming arcs. Those of the abaxial main conducting system are alternate with large air canals, which are traversed internally by trabeculae. Toward the margin, the vascular bundles seem to arrange themselves in a single row. Both the upper and the lower epidermis are uniseriate, consisting of rectangular cells.

Rhizome

The rhizome is the underground stem of turmeric, which can be divided into two parts, the central pear-shaped "mother rhizome" and its lateral axillary branches known as "fingers." Normally, there is only one main axis. Either a complete finger or a mother rhizome is used as planting material. It is also called the "seed rhizome." Normally, the "seed rhizome" produces only one main axis, which develops into the aerial leafy shoot. The base of the main axis enlarges and becomes the first formed unit of the rhizome which ultimately develops into the mother rhizome. Axillary buds from the lower nodes of the "mother rhizome" develop and give rise to the first order of branches, often called the "primary fingers." Their number varies from two to five. Primary branches grow to some length and either develop into an aerial shoot or stop growing further. They grow in a haphazard manner in different directions and in some cases grow up to the ground level with one or two, or even no, leaves. Secondary branches developing at higher nodes of primary branches are diageotropic (Raju and Shah, 1975). Some primary branches after hitting ground level do not form any aerial shoot, but, exhibit positive geotropic growth. Such branches arising from the mother rhizome may be diageotropic, orthogeotropic, plagiotropic (Ravindran et al., 2007). Primary fingers branch further, resulting in secondary and tertiary branches, and these branches do not produce aerial shoots. The majority of them show positive geotropic growth or obliquely downward growth. The C. longa types have more sideward growth, while the C. aromatica types have more downward growth (Ravindran et al., 2007).

Nodes and Internodes

Mature mother rhizomes may have 7–12 nodes, and the intermodal length varies from 0.3 to 0.6 cm. However, the first few internodes at the proximal end are elongated due to which the mother rhizome reaches the ground level (Shah and Raju, 1975). Primary and secondary fingers have longer internodes of about 2 cm length, compared to mother rhizomes. Except the first one or two, all the other nodes in the mother rhizome as well as fingers have axillary buds. The mother rhizome has scale leaves only at the first two to four nodes; the rest of them have sheath leaves and foliage leaves. The secondary and tertiary branches have only scale leaves. The branches with negative geotropic growth have pointed scale leaves or sheath leaves (Ravindran et al., 2007).

Aerial Shoot

The foliage leaves emerge from the buds on the axils of the nodes of the underground bulb and sometimes from the primary finger also. The petiole of the foliage leaf is long and has a thick leaf sheath. The long-leaf sheaths overlap and give rise to the aerial shoot (Ravindran et al., 2007).

Shoot Apex

The apical meristem of the shoot has the tunica–corpus type configuration. The tunica is two-layered, with cells dividing anticlinally, while in the corpus, which is the region proximal to tunica, the cells divide in all directions. The central region underlying the corpus layer is the rib meristem which gives rise to a file of cells, which later become the ground meristem. The central region is surrounded by the flank meristem, which produces the procambrium, cortical region, and leaf primordium (Ravindran et al., 2007).

Roots

Roots emerge from the mother rhizomes and often from fingers, not from the secondary and tertiary fingers. Some of the roots enlarge and become fleshy due to storage of food materials. They serve the function of nutrient and water absorption, anchorage, and storage of assimilated food. In certain species, some of the roots terminate in bulbous tubers (Ravindran et al., 2007). It has been observed that the true seedling progenies of turmeric at their early stages of rhizome formation will produce root tubers as per unpublished data from IISR. Root initials originate from the narrow cell zone separating the inner and outer ground tissues termed the diffuse meristem, which is an extension of the primary elongating meristem and is noticeable below the second or third node. Root meristem originates from the diffuse meristem. The root apex of turmeric shows three sets of initials developing from the diffuse meristem, one each for root cap and plerome and a common zone of dermatogens and plerome. The root cap has two regions, namely a columella in the middle and a calyptras at the periphery. The columella consists of 5–7 layers of vertical files of cells, which divide mostly peridermally. The cells in the peripheral region of calyptras undergoes kappa-type divisions followed by cell enlargement resulting in broadening of this region toward the distal end (Raju and Shah, 1975).

Root Epidermis and Cortex Originate from Single Tier of Common Initials

The protoderm–periblem complex, which is composed of 1–7 cells in the horizontal row. Epidermis and cortex are established from this row by the Korper-type divisions (T-divisions). These divisions, followed by cell enlargement, enable the tissue to widen toward the proximal end. The epidermis is differentiated from the outermost layer. The stellar and pith cells are formed from a group of cells located above the epidermis (the cortical initial) (Raju and Shah, 1975). A detailed account of the differentiation of cell layers in turmeric root has been reported by Pillai et al. (1961).

Turmeric Rhizome—its Developmental Anatomy

The turmeric rhizome anatomy and its development has been investigated (Ravindran et al., 1998; Sherlija et al., 1999). These investigators provided a detailed description of the rhizome and its different developmental stages based on histology. Transverse sections of rhizomes show an outer zone and an inner zone, separated by intermediate layers. Both have vascular bundles. The vessels show spiral and scalariform perforation plates. The phloem contains sieve tubes and two or three companion cells. Early in rhizome development, when it is about 4–7 mm in diameter, the outer zone is 1.5–2.5 mm, the inner zone is 2.5–3.5 mm, and the intermediate layer is about 0.5 mm in thickness. A mature mother rhizome measures about 2–3 cm across, having an outer zone of about 6–10 mm, inner zone about 10–12 mm, and the intermediate layer about 1–1.5 mm in thickness. At this stage, the primary finger is about 1–2 cm in diameter, outer zone 4–5 mm, inner zone 9–10 mm, and intermediate layer about 1 mm in thickness. Rhizome enlargement initiates through the activity of meristematic cells, present below the young primordial of the developing rhizome. These cells develop into primary thickening meristem (PTM), which is responsible for the initial thickening in the width of the developing cortex by producing primary vascular bundles that are collateral. At the lower levels of the developing rhizome, the PTM becomes primarily a root-producing meristem. After the formation of the primary vascular cylinder, some of the pericycle cells at different places undergo one or more periclinal divisions, forming secondary thickening meristems (STMs), which vary from two to six layers. This meristem produces secondary vascular bundles and parenchyma cells on its inner side. These parenchyma cells become packed with starch grains on maturity. The crowded arrangement of the secondary vascular bundles, which are amphicribral, and their distribution clearly distinguishes them from the primary bundles which are collateral and scattered. The cambrium-like zones (PTM and STM) constitute ray initials and fusiform initials, which are visible in certain loci. In addition to this cambial activity, increase in size of the rhizome is also the result of the activity of ground meristem that divides at many loci, followed by cell enlargement. The ground parenchyma in actively growing regions contains oil canals along with phloem and xylem. Oil canals are formed lysigenously by the disintegration of entire cells (Ravindran et al., 2007).

Inflorescence, Flower, Fruit, and Seed Set

Both the cultivar and the climatic conditions decide pattern of flowering in turmeric. After 109–155 days of planting flowering commences. This time range will vary depending on the specific cultivar. This is only a range. The inflorescence lasts about 1–2 weeks after emergence (Pathak et al., 1960). Inflorescence is a cylindrical spike, 10–15 cm long and 5–7 cm wide, which is terminal on the leafy shoot with the scape partly enclosed by leaf sheaths. The brachts are spirally arranged and closely over-lapped giving the inflorescence a cone-like appearance. The brachts are adnate for less than half of their length and are elliptic-lanceolate and acute, 5–6 cm long and about 2.5 cm wide. The upper 3–7 and lower 5–10 brachts are sterile with no flowers. The upper sterile brachts are white or white-streaked with green, pink-tipped in some cultivars grading to light green brachts lower down. The brachteoles are thin, ellipti-cal, and about 3.5 cm long. The flowers are borne in cincinni of two in the axils and brachts, opening one at a time. The number of flowers per inflorescence ranges from 26 to 35 (Purseglove et al., 1981; Sherlija et al., 2001). Cultivars Rajendra-Sonia and BSR-2 produce 50–100 flowers per inflorescence (IISR personal communication, unpublished data). The flowers are thin-textured and fugacious and are about 5 cm long. The calyx is short, tubular, uniquely toothed, and split nearly halfway down one side. The corolla is tubular at the base with the upper half cup shaped with three unequal lobes inserted on the edge of the cup lip. It is whitish, thin, and translucent with the dorsal lobe hooded. There are two lateral staminodes, elliptic-oblong, which are creamy white in color, and with the inner edges folded under the hood of the dor-sal petal. The lip or labellum is obovate, with a broad thickened yellow band down the center and thinner creamy white side-lobes upcurved and overlapping the stami-nodes. The stamen is epipetalous and attached to the throat of the corolla. The fertile stamen has two anther lobes. The filament of the stamen is short and broad, united to a versatile anther about the middle of the anther sacs, and with a broad, curved large spur at the base. The cylindrical stylodes are about 4 mm long. The ovary is inferior, tricolor, and syncarpus with a slender style passing between the anther lobes and held by them. The placentation is axile (Nazeem et al., 1993; Purseglove et al., 1981). Anthesis is between 7 AM and 9 AM, peaking at 8 AM. Anther dehiscence is between 7:15 AM and 7:45 AM (Nazeem et al., 1993; Rao et al., 2006). Based on cultivation place, slight variation of flower open and anther dehiscence have been noted (Nambiar et al., 1982). Pollen fertility is cultivar-dependent. In three cultivars of *C. longa* it is 45.7–48.5% and in five cultivars of *C. aromatica* it is between 68.6% and 74.5% (Nambiar et al., 1982). Pollen grains are ovoid to spherical, light yellow in color, and slightly sticky (Nazeem et al., 1993). Based on acetocarmine staining, these authors investigated pollen fertility in eight cultivars and found that it ranges from 71% in Kodur to 84.5% in Kuchipudi. Though pollen fertility in five acces-sions can range from 53% to 58% based on the above method, actual germination in Brewbaker and Kwack medium containing less than 20% sucrose is less than 10% (Nair et al., 2004). This, probably, is the reason for infrequent seed set in turmeric.

Turmeric is a cross-pollinated crop (Nazeem et al., 1993). Seed set in 11 culti-vars of *C. aromatica*, 6 of *C. longa* was compared by open pollination, where it was found that only 9 of the former set seed, while none in the latter (Nambiar et al., 1982). Poor pollen fertility in the latter, owing to it being triploid, was the reason for the total absence of seed set. Selfing, crossing, and open pollination investigations carried out by Nazeem et al. (1993) using three cultivars of *C. domestica* (synonym of *C. longa*) and five cultivars of *C. aromatica* showed total absence of seed set in self-pollination. Of the 11 cross-combinations, 8 resulted in seed set and the one open pollinated had the maximum seed set. They suggested self-incomptability as the reason for absence of seed setting in self-pollination. Crossing investigations by Renjith et al. (2001) employing two medium-duration and five short-duration cultivars resulted in seed set in three out of twelve cross-combinations, all involving short-duration types. Thus, it appears, some compatibility mechanisms operate in the crosses involving different cultivars, which have to be investigated further. A case of seed setting in cultivated turmeric types was reported by Lad (1993). Nair et al. (2004) obtained seed set in two accessions of *C. longa*, which was later repeated in many germplasm collections.

Turmeric fruit is a trilocular capsule with numerous arillate seeds. The mature fruit will give the appearance of a small garlic bulb and is white in color. The immature seeds are white to light brown in color and mature seeds are brownish black in color. Histological analysis of fruits and seeds by Nair et al. (2004) showed that seeds are attached to the central column inside the fruit. Different seeds derived from the same fruit showed embryos of different developmental stages occasionally. The embryos were clearly monocotyledonary, resembling the embryos of cardamom in structure. Persistence of the nucellus was evident in the mature seed (Nair et al., 2004).

Germination of Seed and Establishment of Seedling Progenies

Nambiar et al. (1982) was the first to report on seed germination and establishment of seedling progenies in turmeric (*C. aromatica*). Seeds matured in 23–29 days after opening of flowers, according to these authors. They germinated within 10–18 days and germination percentage ranged from 30.5 to 62.5 in different cultivars. At germination, seeds absorb moisture and nutrients and enlarge before plumule emergence. The plumule is with two protuberances at the base which later develop into primary roots. The seedling progenies produced mainly roots and root tubers in the first year of growth. The rhizomes were very small. Normal rhizome development occurred in the second year. In the southern state of Andhra Pradesh of India, cultivars of *C. longa* flowered very rarely, but viable seeds could be collected from flowering types (Purseglove et al., 1981). Seedlings were found to be tardy in growth and development and rhizome formed was of poor quality. Seedling progenies of many turmeric cultivars from crossed as well as open-pollinated seeds were established by Nazeem et al. (1993). These investigators observed that in seeds from 17 to 26 days after sowing, seed germination commences, and its duration ranged from 10 to 44 days, depending on the crosses. Percentage of germination varied from 17.22 to 100 in different crosses and was 26.48% in the case of the open-pollinated progenies of cultivar Nandyal. Variations in morphological characters of seedling progenies were observed, and it was suggested that there was scope for selection among the progenies. Seedlings produced only one mother rhizome with root tubers in the first year of growth and the weight ranged from 14.18 to 49.4 g. Size of the mother rhizomes progressively increases over the years and full growth is observed in the third year after sowing. With increase in rhizome size, number of root tubers decline.

Seed germination from two accessions from open-pollinated seeds showed that only few seeds germinated within a month of sowing, and majority of seeds germinated after 5 months showing 75% and 3% germination, respectively, in Accession No. 126 and Accession No. 399. Subsequently, more than 250 open-pollinated progenies of 23 *C. longa* genotypes were established at IISR in 2003–04 and their evaluation is underway (Nair et al., 2004).

The *Curcuma* Cytology

It was in 1936 that the first report on *Curcuma* chromosome number was made by Sugiura, who observed a chromosome number of $2n = 64$ (Sugiura, 1936). Since then, several reports on *Curcuma* cytology have appeared in scientific literature, most of them confining just to the chromosome number. Table 2.2 summarizes these reports.

The most commonly reported and generally accepted chromosome number of *Curcuma* is $2n = 63$ (Chakravorti, 1948; Islam, 2004; Ramachandran, 1961). Deviations have also been reported, as detailed in Table 2.2. The basic chromosome number of the genus *Curcuma* was suggested as $x = 21$, which in turn originated by dibasic amphidiploidy from $x = 9$ and $x = 12$ by secondary polyploidy (Ramachandran, 1961, 1969). The above-mentioned authors suggested that turmeric is a triploid and might have originated as a hybrid between tetraploid *C. aromatica* ($2n = 84$) and an ancestral diploid *C. longa* ($2n = 42$) or one of these has evolved from the other through mutation, represented by the intermediate type which is known to occur. The herbaceous perennial habit of this species, its vegetative mode of propagation, and the small size of the chromosomes favor perpetuation of polyploidy (Ramachandran, 1961). This suggestion of the hybrid origin of *C. longa* was later corroborated by Nambiar (1979). His investigations revealed intercellular variation in chromosome number in different cultivars of *C. longa* and *C. aromatica*.

The *Curcuma* Karyomorphology

Karyomorphological investigations have been rather scanty in *Curcuma*. Sato (1948) while investigating the karyotype of Zingiberaceae reported the karyology of *C. longa* as well. He designated chromosomes of Zingiberaceae from A to H based on morphology. He reported that *C. longa* has 32 chromosomes and suggested that the species could be an allotetraploid with a basic number of $x = 8$. The karyotype formula was

Table 2.2 Species Specific Chromosome Number in *Curcuma*

Species	Number of Chromosomes (2n)	References
C. aeruginosa Roxb.	63	Joseph et al. (1999), Paisooksantivatana and Thepsen (2001)
C. alismatifolia Gagnep.	32	Paisooksantivatana and Thepsen (2001)
C. amada Roxb.	42	Raghavan and Venkatasubban (1943), Chakravorti (1948), Sharma and Bhattacharya (1959)
C. amarissima Rosc.	63	Islam (2004)
C. angustifolia Roxb.	42	Chakravorti (1948), Sharma and Bhattacharya (1959), Islam (2004)
C. aromatica Salisb.	42	Raghavan and Venkatasubban (1943), Chakravorti (1948)
C. attenuata Wall.ex Baker	84	
C. aurantiaca Van Zip.	42	
C. borg Val.	63	
C. caesia Roxb.	22	Das et al. (1999)
C. colorata F.	62	
C. comosa Roxb.	42	Joseph et al. (1999)
C. decipiens Dalz.	42	Ramachandran (1961, 1969)
C. elata Roxb.	63	
C. gracillima Gagnep.	24	
C. haritha Mangaly and Sabu	42	Joseph et al. (1999)
C. harmandii Gagnep.	20	Paisooksantivatana and Thepsen (2001)
C. haeyneana Val. & Zip	63	
C. kwangsiensis S.G.Lee & C.F.Liang	84	
C. latifolia Rosc.		
C. longa L.	63	Islam (2004)
	32	Sato (1948)
	48	Das et al. (1999), Nayak et al. (2006)
	61	Nair and Sasikumar (2009)
	62	Raghavan and Venkatasubban (1943)
	63	Chakravorti (1948), Nair and Sasikumar (2009)
	64	Sugiura (1936), Chakravorti (1948)
	84	Nair and Sasikumar (2009)
	93	Nair and Sasikumar (2009)
C. malabarica Vel. et al.	42	Joseph et al. (1999)
C. manga Val. and Zip	42	
	63	

(Continued)

Table 2.2 (Continued)

Species	Number of Chromosomes ($2n$)	References
C. neilgherrensis Wight	42	Chakravorti (1948), Ramachandran (1961, 1969)
C. oligantha Trim.	42	
	40	
C. parviflora Wall.	28	
	30	
	32	
	34	
	36	
	42	Paisooksantivatana and Thepsen (2001)
C. petiolata Roxb.	64	
	42	
C. phaeocaulis Val.	62, 63, 64	
C. purpurrascens Blume	63	
C. raktakanta Mangaly and Sabu	63	Joseph et al. (1999)
	42	
C. rhabdota Sirigusa & M.F. Newman	24	
C. roscoeana Wall.	42	
C. rubescens Roxb.	63	Islam (2004)
	42	Islam (2004)
C. sessilis Gage	84	
	46,92	
C. soloensis Val.	63	
C. viridiflora Roxb.	42	Islam (2004)
C. thorelii Gagnep.	34	
	36	
C. wenyujiri Y.H. Chen & C. Ling	63	
C. zanthorrhiza Roxb.	63	
C. zedoaria Rosc.	42	Paisooksantivatana and Thepsen (2001)
	63	Chakravorti (1948), Ramachandran (1961, 1969)
	64	Chakravorti (1948)
	66	

Source: Modified from Ravindran et al., (2007) and Skornickova et al., (2007).

presented as $2tA^m + 10A^{sm} + 12B^{sm} + 6C^{ot} + 2tC^{ot}$. The m, sm, and ot represent the centromeric positions and t indicates the presence of a satellite. Joseph et al. (1999) reported the karyotype of six species of *Curcuma*, namely *C. aeruginosa* ($2n = 63$), *C. caesia* ($2n = 63$), *C. comosa* ($2n = 42$), *C. haritha* ($2n = 42$), *C. malabarica*

$(2n = 42)$, and *C. raktakanta* $(2n = 63)$. They found symmetrical karyotypes in all these species. The chromosome length ranged from 0.24 to 0.99 μm³ among these species and total chromosome length varied from 16.21 to 33.06 μm³. The average chromosome length varied from 0.39 to 0.52 μm³. Based on the karyomorphological data of these species, they concluded that both numerical and structural variations have operated in the evolution of the genus *Curcuma* (Joseph et al., 1999). Das et al. (1999) reported the karyotypes of *C. amada, C. caesia* and two varieties of *C. longa*. They reported chromosome numbers of $2n = 40$ for *C. amada*, $2n = 22$ for *C. caesia*, and $2n = 48$ for two varieties of *C. longa*, namely Suroma and TC-4, which are drastically different from most of the earlier reports.

The *Curcuma* Meiotic Investigations

Information on chromosome orientation in *C. longa* and a few related species was reported by Ramachandran (1961) and Nambiar (1979). Meiosis in *C. decipiens* $(2n = 42)$ and *C. longa* $(2n = 63)$ was investigated by Ramachandran (1961), and he concluded that meiosis is regular with the formation of bivalents only at metaphase I in the former, while in the latter a high percentage of trivalent associations were produced despite small chromosome size. Nambiar (1979) analyzed meiosis in three cultivars of *C. longa* and five cultivars of *C. aromatica*. Maximum number of quadrivalents and hexavalents were found in *C. longa* and *C. aromatica*, respectively. Bivalents were predominant in all the cultivars of both species. Later stages of meiosis were almost regular in the cultivars of *C. aromatica*, though increased abnormalities were observed in the cultivars of *C. longa*. Microporogenesis and megaporogenesis in *C. aurantiaca* and *C. lorgengii* were reported by Strapradja and Aminali, as cited by Ravindran et al. (2007).

The *Curcuma* Nuclear DNA Content

Researchers have investigated the nuclear DNA content of *Curcuma*, among which Das et al. (1999) are at the forefront. They observed that 4C DNA content of *Curcuma* species varied significantly from 3.12 to 5.26 pg among species/varieties. Interphase nuclear volume varied from 224.56 in *C. caesia* to 422.56 μm³ in *C. longa*. Nayak et al. (2006) observed significant variation in 4C DNA content of 17 cultivars of *C. longa*, all having $2n = 48$. They attributed this variation to the loss or addition of highly repetitive sequences in the genome. Islam (2004), while investigating cytogenetic makeup of *Curcuma* species and accessions from Bangladesh, observed that the 2C DNA content varied from 2.12 to 5.32 pg among species and accessions investigated. The author employed flow cytometry technique. Further, it was observed that there existed variation in nuclear DNA content among populations of *C. zedoaria* as well.

Skornickova et al. (2007) made detailed analysis of chromosome number and genome size variation in Indian *Curcuma* species, where it was noted that six different chromosome counts, namely $2n = 22$, $2n = 42$, $2n = 63$, $2n = >70$, $2n = 77$, and $2n = 105$, existed. The 2C DNA values varied from 1.66 in *C. vamana* to 4.76 pg in *C. oligantha*, which shows a 287% increase, almost a threefold. Three groups of taxa with significantly different homoploid genome sizes and distinct geographical distribution were identified. Five species exhibited intraspecific variation in nuclear DNA content, attaining 15.1% in cultivated *C. longa*. Based on these investigations, these authors suggested that the basic chromosome number of subgenus of *Curcuma* is $x = 7$ and different published reports corresponds to $6x$, $9x$, $11x$, $12x$, and $15x$. They also presented an exhaustive review on chromosome number and DNA content of different turmeric species from various centers around the world. Table 2.2 has already presented these data.

Chromosome Number in *Curcuma* Seedling Progenies

Chromosome number in seedling progenies is an important trait in understanding the genetics of plants. *Curcuma* is no exception. Numerical variation in chromosome number in *Curcuma* germplasm and seedling progenies has been reported by some researchers, among which the investigations of Nair and Sasikumar (2009) are important. They determined chromosome numbers in 22 germplasm collections and 28 open-pollinated seedling progenies. Among the germplasm collections analyzed, twenty had $2n = 63$, one had $2n = 61$, and another one had $2n = 84$. The seedling progenies showed various chromosome numbers as expected, ranging from $2n = 63$ to $2n = 86$, of which $2n = 84$ was the most frequent. These authors attributed the abnormalities to triploid chromosome segregation, which produced gametes with differing chromosome numbers, resulting in chromosome number variation among progenies. These authors also suggested the origin of germplasm collections with varied chromosome number as natural seedling progenies. Mitotic metaphase has shown diploid chromosome number of $2n = 63$ in a cultivar and $2n = 84$ in a seedling progeny of *C. longa*.

Turmeric Crop Improvement

Generally, crop improvement in turmeric was restricted to clonal selection, induced mutation, and subsequent selection. The principal target in crop improvement was yield enhancement, attaining high curing percentage, and curcumin content. Following reports of seed set in *C. longa* and *C. aromatica*, through open pollination and controlled crosses, the possibility of utilizing recombination breeding opened up in genetic improvement of the crop (Nambiar et al., 1982; Nazeem et al., 1993; Sasikumar et al., 1994). Both IISR at Kozhikode (Calicut) in Kerala State and the AICRP scattered in different parts of India, engage in crop improvement programs. Table 2.3 gives salient results with regard to these efforts.

Clonal Selection in Turmeric

Turmeric crop improvement by clonal selection played the most significant role in developing high-yielding varieties. Rare seed set and insufficient knowledge on establishing seedling progenies were the reasons for this. Selection was mainly applied on landraces collected from different turmeric growing regions of India. More improved varieties of *C. longa* were developed as compared to *C. aromatica* or *C. amada*. High-yielding variety Krishna (7.2 t/ha) was developed by Pujari et al. (1986) for general cultivation in Maharashtra, which was found suitable for Konkan region later (Jalgaonkar et al., 1988). Two high-yielding and disease-tolerant lines with high dry recovery and quality, namely PTS-10 and PTS-24, were identified by HARS, Pottangi (Odisha) and were releases as Roma and Suroma (Table 2.3) and recommended for large-scale cultivation to replace local varieties (Ravindran et al., 2007). Evaluation of about 19 high-yielding lines identified at IISR on multilocation testing resulted in PCT-8, PCT-13, and PCT-14, with high yield and curcumin content and were released as Suvarna, Suguna, and Sudharshana (Ratnambal and Nair, 1986; Ratnambal et al., 1992). Suguna and Sudharshana are short-duration varieties and are also field tolerant to rhizome rot (Ravindran et al., 2007). These were better performers in the State of Andhra Pradesh in India (Reddy et al., 1989). Two more high-yielding and high-quality varieties, namely IISR Alleppey Supreme and IISR-Kedaram, were released from IISR through clonal selection (Sasikumar et al., 2005). Maurya (1990) recommended RH-Rajendra-Sonia for the State of Bihar in India. Its yield potential is about 24 t/ha of fresh rhizomes, and 8.4% curcumin content.

Turmeric Improvement by Seedling Selection

Reports on seed set and seed germination in *C. aromatica* (Nambiar et al., 1982) and *C. longa* (Nair et al., 2004; Nazeem et al., 1993) opened new vistas in crop improvement in turmeric. Evaluation of 15 open-pollinated seedling progenies of *C. longa* at IISR, Calicut, India during 1990–1 resulted in short listing of seven lines. These on multilocation trials during 1992–5 threw up two promising lines which were releases for commercial cultivation by IISR as IISR Prabha and IISR Pratibha. Hence, it is evident that seedling progenies have the potential to generate sufficient variability for selection of better genotypes for commercial cultivation.

Mutation and Selection Induced Crop Improvement in Turmeric

X-ray irradiation was used to develop all the mutant cultivars. Three mutants, namely CO-1, BSR-1, and BSR-2, from Erode local were developed by X-ray irradiation and were subsequently released for large-scale commercial cultivation

Table 2.3 Crop Improvement in Indian Turmeric

Variety	Pedigree	Crop Yield (t/ha)	Fresh Recovery (%)	Dry Content (%)	Curcumin Features (%)	Remarks
Krishna	Clonal selection from Tekurpeta (Maharashtra Agricultural University)	240	9.2	16.4	2.8	Moderately tolerant to pests and diseases
Sugandham	Germplasm selection (Gujarat Agricultural University)	210	15.0	23.3	3.1	Moderately tolerant to pests and diseases
Roma	Clonal selection T. Sunder (Orissa University of Agriculture and Technology)	250	20.7	31.0	6.1	Suitable for hilly areas
Ranga	Clonal selection from Rajpuri local (Orissa University of Agriculture and Technology)	250	29.0	24.8	6.3	Bold rhizomes, moderately resistant to leaf blotch and rhizome scales
Rasmi	Clonal selection from Rajpuri local (Orissa University of Agriculture and Technology)	240	31.3	23.0	6.4	Bold rhizomes
Rajendra-Sonia	Germplasm selection (Rajendra Agricultural University)	225	23.0	18.0	8.4	Bold and plumpy
Megha	Selection from Lakadong types (ICAR Research Center Shillong)	300–315	20.0	16.37	6.8	Bold rhizomes
Pant Peetabh	Germplasm selection (GB Pant University of Agriculture and Technology)	–	29.0	18.5	7.5	Resistant to rhizome rot
Suranjana	Germplasm selection (Uttar Bengal Krishi Viswa Vidyalaya)	235	–	21.2	5.7	Tolerant to rhizome rot and leaf blotch, resistant to rhizome scales and moderately resistant to shoot borer
Suvarna	Germplasm selection (Indian Institute of Spices Research)	200	17.4	20.0	4.3	Bright orange-colored rhizomes
Suguna	Germplasm selection (Indian Institute of Spices Research)	190	29.3	20.4	7.3	Short-duration type, tolerance to rhizome rot

Variety	Origin	Duration (days)	Yield (t/ha)	Dry recovery (%)	Oleoresin/essential oil (%)	Special features
Sudharshana	Germplasm selection (Indian Institute of Spices Research)	190	28.8	20.6	5.3	Short-duration type, tolerance to rhizome rot
IISR Alleppey Supreme	Selection from finger (Indian Institute of Spices Research)	210	35.4	19.0	5.55	Tolerant to turmeric leaf blotch
IISR-Kedaram	Germplasm selection (Indian Institute of Spices Research)	210	35.5	18.9	5.9	Tolerant to leaf blotch
Kanthi	Selection from Mydukur (Kerala Agricultural University)	240–270	37.65	20.15	7.18	Big mother rhizomes and bold fingers with short internodes
Sobha	Germplasm selection (Kerala Agricultural University)	240–270	35.88	19.38	7.39	Big mother rhizomes and bold fingers with short internodes
Sona	Germplasm selection (Kerala Agricultural University)	240–270	4.02 dry	18.88	7.12	Field tolerant to leaf blotch
Varna	Germplasm selection (Kerala Agricultural University)	240–270	4.16 dry	19.05	7.87	Bold rhizome with internodes, field tolerant to leaf blotch
IISR Prabha	Open-pollinated seedling selection (Indian Institute of Spices Research)	205	37.47	19.50	6.50	
IISR Pratibha	Open-pollinated seedling selection (Indian Institute of Spices Research)	225	39.12	18.50	6.21	
CO-1	X-ray mutant selection from Erode local (Tamil Nadu Agricultural University)	270	30.5	19.5	3.2	Bold, bright orange rhizomes, suitable for drought-prone areas
BSR-1	X-ray mutant selection from Erode local (Tamil Nadu Agricultural University)	285	30.7	20.5	4.2	Suitable for drought-prone areas
BSR-2	X-ray mutant selection from Erode local (Tamil Nadu Agricultural University)	245	32.7	26.0	6.1	Bold rhizomes, resistant to scale insects
Suroma	X-ray mutant selection from Tsunder (Tamil Nadu Agricultural University)	253	20.0	–	–	Field tolerant leaf blotch, leaf spot, and rhizome scale

Source: Ravindran et al. (2007).

(Balashanmugham et al., 1986; Cheziyan and Shanmugasundaram, 2000). *Suroma* is another mutant selection from Tsunder subsequent to X-ray irradiation. Gamma rays effect on mutation in turmeric has been reported (Rao, 1999). Use of chemical mutagens, such as Colchicine, EMS, and MNG, each at 20, 500, and 1000 ppm on cultivar Mydukur resulted in taller and high-yielding colchiploids (Ravindran et al., 2007).

Hybridization and Selection in Turmeric

Successful hybridization and establishment of hybrid progenies have been reported (Nazeem et al., 1993) in crosses between *C. longa* and *C. aromatica*. Seed sets in *C. longa* crosses were obtained involving short-duration types (Renjith et al., 2001). However, no commercial cultivars have been released by hybridization and selection so far. A systematic investigation on the incompatibility mechanisms and production of inbred progenies in turmeric may help evolve desirable recombinants and heterotic hybrids.

References

Anuntalabhochai, S., Sitthiphrom, S., Thongtaksin, W., Sanguansermsri, M., Cutler, R.W., 2007. Sci. Hortic. 111, 389–393.
Apavatjrut, P., Anuntalabhochai, S., Sirirugsa, P., Alisi, C., 1999. Molecular markers in the identification of some early flowering *Curcuma* (Zingiberaceae) species. Ann. Bot. 84, 529–534.
Baker, J.G., 1890. Scitaminae In: Hooker, J.D. (Ed.), The Flora of British India, vol. VI L. Reeve and Co., London. 1890–1892 pp. 198–264.
Baker, J.G., 1898. Scitamineae in dyer. Flora Trop. Afr. 7, 311.
Balashanmugham, P.V., Chezhiyan, N., Ahmad Shah, H., 1986. BSR-1 turmeric. S Indian Hortic. 41 (3), 152–154.
Burtt, B.L., 1977. The nomenclature of turmeric and other Ceylon Zingiberaceae. Notes R. Bot. Gard., Edinburgh.
Burtt, B.L., Smith, R.M., 1972. Proposal to conserve 1351 *Curcuma* Roxb. (1810) non Linnaeus (1753). Taxon 21, 709–710.
Cao, H., Komatsu, K., 2003. Molecular identification of six medicinal *Curcuma* plants produced in Sichuan: evidence from plastid *trnK* gene sequences. Yao Xue Xue Bao 38, 871–875.
Cao, H., Sasaki, Y., Fushimi, H., Komatasu, K., 2001. Molecular analysis of medicinally used Chinese and Japanese *Curcuma* based on 18S rRNA gene and *trnK* gene sequences. Biol. Pharm. Bull. 24, 1389–1394.
Chakravorti, A.K., 1948. Multiplication of chromosome number in relation to speciation in Zingiberaceae. Sci. Cult. 14 (4), 137–140.
Cheziyan, N., Shanmugasundaram, K.A., 2000. BSR-2—A promising turmeric variety from Tamil Nadu. Indian J. Arecanut Spices Med. Plants 2, 24–26.
Das, A.B., Rai, S., Das, P., 1999. Karyotype analysis and cytophotometric estimation of nuclear DNA content in some members of the Zingiberaceae. Cytobios 97, 23–33.

Das, D., Bhattachargee, A., Biswas, I., Mukherjee, A., 2004. Foliar characteristics of some medicinal plants of Zingiberaceae. Phytomorphology 8, 291–302.

Fischer, C.E.C., 1928. In Gamble: The Flora of the Presidency of Madras. Pt 8. London.

Holtum, R.E., 1950. The Zingiberaceae of Malay Peninsula. Gard. Bull. Singapore 13, 1–249.

Hooker, J.D., 1894. The Flora of British India, vol. VI. L. Reeve and Co., London, p. 792.

Hutchinson, J., 1934. Families of Flowering Plants II Monocotyledons. Oxford University Press, London, p. 243.

Ibrahim, H., 1996. Isoenzyme variation in selected Zingiberaceae spp. In: Proceedings of the Second Symposium on the Family Zingiberaceae. South China Institute of Botany, Guanchou, pp. 142–149.

Islam, M.A., 2004. Genetic diversity of the genus Curcuma in Bangladesh and further biotechnological approaches for in vitro regeneration and long-term conservation of C. longa germplasm. Ph.D. Thesis, University of Hannover, Germany, p. 136.

Islam, M.A., Kloppstech, K., Esch, E., 2005. Population genetic diversity of Curcuma zedoaria (Chrism.) Roscoe—a conservation prioritized medicinal plant in Bangladesh. Cons. Genet. 6, 1027–1033.

Jalgaonkar, J., Patil, M.M.,Rajput, J.C., 1988. Performance of different varieties of turmeric under Konkan conditions of Maharashtra. In: Proceedings National Seminar on Chillies, Ginger and Turmeric. Hyderabad, India. pp. 102–105.

Jayasree, S., Sabu, M., 2005. Anatomical studies on the genus Curcuma L. (Zingiberaceae) in India—Part I. In: Pandey, A.K., Wen, J., Dogra, J.V.V. (Eds.), Plant Taxonomy: Advances and Relevance CBS Publishers and Distributors, New Delhi, pp. 475–492.

Joseph, R., Joseph, T., Joseph, J., 1999. Kayomorphological studies in the genus Curcuma Linn. Cytologica 64, 313–317.

Kress, W.J., Prince, L.M., Williams, K.J., 2002. The phylogeny and a new classification of the gingers (Zingiberaceae): evidence from molecular data. Am. J. Bot. 89 (11), 1662–1696.

Kumar, S., 1991. Turmeric (Curcuma longa L.) and related taxa in Sikkim Himalaya. J. Econ. Tax. Bot., 721–724.

Lad, S.K., 1993. A case of seed-setting in cultivated turmeric types (Curcuma longa Linn.) 1993. J. Soils Crops 3, 78–79.

Linnaeus, C., 1753. Species Plantarum. London.

Mangaly, J.K., Sabu, M., 1993. A taxonomic revision of the South Indian species of Curcuma (Zingibereae). Rheedea 3, 139–171.

Maurya, K.R., 1990. R.H. 10 a promising variety of turmeric to boost farmers' economy. Indian Cocoa Arecanut and Spices J. 13, 100–101.

Nair, R.R., Sasikumar, B., 2009. Chromosome number variation among germplasm collections and seedling progenies in turmeric (Curcuma longa L.). Cytologia 74 (2), 153–157.

Nair, R.R.,Dhamayanthi, K.P.M., Sasikumar, B., Padmini, K., Saji, K.V., 2004. Cytogenetics and reproductive biology of major spices. In: Rema, J. (Ed.), Annual Report 2003–2004 of Indian Institute of Spices Research, Calicut. pp. 38–41.

Nambiar, M.C.,1979. Morphological and cytological investigations in the genus Curcuma Linn. Ph.D. Thesis, University of Bombay, Bombay (now Mumbai), India, p. 95.

Nambiar, M.C., Pillai, P.K.T., Sarma, Y.N., 1982. Seedling propagation in turmeric Curcuma aromatica Salisb. J. Plant. Crops 10 (2), 81–85.

Nayak, S., Naik, P.K., Acharya, L.K., 2006. Detection and evaluation of genetic variation in 17 promising cultivars of turmeric (Curcuma longa L.) using 4C nuclear DNA content and RAPD markers. Cytologia 71 (1), 49–55.

Nazeem, P.N., Menon, R., Valsala, P.A., 1993. Blossom biological and hybridization studies in turmeric (Curcuma spp.). Indian Cocoa Arecanut Spices J. 16, 106–109.

Ngamriabsakul, C., Newman, M.F., Cronck, Q.C.B., 2004. The phylogeny of tribe Zingibereae (Zingiberaceae) based on ITS (nr DNA) and trn L-F (cp DNA) sequences. Edin. J. Bot. 60 (3), 483–507.

Oischi,J., 1996. Studies on growth and flowering behaviours, and on species grouping by isozyme analysis Curcuma. B.S. Thesis (in Japanese with English abstract), Yamaguchi University, Japan.

Paisooksantivatana, Y., Thepsen, O., 2001. Phenetic relationship of some Thai Curcuma species (Zingiberaceae) based on morphological, palynological and cytological evidences. Thai J. Agri. Sci. 34, 47–57.

Pathak, S., Patra, B.C., Mahapatra, K.C., 1960. Flowering behavior and anthesis in Curcuma longa. Curr. Sci. 29, 402.

Pillai, S.K., Pillai, A., Sachdeva, S., 1961. Root apical organization in monocotyledons—Zingiberaceae. Proc. Indian Acad. Sci. Ser. B 53, 240–256.

Pujari, P.P., Patil, R.B., Sakpal, R.T., 1986. Krishna—a high yielding variety of turmeric. Indian Cocoa Arecanut Spices J. 9, 65–66.

Purseglove, J.W., Brown, E.G., Green, C.L., Robin, S.R.J., 1981. Turmeric In: Spices, 2. Longman, Essex, UK, pp. 532–580.

Raghavan, T.S., Venkatasubban, K.R., 1943. Cytological studies in the family Zingiberaceae with special reference to chromosome number and cyto-taxonomy. Proc. Indian Acad. Sci. Ser. B 17, 118–132.

Raju, E.C., Shah, J.J., 1975. Studies in stomata of ginger, turmeric and mango ginger. Flora Bd. 164, 19–25.

Ramachandran, K., 1961. Chromosome numbers in the genus Curcuma Linn. Curr. Sci. 30, 194–196.

Ramachandran, K., 1969. Chromosome number in Zingiberaceae. Cytologia 34 (2), 213–221.

Rao, D.V.R., 1999. Effect of gamma irradiation on growth, yield and quality of turmeric. Adv. Hortic. For. 6, 107–110.

Rao, A.M., Jagadeeshwar, R., Sivaraman, K., 2006. Turmeric. In: Ravindran, P.N., Nirmal Babu, K., Shiva, K.N., Johny, A.K. (Eds.), Advances in Spices Research Agribios, Jodhpur, pp. 433–492.

Ratnambal, M.J., Babu, K.N., Nair, M.K., Edison, S., 1992. PCT 13 and PCT 14—two high yielding varieties of turmeric. J. Plant. Crops 20, 79–84.

Ratnambal, M.J., Nair, M.K., 1986. High yielding turmeric selection PCT-8. J. Planta. Crops. 14, 94–98.

Ravindran, P.N., Remashree, A.B.,Sherlija, K.K., 1998. Developmental morphology of rhizomes of ginger and turmeric. Final Report of Indian Council of Agricultural Research New Delhi Ad-hoc Project, Indian Institute of Spices Research, Calicut, Kerala State, India.

Ravindran, P.N., Babu, K.N., Shiva, K.N., 2007. In turmeric—The genus Curcuma. In: Ravindran, P.N., Babu, K.N., Shiva, K.N. (Eds.), Botany and Crop Improvement of Turmeric CRC Press, Boca Raton, FL, pp. 15–70.

Reddy, M.L.N., Rao, A.M., Rao, D.V.R., Reddy, S.A., 1989. Screening of short duration turmeric varieties/cultures suitable for Andhra Pradesh. Indian Cocoa Arecanut Spices J. 12, 87–89.

Rendle, A.B..1904. The Classification of Flowering Plants I. Gymnosperms and Monocotyledons. Cambridge, p. 403.

Renjith, D., Nazeem, P.A., Nybe, E.V., 2001. Response of turmeric (Curcuma domestica Val.) to in vivo and in vitro pollination. J. Spices Aromatic Crops 10 (2), 135–139.

Sabu, M.A., 1991. Taxonomic and phylogenetic study of South Indian Zingiberaceae. Ph.D. Thesis, University of Calicut, Kerala State, India, p. 322.

Sasaki, Y., Fushimi, H., Komatasu, K., 2004. Application of single-nucleotide polymorphism analysis of the *trnK* gene to the identification of the *Curcuma* plants. Biol. Pharm. Bull. 27, 144–146.

Sasikumar, B., George, J.K., Saji, K.V., Zachariah, T.J., 2005. Two new high yielding, high curcumin, turmeric (*Curcuma longa* L.) varieties—"IISR Kedaaram" and "IISR Alleppey" Supreme. J. Spices Aromatic Crops 14 (1), 71–74.

Sasikumar, B., Ravindran, P.N., George, J.K., 1994. Breeding ginger and turmeric. Indian Cocoa Arecanut and Spices J. 18 (1), 10–12.

Sato, D., 1948. The karyotype analysis in Zingiberaceae with special reference to the protokaryotype and stable karyotype Scientific Papers of the College of General Education, 10. University of Tokyo, Tokyo, Japan, 2, pp. 225–243.

Shah, J.J., Raju, E.C., 1975. General morphology, growth, and branching behavior of the rhizome of ginger, turmeric and mango ginger. New Bot. 11 (2), 59–69.

Sharma, A.K., Bhattacharya, N.K., 1959. Cytology of several members of Zingiberaceae. La Cellule 10, 297–346.

Sherlija, K.K., Unnikrishnan, K., Ravindran, P.N., 1999. Bud and root development of turmeric (*Curcuma longa* L.) rhizomes. J. Spices Aromatic Crops 8, 49–55.

Sherlija, K.K., Unnikrishnan, K., Ravindran, P.N., 2001. Anatomy of rhizome enlargement in turmeric (*Curcuma longa* L.). In Recent Research in Plant anatomy and Morphology. J. Econ. Tax. Bot. Addl. Ser. 19, 229–235.

Sirirugsa, P. 1997. Thai Zingiberaceae: species diversity and conservation. Invited lecture presented at the International Conference on Biodiversity and Bioresources: Conservation and Utilization, Phuket, Thailand, 23–27 November IUPAC 1999, pp. 1–7.

Skornickova, J., Sabu, M., 2005. *Curcuma roscoeana* Wall (Zingiberaceae) in India. Gard. Bull. Singapore 57, 187–198.

Skornickova, J., Sabu, M., Prasanthkumar, M.G., 2004. *Curcuma mutabilis* (Zingiberaceae)—A new species from South India. Gard. Bull. Singapore 56, 43–54.

Skornickova, J., Sida, O., Jarolimova, V., Sabu, M., Fer, T., Travnicek, P., et al., 2007. Chromosome numbers and genome size variation in Indian species of *Curcuma*. Ann. Bot. 100, 505–526.

Skornickova, J., Sida, O., Wijesundara, S., Marhold, K., 2008. On the identity of turmeric: the typification of *Curcuma longa* L (Zingiberaceae). Bot. J. Linn. Soc. 157, 37–46.

Sugiura, T., 1936. Studies on the chromosome number of the higher plants. Cytologica 7 (4), 544–595.

Syamkumar, S., Sasikumar, B., 2007. Molecular marker based genetic diversity analysis of *Curcuma* species from India. Sci. Hortic. 112, 235–241.

Trimen, H., 1887. Hermann's Ceylon herbarium and Linnaeus's "Flora Zeylanica". J. Lin. Soc. Bot. 24, 129–155.

Valeton, T.H., 1918. New notes on Zingiberaceae of Java and Malaya. Bull. Jard Buitenzorg Ser. II 27, 1–8.

Van Rheede, H.A., 1678. Hortus Indicus Malabaricus, vol. 1–12. Amsterdam. 1678–1693.

Velayudhan, K.C., Muralidharan, V.K., Amalraj, V.A., Gautam, P.L., Mandal, S.K., 1999. *Curcuma* Genetic Resources, Scientific Monograph No.4. National Bureau of Plant Genetic Resources, New Delhi, p. 149.

Watt, G.A., 1972. Dictionary of the economic products of India. 1908. Periodic. Experts, India, Rep. 1972, 689.

Xia, Q., Zhao, K.J., Huang, Z.G., Zhang, P., Dong, T.T.X., Li, S.P., et al., 2005. Molecular genetic and chemical assessment of *Rhizoma curcumae* in China. J. Agric. Food. Chem. 53 (15), 6019–6026.

3 Genetics of Turmeric

Turmeric has a rich genetic background. The genus *Curcuma* has about 100 species and Table 3.1 catalogs the most important of them. These varied species originate from South and Southeast Asia. Some of the floristic studies of the *Curcuma* genus are of those described by Gamble (1925), Hooker (1886), and Roxburgh (1832). In addition to *Curcuma longa*, the other economically important ones are *C. aromatica*, used in medicine and toiletry article manufacture. Also, those used in folklore medicines in Southeast Asia, such as *C. kwangsiensis*, *C. ochrorhiza*, *C. pierreana*, *C. zedoaria*, and *C. caesia* are important. *C. alismatifolia*, *C. roscoeana* have floricultural importance. The popular *C. amada* is used extensively for culinary use, such as in pickles and salads. *C. zedoaria*, *C. malabarica*, *C. pseudomontana*, *C. montana*, *C. decipiens*, *C. augustifolia*, *C. rubescens*, *C. haritha*, and *C. caulina* are used in the manufacture of arrowroot powder (Sasikumar, 2005). The other important species are *C. purpurescens*, *C. mangga*, *C. heyneana*, *C. xanthorrhiza*, *C. aeruginosa*, *C. phaeocaulis*, and *C. petiolata*.

Though it is widely believed and acclaimed that *Curcuma* originated in the Indo-Malaysian region, its spread and acclimatization in South and Southeast Asian regions seem to have religious connotations. The Hindu religion might have triggered its spread to different Asian countries during the post-Aryan period. According to Marco Polo, turmeric reached China in AD 700 (Ridley, 1912). Purseglove et al. (1981) stated that the people of Malaysia believed in a Malaysia–Polynesian connection in the origin of turmeric in that country. Burkill (1966) believed that the crop spread to West Africa in the thirteenth century and to East Africa in the seventeenth century. It was introduced to Jamaica in 1983. Its introduction to Central American countries is of recent origin. The genus *Curcuma*, considered to have originated in the Indo-Malaysian region (Purseglove, 1968), has a widespread occurrence in the tropical belt of Asia to Africa and also Australia. Of the 100 or so species reported in the *Curcuma* genus, about 40% is of Indian origin (Velayudhan et al., 1999).

Taxonomy of the *Curcuma* genus is still far from being clearly established. Some investigations on the anatomical and morphological characteristics of the *Curcuma* species and turmeric varieties have been made, but, scarcely, if any, molecular characterization. A few investigations on isozyme polymorphism and identification of species based on 18S rRNA and *trnK* genes have been made. Cytology of 12 species of *Curcuma* is reported. *C. longa*, cultivated turmeric, is now grown in India, China, Pakistan, Bangladesh, Vietnam, Thailand, The Philippines, Japan, Korea, Sri Lanka, Nepal, South Pacific Islands, East and West Africa, Malaysia, Caribbean Islands, and Central America. There is rich diversity of *Curcuma* in India, especially species and cultivar (Sasikumar et al., 1999). About 40 *Curcuma* species, 50 cultivars, and 20 improved varieties of *C. longa* and one improved variety of *C. amada* are available in India. Table 3.2 lists the different species in India.

The Agronomy and Economy of Turmeric and Ginger. DOI: http://dx.doi.org/10.1016/B978-0-12-394801-4.00002-8

Table 3.1 Turmeric Species of Economic Importance

S. No.	Species	Economic Importance
1.	*C. longa* L. syn *C. domestica* Val.	Dye manufacture, perfumery, aroma therapy, insect repellant, and in religious use
2.	*C. amada* Roxb, *C. mangga* Val. and Zijp.	Spice, medicine, pickles, salads
3.	*C. zedoaria* Roxb.	Folklore medicine, arrowroot manufacture
4.	*C. ochrorhiza* Val. and Van Zijp.	Malyasian folklore medicine
5.	*C. pierreana* Gagnep.	Vietnamese folklore medicine
6.	*C. aromatica* Salisb.	Medicine, toiletry articles, and insect repellant
7.	*C. kwangsiensis* S.G. Lec and C.F. Liang syn *C. chuanyujin*, *C. phaeocaulis* Val.	Chinese folklore medicine
8.	*C. caesia* Roxb.	Spice and medicine
9.	*C. comosa* Roxb.	Thai folklore medicine
10.	*C. angustifolia* Roxb., *C. zedoaria* Roxb.	Arrowroot manufacture
11.	*C. caulina* F.Grah., *C. psedomonantana* F.Grah	Arrowroot manufacture
12.	*C. montana* Roxb., *C. rubescens* Roxb.	Arrowroot manufacture
13.	*C. leucorrhiza*, *C. xanthorrhiza* Roxb.	Arrowroot manufacture
14.	*C. decipiens* Dalz., *C. malabarica* Val.	Arrowroot manufacture
15.	*C. raktakanta* Mangaly and Sabu	Arrowroot manufacture
16.	*C. haritha* Mangaly and Sabu	Arrowroot manufacture
17.	*C. aeruginosa* Roxb.	Arrowroot manufacture
18.	*C. alismatifolia* Gagnep., *C. thorelli*	Cut flower production
19.	*C. roscoeana* Wall.	Cut flower production

The Diversity of Turmeric Species

Of the 100 turmeric species, 41 are known to occur in India, of which at least 10 are endemic to the Indian subcontinent. The ecology of the species varies so much that their habitat ranges from the sea level (sandy coastal habitat) to high altitude, such as more than 2000 msl (mean sea level) in the Western Ghats and Himalayas in India. While species such as *C. longa*, *C. zedoaria*, *C. amada*, and *C. aromatica* are found predominantly in the plains; *C. angustifolia*, *C. neilgherrensis*, *C. kudagensis*, *C. thalakaveriensis*, *C. pseudomontana*, and *C. coriacea* are confined to hills at an altitude above 1000–2500 msl (Velayudhan et al., 1999). Species diversity is at its maximum in south and northeast India and the Andaman and Nicobar islands. Taxonomic revision of *Curcuma* genus is underway. The species *C. zedoaria* syn. and *C. xanthorrihza* are considered synonymous and *C. amada* closely resembles *C. mangga* inasmuch as quality attributes are concerned. Quite likely, the number of Indian species may be reduced to just 30. Similarly, it has now been established that the Chinese species *C. albicoma* and *C. chuanyujin* are synonyms of *C. sichuanensis* and *C. kwangsiensis*, respectively. The Chinese species *C. wenyujin* is now recognized as

Table 3.2 Indian *Curcuma* Species

Species	Statewise Geographic Distribution
C. aeruginosa	West Bengal
C. albiflora	Kerala
C. amada	Throughout India
C. amarissima	West Bengal
C. angustifolia	Uttar Pradesh, Madhya Pradesh, Himachal Pradesh, Northeast India
C. aromatica	Kerala, Tamil Nadu, Karnataka, Andhra Pradesh, Odisha, and Bihar
C. caesia	West Bengal
C. caulina	Maharashtra
C. comosa	West Bengal
C. petiolata	West Bengal
C. decipiens	Kerala and Karnataka
C. rubescens	West Bengal
C. ferruginea	West Bengal
C. longa	Throughout India
C. montana	Kerala, Karnataka, Tamil Nadu, and Andhra Pradesh
C. neilgherrensis	Same as above
C. oligantha	Kerala
C. pseudomontana	Kerala, Karnataka, Tamil Nadu, and Andhra Pradesh
C. reclinata	Madhya Pradesh
C. xanthorrhiza	West Bengal
C. zedoaria	Throughout India
C. sylvatica	Kerala
C. aurantiaca	Kerala, Karnataka, Tamil Nadu, and Andhra Pradesh
C. sulcata	Maharashtra
C. indora	Gujarat, Maharashtra, and Karnataka
C. ecalcarata	Kerala
C. soloensis	West Bengal
C. brog	West Bengal
C. haritha	Kerala
C. raktakanta	Kerala
C. kudagensis	Karnataka
C. thalakaveriensis	Karnataka
C. malabarica	Kerala and Karnataka
C. karnatakensis	Karnataka
C. cannanorensis	Kerala
C. vamana	Kerala
C. lutea	Kerala and Karnataka
C. coriacea	Kerala
C. nilamburensis	Kerala
C. leucorhiza	West Bengal

Source: Velayudhan et al. (1999).

a synonym of *C. aromatica*, while *C. phaeocaulis* was misidentified as *C. zedoaria*, *C. caesia*, and *C. aeruginosa* in China in the past (Liu and Wu, 1999). *C. kwang-siensis* var. *puberula* and var. *affinis* are not accepted and the identity of the Taiwan species *C. viridiflora* remains undecided (Liu and Wu, 1999). However, new species, such as *C. rhabdota* are also reported from Southeast Asia (Sirirugsa and Newman, 2000). *C. prakasha* sp. nov. from India (Tripathi, 2001) and *C. bicolor*, *C. glans*, and *C. rhomba* from Thailand (Mood and Larsen, 2001) have also been reported.

Curcuma spp.—Its Characterization

Employing numerical taxonomy, 31 species of *Curcuma* were investigated by Velayudhan et al. (1999). These could be clustered into nine groups in the den-dogram. In general, the sessile tuberizing species were distinct from the species without sessile tubers. They also collected extensive data on distribution, habitat, flowering time, floral characters, quantitative characters of the floral parts, features of above- and below-ground characters of the 31 species. Most of them exhibited distinguishable morphological features.

Molecular Characterization

The molecular characterization of *Curcuma* species is in its infancy. Randomly amplified polymorphic DNA profiling (RAPD) of rhizome DNA was done by Sreeja (2002) on five *Curcuma* species, namely, *C. longa*, *C. zedoaria*, *C. caesia*, *C. amada*, and *C. aromatica*. Three random decamer primers generated 11 polymor-phic bands among the species investigated. A novel attempt to identify the genuine *Curcuma* species traded as drug in China and Japan, based on sequence analysis of the 188 rRNA, *trnK* genes coupled with amplification refractory mutation sys-tem (ARMS) analysis was done (Sasaki et al., 2002). Though designed to identify the spurious *Curcuma* spp. in the marketed drug, this method is also helpful in the molecular taxonomy profiling of *Curcuma*. A polymerase chain reaction (PCR) tech-nique to identify genuine botanical *Curcuma* species in the marketed turmeric pow-der was developed by Sasikumar et al. (2004). RAPD markers to identify *C. longa* and *C. zedoaria* in the marketed turmeric powder have been reported in literature. Apavatjrut et al. (1999) investigated isozyme polymorphism in seven early flower-ing and two unidentified *Curcuma* species of Thailand. Of the 21 isozymes inves-tigated, 8 were found to be polymorphic and the species were grouped into distinct clusters depicting their polygenetic relationships. Comparative isozyme polymor-phism of the cultivated and natural populations from Thailand and Japan revealed that the cultivated population was far more uniform genetically (Paisookasantivatana et al., 2001). K nucleotide sequencing technique to identify six medicinal *Curcuma* species, namely *C. longa*, *C. phaeocaulis*, *C. sichuanensis*, *C. chuanyujin*, *C. chuanhuangjiang*, and *C. chuanezhu* found in Sichuan, in China, was employed by Cao et al. (2003). The new polygenetic analysis of Zingiberaceae based on DNA sequences of the nuclear internal transcribed spacer (ITS) and plastid mat K regions has been recently proposed (Kress et al., 2002). Results suggest that *Curcuma* is paraphyletic with *Hitchenia, Stahlianthus*, and *Smithatris*. A novel method of identification of *C. longa* and *C. aromatica* called "loop-mediated isothermal amplification" (LAMP) has also been reported (Sasaki and Nagumo, 2007).

Isozyme polymorphism in a germplasm collection of *C. longa* was investigated by Shamina et al. (1998). Acid phosphatase, superoxide dismutase, esterase, polyphenol oxidase, peroxidase, and catalase showed good polymorphism in 15 accessions investigated. Though turmeric is predominantly vegetatively propagated, the variability observed in the isozymes indicate the role of natural selection and conscious selection by human hand over the years in evolving locally adapted cultivars.

The Diversity in Turmeric Cultivars

Many local cultivars of turmeric are known mostly by the names of the locality. Moderate genetic variability exists in the crop and the cultivars vary in yield, duration, and quality. About 70 cultivars were identified. Region-specific varieties are cataloged in Table 3.3.

Both collections and species of *Curcuma* differ in their floral characteristics, aerial and rhizome morphology, and chemical constituents (Valeton, 1918; Velayudhan et al., 1999). India boasts more than 70 turmeric types belonging to *C. longa*, with a few belonging to *C. aromatica*. There are many popular cultivars, which are specific to each region where the crop is cultivated. *Duggirala, Armoor, Sugandham, Nandyal, Alleppey, Rajapuri, GL Puram, Bhavanisagar, Gorakpur, and Jabedi* are some of the more popular local cultivars, which have acquired their names from the regions where they are grown extensively (Nair et al., 1980). Existence of wide variability among the existing cultivars in respect of growth parameters, yield attributes, resistance to biotic and abiotic stresses, and quality characteristics was reported by many researchers (Velayudhan et al., 1999). The cultivars are grouped into short-duration, known as "Kasturi" types, medium-duration known as "Kesari" types, and long-duration, with no specific names as in the former cases (Rao and Rao, 1994).

Table 3.3 The Indian Turmeric Cultivars

Indian State	Cultivar Description
Andhra Pradesh	Duggirala, Mydukkur, Armoor Local, Cuddapah, Kodur, Tekurpet, Kasturi, Chaya Pasupu, Armoor, Amdapuram
Karnataka	Kasturi, Mudaga, Balaga, Cuddapah, Rajapuri, Amalapuram, Shillong
Kerala	Alleppey, Moovattupuzha, Wyanad Local, Tekurpetta, Armoor, Duggirala
Madhya Pradesh	Raigarh, Jangir, Bilaspur
Maharashtra	Krishna, Rajapuri, Sugandham
Odisha	Dindigam
Tamil Nadu	Erode, Salem
Northeast	Lakadong

Table 3.4 Indian Popular Local Cultivars

Cultivar Name	Yield (t/ha)	Remarks
Andhra Pradesh		
Kasturi Types—*C. aromatica*		
Kasturi Kothapeta	15–20	Susceptible to leaf spot
Kasturi Tanuku	12–15	Same as above
Kasturi Amalapuram Chaya pasupu	10–12	Same as above
Kesari Types—*C. longa*		
Kesari Duvvur	-	Susceptible to leaf blotch
Amruthapani Kothapeta	25	Same as above
Long-Duration Types		
Duggairala	32	Susceptible to leaf spot
Tekurpeta	-	Tolerant to leaf spot
Mydukur	32	Susceptible to leaf rot and leaf spot
Armoor	25	Susceptible to leaf rot and leaf spot
Sugandam	20–25	Susceptible to leaf blotch
Vontimitta	20	-
Nandyal	-	-
Ayanigadda	15–18	-
Tamil Nadu		
Erode	30–32	-
Salem	-	-
Kerala		
Alleppey	25	-
Mannuthy local	24	-
Assam		
Shillong	40	Tolerant to leaf blotch and leaf rot
Tall Karbi	30–40	Tolerant to leaf rot and leaf spot
Maharashtra and Gujarat		
Rajapuri	20	Resistant to leaf spot and Susceptible to leaf blotch and leaf rot
Eavaigon	45	-

Cultivars Armoor, Tekurpet, and Mydukkur are long-duration types, while cultivar Kothapeta is medium-duration type and Kasturi short-duration type. There is reasonable variation with regard to reaction to pests and diseases. Cultivars Mannuthy local and Kuchipudi are tolerant to shoot borer, while cultivars Mannuthy local, Tekurpets, and Kodur are tolerant to leaf blotch. Cultivars Suguna and Sudarshana are tolerant to the dreaded disease rhizome rot. Both foliar and rhizome diseases affect turmeric. Among the foliar diseases, leaf spot caused by *Colletotrichum capsici* and leaf blotch caused by *Taphrina maculans* are quite serious. Rhizome rot caused by *Pythium gramicolum* is the most serious malady of the crop. An important breeding objective is to identify disease-resistant varieties. The various collections of turmeric germplasm showed high variability in resistance to pests and diseases. Table 3.4 lists the details of popular turmeric cultivars.

Table 3.5 Genetic Variability in Turmeric Types

Type	Genetic Traits	Range	Good Cultivars and Accessions
C. domestica **Val. (*C. longa*)**			
	Dry recovery (%)	13.5–32.4	Ernad, Cl.No.5A, Amrithapani, Kothapetta, Amalapuram, Sel.III
	Oleoresin (%)	10.0–19.0	Chamakuchi, Kayyam, Gudalur, Palani Amalapuram, Sel.III, Rorathong E. Sikkim
	Curcumin (%)	2.8–10.9	Kaziranga, Jorhat, CII.328, Sugadam, Kayyam, Gudalur, Edapalayam, Erathkunnam, Palapally, Trichur
	Oil (%)	4.0–9.5	Ernad, Kakkayam Local, Kaahikuchi
C. aromatica **Salisb**			
	Dry recovery (%)	14.0–28.0	Burahazer, Dibrugarh Hahim Konni
	Oleoresin (%)	9.6–19.2	Konni
	Oil (%)	4.0–9.0	Nadavayal Kasturi, Armoor
	Curcumin (%)	2.3–8.0	Dibrugarh

Genetic Variability in Turmeric

Genetic variability in quality attributes of more than 100 collections of turmeric from India, belonging to both the *C. longa* and *C. aromatica* types, a few exotic ones and some wild collections were investigated with respect to oleoresin and essential oil contents, and dry recovery of harvested turmeric by Ratnambal and Nair (1986). Data on these are reported in Table 3.5. Dry recovery, oleoresin, and curcumin contents determine the quality of turmeric and high variability was observed in these important traits.

Remarkable genetic advancement coupled with high heritability in rhizome yield, crop duration, leaf number, number of primary fingers, yield of secondary fingers, and height of pseudostem was reported (Indiresh et al., 1992; Philip and Nair, 1986; Singh et al., 2003). These authors suggested that superior genotypes may be obtained by selection based on the number and weight of mother, primary, and secondary rhizomes. Turmeric yield is primarily influenced by length of primary fingers, mother rhizome diameter, and length and girth of secondary fingers. When these genetic traits are correlated with yield, the correlation was both positive and significant (Cholke, 1993; Rao, 2000). Leaf number, number of primary fingers, and crop duration had shown positive correlation with rhizome yield, at both genotypic and phenotypic levels (Panja et al., 2002; Reddy, 1987). Curcumin, oleoresin, and essential oils were negatively correlated with rhizome yield. For turmeric improvement, plant height, and leaf number have to be considered as they are important determinants of the potential for the genotypic yield (Narayanpur and Hanamashetti, 2003). From this, it can be concluded that plant height is the single most morphological trait while selecting genotypes for yield.

Genetic divergence investigations (D2 analysis) with 54 genotypes showed wide diversity among genotypes and were grouped in as many as six clusters (Rao, 2000). PCT-13 and Lakadong formed solitary groups and were genetically most distant. The land races of the northeast region of India were almost clustered in low- to moderate-yield groups, while genotypes from the southern region were scattered among different complexes ranging from moderate to high yielders (Chandra et al., 1997, 1999). PCT-10, 13, and 14 of shorter duration, medium yield, and good curcumin content were identified as potential parents in future breeding programs. There is extensive reference to genetic variability in turmeric influencing important yield attributes and yield (Chandra et al., 1999; Hazra et al., 2000; Lynrah et al., 1998). Environmental effects on curcumin content of turmeric varieties have been reported (Zachariah et al., 1998). The curcumin content of turmeric varieties show high location specific-genotypic effects. Turmeric cultivars are classified either based on their crop duration, as short, medium, and long or depending on the curcumin content, as low or high curcumin content varieties (Sasikumar et al., 1994).

The Conservation and Management of Turmeric Genetic Resources

India must be credited with the highest collection of genetic resources of turmeric. Within the country, the Indian Institute of Spices Research (IISR) at Calicut, Kerala State, under the administrative control and funding of the Indian Council of Agricultural Research (ICAR), New Delhi, has a good collection of turmeric germplasms maintained in field gene banks. There are other centers also maintaining gene banks. At IISR, these nucleus germplasms are planted in tubs to maintain purity, as planting in fields would lead to mixing and loss of purity. An *in vitro* gene bank of important genotypes is also maintained at IISR and the National Bureau of Plant Genetic Resources, New Delhi, (NBPGR, Geetha, 2002; Ravindran et al., 2004). *Ex situ* gene banks were also established at IISR, NBPGR, and in a Regional Research Station in Trichur, Kerala State. These gene banks are established by collecting germplasms from all over the country. The turmeric conservatory at IISR has 1040 accessions, 1018 cultivars, 16 acccessons of related taxa, and six exotic collections. The NBPGR Regional Research Station in Trichur has 650 accessions. In addition to the above, the central government-funded All India Coordinated Research Project on Turmeric (AICRPT) also maintains germplasm collection, details of which are given in Table 3.6.

The turmeric research centers under the AICRP on Spices, funded by the ICAR, maintain about 1280 germplasm accessions. Utilizing these germplasms, 22 high-yielding varieties, with good yield attributes and resistance to pests and diseases have been developed and released for widespread cultivation within the country in different agroclimatic conditions.

Management of turmeric genetic resources is a very complex process, starting from the identification of a target gene pool for conservation and their subsequent management, which demands precise attention on specific growth stages and

Table 3.6 AICRPT Germplasm Collections

AICRPT Center	Number of Cultivated Germplasms	Wild and Related Species	Total
Pottangi	197	-	197
Jagtial	273	-	273
Dholi	87	2	89
Raigarh	42	-	42
Kumarganj	126	-	126
Pundibari	136	14	150
Solan	145	-	145
Coimbatore	258	-	258
Total	1264	16	1280

Source: Annual report, AICRP on Spices, 2007.

other mutually dependent stages. Many of these activities not only generate but also require georeferenced data. Analysis of these data with Geographic Information System (GIS) technology can make the process more effective and time efficient. GIS technology has successfully been employed to study the spread of turmeric germplasms and also generate data on the incidence of pests and diseases.

Biodiversity and GIS Technology

Ecosystem diversity, species, and genetic diversity constitute biodiversity. This is a vast area of scientific scrutiny, as the number of parameters involved is quite large. Biodiversity investigations demand cataloging collections to establish relationships among varied data bases. GIS differentiation of plant populations reflects the dynamics of gene flow and natural selection. Sampling of geographically distinct populations is a practical approach to understand biodiversity of genetic variation. Sampling among distinct and diverse ecogeographic regions is specifically recommended for conserving genetic diversity in the case of rare and wild species. The application of geographic analysis (GA) of the collected germplasms helps predict the occurrence of species naturally, or artificially and where successfully introduced. GA is based on the deployment of three basic technologies, namely, global positioning system (GPS), remote sensing devices (RSD), and GIS technology. The analysis of spatial information with GIS technology introduces new strategies for understanding and exploiting patterns of geographic diversity and can be carried out efficiently with personal computers and GIS software. GIS is a mapping software that links information about where things are with information about what things are like. A GIS map can combine many layers of information to provide a full view of the crop in question. Hence, if the biodiversity data obtains the assistance of GIS to arrange its collection data, then it will be facile to accommodate all the necessary information at one place. A GIS analysis makes it possible to link, or integrate, information that is difficult to associate through any other means. Hence, a GIS can use combinations of mapped variables to build and analyze new variables.

A common strategy for sampling intraspecific genetic diversity is to maximize the sampling of geographically distinct populations. A GIS is a computer-based tool to map and analyze geographic phenomena that exist and events that occur on the Earth. Overall, GIS should be viewed as a technology, not simply as a computer system. It is an integrated set of hardware and software tools used for the manipulation and management of digital spatial (geographic) and related attribute data. Habitat loss and fragmentation are among the most common threats facing endangered species, making GIS-based evaluations an essential component of population viability analyses. Agricultural crop distribution is rarely limited to a crop's native range. Introductions of crops to new areas largely result in crop range. This may not provide optimum growing conditions. For sustainable agricultural production, land suitability analysis is an important prerequisite. It is an interdisciplinary approach, which includes information from different domains, such as crop and soil science, meteorology, social science, economics, and management. Being interdisciplinary, it deals with information that is measured in different scales, including ordinal, nominal, and ratio scales.

The turmeric crop can grow in diverse tropical conditions, from 1500 msl elevation, at a temperature range of 20–30°C, with an annual rainfall of 1500 mm or more, under rainfall or irrigated conditions. It grows in diverse soils, but does best in well-drained sandy or clay loams. Based on these facts, a land suitability map for turmeric has been prepared employing the ecocrop model of DIVA GIS (Utpala et al., 2007). It shows that suitable places like Assam, parts of Andhra Pradesh, and Bihar can grow turmeric of high quality. Assamese farmers grow excellent local varieties, such as Nowgam, which has 20% dry recovery, Hajo with 21% dry recovery, Barhola Jorhat with 25% dry recovery, Dardra Gauhati with 23.2% dry recovery, and Maran with 26% dry recovery (Ratnambal and Nair, 1986).

Turmeric and Intellectual Property Rights

The Intellectual Property Rights (IPR) issue is one of the most vexing questions internationally. A decade ago, a global trade row was kicked up on account of the infringement of IPR rights, and India had to fight its way through to settle the issues. The primary reason for this is the very high medicinal value of turmeric. The International Treaty on Plant Genetic Resources for Food and Agriculture recognizes the great importance of plant genetic resources and their proper conservation. The Treaty was adopted by the FAO Conference on November 3, 2001 and came into force on June 29, 2004. The sheer time span spent between the initial date of introduction of the Treaty and its final official recognition speaks volumes for the kind of pressures and/or counter-pressures in streamlining a very important Treaty, which will have far-reaching consequences to humankind. Article 5 relating to the Conservation, Exploration, Collection, Characterization, Evaluation, and Documentation of plant genetic resources says: "Each Contracting Party shall, in cooperation with other Contracting Parties, promote an integrated approach to the exploration, conservation and sustainable use of plant genetic resources for food and agriculture." And yet, one notes with utter dismay the kind of poaching and pirating of rare plant species around the developing world, through overt or covert collaboration of the natives, often by scientists and officials in responsible positions, and often through the sheer ignorance of the natives, as for instance, in the case of the very many tribal communities who cultivate these rare plant species and who are ignorant of the immense potential value of the rare plant genetic resources in the developed world, for pecuniary benefits. Turmeric is a classical example of this biopiracy.

India is a party to the Convention on Biological Diversity (CBD, 1992). CBD recognizes the sovereign rights of States to use their own biological resources. It expects the parties concerned to facilitate access to genetic resources by other Parties, subject to national legislation and on mutually agreed upon terms (Article 3 and 15 of CBD). Article 8(j) of the CBD recognizes the contributions of local and indigenous communities to the conservation and sustainable utilization of biological resources through traditional knowledge, practices, and innovations and provides for equitable sharing of the benefits that accrue with such people, arising from the utilization of their knowledge, practices, and innovations. However, the actual practice of these provisions is an exception to the rule, as has been exemplified by the Indian Basmati rice IPR fiasco, Neem and Turmeric IPR controversies, to name a few.

Controversial Patent Cases Involving Turmeric and Traditional Knowledge

Turmeric has been domesticated in Southeast Asian countries since very early times. It has properties which make it a very effective pharmaceutical and nutraceutical ingredient, cosmetics, and as a color dye. The following is a glaring example with regard to breach of IPR.

In 1995, two Indian nationals at the University of Mississippi Medical Center were granted US Patent No. 5401, 5504 on "use of turmeric in wound healing." The Council of Scientific and Industrial Research (CSIR) in New Delhi, India, requested the US Patent and Trademark Office (USPTO), to reexamine the granting of the patent. CSIR argued that turmeric has been in use for thousands of years in India as an *Ayurvedic* preparation (Indian traditional medicine) in formulating medical preparations, as it shows rare properties in healing wounds and rashes, and therefore, its medicinal value, as claimed in the US Patent was not original. The CSIR claim was supported by documentary evidence of traditional knowledge, including an ancient *Sanskrit* text and a research paper published in 1953 in the Journal of Indian Medical Association. Despite arguments by the patentees to the contrary, the USPTO upheld the objections raised by the CSIR and revoked the patents granted (Anon, 2007).

Protection of Plant Varieties

Another important aspect of the IPR is the Trade Related Intellectual Property Rights (TRIPS) agreement. After having the ratified TRIPS, India had to give effect to para 27(3) of Part II of TRIPS agreement in relation to plant varieties and would imply that India adopts an effective *sui generis* plant variety protection law. Registration and protection of plant varieties in India are covered under "The Protection of Plant Varieties and Farmers Rights Act, 2002," a *sui generis* system of plant variety protection. The idea behind this Act is to establish an effective and efficient, fool proof, system of protecting plant varieties, farmers' rights and the right of plant breeders to develop new plant varieties. Conditions of registration are that the material is "new" (novel), distinct, uniform, and stable for new varieties and distinct (at least in one plant trait), uniform, and stable for extant varieties. The guidelines for Distinctness, Uniformity, and Stability (DUS) testing in turmeric are prepared by Protection of Plant Varieties and Farmers Rights Authority (PPF & FRA), New Delhi, India, in collaboration with IISR and notified the public. As the DUS guidelines are notified, varieties are eligible for registration under this Act in India.

Geographical Indications

To protect the unique origin of products and/or organisms, the geographical indication (GI) concept has been developed. It is in this context that GI becomes relevant. For instance, "Malabar Pepper" is unique not just to the specific "Malabar" region in Kerala State, India, and hence, pepper grown elsewhere in the country cannot usurp this GI. The same holds good for Scotch whisky of Scotland or champagne of France. A GI is a sign for goods or brands that have a specific geographical origin and possesses qualities or reputation that are due to only that place of origin. Most commonly, a GI consists of the name of the place of origin of the good in question. Agricultural products typically have qualities that derive from their place of production and are influenced by specific local factors, such as climate and soil. Whether a sign functions as a GI is a matter of national law and consumer protection. GIs may be used for a wide variety of agricultural and horticultural products. One comes across hundreds of such products that bear a specific GI—Basmati rice, Benares silk, Ceylon (now Sri Lanka) tea, Indian curry, Swiss cheese, Russian Vodka, etc. Indian turmeric possesses unique features, and precise cataloging of these will help protect them through GI indication, which will eliminate unfair trade practices. Examples of possible Indian GI in turmeric are Lakadang Turmeric or Alleppey Finger Turmeric. Both command a huge market within the country and overseas.

References

Anon, 2007. Traditional knowledge and geographical indications. <http://www.iprcommission.org/graphic/documents/final_report.htm/>

Apavatjrut, P., Anuntalabhochai, P., Sirirugsa, P., Alisi, C., 1999. Molecular markers in the identification of some early flowering *Curcuma* L. (Zingiberaceae) species. Ann. Bot. 84, 529–534.

Burkill, T.H., 1966. A Dictionary of Economic Products of the Malay Peninsula. Ministry of Agriculture and Co-operatives, Kualalumpur.

Cao, H., Komatsu, K., 2003. Molecular identification of six medicinal *Curcuma* plants produced in Sichuan: evidence from plastid trnK gene sequences. Yao Xue Xue Bao 38, 871–875.

Chandra, R., Desai, A.R., Govind, S., Gupta, P.N., 1997. Metroglyph analysis in turmeric (*C. longa* L.) germplasm in India. Sci. Hortic. 70, 211–233.

Chandra, R., Govind, S., Desai, A.R., 1999. Growth, yield and quality performance of turmeric (*C. longa* L.) genotypes in mid altitudes of Meghalaya. J. Appl. Hortic. 1, 142–144.

Cholke, S.M., 1993. Performance of turmeric (*Curcuma longa* L.) cultivars. MSc (Ag) Thesis, University of Agricultural Sciences, Dharvad, India.

Gamble, J.S., 1925. Flora of the Presidency of Madras II. Botanical Survey of India, Calcutta, India.

Geetha, S.P., 2002. *In vitro* technology for genetic conservation of some genera of Zingiberaceae. PhD Thesis, Calicut University, Calicut, Kerala State, India.

Hazra, P., Roy, A., Bandhopadyay, A., 2000. Growth characters as rhizome yield components of turmeric. Crop. Res. 19, 235–246.

Hooker, J.D., 1886. The Flora of British India, vol. V. Reeve L and Co, London (Rep), pp. 78–95.

Indiresh, K.M., Uthaiiah, B.C., Reddy, M.J., Rao, K.B., 1992. Genetic variability and heritability studies in turmeric (*Curcuma longa* L.). Indian Cocoa Arecanut Spices J. 27, 52–53.

Kress, W.J., Prince, L.M., Williams, K.J., 2002. The phylogeny and a new classification of the gingers (Zingiberaceae): evidence from molecular data. Am. J. Bot. 89, 1682–1696.

Liu, N., Wu, T.L., 1999. Notes on *Curcuma* in China. J. Trop. Subtrop. Bot. 7, 146–150.

Lynrah, P.G., Barua, P.K., Chakrabarti, B.K., 1998. Pattern of genetic variability in a collection of turmeric genotypes. Indian J. Genet. Plant Breed. 28, 201–207.

Mood, J., Larsen, K., 2001. New *Curcuma* species from South East Asia. New Plantsman 8, 207–217.

Nair, M.K., Nambiar M.C., Ratnambal, M.J., 1980. Cytogenetics and crop improvement of ginger and turmeric. In: Proceedings of the National Seminar on Ginger and Turmeric, Calicut, India, pp. 15–23.

Narayanpur, V.B., Hanamashetti, S.I., 2003. Genetic variability and correlation studies in turmeric (*Curcuma longa* L.). J. Plant. Crops 31 (2), 48–51.

Paisookasantivatana, Y., Kako, S., Seko, H., 2001. Isozyme polymorphism in *Curcuma alismatifolia* Gagnep. (Zingiberaceae) populations from Thailand. Sci. Hortic. 88, 244–307.

Panja, B.N., De, D.K., Basak, S., Chattopadhyay, 2002. Correlation and path analysis in turmeric (*Curcuma longa* L.). J. Spices Aromatic Crops 11, 70–73.

Philip, J., Nair, P.C.S., 1986. Studies on variability, heritability and genetic advance in turmeric. Indian Cocoa Arecanut Spices J. 10, 29–30.

Purseglove, J.W., 1968. Tropical Crops: Monocotyledons. Longman, London.

In: Purseglove, J.W. Brown, E.G. Green, C.L. Robin, S.R. (Eds.), Spices, vol. II Longman, New York, NY, pp. 532–580.

Rao, A.M., 2000. Genetic variability, yield and quality studies in turmeric (*Curcuma longa* L.). PhD Thesis, ANGRAU, Rajendranagar, India.

Rao, M.R., Rao, D.V.R., 1994. Genetic resources of turmeric In: Chadha, K.L. Rathinam, P. (Eds.), Advances in Horticulture: Plantation Crops and Spices, vol. 9 Malhotra Publishing House, New Delhi, pp. 131–150.

Ratnambal, M.J., Nair, M.K., 1986. High yielding turmeric selection PCT 8. J. Plantation Crops 14, 91–98.

Ravindran, P.N., Nirmal Babu, K., Saji, K.V., Geetha, S.P., Praveen, K., Yamuna, G., 2004. Conservation of Spices genetic resources in in-vitro gene banks. ICAR Project Report. Indian Institute of Spices Research, Calicut, Kerala State, India, p. 81.

Reddy, M.L.N., 1987. Genetic variability and association in turmeric (*Curcuma longa* L.). Prog. Hortic. 19, 83–86.

Ridley, H.N., 1912. Spices. Macmillan, London.

Roxburgh, W., 1832. Flora Indica or Description of Indian Plants. Serampur, India, Carey, W. (Ed.).

Sasaki, Y., Nagumo, S., 2007. Rapid identification of *Curcuma longa* and *C. aromatica* by LAMP. Biol. Pharm. Bull. 30 (11), 2229–2230.

Sasaki, Y., Fushimi, H., Hui, C., Shao Qing, C., Komatsu, K., 2002. Sequence analysis of Chinese and Japanese *Curcuma* drugs on 18S rRNA genes and *trnK* gene and the application of amplification refractory mutation system analysis for their authentication. Biol. Pharm. Bull. 25, 1593–1599.

Sasikumar, B., 2005. Genetic resources of Curcuma: diversity, characterization and utilization. Plant Genet. Resour. 3 (2), 230–251.

Sasikumar, B., Ravindran, P.N., George, J.K., 1994. Breeding ginger and turmeric. Indian Cocoa Arecanut Spices J. 18, 10–12.

Sasikumar, B., Krishnamoorthy, B., Johnson, K.G., Saji, K.V., Ravindran, P.N., Peter, K.V., 1999. Biodiversity and conservation of major spices in India. Plant Genet. Resour. Newslet. 118, 19–26.

Sasikumar, B., Syamkumar, S., Remya, R., Zachariaj, T.J., 2004. PCR based detection of adulteration in market samples of turmeric powder. Food Biotechnol. 18, 299–306.

Shamina, A., Zachariah, T.J., Sasikumar, B., Johnson, K.G., 1998. Biochemical variation in turmeric (*C. longa*) accessions based on isozyme polymorphism. J. Horticult. Sci. Biotechnol. 73, 479–483.

Singh, G., Singh, O.P., de Lampasona, M.P., Catalan, C., 2003. *Curcuma amada* Roxb. chemical composition of rhizome oil. Indian Perfumer 47, 143–148.

Sirirugsa, P., Newman, M.A., 2000. A new species of *Curcuma* L. (Zingiberaceae) S.E. Asia. New Plantsman 7, 196–199.

Sreeja, S.G., 2002. Molecular characterization of *Curcuma* species using RAPD markers. MSc (Biotech) Thesis. Periyar University, Tamil Nadu, India.

Tripathi, S., 2001. *Curcuma prakasha* sp. nov. (Zingiberaceae) from North Eastern India. Nordic. J. Bot. 21, 549–550.

Utpala, P., Johny, A.K., Jayarajan, K., Parthasarathy, V.A., 2007. Site suitability for turmeric production in India—A GIS interpretation. Nat. Product Radiance 6 (2), 142–147.

Valeton, T.H., 1918. New notes on the Zingiberaceae of Java and Malaya. Bull. Jard Buitenzorg Ser II 27, 1–81.

Velayudhan, K.C., Muralidharan, V.K., Amalraj, V.A., Gautam, P.L., Mandla, S., Kumar, Dinesh, 1999. *Curcuma* genetic resources Scientific Monograph No. 4. National Bureau of Plant Genetic Resources, New Delhi.

Zachariah, T.J., Sasikumar, B., Nirmal Babu, K., 1998. Diversity for quality components in ginger and turmeric germplasm and interaction with environment. In: Sasikumar, B., Krishnamoorthy, B., Rema, J., Ravindran, P.N., Peter, K.V. (Eds.), Proceedings of the Golden Jubilee National Symposium on Spices, Medicinal and Aromatic Plants— Biodiversity, Conservation and Utilization Indian Institute of Spices Research, Calicut, Kerala State, India, pp. 16–120.

4 The Chemistry of Turmeric

There are various products obtained from turmeric, which have tremendous potential as commercial products. Turmeric, like other spices such as black pepper and cardamom, is available in the market as whole, ground, or as oleoresin. The institutional sector in western countries buys turmeric oleoresins, whereas in the industrial sector demand is more for whole turmeric; value-added products from turmeric include dehydrated turmeric powder, turmeric oils, oleoresins, and curcuminoids.

Turmeric Oil

The volatile turmeric oil is extracted from dried rhizomes, containing about 5–6% oil, and leaves about 1.5%. It is generally extracted by steam distillation, and processing and extraction techniques play a vital role in maximizing oil yield, pigments, and their constituents (Chempakam and Parthasarathy, 2008). Generally, the volatile oil is extracted by steam or hydrodistillation. Supercritical extraction, using carbon dioxide, is also used to extract volatile oil and oleoresin. The characteristic turmeric aroma is imparted by ar-turmerone, the major aroma principle in the oil (Chempakam and Parthasarathy, 2008).

Turmeric Oleoresin

Turmeric oleoresin is the organic extract of turmeric and is added to food items as a coloring agent. This is obtained by the extraction of the ground spice with organic solvents, such as acetone, ethylene dichloride, or ethanol, for 4–5h (Krishnamurthy et al., 1976). Supercritical CO_2 extraction and molecular distillation are the latest technologies followed to extract and separate turmeric oleoresin and to get a high-quality product. Supercritical fluid extraction (SFE) does not result in thermal degradation products or contamination of the solvent. The extraction rate of turmeric oil in supercritical CO_2 is a function of pressure, temperature, and flow rate at constant extraction time. The total yield decreased with temperature at constant flow rate, but increased with flow rate at constant pressure and temperature. The optimum pressure for the extraction was found to be 22.5MPa (Gopalan et al., 2000). Turmeric oleoresin is orange red in color. Its yield ranges from 5% to 15%. Curcumin, the principal coloring matter, forms about a third of the good-quality oleoresin. Its use is mainly in institutional cooking of meat and fish products, and also certain products, such as mustard, pickles, butter, and cheese.

Oleoresin is composed of 30–45% of curcuminoid pigments and 15–20% of volatile oil. The latter contains 60% turmerone, 25% zingiberene, and small quantities of

The Agronomy and Economy of Turmeric and Ginger. DOI: http://dx.doi.org/10.1016/B978-0-12-394801-4.00004-1

D-α-phellandrene, D-sabinene, cineole, and farnesol. It undergoes oxidative degradation. The pigments decompose on exposure to O_2. The hydroxyl groups of pigments
are converted to unstable ketones, which are further decomposed to colorless compounds with a short-carbon skeleton. Microencapsulation overcomes these problems.

Microencapsulation of Oleoresin

The specific advantages of turmeric oleoresin over ground turmeric are offset by
their sensitivity to light, heat, oxygen, and alkaline conditions. This can be overcome
by microencapsulation, which is defined as the technique of embedding minute particles of core material with a continuous polymer film that is designed to release its
core material under predetermined conditions. This also protects the flavors from
undesirable interactions (Amol et al., 2009). These investigators used gum arabic and maltodextrin to evaluate the appropriate wall material for encapsulation of
turmeric oleoresin by spray drying. To prepare the microencapsulated free-flowing
powder, the wall material was dissolved in water and oleoresin was dispersed
therein to obtain a dispersion or emulsion that has a continuous phase containing
the film-former, and a discontinuous phase containing the oleoresin. The emulsion
was stabilized using a stabilizer and/or emulsifying agent under continuous vigorous agitation. Gum acacia, dextrinized starch, maltodextrin, pectin, alginate, or a
proteinaceous material such as gelatin or casein is used as a stabilizer. Alternatively,
nonpolymeric emulsion stabilizers such as fatty acid partial esters of sorbitol anhydride (sorbitan or Span) and polyoxyethylene derivatives of fatty acid partial esters
of sorbitol anhydride (tween) are used. The dispersions are then emulsified by using
a homogenizer, which can then be spray-dried (Amol et al., 2009).

Volatiles of Turmeric

Dried and cured *C. longa* generally yields 1.5–3.0% volatile oil. *C. aromatica* is
comparatively much higher in volatile oil (4–8%) and low in curcuminoids
(1.5%). Turmeric owes its aromatic taste to the volatile oils present in the rhizome
(Chempakam and Parthasarathy, 2008).

The Constituents of Volatile Oils

Fresh turmeric oil from rhizomes obtained from French Polynesia is reported to
contain 20 components with zingiberene (16.7%), ar-turmerone (15.5%), and
α-phellandrene (10.6% as the major components; Lechat et al., 1996). GC and
GC–MS analysis of the rhizome oils from turmeric grown in Bhutan showed ar-turmerone content of 16.7–25%, α-turmerone (30.1–30.2%), and β-turmerone (14.7–
18.4%) as chief constituents (Sharma et al., 1997). Leaf oil from turmeric grown in
North Indian plains showed 20 compounds by GC–MS; dominated by monoterpenes
(57%), sesquiterpene hydrocarbons, and oxygenated mono- and sesquiterpenes contributed 10%, 3.3%, and approximately 0.1%, respectively. The major compounds
identified were: *p*-cymene (25.4%), 1,8-cineole (18%), *cis*-sabinol (7.4%), and

β-pinene (6.3%) (Garg et al., 2002). In another investigation, the rhizome oil of Chinese-origin turmeric indicated 17 chemical constituents of which turmerone 24%, ar-turmerone 18%, and germacrone 11% were considered the major constituents (Zhu et al., 1995).

Green leaves of *C. longa* from India for monoterpenes showed high concentration of α-phellandrene and terpinolene. The rhizome oil did not contain α-pinene, but included car-3-ene, α-terpinene, θ-terpinene, and terpinolene (McCarron et al., 1995). GC–MS analysis of the hydrodistilled oil from three *Curcuma* species, namely *C. longa*, *C. aromatica*, and *C. xanthorhiza*, indicated α, β content of 19–24% and 11–36% of turmerone, respectively and 4–14% of ar-turmerone (Kojima et al., 1998). Gopalan et al. (2000) compared the composition of steam-distilled essential oil and the oil extracted by supercritical CO_2 by GC–MS technique and found that of the 21 components identified, ar-turmerone and turmerone constituted nearly 60% of the total oil extracted. Analysis of the cyclohexane extract of turmeric by GC–MS coupled with pseudo-Sadtler retention indices revealed a series of saturated and unsaturated fatty acids along with sesquiterpenes. The fatty acids reported were tetradecanoic acid, *cis*-9-hexadecenoic acid, hexadecanoic acid, *cis–cis*-9, 12-octadecenoic acid, *cis–trans*-9-octadecenoic acid, octadecanoic acid, and eicosa-decanoic acid.

The constituents of essential oil from leaves, flowers, rhizomes, and roots of turmeric showed that the oils from rhizomes and roots were similar when compared to that from leaves and flowers. This indicates the presence of biologically linked traits (Leela et al., 2002). Volatile oils from flowers and leaves were dominated by monoterpenes, while major part of the oil from roots and rhizomes contained sesquiterpenes (Table 4.1). Bansal et al. (2002) analyzed oil from leaf lamina, petiole, stem, inflorescence, primary and secondary rhizomes, and rhizoids. Essential oil from the stem, rhizome, and rhizoid were similar in composition. Essential oils in rhizome were dominated by α- and ar-turmerones (40.8%), myrcene (12.6%), 1,8-cineole (7.7%), and *p*-cymene (3.8%). Essential oils in leaf, petiole, and lamina were dominated by myrcene (35.9%), 1,8-cineole (12.1%), and *p*-cymene (21.7%). Essential oil from the leaf of turmeric from Nigeria showed α-phellandrene (47.7%) and terpinolene (28.7%) as the major constituents (Table 4.2).

Turmeric Turmerones

The unique aroma of turmeric is derived from two major ketonic sesquiterpenes, namely ar-turmerone and turmerone ($C_{15}H_{20}O$ and $C_{15}H_{22}O$), which were identified in 1934. Additionally, *p*-cymene, β-sesquiphellandrene, and sesquiterpene alcohols have also been reported (Chempakam and Parthasarathy, 2008). Golding et al. (1982) demonstrated the structure of the two turmerones and defined them as 2-methyl-6-(4-methylcyclohexa-2,4-dien-1-yl)-hept-2-en-4-one (α-turmerone) and 2-methyl-6-(4-methylenecyclohex-2en-1-yl)-hept-2-en-4-one   (β-turmerone).The   ratio   of turmerone to ar-turmerone is reported to be 60:40, while analysis of the volatile oils from commercial oleoresin shows a ratio of 80:20 (Rupe et al., 1934; Salzer, 1977). Essential oil from rhizomes of different maturity levels grown in Sri Lanka

Table 4.1 The Composition of Essential Oils in Different Vegetative Parts of
Curcuma Longa L.

Constituent	Concentration (%)			
	Leaf	**Flower**	**Root**	**Rhizome**
α-Pinene	2.1	0.4	0.1	0.1
β-Pinene	2.8	0.1	0.1	Traces
Myrcene	2.3	0.2	Traces	0.1
α-Phellandrene	32.6	–	0.1	0.1
ω-3-Carene	1.1	0.6	–	–
α-Terpinene	1.3	0.1	–	–
p-Cymene	5.9	1.6	3.3	3.0
β-Phellandrene	3.2	Traces	–	Traces
1,8-Cineole	6.5	4.1	0.7	2.4
Z-β-Ocimene	0.2	–	–	–
E-β-Ocimene	0.4	–	–	–
r-Terpinene	1.5	–	–	–
Terpinolene	26.0	7.4	0.1	0.3
Linalool	0.7	1.1	0.1	–
1,3,8-Paramenthatriene	0.2	0.3	–	–
p-Methyl acetophenone	0.1	0.3	Traces	Traces
p-Cymen-8-ol	0.8	26.0	1.5	0.3
α-Terpineol	0.4	1.1	0.1	0.2
Thymol	0.3	–	0.1	–
Carvacrol	0.1	–	0.3	0.1
r-Curcumene	0.1	Traces	0.4	0.1
ar-Curcumene	0.2	1.9	7.0	6.3
α-Zingiberene	0.5	0.8	Traces	Traces
β-Bisabolene	–	0.9	2.3	1.3
β-Sesquiphellandrene	0.3	1.1	Traces	2.6
E-Nerolidal	0.1	1.1	–	–
Dehydrocurcumene	–	–	4.3	2.2
ar-Turmerone	0.1	1.2	46.8	31.1
Turmerone	0.9	1.0	–	10.0
Curlone	0.2	0.3	0.6	10.6
Curcuphenol	Traces	Traces	0.6	0.5
6S-7R-Bisabolene	0.1	0.4	1.2	0.9
Others	9.0	48.0	30.3	27.8

Source: Leela et al. (2002).

indicated that the ar-turmerone content ranged from 24.7% to 48.9%, while turmerone varied from 20% to 39%, both of which were the major constituents (Cooray et al., 1988). Ar-turmerone has been proved to be a potential antivenom agent (Ferreira et al., 1992). It is because of the biological importance of turmerones that attempts were made to isolate them in pure form. Negi et al. (1999) isolated turmerones from the mother liquor of turmeric oleoresin. Extraction of mother liquor

Table 4.2 Composition of Essential Oil of Curcuma Longa L. Leaves

Components	Composition (%)	Identification
α-Pinene	2.2	GC–MS, ^{1}H-NMR
β-Pinene + Myrcene	6.3	^{1}H-NMR
α-Phellandrene	47.7	RGC–MS, ^{1}H-NMR and ^{13}C-NMR
ω-3-Carene	1.2	GC–MS, ^{1}H-NMR
α-Terpinene	1.8	GC–MS–^{1}H-NMR
p-Cymene	1.2	GC–MS–^{1}H-NMR
Limonene + 1,8-cineole	6.0	GC–MS–^{1}H-NMR
r-Terpinene	2.0	GC–MS–^{1}H-NMR
Terpinolene	28.9	GC–MS–^{1}H-NMR
Sesquiphillandrene	0.8	GC–MS
α-Terpineol	<0.2	GC–MS
4-Terpineol	<0.2	GC–MS
Sabinol	<0.2	GC–MS

Source: Oguntimein et al. (1990).

with hexane followed by column chromatography using silica gel and elution with 5% ethyl acetate yields mainly ar-turmerone, turmerone, and curlone. Chang et al. (2006) isolated ar-turmerone from the supercritical CO_2 extracted turmeric oil by normal phase-silica-gel column chromatography. Kao et al. (2007) were able to enrich α-, β-, and ar-turmerone in oil up to 91.8%, using SFEs followed by solid–liquid column partition fractionation. Rhizome oils from five different *Curcuma* species showed ar-turmerone content in the range of 2.6–70.3%, ar-turmerone content in the range of trace to 46.2%, and zingiberene content in the range of trace to 36.8%. A number of sesquiterpenes have been identified in volatile oils of turmeric rhizome. Turmeronol A and B, two sesquiterpene ketoalcohols isolated from the dried rhizomes of turmeric, inhibited soybean lipoxygenase activity (Imai et al., 1990). Five sesquiterpenes, namely germacrone-1 3-al, 4-hydroxybisabola-2, 10-diene-9-one, 4-methoxy-5-hydroxy-bisabola-2,10-diene-9-one,2,5-dihydroxybisabola-3,10-diene, and procurcumadiol, were isolated and identified by nuclear magnetic resonance (NMR) (^{1}H and ^{13}C) spectroscopy. The investigation revealed more of bisabolene-type sesquiterpenes in turmeric (Oshiro et al., 1990).

Turmeric Curcuminoids

The investigator Vogel, as early as 1815 and Pelletier in 1818 (Govindarajan, 1980), isolated curcumin ($C_{21}H_{20}O_6$, mp 184–185°C). It is insoluble in water, but soluble in ethanol and acetone. The physicochemical properties of the three curcuminoids are detailed in Table 4.3.

The structure of curcumin as diferuloylmethane and its synthesis was confirmed by Lampe et al. (1910) and also by Majeed et al. (1995). The main yellow-colored substance in the rhizome is curcumin (1,7-bis-(4-hydroxy-3-methoxy

Table 4.3 Physicochemical Properties of Curcuminoids

	Name of Chemical		
	Diferuloylmethane	**4-Hydroxy Cinnamoyl (Feruloyl) Methane**	**Bis-4-Hydroxy Cinnamoyl Methane**
Common name	Curcumin	Demethoxycurcumin	Bis-demethoxycurcumin
Molecular weight	368.4	338.0	308.1
Melting point (°C)	184–186	172.5–174.5	224
Solubility			
Hexane or ether	Insoluble	Insoluble	Insoluble
Alcohol/acetone	Soluble	Soluble	Soluble
Relative Rf			
Chloroform: Ethanol (25:1)	1	0.66	0.41

prenyl)-1,6-heptadiene-3,5-dione) and two related demethoxy compounds, demethoxy curcumin and bis-demethoxycurcumin, which belong to the group of diarylheptanoids. Curcumin gives vanillic and ferulic acids on boiling with alkali. The UV-absorption spectra of these three curcuminoids vary slightly, with curcumin at 429 nm, demethoxy curcumin at 424 nm, and bis-demethoxy curcumin at 419 nm. The three curcuminoids also exhibit fluorescence under UV light. Using thin-layer plates, the curcuminoids can be directly estimated by fluorescence–densitometer by irradiation at 350 nm (Jentzsch et al., 1959). The fluorescence spectra of curcuminoids exhibited distinct excitation at 435 nm, and emission at 520 nm (Maheshwari and Singh, 1965). The percentage contribution of curcumin, demethoxy curcumin, and bis-demethoxy curcumin in oleoresin was estimated to be 60:30:10, 47:24:29, and 49:39:22 by different investigators (Jentzsch et al., 1959; Perotti, 1975; Roughley and Whiting, 1973).

Curcumin or curcuminoids concentrate for use in coloring food products is not considered a regular article of commerce, because for most of the current uses the cheaper substitute, oleoresin, has been found suitable. Curcumin is included in the list of colors with a restricted use because it has low acceptable daily intake (ADI) of 0–1.0 mg/kg of body weight/day. Curcumin gives a bright yellow color even at doses as low as 5–200 ppm. A variety of blends are available to suit the color of the product (Henry, 1998). Application of iron, manganese, and zinc influenced the curcumin content of turmeric genotypes (Jirali et al., 2007). Bos et al. (2007) determined the curcuminoid contents in four *Curcuma* species, namely *C. mangga*, *C. heyneana*, *C. aeruginosa*, and *C. soloensis*, which were found to vary between 0.18–0.47%, 0.98–3.21%, 0.02–0.03%, and 0.40%, respectively.

In addition to the three forms of curcuminoids, three minor constituents have also been identified which are considered as geometrical isomers of curcumin. Heller (1914) isolated an isomer of curcumin with a diketone structure. Utilizing liquid chromatography–electrospray ionization tandem–mass spectrometry

(LC–ESI–MS/MS) coupled with DAD analysis, 19 diarylheptanoids were identified from fresh turmeric rhizome extracts (Jiang et al., 2006). Kiuchi et al. (1993) isolated cyclocurcumin, possessing nematicidal activity. It had the same molecular formula of curcumin, but differed in structure by an intramolecular Michael addition of the enol-oxygen to the enone group.

Specific Beneficial Properties of Curcuminoids

The human body is adversely affected by external pollutants and mutagens, such as smoke, contaminated air, and water. Curcumin provides a body protection against these. Recent investigations indicate that curcuminoids are active in the external treatment of certain cancerous conditions, and this is related to the cytotoxicity of these substances. Abhishek and Dhan (2008) reported that curcumin possessed higher superoxide anion scavenging activity compared to demethoxycurcumin or bis-demethoxycurcumin. Curcumin acts as a prooxidant in the presence of transition metal ions, such as copper and iron, and is a potent bioprotectant with a wide range of therapeutic applications. Erika et al. (2003) investigated the stability of three curcuminoids and found that curcumin I (curcumin) was least stable and curcumin III (bis-demethoxycurcumin) is the most stable.

Despite the surprisingly wide range of chemopreventive and chemotherapeutic properties of curcumin, the clinical application is limited by its chemical instability and poor metabolic property. The reactive α-diketone moiety of curcumin is responsible for the pharmacokinetic limitation. The stability and pharmacokinetic profiles of curcumin were significantly improved by synthesizing monocarbonyl analogs of curcumin (Liang et al., 2009). Limtrakul et al. (2004) reported that bis-demethoxycurcumin is the most active of the curcuminoids present in turmeric for modulation of *MDR-1* gene in multidrug-resistant human cervical carcinoma cell line, KV-VI. The modulation of *MDR-1* expression may be an attractive target for new chemosensitizing agents.

Extraction and Estimation of Curcumin

A popular turmeric variety, "Alleppey," grown in Kochi district (central Kerala State, India), was found to contain 5.4% curcumin, and was found to be the ideal raw material for oleoresin extraction (Mathew, 1989). As curcuminoids have closely related physical and chemical properties, isolation of curcumin by usual chromatographic methods results in simultaneous crystallization of the three forms. Yet, several methods on isolation of curcuminoids have been reported. Some of these are listed below.

Using acetone for selective extraction of curcuminoids from *C. longa*, a microwave-assisted extraction (MAE) technique was developed by Dandekar and Gaikar (2002). These investigators succeeded in extracting up to 60% curcuminoids of

75% purity by this technique. The same investigators developed a hydrotropy-based extraction technique for selective extraction of curcuminoids from *C. longa* (Dandekar and Gaikar, 2003). The two-step process involving hydrotropic solubilization followed by dilution using water with or without pH adjustment, resulted in good curcuminoid yield of high purity. Sodium cumene sulfonate (Na–CuS) was found to be an effective hydrotropic extractant for curcuminoids. Though there are several methods reported for estimation of curcuminoids, the high precision liquid chromatography (HPLC) method is the most reliable to determine individual curcuminoid. Bos et al. (2007) determined curcumin, demethoxycurcumin, and bis-demethoxy curcumin in rhizomes of *Curcuma* species by the HPLC method with photodiode array detector. Separation and purification of individual curcuminoid was accomplished employing standard high-speed counter-current chromatography (CCC), as well as by pH-zone-refining CCC. The latter technique was able to separate multigram quantities of curcumin and other curcuminoids from crude curcumin and turmeric powder, while maintaining high purity (Patel et al., 2000). Inoue et al. (2003) evaluated the curcuminoids in food products using liquid chromatography (LC) or mass spectrometry (MS) coupled with an electrospray ionization interface. The detection limit of curcumin by this method is 1.0 ng/ml.

Simple two-directional high-performance thin-layer chromatography (HPTLC) method for simultaneous determination of curcumin, metanil yellow, and Sudan dyes in turmeric, chili, and curry powders was developed by Sumita et al. (2008). The method offers resolution of curcumin, demethoxy curcumin, and bis-demethoxy curcumin using the mobile phase chloroform–methanol (9:1), followed by second-directional mobile phase toluene–hexane–acetic acid (50:50:1). Pathania et al. (2006) developed a rapid, simple HPTLC method for quantification of curcuminoids employing a mobile phase chloroform–methanol (98:2). Simultaneous determination of curcuminoids in *Curcuma* samples using HPTLC was developed by Gupta et al. (1999). Paramasivam et al. (2008, 2009) also evaluated the effectiveness of HPTLC technique to separate curcuminoids using precoated silica gel—60 GF 254 thin-layer chromatography (TLC) plates and chloroform:methanol (48:2, v/v). Goren et al. (2009) developed the ^{1}H-NMR technique for the quantification of curcumin in the rhizome of *C. longa*.

Biosynthesis of Curcuminoids

Since time immemorial, turmeric has received much attention in India on account of its wide-ranging properties as a culinary spice and a medicinal one as well. Initial investigations into the biosynthesis of curcuminoids were carried out a quarter century ago. Radio-tracer feeding investigations suggested that these compounds are derived from intermediates in the phenylpropanoid pathway that are condensed with other molecules, derived in turn from the acetate and short- and medium-chain fatty acid pathways. The structure established for curcumin suggested the reasonable biosynthetic mechanism involving two cinnamoyl units (Geissman and Crout, 1969; Pabon, 1964). The proposed scheme was tested by Roughley and Whiting (1973) by incorporating labeled precursors. They concluded that an alternate scheme might exist for curcumin biosynthesis involving a cinnamate starter extending by five acetate or malonate units.

Cyclization of this chain would give the second aromatic ring. Subsequently, the biosynthesis would be completed by hydroxylation and methylation. Tracer investigations by the same authors using tritated cinnamic acids (ferulic, coumaric, and caffeic acids) along with ^{1-14}C-phenylalanine revealed higher incorporation of phenylalanine into curcumin. They also found that caffeic acid was relatively unacceptable as a precursor, supporting their proposed scheme.

Phenylalanine ammonia lyase (PAL), the first enzyme of the biosynthetic sequence, and a flavonoid glucosyltransferase, appear to be located in the lumen of the membranes. Cinnamate-4-hydroxylase is membrane embedded, while other enzyme activities appear to be weakly associated with the cytoplasmic phase of endoplasmic reticulum membranes (Hrazdina and Wagner, 1984). PAL, which initiates the series of reactions leading to curcumin synthesis, was investigated during the early germination phase. The activity was maximum in leaves compared to roots, rhizomes, and pseudostem, indicating that the conversion of phenylalanine to cinnamic acid mostly takes place in leaves. Investigations on the localization of PAL activity in various cell fractions showed maximum activity in microsomal fraction. Employing tracer technique with labeled phenylalanine and malonate, it was seen that the cinnamate pathway might lead to curcumin synthesis, ruling out the alternate pathway involving acetate (Neema, 2005).

Recently, curcuminoid synthase has been found capable of forming the curcuminoids in turmeric (Maria et al., 2006). This activity required malonyl-CoA and phenylpropanoid pathway derived hydroxycinnamoyl-CoA esters as substrates, suggesting that the corresponding protein is a polyketide synthase or an enzyme which is closely related. It is postulated that this activity could be the result of single enzyme or of multiple enzymes in sequence.

A novel type III polyketide synthase named curcuminoid synthase, which synthesizes bis-demethoxycurcumin via a unique mechanism from two 4-coumaroyl-CoAs and one malonyl-CoA has been found (Katsuyama et al., 2007). The reaction begins with the thioesterification of the thiol moiety of Cys-174 by a starter molecule, 4-coumaroyl-CoA. Decarboxylative condensation of the first extender substrate, malonyl-CoA, onto the thioester of 4-coumarate results in the formation of a diketide-CoA intermediate. Subsequent hydrolysis of the intermediate yields an α-keto acid, which in turn, acts as the second extender substrate. The α-keto acid is then joined to the Cys-174-bound 4-coumarate by decarboxylative condensation to form bis-demethoxycurcumin. The reaction violates the traditional head-to-tail model of polyketide assembly; the growing diketide intermediate is hydrolyzed to a β-keto acid which subsequently serves as the second extender to form curcuminoids. Curcuminoid synthase appears capable of synthesizing not only diarylheptanoids but also gingerol analogs, because it synthesized cinnamoyl (hexanoyl) methane, a putative intermediate of gingerol, from cinnamoyl-CoA and 3-oxo-octanoic acid. ^{13}C- incorporation investigations by Kita et al. (2008) indicated that the possibility of utilization of two cinnamoyl-CoAs and one malonyl-CoA in the formation of curcumin.

The pathway proposed by Katsuyama et al. (2009) for biosynthesis of curcuminoids includes two type III polyketide synthases, namely diketide-CoA synthase (DCS) and curcumin synthase (CURS). According to this investigator, DCS synthesizes feruloyldiketide-CoA, and CURS, which then converts the diketide-CoA esters into a curcuminoid scaffold. Diketide-CoA synthase catalyzes the formation feruloyldiketide-CoA by condensing feruloyl-CoA and malonyl-CoA. Curcumin synthase catalyzes the *in vitro* formation of curcuminoids from cinnamoyldiketide-*N*-acetylcysteamine and feruloyl-CoA. Co-incubation of DCS and CURS in the presence of feruloyl-CoA and malonyl-CoA yielded curcumin at high efficiency, although CURS itself possessed only low activity for the synthesis of curcumin from feruloyl-CoA and malonyl-CoA. Detailed enzyme systems involved in the biosynthetic pathway is yet to be elucidated.

Turmeric, well known for its coloring and flavoring properties, has unique biological properties and their beneficial effects have been correlated with its chemical constituents. This naturally enhances the position of the crop from a mere food additive to one as a substantial growth-enhancing product. Key to tapping its full potential as perhaps the world's most important medicinal spice, will be in focused modern research on the above lines, so that turmeric comes to enjoy its real preeminent position as a boon to human health, amply and unequivocally recorded in the Indian scriptures.

References

Abhishek, N., Dhan, P., 2008. Chemical constituents and biological activities of turmeric constituents (*Curcuma longa* L.)—a review. J. Food Sci. Technol. 45 (2), 109–116.

Amol, C.K., Vishwajeet, B.Y., Sarkar, A., Rekha, S.S., 2009. Efficacy of pullulan in emulsification of turmeric oleoresin and its subsequent microencapsulation. Food Chem. 113, 1139–1145.

Bansal, R.P., Bahl, J.R., Garg, S.N., Naqvi, A.A., Kumar, S., 2002. Differential chemical composition of the essential oils of the shoot organs, rhizomes and rhizoids in the turmeric *Curcuma longa* grown in Indo-Gangetic Plains. Pharm. Biol. 40 (5), 384–389.

Bos, R., Windono, T., Woerdenbag, H.J., Boersma, Y.L., Koulman, A., Kayser, O., 2007. HPLC-photodiode array detection analysis of curcuminoids in *Curcuma* species indigenous to Indonesia. Phytochem. Anal. 18 (2), 118–122.

Chang, L.H., Jong, T.T., Huang, H.S., Nien, Y.F., Chang, C.M., 2006. Supercritical carbon dioxide extraction of turmeric oil from *Curcuma longa* Linn. and purification of turmerones. J. Sep. Purif. Technol. 47 (3), 119–125.

Chempakam, B., Parthasarathy, V.A., 2008. Turmeric. In: Parthasarathy, V.A., Chempakam., B., John Zacharraiah, T. (Eds.), Chemistry of Spices CABI Pub., UK, pp. 97–123.

Cooray, N.F., Jansz, S.R., Ranatnga, J., Wimalasena, S., 1988. Effect of maturity on some chemical constituents of turmeric (*Curcuma longa* L.). J. Nat. Sci. Counc. Sri Lanka 16, 39–51.

Dandekar, D.V., Gaikar, V.G., 2002. Microwave assisted extraction of curcuminoids from *Curcuma longa*. Sep. Sci. Technol. 37 (11), 2669.

Dandekar, D., Gaikar, V., 2003. Hydrotropic extraction of curcuminoids from turmeric. Sep. Sci. Technol. 38 (5), 1185.

Erika, P., Simone, H., Aniko, M.S., Manfred, M., 2003. Studies on the stability of turmeric constituents. J. Food Eng. 56, 257–259.

Ferreira, L.A.F., Hneriques, O.B., Andreoni, A.A.S., Vittal, G.R.F., Campos, M.M.C., Habermehl, G.G., 1992. Antivenom and biological effects of ar-turmerone isolated from *Curcuma longa* (Zingiberaceae). Toxicon 30, 1211–1218.

Garg, S.N., Mengi, N., Patra, N.K., Charles, R., Kumar, S., 2002. Chemical examination of the leaf essential oil of *Curcuma longa* L. from North Indian plains. Flavour Frag. J. 17, 103–104.

Geissman, T.A., Crout, D.H.G., 1969. Organic Chemistry of Secondary Plant Products. Freeman Cooper, San Francisco, CA.

Golding, B.T., Pombo, E., Christopher, J.S., 1982. Tumerones: isolation from turmeric and their structure determination. J. Chem. Soc. Chem. Commun. 16, 363–364.

Gopalan, B., Goto, M., Kodama, A., Hirose, T., 2000. Response surfaces of total oil yield of turmeric (Curcuma longa) in supercritical carbon dioxide. Food Res. Int. 33, 341–345.

Goren, A.C., Cykrykcy, S., Cergel, M., Bilsel, G., 2009. Rapid quantification of curcumin in turmeric via NMR and LC-tandem mass spectrometry. Food Chem. 113 (4), 1239–1242.

Govindarajan, V.S., 1980. Turmeric chemistry technology and quality. Crit. Rev. Food Sci. Nutr. 12 (3), 199–301.

Gupta, A.P., Gupta, M.M., Kumar, S., 1999. Simultaneous determination of curcuminoids in Curcuma samples using high performance thin layer chromatography. J. Liq. Chromatogr. Related Technol. 22 (10), 1561–1569.

Heller, G., 1914. Ber. Dtsch. Chem. Ges. 47, 2998. Cited by Govindarajan (1980).

Henry, B., 1998. Use of capsicum and turmeric as natural colors. Indian Spices 35 (3), 7–14.

Hrazdina, G., Wagner, G.J., 1984. Metabolic pathways as enzyme complexes: evidence for the synthesis of phenylpropanoids and flavonoids on membrane associated enzyme complexes. Arch. Biochem. Biophys. 237 (1), 88–100.

Imai, S., Morikiyo, M., Furihata, K., Hayakawa, Y., Seto, H., 1990. Turmeronol A and turmeronol B, new inhibitors of Soybean lipooxygenase. Agric. Biol. Chem. 54, 2367–2371.

Inoue, K., Hamasaki, S., Yoshimura, Y., Yamada, M., Nakamura, M., Yoshio, I., et al., 2003. Validation of LC/electroscopy–MS for determination of major curcuminoids in foods. J. Liq. Chromatogr. Related Technol. 26 (1), 53–62.

Jentzsch, K., Gonda, T., Holler, H., 1959. Paper chromatographic and pharmacological investigation on Curcuma pigments. Pharm. Acta Helv. 34, 181–188.

Jiang, H.L., Barbara, N.T., Gang, R.D., 2006. Use of liquid chromatography-electrospray inonization tandem, mass spectrometry to identify dairylheptanoids in turmeric rhizome. J. Chromatogr. A 1111 (1), 21–31.

Jirali, D.I., Hiremath, S.M., Chetti, M.B., Patil, S.A., 2007. Biophysical, biochemical parameters, yield and quality attributes as affected by micronutrients in turmeric. Plant Arch. 7 (2), 827–830.

Kao, L., Chen, C.R., Chang, C.M., 2007. Supercritical CO_2-extraction of turmerones from turmeric and high-pressure phase equilibrium of CO_2+turmerones. J. Supercritical Fluids 43 (2), 267–282.

Katsuyama, Y., Matsuzawa, M., Funa, N., Horinouchi, S., 2007. In vitro synthesis of curcuminoids by type III polyketide synthase from Oryza sativa. J. Biol. Chem. 282 (52), 37702–37709.

Katsuyama, Y., Kita, T., Funa, N., Horinouchi, S., 2009. Curcuminoid biosynthesis by two type III polyketide synthases in the herb Curcuma longa. J. Biol. Chem. 284 (17), 11160–11170.

Kita, T., Imai, S., Sawada, H., Kumagai, H., Seto, H., 2008. The biosynthetic pathway of curcuminoid in turmeric (Curcuma longa) as revealed by ^{13}C-labeled precursors. Biosci. Biotechnol. Biochem. 72 (7), 1789–1798.

Kiuchi, F., Goto, Y., Sugimoto, N., Akao, N., Kondo, K., Tsuda, Y., 1993. Nematicidal activity of turmeric: synergistic action of curcuminoids. Chem. Pharm. Bull. 41 (9), 1640–1643.

Kojima, H., Yanai, T., Toyota, A., Hanani, E., Saiki, Y., 1998. Essential oil constituents from Curcuma aromatica, C. longa, and C. xanthorrhiza rhizomes. In: Ageta., H., Aimi., N., Ebizaka., Y., Fujita, T., Honda, G. (Eds.), Towards Natural Medicine Research in the 21st Century Elsevier, Amsterdam, pp. 531–539.

Krishnamurthy, N., Mathew, A.G., Nambudiri, E.S., Shivasankar, S., Lewis, Y.S., Natarajan, C.P., 1976. Oil and oleoresin of turmeric. Trop. Sci. 18 (1), 37.

Lampe, V., Milobedeska, J., Kostanecki, V., 1910. Synthese von *p, p'*-dioxy-and *p*-oxy dicinnamolymethane. Ber. der Deutsch. Chem. Gessellschaft 43, 2163.

Lechat, V.I., Menut, C., Lamaty, G., Bessiere, J.M., 1996. Huiles essentielles de Polynesia Franqais Rivista Italiana Eppos (Numero Speciale), 627–638.

Leela, N.K., Aldo, T., Shaji, P.M., Sinu, P.J., Chempakam, B., 2002. Composition of essential oils from turmeric (*Curcuma longa* L.). Acta Pharm. 52, 137–141.

Liang, G., Shao, L., Wang, Y., Zhao, C., Chu, Y., Xiao, Y., et al., 2009. Exploration and synthesis of curcumin analogues with improved structural stability both *in vitro* and *in vivo* as cytotoxic agents. Bioorg. Med. Chem. 17 (6), 2623–2631.

Limtrakul, P., Anuchapreeda, S., Buddhasukh, D., 2004. Modulation of human multidrugresistance *MDR-1* gene by natural curcuminoids. BMC Cancer 4, 13–16.

Maheshwari, P., Singh, U., 1965. Dictionary of Economic Plants in India. Indian Council of Agricultural Research, New Delhi.

Majeed, M., Badmaev, V., Shivakumar, U., Rajendran, R., 1995. Curcuminoids—Antioxidant Phytonutrients. Nutriscience Publishers Inc., Piscataway, NJ.

Maria, C.R. del, Barbara, N.T., David, R.G., 2006. Biosynthesis of curcuminoids and gingerols in turmeric (*Curcuma longa*) and ginger (*Zingiber officinale*): identification of curcuminoid synthase and hydroxycinnamoyl-CoA thiosterases. Phytochemistry 67, 2017–2029.

Mathew, A.G., 1989. Oil and oleoresin from Indian spices. Proceedings 11th International Congress of Essential Oils, Fragrances, and Flavours. New Delhi, India, 12–16 November, 1989, vol. 4. Chemistry—Analysis and Structure. Aspect Publishing, London, 1990, pp. 189–195.

McCarron, M., Mills, A.J., Whittakar, D., Sunny, T.P., Verghese, J., 1995. Comparison of the monoterpenes derived from green leaves and fresh rhizomes of *Curcuma longa* L. from India. Flavour Frag. J. 10, 355–357.

Neema, A., 2005. Investigations on the biosynthesis of curcumin in turmeric (*Curcuma longa* L.). Ph.D. Thesis, Calicut University, Calicut, Kerala State, India, p. 171.

Negi, P.S., Jayaprakasha, G.K., Rao, L.J., Sakariah, K.K., 1999. Antibacterial activity of turmeric oil: a byproduct from curcumin manufacture. J. Agric. Food Chem. 47 (10), 4297–4300.

Oguntimein, B.O., Weyerstahl, P., Marshall, H., 1990. Essential oil of *Curcuma longa* L. leaves. Flavour. Frag. J. 5, 89–90.

Oshiro, M., Kuroyanagi, M., Ueno, A., 1990. Structures of sesquiterpenes from *Curcuma longa*. Phytochemistry 29, 2202–2205.

Pabon, H.J.J., 1964. Synthesis of curcumin and related compounds. Rec. Trav. Chim. Pays-Bas. 83, 379.

Paramasivam, M., Rajalakshmi, P., Banerjee, H., 2008. Quantitative determination of curcuminoids in turmeric powder by HPTLC technique. Curr. Sci. 95 (11), 1529–1531.

Paramasivam, M., Poi, R., Banerjee, S., Bandopadhyaya, A., 2009. High-performance thin layer chromatographic method for quantitative determination of curcuminoidsin *Curcuma longa* germplasm. Food Chem. 113 (2), 640–644.

Patel, K., Krishna, G., Sokoloski, E., Yoichiro, I., 2000. Preparative separation of Curcuminoids from crude curcumin and turmeric powder by pH-zone refining counter current chromatography. J. Liq. Chromatogr. R. T. 23 (14), 2209–2218.

Pathania, V., Gupta, A.P., Singh, B., 2006. Improved HPTLC method for determination of curcuminoids from *Curcuma longa*. J. Liq. Chromatogr. Related Technol. 29, 877–887.

Perotti, A.G., 1975. Curcumin—a little known but useful vegetable colour. Ind. Aliment Prod. Vegetables 14 (6), 66.

Roughley, P.J., Whiting, D.A., 1973. Experiments in the biosynthesis of curcumin. J. Chem. Soc. Perkin Trans. I, 18, 2379.

Rupe, H., Clar, G., Pfau, A., Plattner, P.C., 1934. Volatile plant constituents II. Turmerone, the aromatic principle of turmeric oils (in German). Helv. Chim. Acta 17, 372.

Salzer, U.J., 1977. The analysis of essential oils and extracts (oleoresins) from seasonings—a critical review. Crit. Rev. Food Sci. Nutr. 9 (4), 345.

Sharma, R.K., Misra, B.P., Sarma, T.C., Bordoloi, A.K., Pathak, M.G., Leclercq, P.A., 1997. Essential oils of *Curcuma longa* L. Bhutan. J. Essent. Oil Res. 9, 589–592.

Sumita, D., Subash, K.K., Mukul, D., 2008. A simple 2-directional high performance thin layer chromatographic method for the simultaneous determination of curcumin, metanil yellow and Sudan dyes in turmeric, chilli and curry powders. J. AOAC Int. 91 (6), 1387–1396.

Vogel, Pelletier, 1818. J. Pharma. Sci. 2, 50. Cited in Govindarajan, (1980).

Zhu, L.F., Li, Y.H., Li, B.L., Ju, B.Y., Zhang, W.L., 1995. Aromatic Plants and Essential Constituents (Suppl.1): South China Institute of Botany. Chinese Academy of Sciences, Hai Feng Publ. Co. Distributed by Peace Book Co. Ltd., Hong Kong.

5 The Biotechnology of Turmeric

Biotechnological interventions in turmeric can be summarized in the following sections.

Tissue Culture

True to type multiplication in turmeric is necessary for the following reasons: (i) low rhizome multiplication and difficulty in storage; (ii) limited availability of high-yielding genotypes; and (iii) expensive field maintenance and high susceptibility to pests and diseases. Tissue culture is one of the most important techniques to meet some of the above requirements. Tissue culture enables large-scale propagation in turmeric. Research in tissue culture is quite exhaustive. A number of researchers from all over the world have reported very effective *in vitro* culture protocols for direct and indirect regeneration from turmeric plant. However, only limited investigations refer to genetic fidelity of micropropagated plantlets using molecular tools.

Explants and Media

In vitro cultures from underground rhizomes are the main methodology employed by most investigators, where contamination has been a common serious problem (Islam et al., 2004). Almost 50% contamination has been reported by Balachandran et al. (1990). To contain contamination, mercuric chloride $HgCl_2$ is used as a disinfectant. Rahman et al. (2004) used a pretreatment with Savlon (a patented disinfectant in the Indian market) at 5% before treating with $HgCl_2$. Sodium hypochlorite and Tween 80 (another patented disinfectant) were also used for treatment of terminal bud explants to reduce contamination (Prathanturarug et al., 2005).

For multiple shoot production, mostly vegetative bud explants (Salvi et al., 2002), sprouting buds from rhizomes (Balachandran et al., 1990), axillary buds from unsprouted rhizomes (Panda et al., 2007), pseudostem (Gayatri et al., 2005), leaf (Salvi et al., 2001), shoot tips (Salvi et al., 2001), immature inflorescence (Salvi et al., 2000), and nodal explants (Roy and Ray Chaudhuri, 2004) were used.

A number of investigators used full-strength Murasighe and Skog (MS) media, while Sit and Tiwari (1996) reported suitability of half-strength MS medium for better micropropagation. Cultures were maintained at 2000–3000 lux with a photoperiod of 16/8 (light/dark) cycle at 22–25°C and 55–60% relative humidity. In most cases, MS medium is used except in investigations by Nadgauda et al. (1978, Smith medium), Mukhri and Yamaguchi (1986, Ringe and Nitsch media), Nasirujjaman et al. (2005, woody plant medium (WPM)), and Gayatri et al. (2005, Linsmaier and Skoog (LS) medium).

The Agronomy and Economy of Turmeric and Ginger. DOI: http://dx.doi.org/10.1016/B978-0-12-394801-4.00005-3

Within 5–10 days of inoculation, shoot initiation took place. Phytohormones such as 6-benzylaminopurine (BAP), 1,-naphthaleneacetic acid (NA), and kinetin were the most commonly used by majority of the workers for shoot initiation, multiplication, and development, but in some cases gibberellic acid (GA) was used for initial establishment of cultures (Meenakshi et al., 2001). Some investigations point out use of thidiazuron (TDZ) containing media to improve the multiple shoot-inducing ability (Prathanturarug et al., 2005). Salvi et al. (2000) have reported use of indoleacetic acid (IAA) along with TDZ for direct shoot development from immature inflorescence explants. Praveen (2005) reported direct regeneration from leaf explants on TDZ and BAP media. A perusal of the results of investigations indicates that BAP is the most preferred phytohormone for shoot elongation and multiplication, used singly (Balachandran et al., 1990) irrespective of the kind of basal media employed. Majority of the investigators had employed a combination of BAP and NAA (Nasirujjaman et al., 2005), while some used a combination of kinetin and BAP (Balachandran et al., 1990), while Sunitabala et al. (2001) reported use of kinetin and NAA. Use of adjuvants such as coconut water (Nadgauda et al., 1978) has been found to be more preferable, resulting in better bud elongation and generation of multiple shoots.

Salvi et al. (2002), in an elaborate investigation on the effect of various physical and biochemical factors on *in vitro* response of turmeric, employed young rhizome buds and shoot tips to evaluate a range of plant growth regulators and found that adenine sulfate is also equally effective for shoot multiplication. The effect of various cytokinins on shoot multiplication was investigated by culturing the young shoot buds from sprouting rhizomes on MS liquid medium supplemented with benzyladenine, benzyladenine riboside, kinetin, kinetin riboside, zeatin, 6-yy-dimethylallylaminopurine, adenine, adenine sulfate, or metatopolin in combination with NAA. Among the different carbon sources examined, fructose, glucose, sugar cubes, maltose, levulose, and market sugar were all found to be equally effective in shoot multiplication. However, xylose, rhamnose, lactose, and soluble starch were found to be inhibitory. Tyagi et al. (2007) reported no adverse effect of commercial or market sugar on regeneration and *in vitro* conservation. Up to 73% reduction in the cost of making media could be achieved by using cheap alternative sources of carbon and gelling agent.

Most of the reports suggest use of 3% sucrose for multiple shoot induction with some exceptions. Keshavchandran and Khader (1989) and Shetty et al. (1982) report a higher level of 4% sucrose for multiple shoot induction and Shirgurkar et al. (2001) report 2% sucrose for bud elongation. Semisolid (0.8% agar) media is reported to be as appropriate for multiple shoot induction by most of the investigators. However, Salvi et al. (2002) report multiple shoots on low-agar (0.4–0.6%) medium. Prathanturarug et al. (2005) also used 0.55% agar gel for micropropagation from dissected terminal buds. As a cheap alternative to agar, isabgol was identified by many investigators like Praveen (2005) and Anuradha et al. (2008), who reported no adverse effect on plantlet regeneration. A better survival of plantlets from 33% to 44% was observed after 12 months in the presence of isabgol than on agar medium (Anuradha et al., 2008). Some investigators have tried media without agar and observed better results. Shirgurkar et al. (2001) and Salvi et al. (2003) report better induction of multiple shoots on liquid MS and Nadgauda et al. (1978) on liquid Whites media.

Production of an increased number of multiple shoots on liquid MS medium has been reported by Chang and Thong (2004). Ali et al. (2004), Winnaar (1989), and Rahman et al. (2004) corroborate the above findings through their own investigations on *C. longa*. Prathanturarug et al. (2005) report a preculture on liquid MS with TDZ prior to transfer onto plain MS media to enhance multiple shootlet formation and also the application of a temporary immersion culture system to reduce the cost of plantlet production during mass multiplication. Adelberg and Cousins (2006) report that use of liquid media in large culture vessel with gentle tilting agitation can give bigger plantlets, while small vessels on a shaker gave higher number of plantlets. Increased biomass formation was thus reported in liquid medium as compared to solid medium.

The number of shoots regenerated per explants ranged from 2 to 25 plantlets per explants, when all the different results of the investigation were compared. The multiple shoot regeneration reported by various investigators are 10–12 (Shetty et al., 1982); 18 and 11 in 1 year (Prathanturarug et al., 2005); 25 (Praveen, 2005); and as low as 2 (Keshavchandran and Khader, 1989) and 4 (Tule et al., 2005).

Rooting in turmeric was observed to be either spontaneous or induced as reported by several investigators. Mostly NAA alone up to 1 mg/l is employed (Salvi et al., 2000). A combination of NAA with BAP in the initial culture media (Nasirujjaman et al., 2005) was also observed to induce rooting. Meenakshi et al. (2001) also tried IBA as rooting medium. Rahman et al. (2004) reported that NAA, IBA, IAA in the range 0.1–1 mg/l is effective for root induction. It was observed that IBA was best for rooting. These reports are substantiated by Tule et al. (2005), who suggest that IBA up to 2 mg/l to give maximum number of roots. Tule et al. (2005) report that NAA or 2,4-D can be used along with BAP for induction of rooting in vegetative bud explants derived multiple shoots.

Spontaneous rooting was reported by Prathanturarug et al. (2005) in regenerants derived from bud explants, on MS liquid supplemented with TDZ and also by Gayatri et al. (2006) on LS medium containing 2,4-D, where complete plantlets were regenerated. Nasirujjaman et al. (2005) reported spontaneous rooting in the presence of BAP and NAA. Mukhri and Yamaguchi (1986) reported development of complete plantlets on Ringe and Nitsch medium with BAP. Similarly, Nadgauda et al. (1978) also reported root induction in the presence of BA and CM. Nirmal Babu et al. (1997) and Balachandran et al. (1990) also report simultaneous shooting and rooting in the presence of BA alone.

Field survival in the range of 95–100% can be considered as good survival of hardened plantlets. A sand–soil–farmyard mixture in the ratio of 1:1:1 was found to give 95% survival after hardening, according to most investigators. Up to 95% survival was obtained in sterilized soil (Salvi et al., 2002). Sand–rice shell ash in the ratio of 1:1 was also found to be good for hardening (Prathanturarug et al., 2003). Ali et al. (2004) and Prathanturarug et al. (2005) reported 100% survival on sand:clay:compost mixture in the ratio of 1:1:1. Micropropagated plants showed increase in shoot length, number of tillers, length and number of leaves, number of fingers, and total fresh rhizome compared to conventional propagated ones. Also, the former plants were reported to show no change in morphological characters (Balachandran et al., 1990), while Salvi et al. (2002) reported that among 48 plants, 2 showed variegated leaves on the tillers. Genetic fidelity was confirmed by some investigators who reported no variation in the RAPD profiles (Salvi et al., 2002). Panda et al. (2007) have confirmed genetic fidelity of micropropagated plantlets, both through cytometry and RAPD. Metabolic profiling, using GC–MS and liquid chromatography–electrospray ionization tandem–mass spectrometry (LC–ESI–MS), to analyze differences in curcuminoids and mono- and sesquiterpenoids indicated no significant differences between conventional greenhouse-grown and *in vitro* propagation-derived plants (Ma and Gang, 2006).

Callus Induction

The first investigation on callus induction from stem tip explants was that of Mukhri and Yamaguchi (1986). The callus was mostly induced from slices of rhizome (Yasuda et al., 1988), inflorescence (Salvi et al., 2000), leaf base (Salvi et al., 2001), shoot tips (Sunitabala et al., 2001), leaf (Praveen, 2005), vegetative buds (Praveen, 2005), pseudostems (Gayatri et al., 2005), and stem explants (Nadgauda and Masacarenhas, 1986). Both MS medium and Rige Nitsch media are reported for callus induction (Mukhri and Yamaguchi, 1986).

Dark incubation is suggested by most of the investigators (Nadgauda and Masacarenhas, 1986; Yasuda et al., 1988) for effective callus induction. Most investigators found 2,4-D up to 3 mg/l as the best rate (Nirmal Babu et al., 1997). Dicamba up to 5 mg/l (Salvi et al., 2001), picloram up to 1 mg/l (Salvi et al., 2001), NAA up to 5 mg/l (Sunitabala et al., 2001), and low levels of kinetin up to 0.5 mg/l (Shetty et al., 1982) also could induce callus from various explants. Tule et al. (2005) reported a medium width 40% sucrose supplemented with NAA, 2,4-D, and BAP, for callusing from vegetative bud explants. Praveen (2005) observed embryogenesis on MS with low levels of TDZ. Nadgauda and Masacarenhas (1986) reported use of a single plant growth regulator, either BAP or kinetin, for callus induction from stem base, while Zapata et al. (2003) and Sumathi (2007) reported use of both BAP and 2,4-D as best for callus induction. Sumathi (2007) observed embryogenic callus on media with BAP and 2,4-D and generation of nonembryogenic calli on media with 2,4-D alone. Histological studies confirmed both organogenic and embryogenic pathways.

To induce callus in case of turmeric from rhizome bud explants, a combination of NAA up to 1 mg/l and BAP or kinetin up to 2 mg/l was found good (Sumathi, 2007; Sunitabala et al., 2001). Praveen (2005) reported the use of TDZ up to 0.11 mg/l for effective plantlet regeneration from vegetative bud- and leaf-derived calli. Sunitabala et al. (2001) advocated transfer of calli to a kinetin (up to 1 mg/l) supplemented media as an effective method for plantlet regeneration. Sunitabala et al. (2001) and Salvi et al. (2001) also reported use of BAP up to 5 mg/l for callus induction from turmeric. Sumathi (2007) observed use of NAA and BA to be the best. For better plantlet regeneration, transfer of leaf base calli onto BA in combination with TIBA is appropriate (Salvi et al., 2001), while Zapata et al. (2003) reported use of kinetin and IAA for organogenesis. Zapata et al. (2003) also suggest a hydroponic system for better survival (up to 90%) of callus-regenerated plantlets while hardening.

The number of shoots regenerated per explants ranged from 8 to 10 in 40 days of the culture (Nirmal Babu et al., 1993). Maximum plantlet regeneration per explants reported was 27.1 and 28.3 from vegetative bud derived from calli and the leaf explants, respectively (Praveen, 2005).

Variations were reported in callus-derived plantlets by most of the investigators. Nadgauda et al. (1978) observed variants for high curcumin content. Sunitabala et al. (2001) isolated some somaclonal variants (SCV) from rhizome buds and shoot tip calli-derived regenerants and Salvi et al. (2001) from leaf base and inflorescence-derived calli.

Distinct variation in patterns of rhizome development, orientation and number of primary, secondary, and tertiary rhizomes, and intermodal and nodal distances were observed by Praveen (2005). Horn-like primaries and longer intermodal distances and also dark pink-colored rhizomes were observed under field conditions and variations were confirmed by RAPD. The same investigator (Praveen 2005) also observed cytotypes and morphological variants and variegated plant types in 1% cultures. Salvi et al. (2001) compared the RAPD profiles and found that callus-derived plants showed variations in RAPD profiles compared to those from shoot tips. Sumathi (2007) observed variation in plant height, leaf breadth, yield, and nodes in primary and secondary fingers and RAPD profiles in SCV. Stomatal frequency (SF) of SCVs were found to increase in number and decrease in size. Variations in chromosome number and structural aberrations were also reported in comparison to the mother plant (Sumathi, 2007).

Metabolic profiling using GC–MS and LC–ESI–MS among micropropagated plants and conventional greenhouse-grown plants revealed no difference in the major chemical constituents of the curcuminoids and mono- and sesquiterpinoids, as reported by Ma and Gang (2006).

In Vitro Screening

To isolate useful variants in turmeric through induced mutations is bedeviled by many problems and these are well recorded in scientific literature. *In vitro* screening techniques were employed once callus cultures were developed. Nadgauda and Masacarenhas (1986) attempted to develop an *in vitro* screening method to select high curcumin in turmeric lines by estimating the curcumin content from the "rhizome-like" portions of the *in vitro* grown plants and later correlating this value with that of the field grown plants. However, this work failed to take off owing to high variability in variety–environment interaction. Isolation of *Pythium gaminicolum* cell lines from *C. longa* variety Suguna by incorporating culture filtrate of *P. graminicolum* in the culture medium (Gayatri et al., 2005). *Pythium*-tolerant clones were regenerated through continuous and discontinuous *in vitro* selection and *in vitro* sick plot techniques. The plants derived from tolerant cell lines exhibited better disease tolerance as compared to the control. Even though the above-mentioned authors reported stable genetic make-up of the clones, such investigations have been more of an academic nature, as observed in the case of ginger and black pepper, than of much practical utility, an important reason being selection of no stable epigenetic cell lines and their subsequent reversion or suppression.

Praveen (2005) developed efficient cell suspension systems suitable for *in vitro* mutant selection. A good cell suspension culture of small cell aggregates of turmeric was established by subculturing at 200 rpm at weekly intervals. The cell suspension has to be aseptically diluted to 1:3 in fresh medium to maintain properly. In petriplate (90 mm) containing solidified MS + BA for further growth and development was plated 2 ml of the diluted suspension. The plating efficiency was high with 40–60 well-developed microcalli. These microcalli were further proliferated on callus-proliferating medium.

Exchange of Turmeric Germplasm and *In Vitro* Conservation

To help increase intervals between subculturing and thereby extending storage period of *in vitro* cultures, chemical and physical manipulation during culturing has to be done. This also reduces the risk of contamination at each transfer level, inputs in terms of labor and consumables. A number of investigators (Nirmal Babu et al., 1994; Geetha et al., 1995) have investigated methods to improve the *in vitro* conservation in turmeric. Balachandran et al. (1990) described a short-term conservation of *in vitro* raised plants of different *Curcuma* species as an *in vitro* conservation strategy. Subculture intervals for *in vitro* conservation of *C. longa* were standardized (Tyagi et al., 1998). Conservation of germplasm in turmeric by slow growth was reported by a number of workers (Balachandran et al., 1990; Nirmal Babu et al., 1999). The cultures could be stored up to 1 year without subculture in half-strength MS medium with sucrose and mannitol in sealed culture tubes (Geetha et al., 1995). Under the slow-growth mode in presence of sucrose, a 90% survival rate could be observed in a period of 360 days that resulted in 90% field establishment (Nirmal Babu et al., 1996). Valuable accessions of *C. longa* are currently maintained in the National Facility for Plant Tissue Culture Repository at the National Bureau of Plant Genetic Resources (NBPGR) in New Delhi, India.

Production of synthetic seeds in turmeric under *in vitro* condition as a safe means of germplasm exchange and conservation is highlighted by Sajina et al. (1996). Gayatri et al. (2005) investigated the effect of two types of hydrogels in encapsulating the nonembryogenic vegetative shoot bud propagules of *C. longa* var. Suguna. It was observed that among the two encapsulating agents tried, namely sodium alginate and carboxymethyl cellulose, sodium alginate (4%) containing LS medium fortified with BAP, dipped in 50 mM calcium chloride solution and incubated for 30 min in an orbital shaker was found to be the best matrix and complexing agent for encapsulation and regeneration of *in vitro* buds.

Cryopreservation

Cryopreservation technique is the only available technique to conserve turmeric species for long term. Plant germplasm stored in liquid N ($-196°C$) does not undergo cellular divisions. In addition, all metabolic and physical processes are stopped at this temperature. As such, plants can be stored for very long periods, avoiding both genetic instability and the risk of losing accessions, due to contamination or human error during subculturing. Most cryopreservation attempts deal with recalcitrant seeds and *in vitro* tissues from vegetatively propagated crops.

Successful cryopreservation of turmeric shoot tips was done with 80% recovery using vitrification technique. When encapsulated shoot tips were dehydrated in progressive sucrose concentration increase in the medium with 4–8 h of desiccation, the recovery rate was only 40% (Peter et al., 2002).

In Vitro Mutagenesis

Because turmeric can be vegetatively propagated, the crop improvement through mutation breeding was attempted. *In vitro* mutation induction in *C. longa* varieties Suvarna and Prabha using ethylmethane sulfonate (EMS) and physical mutagens (gamma radiation) resulted in cytotypes with distinct chromosome numbers and variations in RAPD profiles (Praveen, 2005).

In Vitro Pollination

Seed set was obtained through *in vitro* pollination using flowers on the day of anthesis and 1 day prior to anthesis (Renjith and Valsala, 2007). Placental pollination was found to be the best method. The culture medium composed of half-strength MS supplemented with BAP and kinetin and NAA along with sucrose, CW, and casein hydrosylate used to support ovule development and *in vitro* pollination. In *in vitro* developed seed, endosperm development was not complete.

The first attempt to integrate *in vitro* and *in vivo* methods to accelerate turmeric breeding was made by Vijayasree and Valsala (2007). Seeds obtained through *in vivo* fertilization were cultured *in vitro* for rapid multiplication to establish a population of six plantlets/seedling within a period of 10 months. They crossed between two short-duration types VK 70 and VK 76 to develop a high-yielding variety with high curcumin content and curing percentage.

An attempt to develop *in vitro* pollinated ovule to seed was made by Renjith and Valsala (2007). Presence of half-strength MS and NAA and BAP showed maximum ovule swelling, while a combination of BAP and IAA produced callusing instead of ovule swelling.

Microrhizomes

Induction of microrhizomes (mr) in turmeric was reported by a number of investigators. Plantlets generated from rhizome buds were used for this purpose. Factors such as concentration of sucrose and BA in the medium and photoperiod and their interaction were found to have a significant effect in the induction of mr and sucrose was most effective in rhizome formation, followed by photoperiod and BA in the medium (Nayak and Naik, 2006). Low sucrose is reported to decrease the size of mr (Shirgurkar et al., 2001). Nayak and Naik (2006) and Sunitabala et al. (2001) have obtained optimum mr production at a rate of 6–9% sucrose. However, Geetha (2002) has reported 12% sucrose to be optimum.

A reduced photoperiod of 4 h was found to favor mr production (Nayak and Naik, 2006). But 0 h dark incubation reduced sprouting. However, incubation under fully dark condition was found to be favorable (Shirgurkar et al., 2001). In general, the phytohormones used were BA (up to 3 mg/l), NAA (up to 0.1 mg/l), and kinetin (up to 1 mg/l). Shirgurkar et al. (2001) and Peter et al. (2002) observed induction of mr on media containing BA alone, while Rajan (1997) reported a combination of BA and NAA along with low levels up to 0.5 mg/l of ancymidol for the same effect. Sunitabala et al. (2001) and Shirgurkar et al. (2001) reported use of kinetin in combination with NAA and BA, respectively for mr induction. Cousins and Adelberg (2008) used methyl jasmonate up to 0.45 mg/l along with BA (up to 0.45 mg/l) for induction of mr. Meja reduced root and rhizome biomass and BA had a positive effect on biomass accumulation. These reports contradict those of Shirgurkar et al. (2001), who observed that at 7.93 mg/l of BA, total inhibition of mr induction was observed.

Better recovery of curcumin vis-à-vis seed weight was the main advantage of in vitro mr induction. Also, high antioxidant activity (about 30-fold more than that of standard food preservative butylated hydroxyl toluene, BHT) was observed. They are convenient for packing and transport, besides aiding in germplasm conservation and exchange. The salient observations made by various investigators suggest that mr production depended on the size of the multiple shoots used (Geetha, 2002; Shirgurkar et al., 2001). Though this technology has to be cost-sensitive for commercial viability, no precise information is available on the cost reduction involved. Shirgurkar et al. (2001) used half-strength MS for optimum production of mr, while Nayak and Naik (2006) used liquid MS for mr induction. Lower MS strength resulted in lower number of smaller mr, while full strength produced larger sized mr through a lower number (Islam et al., 2004; Nayak and Naik, 2006). Cousins and Adelberg (2008) also reported that liquid media was more effective for mr induction, which can lead to cost reduction of gelling agent.

Those mr with about 2 g fresh weight can be planted directly in the field without hardening, showing about 80% success (Geetha, 2002). Field evaluation reports suggest a higher number of mother rhizomes per plant and reasonably large rhizomes, up to 900 g per plant yield (Nirmal Babu et al., 2003). Comparison of field performance of conventionally raised seed propagated plants and those micropropagated plants revealed relative improvement in yield in the case of the latter through in vitro induction (Nayak and Naik, 2006). Harvested mr can be stored in moist condition for convenience and sprouting is observed after 2 months of storage at room temperature (Nayak and Naik, 2006).

Investigational reports on mr of *Curcuma* species are rather limited. This is principally because there is a great need to improve the protocol which is essential to obtain more and larger mr, as the survival of small mr is very low and small rhizomes normally produce unhealthy and stunted plants, while bigger ones are more vigorous and grow better and faster in the field.

Molecular Markers

Molecular Characterization and Diversity Investigations

There are, but limited, investigations on molecular biology of *Curcuma* species. There is yet no published information on a global taxonomic revision of the genus. Primary constraints in this aspect are the lack of specimens, illustrations of the old species, lack of protologs with finer details of the earlier gathered literature, absence of important floral parts in the herbarium specimens, incomplete description of the rhizome features in the herbarium sheets, and fleshy and perishable aerial portions (Sasikumar, 2005). A brief account of this aspect has been presented in an earlier chapter in this book on botany (Chapter 2) and breeding of *Curcuma*. Relying much on the morphological traits alone in species delimitation has its own limitations in the genus. Molecular marker techniques like ISSR/RAPD/SSR markers thus assume significance.

Among the researchers in this line of research, it was Shamina et al. (1998) who, for the first time, used biochemical markers to identify variability in 15 *C. longa* accessions collected from different geographical locations in India with respect to six isozymes, namely, acid phosphatase, superoxide dismutase, esterase, polyphenol oxidase, peroxidase, and catalase. Acid phosphatase produced the maximum number of bands among the accessions, while polyphenol oxidase and peroxidase were more consistent and reproducible. A high degree of polymorphism (63.8–96%) was observed in the population investigated for these isozymes. Accessions collected from the same geographical area showed highest similarity (above 90%) and two true turmeric seedlings included in the study stood out distinctly from the rest of the dendrogram. Turmeric being a rhizome-propagated crop, all the varieties/cultivars are not easily discernible based on aerial and/or rhizome morphology. This could give rise to unscrupulous seed trade practice in this crop. Hence, protocols were optimized for isolation of DNA from rhizomes (Syamkumar et al., 2003).

RAPD profiles of DNA isolated from fresh rhizomes of *C. longa* were compared with five other species of *Curcuma*, and it was observed that eleven polymorphic bands were produced in the five species investigated using three RAPD markers (Sreeja, 2002). Kress et al. (2002) proposed a new phylogenetic analysis of Zingiberaceae based on DNA sequences of the nuclear ITS and plastid mat K regions. The results suggest that *Curcuma* is paraphyletic with Hitchenia, Stahlianthus, and Smithatris. A phylogenetic analysis of the tribe Zingiberae (*Zingiberaceae*) was done by Ngamriabsakul et al. (2003) using nuclear ribosomal DNA (ITS1, 5.8S, and ITS2) and chloroplast DNA (trnL (UAA) 5′) exon to trnF (GAA) and suggested that the genus *Curcuma* is monophyletic. Cao et al. (2003) used trn K nucleotide sequencing for identification of six medicinal species of *Curcuma*, including *C. longa*.

Genetic diversity analysis of 20 accessions of *C. longa* from different parts of Brazil using RAPD markers produced 45 polymorphic loci and the dendrogram produced by UPGMA grouping using Jaccard's Index of similarity formed 2 groups. Among the groups, 44.4% genetic variability was observed and most of the variation was found within the groups (Nayak and Naik, 2006). Nayak et al. (2006) carried out 4C nuclear DNA content and RAPD analysis of 17 promising cultivars of turmeric (*C. longa*) from India. Significant variation in the 4C DNA content and RAPD profiles was observed among the cultivars. In RAPD, the intercultivar polymorphism ranged from 35.6% to 98.6% and the amplification fragments per primer ranged from 4 to 17 with fragment size ranging from 0.4 to 3 kb. Two primers OPN 06 and OPA 04 could clearly distinguish all the cultivars. Syamkumar (2007) investigated over 36 Indian cultivars and related species of turmeric using RAPD and ISSR (Indian Institute of Spices Research, Calicut, Kerala State, India) profiling for establishing their interrelationships.

Most of the improved varieties clustered distinctly from the land races/cultivars. Further, most of the land varieties, collected mainly based on vernacular names from one geographical region clustered together along with few released varieties which were evolved through germplasm selection of material collected from the same region. In all the dendrograms constructed using separate or combined RAPD/ISSR markers, irrespective of the grouping procedure adopted, the popular varieties, such as Pratibha and Alleppey Supreme, clustered together and showed maximum similarity within the group. Cultivars Amruthapani and Armoor and/or Amalapuram, Amruthapani, and Armoor clustered together showing maximum similarity. This suggests that these cultivars are genetically very similar or got collected as distinct accessions based on vernacular names.

Evaluation of genetic diversity employing ISSR and RAPD markers among *C. longa* and 14 *Curcuma* species (Syamkumar and Sasikumar, 2007) points out the inadequacies of conventional taxonomic tools for resolving taxonomic confusion prevailing in the genus and suggests the need for molecular markers in conjunction with morphotaxonomic and cytologic investigations while revisiting the genus. Dendrograms constructed based on the unweighted pair group method using arithmetic averages placed the 15 species into 7 groups which is somewhat congruent with classification based on morphological characters proposed by the earlier investigators. Syamkumar (2007) also reports use of RAPD and RAPD plus ISSR markers to group OP lines and their mother lines. Three groups were observed in RAPD, while ISSR dendrograms showed five different groups among the OP lines and their mother parents. A definite pattern or clustering of the OP lines could not be observed in the present case. Perhaps, the very few number of progenies investigated may be the reason for this behavior of the progenies. The morphological characterization also showed a similar trend.

In yet another investigation on genetic variability in starchy *Curcuma* species, including *C. longa*, using RAPD, the banding profiles generated by 20 primers revealed high degree of polymorphism (Angel et al., 2008). A total of 274 bands were generated of which 264 were polymorphic. All the species were separated into three clusters using UPGMA. As it is difficult to distinguish different species by leaf morphology, the RAPD pattern has high utility in identification of *C. longa* from other species investigated.

Expressed sequence tags (ESTs) from turmeric were used to screen type and frequency of Class I (hypervariable) simple sequence repeats (SSRs). A total of 231 microsatellite repeats could be detected from 12,593 EST sequences of turmeric after redundancy elimination. The average density of Class I SSRs accounts to one SSR per 17.96 kb of EST. Mononucleotides was the most abundant class of motifs followed by trinucleotides. A robust set of 17 polymorphic EST–SSRs were developed and used to evaluate 20 turmeric accessions. The number of alleles detected ranged from 3 to 8 per loci. The developed markers were also evaluated in 13 related species of *C. longa* confirming the highest rate (100%) of cross-species transferability (Siju et al., 2009). The polymorphic microsatellite markers will be useful in genetic diversity analysis, resolving the taxonomic confusion prevailing in the genus.

Checking Adulteration and Purity Assessment

Remya et al. (2004) described an efficient protocol for the isolation of high-molecular-weight DNA from dry turmeric powder. This will help in PCR-based detection of adulteration in marketed turmeric powders. Detection of extraneous contamination of powdered turmeric by RAPD technique was explained by Sasikumar et al. (2004). Analysis of three market samples of turmeric powder from Kerala State, India showed the prevalence of *C. zedoaria* (wild species) powder admixture with *C. longa* (the economically important species which is generally used). However, the curcumin levels in the samples conformed with prevalent standards. A similar investigation was carried out by Cao et al. (2001) and Sasaki et al. (2004). The former investigators reported that 18S rRNA gene sequence of *Curcuma* species was found to be 1 8 1 0 bp long and is a conserved region among the six *Curcuma* species. Only one base transversion substitution from cystosine to thymine was observed at nucleotide positions 234 in *C. kwangsiensis*, when compared to the common species of *C. longa*, *C. phaeocaulis*, *C. wynujin*, and *C. aromatica*. This investigation also indicated that the length of the *trnK* gene varied from 2698 to 2705 bp and the gene is highly conserved in all the *Curcuma* species, particularly in the *mat K* region. Sasaki et al. (2002) used sequence analysis of Chinese and Japanese *Curcuma* drugs on the 18S rRNA gene and *trnK* gene and the application of ARMS analysis for their identification and authentication.

Application of single nucleotide polymorphism (SNP) analysis based on species-specific nucleotide sequence was developed by Sasaki et al. (2004) to identify the plants and drugs derived from *Curcuma* species, including *C. longa*. Based on the difference in the nucleotide positions at 177, 645, 724, and a 4-base indel on the *trnK* gene obtained using three different lengths of (26 mer, 30 mer, and 34 mer) reverse primers helped to identify the four *Curcuma* species investigated. The SNP analysis method is a useful method for the identification of botanical origins of *Curcuma* drugs used in Chinese medicine, which is difficult to identify morphologically and phytochemically. Xia et al. (2005) used molecular (5S rRNA spacer domains) and chemical fingerprints for quality control and authentication of *Rhizoma curcumae*, a traditional Chinese medicine used to remove blood stasis and alleviate body pain.

Molecular analysis based on polymorphisms of the nucleotide sequence of chloroplast DNA (cpDNA) was performed in order to distinguish four *Curcuma* species including *C. longa*. Nineteen regions of cpDNA were amplified successfully via PCR using total DNA. Using the intergenic spacer between trnS and trnfM (tranSfM), all of the four *Curcuma* species were precisely identified. Additionally, the number of AT repeats in the trnSfM region was predictive of the curcumin content in the rhizome of *C. longa* (Minami et al., 2009).

A simple and rapid method to identify six *Curcuma* medicinal species, namely *C. longa, C. phaeocaulis, C. sichuanensis, C. chuanyujin, C. chuanhuangjiang,* and *C. chuanezhu,* from Sichuan Province of China employing the chloroplast *trnK* nucleotide sequencing technique was developed by Komatsu and Cao (2003). They sequenced entire chloroplast *trnK* gene region spanning 2699–2705 bp. The *matK* gene (an intron embodied in *trnK* gene) sequence and the intron spacer region of the *trnK* gene have great diversity within these six medicinal *Curcuma* species. There were six single-base substitutions between trnK coding region and matK region, the 9 bp deletion and 4 bp or 14 bp insertion repeat at some sites of matK region in each taxon. These relatively variable sequences were potentially informative in the identification for these six *Curcuma* species at the DNA level.

Genetic Fidelity and Identification of SCV

The morphological and molecular variations among micropropagated and callus-regenerated plants were investigated by Nirmal Babu et al. (2003) and they found variations in both but with a higher percentage of variation in callus-regenerated somaclones. *In vitro* plants developed through mr have exhibited the least variations. The authors inferred that this is due to the accumulated vegetative mutations (mosaic) in turmeric.

Genetic fidelity investigations of turmeric germplasms conserved in an *in vitro* gene bank using RAPD profiling showed their genetic integrity (Geetha, 2002). Salvi et al. (2001, 2003) and Praveen (2005) using RAPD analyzed turmeric somaclones and concluded that plants regenerated from vegetative buds showed uniform banding pattern, whereas callus- and inflorescence-derived plants showed polymorphism in a banding pattern when compared with conventionally propagated plants. Genetic stability of 12-month-old *in vitro* conserved *C. longa* cv. Pratibha plants was assessed using 25 RAPD primers (Tyagi et al., 2007). No significant variation was observed in RAPD profiles of mother plants and *in vitro* conserved plantlets. Micropropagated, callus-regenerated, EMS-treated, and gamma-irradiated turmeric somaclones were screened for polymorphic difference using RAPD for indexing genetic variability. The molecular profiles did not show genetic variation among the micropropagated plants. The callus-regenerated plants showed difference in profiles in RAPD. EMS-treated and gamma-irradiated plants also showed profile differences in RAPD (Praveen, 2005).

In another investigation carried out by Salvi et al. (2002), it was observed that RAPD analysis of 11 regenerated plants using 16 10-mer primers did not show any polymorphism. Salvi et al. (2003) analyzed 10 turmeric plants using RAPD and found that plants regenerated using shoot tips showed uniform banding pattern, whereas callus- and inflorescence-derived plants showed polymorphism. Salvi et al. (2001) reported 38 novel bands and also confirmed absence of about 51 bands in comparison to a control when leaf base callus-derived regenerants were analyzed by RAPD and PAGE. A comparative RAPD investigation conducted by Sumathi (2007) on micropropagated, mr- and callus-regenerated plants revealed monomorphic banding pattern in the first two, while the last one showed polymorphic bands. Genetic uniformity of micropropagated turmeric plants was confirmed by RAPD analysis (Panda et al., 2007).

Early Flowering

Association of a few isozymes markers in the identification of some of the early flowering *Curcuma* species has been reported by Pimchai et al. (1999). Eight isozymes showed reliable polymorphism to distinguish between the taxa analyzed. Patterns from isozyme data were analyzed using cluster analysis and UPGMA to produce a dendrogram depicting the degree of relationship among the species.

Genetic Transformation

An efficient method for stable transformation of turmeric, *C. longa* L. was developed using particle bombardment. Callus cultures initiated from shoots were bombarded with gold particles coated with plasmid pAHC25 containing the *bar* and *gusA* genes, each driven by the maize ubiquitin promoter. Transformants were selected on medium containing glufosinate. Transgenic lines were established on selection medium from 50% of the bombarded calluses. Transgenic shoots regenerated from these were multiplied and stably transformed plantlets were produced. PCR and histochemical GUS assay confirmed the stable transformation. Transformed plantlets were resistant to glufosinate (Shirgurkar et al., 2006).

References

Adelberg, J., Cousins, M., 2006. Thin film of liquid media for heteromorphic growth and storage organ development: Turmeric (*Curcuma longa*) as a model plant. Hortic. Sci. 41, 539–542.
Ali, A., Munawer, A., Siddiqui, F.A., 2004. *In vitro* propagation of turmeric (*Curcuma longa* L.). Int. J. Biol. Biotech. 1, 511–518.
Angel, G.R., Makeshkumar, T., Mohan, C., Vimala, B., Nambisan, B., 2008. Genetic diversity analysis of starchy *Curcuma* species using RAPD markers. J. Plant Biochem. Biotechnol. 17.
Anuradha, A., Mahalakshmi, C., Tyagi, R.K., 2008. Use of commercial sugar, isabgol and ordinary water in culture medium for conservation of *Curcuma longa* L. J. Plant Biochem. Biotechnol. 17, 85–89.

Balachandran, S.M., Bhat, S.R., Chandel, K.P.S., 1990. *In vitro* multiplication of turmeric (*Curcuma* sp.) and ginger (*Zingiber officinale* Rosc.). Plant Cell Rep. 8, 521–524.

Cao, H., Sasaki, Y., Fushii, H., Komatsu, K., 2001. Molecular analysis of medicinally used Chinese and Japanese *Curcuma* based on 18S rRNA and *trnK* gene sequences. Biol. Pharm. Bull. 24, 1389–1394.

Cao, H., Komatsu, K., Xue, Y., Bao, X., 2003. Molecular identification of six medicinal *Curcuma* plants produced in Sichuan: evidence from plastid *trnK* gene sequences. Biol. Pharm. Bull. 11, 871–875.

Chang, L.K., Thong, W.H., 2004. *In vitro* propagation of Zingiberaceae species with medicinal properties. J. Plant Biotechnol. 6, 181–188.

Cousins, M.M., Adelberg, W.J., 2008. Short-term and long-term time course studies in turmeric (*Curcuma longa* L.) microrhizome development *In Vitro*. Plant Cell Tissue and Organ Cult. 93, 283–293.

Gayatri, M.C., Roopadarshini, V., Kavyashree, R., 2005. Selection of turmeric callus tolerant to culture filtrates of *Pythium graminicolum* and regeneration plants. Plant Tissue and Organ Cult. 83, 33–40.

Gayatri, M.C., Roopadarshini, V., Kavyashree, R., 2006. Indirect organogenesis through pseudostem callus in turmeric variety *Suguna* of *Curcuma longa* L. concepts in tropical agriculture. In: Kumar, A. (Ed.), Daya Publishers, New Delhi India, pp. 28–35.

Geetha, S.P., 2002. *In vitro* Technology for Genetic Conservation of Some Genera of Zingiberaceae. Ph.D. Thesis, Calicut University, Calicut, Kerala State, India, p. 325.

Geetha, S.P., Manjula, C., Sajina, A., 1995. *In vitro* conservation of genetic resources of spices. In: Proceedings of the Seventh Kerala Science Congress. 27–29 January, Palakkad, Kerala State, India, pp. 12–16.

Islam, K., Kloppstech, M.A., Jacobsen, H.J., 2004. Efficient procedure for *in vitro* microrhizome induction in *Curcuma longa* L. (Zingiberaceae)—a medicinal plant of tropical Asia. Plant Tissue Cult. 14, 123–134.

Keshavchandran, E., Khader, M.A., 1989. Tissue culture propagation of turmeric. Indian Hortic. 37, 101–102.

Komatsu, K., Cao, H., 2003. Molecular identification of six medicinal *Curcuma* plants produced in Sichuan: evidence from plastid trnK gene sequences. Acta. Pharm. Sin. 38, 871–875.

Kress, W.J., Prince, L.M., Williams, K.J., 2002. The phylogeny and a new classification of the gingers (Zingiberaceae): evidence from molecular data. Am. J. Bot. 89, 1682–1696.

Ma, X., Gang, D.R., 2006. Metabolic profiling of turmeric (*Curcuma longa* L.). Plants derived from *in vitro* micropropagation and conventional greenhouse cultivation. J. Agric. Food. Chem. 54, 9573–9583.

Meenakshi, N., Suliker, G.S., Krishnamoorthy, V., Hegde, R.V., 2001. Standardization of chemical environment for multiple shoot induction of turmeric (*Curcuma longa* L.) for *in vitro* clonal propagation. Crop Res. 22, 449–453.

Minami, M., Nishio, K., Ajioka, Y., Kyushima, H., Shigeki, K., Kinjo, K., et al., 2009. Identification of *Curcuma* plants and curcumin content level by DNA polymorphisms in the trn S-trn fM intergenic spacer in chloroplast DNA. J. Nat. Med. 63, 75–79.

Mukhri, Z., Yamaguchi, H., 1986. *In vitro* plant multiplication from rhizomes of turmeric (*Curcuma domestica* Val.) and Temeo Lawak (*C. xanthorhiza* Rodb.). Plant Tissue Cult. Lett. 3, 28–30.

Nadgauda, R.S., Masacarenhas, A.F., 1986. A method for screening high curcumin-containing turmeric (*Curcuma longa* L.) cultivars *in vitro*. J. Plant Physiol. 124, 359–364.

Nadgauda, R.S., Mascarenhas, A.F., Hendre, R.R., Jagannathan, V., 1978. Rapid multiplication of turmeric (*C. longa*) plants by tissue culture. Indian J. Exp. Biol. 16, 120–122.

Nasirujjaman, K., Uddin, M.S., Zaman, S., Reza, M.A., 2005. Micropropagation of turmeric (*Curcuma longa* Linn.) through *in vitro* rhizome bud culture. J. Biol. Sci. 5, 490–492.

Nayak, S., Naik, P.K., 2006. Factors affecting *In Vitro* microrhizome formation and growth in *Curcuma longa* L. and improved field performance of micropropagated Plants. Sci. Asia 32, 31–37.

Nayak, S., Naik, P.K., Acharya, L.K., Patnaik, A.K., 2006. Detection and evaluation of genetic variation in 17 promising cultivars of turmeric (*Curcuma longa* L.) using 4C nuclear DNA content and RAPD markers. Cytologia 71, 49–55.

Ngamriabsakul, C., Newman, M.F., Cronck, Q.C.B., 2003. The phylogeny of tribe Zingberaceae (Zingiberaceae) based on its (nr DNA) and trnl-f (cpDNA) sequences. Edinburgh J. Bot. 60, 483–507.

Nirmal Babu, K., Sasikumar, B., Ratnambal, M.J., George, K.J., Ravindran, P.N., 1993. Genetic variability in turmeric (*Curcuma longa* L.). J. Genet. Plant Breed. 53 (1), 91–93.

Nirmal Babu, K., Rema, J., Ravindran, P.N., 1994. Biotechnology research in spice crops. In: Chadha, K.L., Rethinam, P. (Eds.), Advances in horticulture plantation and spice crops vol. 9 (Part I). Malhotra Publishig House, New Delhi, India, pp. 635–653.

Nirmal Babu, K., Samsudeen, K., Raveendran, P.N., 1996. Biotechnological approaches to crop improvement in ginger, *Zingiber officinale* Rosc. In: Ravishankar, G.A, Venkataraman, L.V (Eds.), Recent Advances in Biotechnological Applications of Plant Tissue and Cell Culture IBH Publishing Co, New Delhi, pp. 321–332.

Nirmal Babu, K., Ravindran, P.N., Peter, K.V., 1997. Protocols for Micropropagation of Spices and Aromatic Crops. Indian Institute of Spices Research, Calicut, Kerala State, p. 35.

Nirmal Babu, K., Geetha, S.P., Minoo, D., Ravindran, P.N., Peter, K.V., 1999. *In vitro* conservation of germplasm. In: Ghosh, S.P (Ed.), Biotechnology and its Application in Horticulture Narosa Publishing House, New Delhi, pp. 106–129.

Nirmal Babu, K., Ravindran, P.N., Sasikumar, B., 2003. Field Evaluation of Tissue Cultured Plants of Spices and Assessment of their Genetic Stability Using Molecular Markers. Final Report Submitted to the Department of Biotechnology, Government of India, New Delhi, p. 94.

Panda, M.K., Sujata, M., Enketeswar, S., Laxmikanta, A., Saghamitra, N., 2007. Assessment of genetic stability of micropropagated plants of *Curcuma longa* by cytophotometry and RAPD analyses. Int. J. Integr. Biol. 1, 189–195.

Peter, K.V., Ravindran, P.N., Nirmal Babu, K., Sasikumar, B., Minoo, D., Geetha, S.P., et al., 2002. Establishing *In Vitro* Conservatory of Spices germplasm. Indian Institute of Spices Research, Calicut, Kerala State, (ICAR Project Report) p. 131.

Pimchai, A., Somboon, A.I., Puangpen, S., Chiara, S., 1999. Molecular markers in the identification of some early flowering *Curcuma* L. (Zingiberaceae) species. Ann. Bot. 84, 529–534.

Prathanturarug, S., Soornthornchareonnon, N., Chuakul, W., Phidee, Y., Saralamp, P., 2003. High frequency shoot multiplication in *Curcuma longa* L using thidiazuron. Plant Cell Rep. 21, 1054–1059.

Prathanturarug, S., Soornthornchareonnon, N., Chuakul, W., Phaidee, Y., Sarakamp, P., 2005. Rapid micropropagation of *Curcuma longa* using bud explants pre-cultured in thidiazuron-supplemented liquid medium. Plant Cell Tissue Organ Cult. 80, 347–351.

Praveen, K., 2005. Variability in somaclones of turmeric (*C.longa* L.) Ph.D Thesis. Indian Institute of Spices Research, Calicut University, Calicut, Kerala State, India, pp.131.

Rahman, M.M., Amin, M.M., Jahan, H.S., Ahemed, R., 2004. *In vitro* regeneration of plantlets of *Curcuma longa* L., a valuable spice plant in Bangladesh. Asian J. Plant Sci. 3, 306–309.

Rajan, V.R., 1997. Micropropagation of turmeric (*Curcuma longa* L.) by *in vitro* microrhizome. In: Edison, S, Ramana, K.V., Sasikumar, B., Babu, K.N., Eapen, S.J. (Eds.), Biotechnology of Spices, Medicinal and Aromatic Plants Indian Society of Spices, Calicut, Kerala State, pp. 25–28.

Remya, R., Syamkumar, S., Sasikumar, B., 2004. Isolation and amplification of DNA from turmeric powder. Br. Food J. 106, 673–678.

Renjith, D., Valsala, P.A., 2007. Optimisation of media components for seed development in turmeric after *in vitro* pollination. In: Keshavachandran, R. (Ed.), Recent Trends in Horticultural Biotechnology New India Publishing Agency, New Delhi, India, pp. 451–455.

Roy, S., Ray Chaudhuri, S.S., 2004. *In vitro* regeneration and estimation of curcumin content in four species of *Curcuma*. Plant Biotechnol. 21, 299–302.

Sajina, A., Minoo, D., Geetha, S.P., Samsudeen, K., Rema, J., Babu, K.N., et al., 1996. Production of synthetic seeds in spices crops. In: Edison, S, Ramana, K.V., Sasikumar, B., Babu, K.N., Eapen, S.J. (Eds.), Proceedings National Seminar on Biotechnology of Spices, Medicinal and Aromatic Plants Indian Institute of Spices Research, Kozhikode, Kerala State, pp. 65–69.

Salvi, N.D., George, L., Eapen, S., 2000. Direct regeneration of shoots from immature inflorescence cultures of turmeric. Plant Cell Tissue Organ Cult. 62, 235–238.

Salvi, N.D., George, L., Eapen, S., 2001. Plant regeneration from leaf base callus of turmeric and random amplified polymorphic DNA analysis of regenerated plants. Plant Cell Tissue Organ Cult. 66, 113–119.

Salvi, N.D., George, L., Eapen, S., 2002. Micropropagation and field evaluation of micropropagated plants of turmeric. Plant Cell Tissue Organ Cult. 68, 143–151.

Salvi, N.D., Geoge, L., Eapen, S., 2003. Biotechnological studies of turmeric (*C. longa* L.) and ginger (*Z. officinale* Rosc.). Adv. Agric. Biotechnol. 11, 32.

Sasaki, Y., Fushimi, H., Cao, H., Cai, S.Q., Komatsu, K., 2002. Sequence analysis of Chinese and Japanese *Curcuma* drugs on the 18S rRNA gene and *trnK* gene and the application of amplification-refractory mutation system analysis for their authentication. Biol. Pharm. Bull. 25, 1593–1599.

Sasaki, Y., Fushimi, H., Komatsu, K., 2004. Application of single-nucleotide polymorphism analysis of the *trnK* gene to the identification of *Curcuma* plants. Biol. Pharm. Bull. 27, 144–146.

Sasikumar, B., 2005. Genetic resources of *Curcuma*: diversity, characterization and utilisation. Plant Genet. Res. 3, 230–251.

Sasikumar, B., Syamkumar, S., Remya, R., Zacharia, T.J., 2004. PCR-based detection of adulteration in the market samples of turmeric powder. Food Biotechnol. 18, 299–306.

Shamina, A., Zachariah, T.J., Sasikumar, B., George, J.K., 1998. Biochemical variation in turmeric based on isozyme polymorphism. J. Hortic. Sci. Biotech. 73, 477–483.

Shetty, M.S.K., Hariharan, P., Iyer, R.D., 1982. Tissue culture studies in turmeric. In: Nair, N.M., Prem Kumar, T., Ravindran, P.N., Sharma, Y.R. (Eds.), Proceedings National Seminar on Ginger and Turmeric, Calicut, Kerala State, India CPCRI, Kasaragod, Kerala State, pp. 39–41.

Shirgurkar, M.V., John, C.K., Nadgauda, R.S., 2001. Factors affecting *in vitro* micro-rhizome production in turmeric. Plant Cell Tissue Organ Cult. 64, 5–11.

Shirgurkar, M., Naik, B.V., von Arnold, S., Nadgauda, R.S., David, C., 2006. An efficient protocol for genetic transformation and shoot regeneration of turmeric (*Curcuma longa*) via particle bombardment. Plant Cell Rep. 5, 112–116.

Siju, S., Dhanya, K., Syamkumar, S., Sasikumar, B., Sheeja, T.E., Bhat, A.I., et al., 2009. Development, characterization and cross species amplification of polymorphic microsatellite markers from expressed sequence tags of turmeric (*Curcuma longa* L.). Mol. Biotechnol. doi: 10.1007/s 12033-009-9222-4.

Sit, A.K., Tiwari, R.S., 1996. Micropropagation in turmeric (*C. longa*). Souvenir National Symposium of Horticulture and Biotechnology, Bangalore, p. 23.

Sreeja, S.G., 2002. Molecular Chracterization of *Curcuma* Species Using RAPD Markers. M.Sc. (Biotech) Thesis, Periyar University, Tamil Nadu, p. 20.

Sumathi, V., 2007. Studies on Somaclonal Variation in Zingiberaceous Crops. Ph.D. Thesis, University of Calicut, Calicut, Kerala State, p. 227.

Sunitabala, H., Damayanti, M., Sharma, G., 2001. *In vitro* propagation and rhizome formation in *Curcuma longa* Linn. Cytobios 105, 71–82.

Syamkumar, S., 2007. Molecular, Biochemical and Morphological Characterization of Selected *Curcuma* Accessions. Ph.D. Thesis, University of Calicut, Calicut, Kerala State, p. 311.

Syamkumar, S., Sasikumar, B., 2007. Molecular marker based on genetic diversity analysis of *Curcuma* species from India. Sci. Hortic. 112, 224–235.

Syamkumar, S., Lawrence, B., Sasikumar, B., 2003. Isolation and amplification of DNA from rhizomes of turmeric and ginger. Plant Mol. Biol. Reptr. 21 171a–171e.

Tule, D., Ghorade, R.B., Mehatre, S., Pawar, B.V., Shinde, E., 2005. Rapid multiplication of turmeric by micropropagation. Ann. Plant Physiol. 19, 35–37.

Tyagi, R.K., Bhat, S.R., Chandel, K.P.S., 1998. *In vitro* conservation strategies of spices crop germplasm—*Zingiber*, *Curcuma* and *Piper* spices. In: Mathew, N.M, Jacob Kuruvilla, C., Licy, J, Joseph, T, Meenattoor, J.R., Thomas, K.K (Eds.), Developments in Plantation Crops Research. Proceedings Twelfth Symposium on Plantation Crops Allied Publishers, New Delhi, pp. 72–82.

Tyagi, R.K., Agrawal, A., Mahalakshmi, C., Hussain, Z., Tyagi, H., 2007. Low-cost media for *in vitro* conservation of turmeric (*Curcuma longa* L.) and genetic stability assessment using RAPD markers. *In vitro* Cell. Dev. Biol. Plant 43, 51–58.

Vijayasree, P.S., Valsala, P.A., 2007. Micropropagation supports heterosis breeding in turmeric (*Curcuma longa* Val.). In: Keshavachandran, R. (Ed.), Recent Trends in Horticultural Biotechnology New India Publishing Agency, New Delhi, India, pp. 421–424.

Winnaar, E.de, 1989. Turmeric successfully established in tissue culture. Inform. Bull. Citrus Subtrop. Fruits Res. Inst. 199, 1–2.

Xia, Q., Zhao, K.J., Huang, Z.G., Dong, T.T., Li, S.P., Tsim, K.W., 2005. Molecular genetics and chemical assessment of *Rhizoma curcumae* in China. J. Agric. Food Chem. 53, 6019–6026.

Yasuda, K., Tsuda, T., Shimizu, H., Sugaya, A., 1988. Multiplication of *Curcuma* species by tissue culture. Planta Med. 54, 75–79.

Zapata, E.V., Morales, G.S., Lauzardo, A.N.H., Bonfil, B.M., Tapia, G.T., Sanches, A.J., et al., 2003. *In vitro* regeneration and acclimatization of plants of turmeric (*Curcuma longa*) in a hydroponic system. Biotechnol. Appl. 31, 20–25.

6 The Agronomy of Turmeric

Soil and Climate Suitability

Turmeric is a crop of the tropical and subtropical regions. It is cultivated from subtropical dry to subtropical wet zone, through dry to wet tropical zone, at an altitude ranging from 1200m above mean sea level (MSL), optimum being 450–900MSL. In India, it is widely grown in warm to hot, per humid to humid ecosubregions. It can grow in rainfall ranging from 64 to 429cm, and moderate rainfall at 150cm at sowing, fairly heavy and well distributed during growing period, and dry weather for about a month before harvest is best for its field performance. Optimum temperature range can be between 18.2°C and 27.4°C. Where rainfall is bimodal, the crop is raised as rainfed, and where rainfall is unimodal and low, it is raised with supplemental irrigation. Fairly heavy rainfall for the first 2 months after planting is essential for sprouting, root and shoot emergence, and enlargement when the crop is grown as a rainfed one. Ibrahim and Krishnamurthy (1955) observed that turmeric growth was stimulated by the receipt of rain even under irrigated conditions. Panigrahi et al. (1987) recorded the air temperatures of 30–35°C, 25–30°C, 20–25°C, and 18–20°C during germination, tilling, rhizome initiation, and bulking, respectively, to be optimum. Rainfall received during the second month after planting and rhizome yield were positively correlated, with a high degree of significance, $r=0.6024$ (Kandiannan et al., 2002). Crop–weather interrelationship in turmeric indicated that mean minimum air temperature, total rainfall, number of rainy days, and mean minimum relative humidity showed positive correlation with yield, while mean evaporation, mean sunshine hours, mean solar radiation, and mean maximum air temperature showed negative correlation (Kandiannan et al., 2002).

A wide variety of soil types, varying in fertility, have been found suitable for turmeric. These soils are grouped under Inceptisols, Entisols, Vertisols, Alfisols, and Ultisols. Soils which have high organic carbon content, base saturation, and major and secondary nutrient content are suitable for turmeric cultivation. Well-drained, deep loamy to clay loam soils, with good organic matter status are well suited for the crop, while very coarse and heavy soils are unsuitable for rhizome development. The crop can thrive well in the soil pH range of 4.3–7.5. The differential performance of the crop based on soil types have been well recorded by Sahu and Mitra (1992).

Propagation of the Crop

Seed Rhizome

For commercial production, turmeric is propagated vegetatively. Rhizomes, also known as "Clump," "Bulb," "Corms," "Set," and "Tuber" in scientific literature, are

The Agronomy and Economy of Turmeric and Ginger. DOI: http://dx.doi.org/10.1016/B978-0-12-394801-4.00006-5

of two types, namely "mother rhizome" and "finger rhizome," also known as "daughter rhizome" (developed from mother rhizome). The fingers are primary, secondary, and tertiary depending on their position. Primary fingers constitute a major share in the clump, while secondary and tertiary are less in quantity. Both mother and finger rhizomes are used in propagation. However, primary fingers are commonly used in planting owing to its availability in large quantities. In India, mother rhizomes are used for planting in Krishna and Guntur districts, Andhra Pradesh, India, while finger rhizomes are exclusively used in Cuddapah district of the state, while in Tamil Nadu, India, both mother and finger rhizomes are used separately (Mudaliar, 1960). In field performance, mother rhizomes have been found to be better than finger rhizomes. Whole mother rhizomes grow rapidly and develop well and are found to be better yielders than finger rhizomes (Aiyadurai, 1966). Also, mother rhizomes yielded more than primary or secondary rhizomes (Rashid et al., 1996). Yothasiri et al. (1997) found that the maximum yield was obtained from whole mother rhizome followed by primary rhizomes with five or six internodes and the half-cut mother rhizome. The primary, secondary, and tertiary rhizomes with 3–4 internodes did not differ from one another in terms of growth and yield. Highest yield (72.27 t/ha) was obtained from half-cut mother rhizome (Zaman et al., 2004), and this yield was statistically at par with that obtained from whole mother rhizome (69.12 t/ha) but was significantly higher than that obtained from other seed rhizomes. The combined effect of half-cut mother rhizome with an application rate of 120–60–120 kg/ha in the ratio of $N:P_2O_5:K_2O$ produced the highest yield of 94.26 t/ha. Menzes et al. (2005) reported that head rhizomes (mother rhizomes) led to 30% higher productivity than when finger rhizomes are planted. But there was no difference in essential oil content. Primary fingers are used as seed stock in most of the turmeric growing regions in India (Rao et al., 2006).

Best planting sequence was with a spacing of 45 cm × 30 cm using mother rhizome. Both mother and primary rhizomes are good planting material. With regard to plant growth characteristics, yield per plant, size of mother rhizome, and primary and secondary finger production, planting mother rhizome and primary fingers was found to be significantly superior to planting secondary fingers (Dhatt et al., 2008). Although mother rhizome and primary finger were at par statistically, in terms of plant growth characteristics, yield, and size of secondary fingers, the former turned out to be a better planting material in terms of the production of mother rhizome and primary finger. Hence, growers must choose either the mother rhizome or the primary finger as planting material when they target high turmeric yield.

Size of the Seed

Broken fingers of 5–10 cm length and weighing 50–100 g obtained from longitudinally split mother rhizomes into two halves, constitute the planting material (Aiyadurai, 1966). Highest yield could be obtained from seed rhizome pieces with 2–3 eyes, and primary rhizome weighing 30–40 g from larger diameter mother rhizome weighing 25–34 g have resulted in highest yield (Philip, 1983). Although the mother rhizomes are better yielders, Tayde and Deshmukh (1986) suggested that during storage of mother rhizomes, secondary rhizomes could be used successfully with additional manuring to obtain high yield. Aoi (1992) found that increasing tuber size from 20 to 29 g to that of 80–89 g increased yields substantially. Whole mother rhizomes weighing 70–80 g planted at a spacing of 50 cm × 20 cm resulted in highest yields in Haryana State, India (Singh et al., 2000). Planting full mother rhizomes weighing 80–100 g led to minimum blotch incidence caused by *Colletotrichum capsici*, which led to a maximum yield of 22 t/ha, which was followed by half mother rhizomes weighing 50–80 g (Archana et al., 2000). Plants from daughter rhizomes weighing 30, 40, and 50 g resulted in significantly larger shoot biomass and higher yield than those from smaller daughter rhizomes, in both greenhouse conditions and field experiments. Shoot biomass and yield are highest in plants grown directly from mother rhizomes, when compared with plants from daughter rhizomes attached to mother rhizomes. This investigation further substantiated the fact that turmeric seed rhizome should weigh 30–40 g, have a large diameter, and seed mother rhizome should be free from daughter rhizomes. Rapid multiplication was obtained from "minisett" weighing 5 g in Nigeria (Okoro et al., 2007).

Rate of the Seed

When mother rhizome was used, the seed rate was 1800 kg/ha, whereas in the case of finger rhizome, it was 1200 kg/ha, while the general seed rate varied from 1000 to 1200 kg/ha (Rao, 1957). Both plant spacing and type of rhizome used decided the seed rate. In fact, in literature one would find that investigators have recommended varying seed rates: 2000–3000 kg/ha (Adhate, 1958), 1000 kg/ha of finger rhizome (Rao et al., 1975), 1500 kg/ha at a plant spacing of 30 cm × 20 cm (Rao, 1978), and 3000 kg/ha (Govinden and Cheong, 1995).

Transplantation

In turmeric cultivation, transplanting of sprouted buds has been practiced (Aiyadurai, 1966). Anjaneyulu and Krishnamurthy (1979) suggested that under delayed planting, nursery raising followed by transplanting, rather than direct planting in the main field was found to be a better practice. Planting whole mother rhizomes led to highest average yield, followed closely by transplanting 30-day-old cut rhizome seedlings (Patil and Borse, 1980; Umarani et al., 1982a,b). Randhawa et al. (1984) reported that direct planting was much better than transplanting, the former practice giving 49.7% more yield. Plants with higher number, longer, and higher fresh and dry weight of mother, primary, and secondary rhizomes, leading to 27% more yield of fresh rhizomes (40 t/ha) were obtained by planting 14-day-old sprouted rhizomes (Shanmugam et al., 2000). Curcumin, oleoresin, and oil yield per hectare were also the highest from the 14-day-old sprouted rhizomes.

Crop Season

Because turmeric is an annual crop, its duration can vary from about 7 to 10 months. In India, the most important monsoon, which is the southwest monsoon (SW monsoon), breaks out in the early part of June, starting from the State of Kerala, and spreads gradually to the central, north, west, and east of India. Hence, important crops like rice and maize are sown coinciding with the SW monsoon, and turmeric is no exception. In fact, planting can start from as early as April, when summer showers arrive, and last up to August, when the SW monsoon can be in full fury. Planting on May 1 and May 10 proved to be better than planting on other dates (Om et al., 1978). Planting on these two dates resulted in better crop growth and consequently better rhizome yield. Planting at the end of April resulted in the highest yield in Andhra Pradesh, India (Anjaneyulu and Krishnamurthy, 1979). Nair (1982) reported that planting from mid-March to early April in the State of Punjab, India, and between mid-April and mid-May in the State of Kerala, India, gave the best rhizome yield. Even though planting later than the normal date led to decreased incidence of leaf blotch (*Taphrina maculans*), corresponding yield reduction would be high. When planting was delayed from April 20 to April 30, both fresh and dry yield of rhizomes increased, but planting beyond first week led to decreased yield (Shanmugam et al., 2000). In Odisha State and Tamil Nadu, India, the optimum time to plant was first week of May in the former case and June in the case of the latter (Panigrahi et al., 1987). In Japan, planting starting from the first fortnight of May to the first fortnight of July was ideal to contain disease incidence and obtain high rhizome yield (Aoi, 1992). Barholia et al. (1992) obtained higher fresh yield of rhizomes with planting on May 16 in Madhya Pradesh, India. In South Africa, planting in September, spaced $0.45 \, \text{m} \times 0.10 \, \text{m}$ was found to be optimum, as advancement of the date beyond September can affect the crop due to sun scorch, as the country enjoys the polar climate, where summer starts from late October and intensifies by December (Nel, 1988). In Brazil, planting is done during the months of October to December, and higher rhizome yield was obtained with November 20 planting with $0.30 \, \text{m} \times 0.30 \, \text{m}$ spacing (Cecilo et al., 2004). The emergence pattern, growth and yield of turmeric plant, and weed growth in field experiments in Okinawa, Japan, suggested that turmeric is best planted in April, followed by March planting (Ishimine et al., 2004). In India, May 15 planting was found better than later planting, when the crop is irrigated (Kandiannan and Chandragiri, 2008). Planting is best when conditions such as water availability, through either the monsoon or irrigation source, and ambient temperature are optimum and delayed planting invariably leads to yield reduction.

Planting

Treating the Seed Stock

Seed stock prior to planting should be dipped in fungicide solution (captan 0.2% and quintozene 0.5%—Indian trade mark, Singh et al., 1976), air dried, and subsequently

kept in aerated pits under shade. The fungicide treatment is to control disease incidence, enhance germination, and lead to higher rhizome yield. Hot water (50°C)
treatment can follow to control any possible fungal attack and to ensure optimum
germination.

Preparatory Tillage

Turmeric needs a fine til and this can only be achieved through adequate number of
plowings after the onset of monsoon or when absent with irrigation. Land preparation depends on the type of soil. Mudaliar (1960) reported that the number of plowing varies between two in the Mydudukur region and sixteen in the Rajampet area,
both in the Cudappah district, Andhra Pradesh, India, and on average works out to
be 6–8. In earlier times, bullock- or cattle-drawn plows were used. Farmers have
now switched over to tractor-drawn plows, as maintenance of cattle/bullocks has
become an expensive affair due to inadequacy of fodder and other expenses. There
is yet no research on the impact of soil compaction on the turmeric plant. It is a common practice to apply organic manure (farm compost or farmyard manure, which,
when cattle are available, is a mixture of cow dung and other vegetative remnants
left over after harvest). And this is thoroughly mixed with the soil. On sloped lands,
beds are prepared as plowing is impossible. In areas where the soil pH is low, especially in high rainfall regions like Kerala State, India, hydrated lime is applied prior
to planting to neutralize acidity and raise soil pH. But this is not a common practice.
Availability of lime and its cost is also a constraint.

Planting Method

Based on soil type, rainfall, irrigation mode, and the cropping system followed, planting method can vary from place to place (Randhawa et al., 1984). In the Malabar
coast of Kerala State, India, where rainfall intensity is very high during the SW monsoon period, planting is done on raised beds of 1–1.5 m width and 15 cm height and
of convenient length with spacing of 50 cm between beds. Small pits at a distance of
20–25 cm apart on either side are dug on the beds and filled with well-decomposed
cattle manure or farmyard compost and the seed set is placed over it and covered with
soil. These raised beds allow free drainage. If the field is irrigated, this system is not at
all practiced. Planting is done on ridges and furrows, and ridges are generally broad.
In loamy soils, the ridges are made 1–1.5 ft apart and planting is done on the crest of
the ridges. In clayey soils, the ridges are 2–2.5 ft apart and planting is done on both
the sides of the ridges (Mudaliar, 1960). The author also noted that sometimes the
rhizomes are scattered evenly on the field and pressed into the soil by hand. This is
done to save on the cost of manual labor, as making ridges and planting is more labor
intensive. In yet another method of planting, at first plow furrows are opened and rhizomes are dropped into them and covered with soil by hand. This method of planting
is impossible when tractors are employed. Li et al. (1997) found that dibble planting
proved more effective in raising yield than drilling and planting. In dibble planting, the
quantity of rhizomes used has no effect on the resultant tuber size. In Meghalaya, India,

turmeric is grown by the "*Bun*" method, wherein a series of beds (called "*Buns*") are formed along hill slopes and planting is done on them (Ghadge et al., 2001; Ngachan and Deka, 2008). Gill et al. (2004) found that flat beds are preferred in red soils and raised beds on black clay loam in Punjab State, India. Turmeric fresh rhizome yield was significantly higher in the ridge planted method as compared to that obtained in flat planting. Ridge planted rhizomes produced significantly taller plants and higher in number and weight of primary and secondary rhizomes in Ludhiana, Punjab State, India (Gill et al., 2004). Planting primary fingers 20 cm apart on broad ridges of 80 cm (long) × 10 cm (wide) × 20 cm (high) with 30 cm wide irrigation channels resulted in 52% more yield, than those planted in conventional ridge and furrow system (Anjaneyulu and Krishnamurthy, 1979). In Japan, turmeric is planted in a triangular pattern to obtain higher yield of shoot and rhizomes (Hossain et al., 2005). Singh et al. (2000) reported that flat bed planting followed by earthing up subsequently as the best method to obtain higher rhizome yield. It is essential to provide good drainage, because in its absence, soil infection will lead to onset of diseases on the planted seed material. This happens because of waterlogging, which is conducive to the buildup of soil pathogens. Wakhare et al. (2007) found that there was no significant difference in yield due to differences in planting method. The ridge and furrow system of planting is conventional and easiest to manage inasmuch as labor, irrigation, and subsequent cultural practices are concerned (Anandraj et al., 2008).

Depth of Planting

Depth of planting, on the whole, had little effect on rhizome yield. Going by experience, it is seen that a depth of 5 cm appeared to be optimum in conditions prevailing in Bangladesh (Amin and Hoque, 1989; Rahman and Faruque, 1974). No systematic investigations of planting depth have been carried out anywhere else. Mishra et al. (2000) found that germination, growth, crop yield, yield attributes, and fresh rhizome yield were positively influenced by increasing planting depth of rhizome. Planting at 8–10 cm depth in dark red soil in Japan was found better to obtain higher yield and reduce weed competition (Ishimine et al., 2003). Rhizome development was also earlier, coupled with higher shoot biomass and rhizome yield, when planted 8, 12, and 16 cm deep than at 4 cm. Hossain (2005) reported that harvest from deep planted crop was difficult to manage.

Plant Geometry

Plant geometry is an important factor that decides plant yield. Many factors influence plant geometry and the important ones are soil type, fertility, crop season, irrigation schedule, rainfall distribution, and, above all, the cultivar used. Planting is done in small pits 25 cm × 30 cm spaced and covered with soil or dry powdered cattle manure. For flat bed method planting, optimum spacing is 25 cm × 25 cm, while for the ridge and furrow method, it is 45–60 cm between rows and 25 cm between plants (Aiyadurai, 1966). A planting distance of 15 cm at a depth of 5 cm seems to be the common plant geometry. When turmeric is planted at a closer spacing, rhizome enlargement does not take place properly, which leads to the production of smaller sized yields. Sundarraj and Thulasidas (1976) suggested 15 cm × 20 cm to be the optimum spacing. Anjaneyulu and Krishnamurthy (1979) recommended planting whole mother rhizomes instead of half ones spaced 30 cm × 22.5 cm to be the maximum yield. Different spacings have been suggested, for instance, 30 cm × 20 cm (Rao, 1978), 45 cm × 20 cm (Ponnuswamy and Muthuswami, 1981), 30 cm × 45 cm (Rajput et al., 1982), and 30 cm × 15 cm (Nair, 1982).

Shashidhar and Sulikeri (1996) reported that curing percentage was higher at a medium spacing of 44 cm × 22.5 cm. But Valsala et al. (1998) observed that planting at different spacings did not affect either the curing percentage or the curcumin content. Plant spacing of 20 cm was optimum for higher rhizome yield (Carvalho et al., 2001). Though a spacing of 30 cm × 30 cm and 30 cm × 45 cm gave the maximum plant height, tiller number, and leaves per clump, the highest fresh weight of rhizomes and cured yield was obtained with a spacing of 30 cm × 15 cm (Manjunathagoud et al., 2002). Planting at a spacing of 45 cm × 45 cm (49,383 plants per hectare) significantly reduced the intensity of leaf blotch without adversely affecting the yield (Prasadji et al., 2002). The yield was 25 t/ha in the 0.05 m intra-row spacing, which reduced to 18 t/ha in the 0.40 m intra-row spacing. Higher yield due to lower intra-row spacing (<10 cm) did not compensate the higher seed cost (Silva et al., 2004). Kandiannan and Chandragiri (2006) recorded that closer spacing of 30 cm × 15 cm gave maximum yield compared to wider ones (45 cm × 15 cm and 60 cm × 15 cm) and that spacing did not affect the curcumin content or its recovery. Wakhare et al. (2007) reported that a spacing of 45 cm × 15 cm produced better growth parameters and yield attributes compared to other spacing treatments tried in the investigation. The weight of fresh rhizome per clump was maximum at a spacing of 30 cm × 30 cm × 45 cm (Ram and Singh, 2007).

All of the above observations point to the crucial fact, while plant geometry, as a whole, did influence rhizome yield and plant growth parameters, no single spacing can be recommended as universally applicable. The findings have been as varied as the situations are.

Intercultivation Practices

Mulching Practices

Mulching and growing crops is a common feature of tropical agriculture. This is especially so in rainfed situations, the principal purpose being to conserve soil moisture. There are several advantages of mulching. First of all, it protects the rhizome from getting exposed to sun scorch, because the mulch material splashes the rain drops onto the principal crop. The mulches also provide carbon for many beneficial soil-borne organisms. Mulching with Sunhemp and Dhaincha (two popular green manure crops) leaves was also found to be useful (Aiyadurai, 1966). Mulching was found to increase the organic carbon content of the soil and also hastened germination by a week while suppressing the weed growth. Hussain et al. (1969) found mulch to enhance germination rate and led to increase in rhizome yield in Punjab State, India. First, mulching is done with green leaves at the rate of 12–15 t/ha and the operation is repeated after a month and a half at the time of top dressing of fertilizers. Early planting (May 23) combined with mulching with *Dalbergia sissoo* leaves gave the highest yield of fresh rhizomes coupled with the highest returns on nonirrigated calcareous soil compared to mulching with mango leaves or rice straw (Jha et al., 1983). Application of paddy husk and wheat straw increased the rhizome yield (Mahey et al., 1986). Singh and Randhawa (1988) reported that application of straw mulch was more beneficial than intercropping turmeric with pigeon pea, maize, or green gram in terms of net returns, and no improvement in the soil or curcumin content was noticed by mulch application. The maximum yield and cost–benefit ratio were obtained with dry forest leaf litter mulch + intercropping with French bean, followed by dry forest leaf litter mulch alone (Mohanty et al., 1991). In the Odisha state, India, 12.5 t/ha green mulch material is commonly applied thrice in a cropping season (at planting time, 45 days after planting, and 90 days after planting) to obtain maximum yield (Mishra et al., 2000), but no significant difference in rhizome yield and growth due to application of different types of mulch was observed. Application of wheat straw as mulch improved both growth and yield (Gill et al., 1999). Alam et al. (2003) noted that sungrass mulch gave significantly higher yield compared to other mulches.

Kumar et al. (2003) reported that application of mulch at the rate of 10 t/ha conserved more moisture and increased the yield of turmeric by 12% and paddy straw mulch increased the yield by 18%, both compared to *Gliricidia maculata*, a common green manure plant (Kumar et al., 2003). Hossain (2005) noted that mulching suppressed weed growth and improved the yield. Menzes et al. (2005) reported that mulching doubled the rhizome productivity and had no effect on the essential oil content of the crop. Dinesh (2006) found that the highest organic matter content (5.68%) and nitrogen content (69.40 kg/ha) in the surface layer during rhizome formation were recorded with 10 t mulch/ha. The different types of mulch also had significant effects on soil fertility status. The maximum organic matter content (6.37%) and maximum phosphorus content (76.0 kg/ha) in the surface layer during rhizome formation were obtained with rice straw mulch. However, the nitrogen content was highest (73.81 kg/ha) with *G. maculata* mulch. The potassium content of the subsurface layer during the rhizome formation stage was highest (302.33 kg/ha) with rice straw mulching. Annu and Sarnaik (2006) noted that mulching using paddy straw resulted in obtaining the tallest turmeric plants, with maximum number of leaves, both of which were very positively reflected in enhanced rhizome yield. Application of paddy straw mulch at the rate of 22.5 t/ha increased rhizome yield by 62.2% as compared to no mulching (Swain et al., 2007).

Hilling

"Hilling" or earthing up is yet another important cultural operation which is normally carried out twice or thrice during the crop season, accompanied by weeding and side dressing the crop with fertilizers, under irrigated conditions, whereas in rainfed conditions, weeding, earthing up, and mulching are carried out simultaneously. Usually, earthing up is done during 45–60 days after planting (DAP), 90–105 DAP, and 120–135 DAP. This intercultural operation helps to form and enlarge finger rhizomes and also ensures adequate aeration to roots. It also protects rhizome from the attack by scale insects and check weed growth (Panigrahi et al., 1987). Flat bed, followed by earthing up, was found to be the best practice to maximize yield (Ajai et al., 2002).

Weed Management

Rashid et al. (1992) reported that a survey of turmeric fields in the Banu district, Pakistan, indicated the presence of 83 weed species which belong to 73 genera. Of the 34 families, 5 were monocotyledonous and 29 were dicotyledonous. The most dominant mono- and dicotyledonous families were the Poaceae (Gramineae) with 21 species, and the Asteraceae (Compositae) with 8 species. Gill et al. (2000) from Punjab observed that *Digitaria ischaemum, Cynodon dactylon, Cyperus rotundus, Eleusine aegypticum (Dactyloctenium aegypticum), Euphorbia hirta, Commelina benghalensis,* and *Eragrostis pilosa* as dominant species. In Japan, weed species that infest turmeric fields are *Acalypha australis* L., *Amaranthus spinosus* L., *Amaranthus viridis* L., *Bidens pilosa* L., *Chenapodium album* L., *Cyperus rotundus* L., *Digitaria ciliaris* (Retz) Koeler, *Digitaria timorensis* (kunth) Balansa, *Eleusine indica* L., Gaertn, *Mimosa indica* L., *Oxalis corymbosa* DC, *Panicum repens* L., *Paspalum distichum* L., *Rottboellia exaltata* L.f., *Solanum nigrum* L., *Sonchus asper* (L.) Hill, and *Sonchus oleraceus* L. (Hossain et al., 2005; Ishimine et al., 2004). The above information clearly establishes the fact that weed flora varies largely depending on geographic locations where the turmeric crop is grown.

Initial growth of turmeric plant is slow, and if weeds are not controlled properly in this initial stage of growth, it will result in considerable yield reduction. Rethinam et al. (1977) reported that application of the weedicide Alachlor (a patented product easily available in India) at the rate of 2 kg/ha led to effective weed control, as compared to manual weeding, leading to good cost–benefit ratio. Preemergence application of Lasso (another patented product) at the rate of 2 kg active ingredient per hectare in a turmeric field intercropped with pigeon pea or maize was more effective as compared to manual weeding (Mishra and Mishra, 1982). Balashanmugam et al. (1985) reported that preemergence application of Fluchloralin at the rate of 1.0 and 1.5 kg/ha, Oxidiazon at the same rate as above (Oxyfluorfen at 0.15 and 0.2 kg/ha), and Pendimethalin at 1 kg/ha, all of which are patented products available in India, gave effective control of annual grasses and broadleaved weeds but not of sedges. Oxyfluorfen at 0.15 kg/ha led to maximum rhizome yield followed by Oxadiazon at 1 kg/ha, Fluchloralin at 1.5 kg/ha, and Pendimethalin at 1 kg/ha, in this order. The weedicides had no deleterious effect on the succeeding groundnut crop. Application of weedicide + one manual weeding was more economic than the common practice of farmers, which is two manual weedings. Mohanty et al. (1991) reported that mulching reduced weed growth in turmeric fields. Herbicide treatment alone did not provide season-long weed control, but integrated treatments achieved similar levels of weed control by applying 0.7 kg/ha Metribuzin followed by the application of 1 kg/ha of Diuron (Gill et al., 2000). Both are patented products.

Weedicide applications are effective in controlling weed growth significantly and enhancing rhizome yield as compared to fields where no control was there (Anil Kumar and Reddy, 2000). Ajai et al. (2002) observed that Pendimethalin and Oxyflurofen, followed by manual weeding, resulted in 45% and 35% more rhizome yield, compared to unweeded control treatments. Early weed establishment was found in shallow planted turmeric (at 4 cm depth), causing nutrient deficiency and reduced rhizome yield (Ishimine et al., 2003). They also observed significantly reduced weed infestation at the second and third weeding with turmeric planted at depths of 8, 12, and 16 cm, because of better canopy structure compared to shallow planted crops. Weed biomass was reduced significantly with turmeric grown from seed bits of 30–40 g because larger seed bits provided higher shoots and better canopy (Alam et al., 2003; Ishimine et al., 2003). In Japan, Ishimine et al. (2004) found that fields of turmeric planted in February and March required additional weeding before emergence because winter and spring weeds emerged earlier and grew vigorously. The order of total weed dry weight was as follows: February planting > March planting > April planting > May planting > June planting. The emergence pattern, growth and yield of turmeric plants, and weed growth in the field experiment suggested that turmeric planting should be done in April followed by March in Okinawa, Japan (Ishimine et al., 2004). Hossain (2005) concluded that seed rhizomes weighing 30–40 g and/or mother rhizome could be planted in a 30 cm triangular pattern at a depth of 8–12 cm on two ridges spaced 75–100 cm apart from March to April in order to reduce weed infestation and obtain higher yield in Japan; besides, mulching also reduced weed growth and improved rhizome yield. He suggested integrating biological weed management practices using rabbits, goats, sheep, ducks, cover crops, or intercrops to control weeds in turmeric fields. The investigation by Hossain et al. (2008) indicated that for reducing weed interference, and obtain high yield, turmeric should be planted in a 30 cm triangular pattern, on two-row ridge spaced 75–100 cm apart. Turmeric when planted in this triangular geometry effectively utilizes all the available space for the growth and enlargement of the rhizome, which ensures a higher weed control (9%) and higher rhizome yield (11%), without any additional cost. Weed infestation did not vary with the planting patterns until about 50 DAP, as all the plants cannot emerge during this period. Hossain et al. (2008) reported that purple nut sedge (<3000 plants per m^2) did not significantly reduce turmeric yield, whereas the combined weed species reduced yield by as much as 40%. Crop interference by purple sedge was not high, and other weeds could be removed during 60–70 DAP, ensuring reduced labor input cost, commensurate with higher rhizome yield. Weed control at the early vegetative phase is a must in turmeric cultivation, and paucity of labor is an emerging phenomenon in many countries, including India, which leaves the option for herbicide use as an effective means to control weeds.

Cropping Pattern

In many tropical cropping systems, turmeric fits in very well as a field crop, a horticultural crop, and a plantation crop. Turmeric can be well intercropped. In coconut, arecanut, and mango gardens, turmeric forms an understory (low tier) crop. It can also be raised as a mixed crop with red chilies, colocasia, onion, brinjal, and cereals like maize, bajra, and ragi (*Eleusine coracana*—a poor man's highly nutritious minor millet crop grown especially in rainfed areas of Karnataka State, India). Turmeric + maize/red gram is a common intercropping system. Growing turmeric + maize in the ration of 2:1 brought in high monetary returns (Shankaraiah et al., 1987). Intercropping turmeric with pigeon pea, maize, or green gram reduced the availability of incident light, which in turn adversely affect rhizome formation and enlargement (Singh and Randhawa, 1988). However, intercropping was invariably more profitable than monocropping. Rao and Reddy (1990) observed that one row of maize alternating every other interrow space of turmeric, maintaining 100% maize density and application of supplemental fertilizers, led to highest yield of both the principal turmeric crop and the intercrop maize. Maize also provided the requisite shade for the turmeric crop in its initial stages of growth. Reduction in turmeric rhizome yield when intercropped with maize has been reported (Anil Kumar and Reddy, 2000; Sivaraman and Palaniappan, 1994).

In rubber plantations, turmeric has been found to be a good intercrop (Sreenivasan et al., 1987). Jaswal et al. (1993) found that turmeric cultivation was more remunerative comparatively than ginger, as an intercrop with poplar trees, at a spacing of 5 m × 5 m. Turmeric rhizome yield under partial shade was higher, but the highest cured yield was obtained when the crop was grown under fully open conditions as a monocrop, obviously due to the positive effect of fully incident light enhancing dry matter accumulation. Higher solar energy input under fully open conditions helps in higher crop growth rate during bulking of the rhizomes (Latha et al., 1995). *Leucaena leucocephala* and *Eucalyptus camaldulensis* are the most compatible agroforestry systems for turmeric intercropping (Mishra and Pandey, 1998). Pal et al. (2000) observed increased rhizome yield in a sole turmeric crop compared to alley cropping within a 2 m distance from a 5-year-old established plantation of *L. leucocephala*. Pigeon pea–turmeric intercropping in a 10-year-old poplar (*Populus deltodes*) tree plantation is also an economically viable system (Chaturvedi and Pandey, 2001). Turmeric crop is a component in an arecanut-based high density multispecies cropping system (HDMSCS) (Ray and Reddy, 2001). Multiple cropping systems in Assam involving coconut + betel vine + banana + Assam lemon + turmeric + colacasia improved soil fertility, thereby enhancing plant nutrient availability, beneficial microflora of *Azotobacter* population, and the net income per hectare (Sarma and Chowdhury, 2002). Singh and Rai (2003) reported high yields from turmeric–mango intercropping. Pradhan et al. (2003) recorded maximum rhizome yield of turmeric when grown within the rows of *Leucaena* planted 4 m × 1.0 cm apart and pruned 5 times/year. The yield of intercrop per unit area slightly increased with increased tree spacing and frequency of pruning, whereas the yield of *Leucaena* decreased with increased spacing and also the frequency of pruning. Yield traits, such as the number of tillers per clump and fresh rhizome yield, were lowest under heavy shade, as compared to the turmeric crop cultivated in open space without any shade. Severity of *Colletotrichum* leaf spot was significantly lower at 1.8% in heavy shade and at 4.8% in partial shade, as compared to open cultivation (without any shade) where it was found to be the highest at 23.7% (Singh and Edison, 2003). Turmeric is grown as a successful intercrop in a silvi–horti system (silviculture–horticulture system) with fodder trees, such as bhimal (*Grewia optiva*), khark (*Celtis australis*), banj (*Quercus leucotrichophora*), and kachnar (*Bauhinia variegata*) in India. The highest rhizome yield was recorded in the case of *Q. leucotrichophora* (Bisht et al., 2004). Khaunkuab et al. (2008) found that yield of turmeric under 1-year-old para rubber plants was significantly higher than in monocropping. However, the intercropping system between turmeric, *Citrus reticulata*, and the 3-year-old rubber plant was not significantly different in terms of growth and yield.

Both market access and choice by farmers decide the choice of the farmers. Kandiannan et al. (2005) have summarized the scope of intercropping turmeric in India. Intercropping with crops such as onion, okra, black gram (*Phaseolus mungo* L.), and green gram (*Phaseolus aureus* L.) increased rhizome yield, especially in the case of the latter two crops, which can fix a considerable amount of atmospheric N and thereby enhance the soil fertility status, unlike maize and finger millet (*Eleusine coracana*) which reduced turmeric yield, as these compete with the main turmeric crop for both water and plant nutrients from the soil. Also income from the latter combination was lower (Rethinam et al., 1984). Growing coriander (*Coriandrum sativum*) and soybean (*Glycine max*) as companion crops to turmeric reduced turmeric yield, more than when garlic and onion were grown as companion crops. It was concluded that coriander, onions, garlic, and soybean should not be intercropped with turmeric (Narayanpur and Sulikeri, 1996). Prasad et al. (2004) reported that turmeric + rice + peas for green pod was the most profitable and preferable intercropping combination, followed by turmeric + maize + peas. Singh et al. (2006) found that okra gave the highest yield as an intercrop. However, the highest rhizome yield (12.7t/ha) and net economic return was when turmeric was intercropped with the vegetable, bitter gourd or round melon. Both combinations proved better than monocropping of turmeric in India. When French bean was compared with cowpea, as an intercrop with turmeric, the former combination proved to be better both in terms of rhizome yield and cost–benefit ratio (Yamgar et al., 2006).

Harvesting

Drying up of the entire plant, including its basal portion, is indicative of maturity in turmeric. This takes place in about 7–9 months, of course, depending on the variety cultivated, with early varieties ready for harvest by about 7–8 months, medium ones in about 8–9 months, and late ones after about 9 months. Harvesting time is normally from January to April, depending on the location in India. Market requirements might prepone harvest by about a month to December. Late harvest can also be resorted to, when a need arises, because the underground rhizome does not deteriorate when left in the field for some months more after full maturity. Irrigation is stopped a month in advance before harvest, allowing the tops to dry. Leaves and stem are cut to the ground level, and the rhizomes are dug out by either plows or using hand hoes. To obtain maximum yield, the turmeric crop should be harvested in about 8–9 months (Govind, 1987; Umarani et al., 1982a,b). When the crop is intended for the extraction of curcumin, it should be harvested in about 7–8 months. Power tiller-based and tractor-drawn turmeric harvesters are available to ensure efficient harvest and minimize on labor cost (Viswanathan et al., 2008).

Storage of Seed Rhizome

Rhizomes, when required as seed, should be stored by heaping in well-ventilated enclosures and covered with turmeric leaves. Seed rhizome can also be stored in pits filled with sawdust (obtained from timber mills), sand, leaves of *Glycosmis pentaphylla* and *Strychnos nux-vomica*. The pits must be covered with wooden planks, with one or two holes made on their top for proper aeration and incidence of sunlight so that no rotting of the rhizomes takes place inside the pits. The rhizomes should be dipped in 0.075% of Quinalphos (a patented Indian fungicide) when scale infestations are observed and dipped in 0.3% Mancozeb (also a patented Indian fungicide) to contain storage losses due to fungal infestation. Venkatesha et al. (1997) found that rhizomes stored at 10°C, unlike in the case of all other treatments, did not rot, nor was there any sprouting. However, these rhizomes showed low sprouting (33.05%) in the field. The most effective method for storage was to keep the rhizomes in 100-gauge polyethylene bags with 3% ventilation which ensures almost 100% (98.88% precisely) recovery of healthy rhizomes, which when subsequently planted in the field showed up to 91.9% sprouting. The above treatment showed <20% sprouting and rooting during storage and ensured almost negligible rotting.

References

Adhate, S., 1958. Turmeric. Farmer 9, 21–27.
Aiyadurai, S.G., 1966. A Review of Research on Spices and Cashew Nut in India. Regional Office (Spices and Cashew), ICAR, Ernakulam, Kerala, p. 209.
Ajai, S., Bajrang, S., Vaishya, R.D., 2002. Integrated weed management in turmeric (*Curcuma longa*) planted under poplar plantation. Indian J. Weed Sci. 34 (3/4), 329–330.
Alam, M.K., Islam, Z., Rouf, M.A., Alam, M.S., Mondal, H.P., 2003. Response of turmeric to planting material and mulching in the hilly region of Bangladesh. Pak. J. Biol. Sci. 6 (1), 7–9.
Amin, M.R., Hoque, M.M., 1989. Effect of number of eyes in a seed and depth of planting on the yield of turmeric. Bangladesh Hortic. 17 (2), 42–43.
Anandraj, M., Johy, A.K., Kandiannan, K., 2008. Technologies for sustainable production of turmeric. In: Krishnamurthy, K.S., Prasath, D., Kandiannan, K., Suseela Bai, R., Johnson George, K., Parthasarathy, V.A. (Eds.), National Workshop on Zingiberaceous Spices–Meeting the Growing Demand Through Sustainable Production Indian Institute of Spices Research, Calicut, Kerala State, pp. 51–63.
Anil Kumar, K., Reddy, M.D., 2000. Integrated weed management in maize + turmeric intercropping system. Indian J. Weed Sci. 32 (1/2), 59–62.
Anjaneyulu, V.S.R., Krishnamurthy, D., 1979. Efficacy of type of seed materials for different times of planting turmeric under Dugirala conditions. Indian Arecanut Spices Cocoa J. 2 (4), 115–116.

Annu, V., Sarnaik, D.A., 2006. Effect of different types of mulches on growth and yield of tur-
    meric (*Curcuma longa* L.). Int. J. Agric. Sci. 2 (2), 425–426.
Aoi, K., 1992. The characteristics and cultivation methods of the medicinal plant *Curcuma
    domestica*. Agric. Hortic. 67 (5), 507–511.
Archana, K., Rai, B., Jha, M.M., 2000. Effect of different size of rhizomes on the severity of
    leaf blotch disease and yield of turmeric. J. Hortic. Sci. 31, 302.
Balashanmugam, P.V., Ali, A.M., Chamy, A., 1985. Annual grass and broad leaved weed
    control in turmeric Annual Conference of Indian Society of Weed Science. Gujarat
    Agricultural University, Gujarat State, p. 25.
Barholia, A.K., Bisen, A.L., Mishra, A.K., 1992. Growth and yield of turmeric as influenced
    by time and planting and planting material. Gujarat Agric. Univ. Res. J. 17 (2), 172–174.
Bisht, J.K., Chandra, S., Singh, R.D., 2004. Performance of taro (*Colacasia esculenta*) and
    turmeric (*Curcuma longa*) under fodder trees in a agri–silvi–horti system of Western
    Himalaya. Indian J. Agric. Sci. 74 (6), 291–294.
Carvalho, C.M., Souza, R.J., de Cecilio, A.B., Filho, 2001. Yield of turmeric (*Curcuma longa* L.)
    grown at different planting densities. Cienciae Agrotecnologia 25 (2), 330–335.
Cecilo, F.A.B., Souza, R.J., de Faquin, V., Carvalho, C.M.de, 2004. Time and density of plan-
    tation on the turmeric production. Ciencia Rural 34 (4), 1021–1026.
Chaturvedi, O.P., Pandey, I.B., 2001. Yield and economics of *Populus deltoides* G3 Marsh
    based inter-cropping system in Eastern India. Forests Trees Livelihoods 11, 207–216.
Dhatt, A.S., Sidhu, A.S., Garg, N., 2008. Effect of planting material on plant growth, yield and
    rhizome size of turmeric. Indian J. Hortic. 65 (2), 193–195.
Dinesh, K., 2006. Effect of organic mulches on soil fertility in turmeric field under rainfed
    conditions of Orissa. Orissa J. Hort. 34 (2), 52–56.
Ghadge, S.V., Agarwal, K.N., Singh, R.K.P., Satpathy, K.K., 2001. Turmeric in Meghalaya—a
    case study of Shangpung village Jaintia Hills. Indian J. Hill Farm. 14 (2), 99–104.
Gill, B.S., Randhawa, R.S., Randhawa, G.S., Singh, J., 1999. Response of turmeric (*Curcuma
    longa*) to nitrogen in relation to application of farm yard manure and straw mulch.
    J. Spices Aromat. Crops 8 (2), 211–214.
Gill, B.S., Randhawa, G.S., Saini, S.S., 2000. Integrated weed management studies in turmeric
    (*Curcuma longa* L.). Indian J. Weed. Sci. 32 (1/2), 114–115.
Gill, B.S., Kaur, S., Saini, S.S., 2004. Influence of planting methods, spacing and farm yard
    manure on growth, yield and nutrient content of turmeric (*Curcuma longa* L.). J. Spices
    Aromat. Crops 13 (2), 117–120.
Govind, S., 1987. Studies on optimum harvesting time of turmeric. Haryana J. Hortic. Sci.
    16 (3–4), 257–263.
Govinden, N., Cheong, W.Y.K., 1995. Planting material and optimum planting rate for tur-
    meric. Revue Agricole et Sucriere de l'Ile Maurice 74 (3), 1–8.
Hossain, M.A., 2005. Agronomic practices for weed control in turmeric (*Curcuma longa* L.).
    Weed Biol. Manag. 5 (4), 166–175.
Hossain, M.A., Ishimine, Y., Motomura, K., Akamine, H., 2005. Effects of planting pattern
    and planting distance on growth and yield of turmeric (*Curcuma longa* L.). Plant Prod.
    Sci. 8 (1), 95–105.
Hossain, M.A., Yamawaki, K., Akamine, H., Ishimine, Y., 2008. Weed infestation in turmeric
    in Okinawa, Japan. Weed Technol. 22 (1), 56–62.
Hussain, A., Washeeda, A., Zafar, M.A., 1969. Effect of mulches on germination, growth and
    yield of turmeric. W. Pak. J. Agric. Res. 7, 153–157.
Ibrahim, S., Krishnamurthy, N.H.V., 1955. Preliminary studies in turmeric in Godaveri delta.
    Andhra Agric. J. 2 (5), 241–246.

Ishimine, Y., Hossain, M.A., Murayana, S., 2003. Optimal planting depth for turmeric (*Curcuma longa* L.) cultivation in dark red soil in Okinawa Island, Southern Japan. Plant Prod. Sci. 6 (1), 83–89.

Ishimine, Y., Hossain, M.A., Motomura, K., Akamine, H., Hirayama, T., 2004. Effects of planting date on emergence, growth and yield of turmeric (*Curcuma longa* L.) in Okinawa prefecture, Southern Japan. Jap. J. Trop. Agric. 48 (1), 10–16.

Jaswal, S.C., Mishra, V.K., Verma, K.S., 1993. Intercropping ginger and turmeric with poplar (*Populus deltoides* "G-3" Marsh.). Agroforest. Syst. 22 (2), 111–117.

Jha, R.C., Sharma, N.N., Maurya, K.R., 1983. Effect of sowing dates and mulching on the yield and profitability of turmeric (*Curcuma longa*). Bangladesh Hortic. 11, 1–4.

Kandiannan, K., Chandragiri, K.K., 2006. Influence of varieties, dates of planting, spacing and nitrogen levels on growth, yield and quality of turmeric (*Curcuma longa*). Indian J. Agric. Sci. 76 (7), 432–434.

Kandiannan, K., Chandragiri, K.K., 2008. Monetary and non-monetary inputs on turmeric growth, nutrient uptake, yield and economics under irrigated condition. Indian J. Hortic. 65 (2), 209–213.

Kandiannan, K., Chandragiri, K.K., Sankaran, N., Balasubramanian, T.N., Kailasam, C., 2002. Crop-weather model for turmeric yield forecasting for Coimbatore district, Tamil Nadu. India Agric. Forest Meteorol. 112, 133–137.

Kandiannan, K., Thankamani, C.K., Shiva, K.N., 2005. Intercropping with spices—a special reference to turmeric. Spice India XVIII (9), 15–20.

Khaunkuab, N., Suwunnalert, S., Chotiphan, R., Namphech, L., Sinphai, J., Chaiburi, J., 2008. Study on intercropping system between turmeric (*Curcuma longa* Linn.) in para rubber and fruit plantation on upper southern area. In: Proceedings of the 46th Kasetsart University Annual Conference. 29 January–1 February, Kasetsart University, Bangkok, Thailand, pp. 373–379.

Kumar, D., Singh, R., Gadekar, H., Patnaik, U.S., 2003. Effect of different mulches on moisture conservation and productivity of rainfed turmeric. Indian J. Soil Conserv. 31 (1), 41–44.

Latha, P., Giridharan, M.P., Naik, J., 1995. Performance of turmeric (*Curcuma longa* L.) cultivars in open and partially shaded conditions under coconut. J. Spices Aromat. Crops 4 (2), 139–144.

Li, L., Qin, S., Yang, H., 1997. Effect of cultivating measures on the tuber yield of *Curcuma longa* L. China J. Chin. Mat. Med. 22 (2), 77–78.

Mahey, R.K., Randhawa, G.S., Gill, S.R.S., 1986. Effect of irrigation and mulching on water conservation, growth, and yield of turmeric. Indian J. Agron. 31 (1), 79–93.

Manjunathagoud, B., Venkatesha, J., Bhahavantagoudra, K.H., 2002. Studies on plant density and levels of NPK on growth, yield and quality of turmeric cv. Bangalore local Mysore. J. Agric. Sci. 36 (1), 31–35.

Menzes, J.A., Borella, J.C., Franca, S.C., Masca, M.G.C.C., 2005. Effects of type of rhizome used to proliferation and mulching on growth and productivity of turmeric (*Curcuma longa* L.). Revista Brasileira de Plantas Medicinais 8 (1), 30–34.

Mishra, M., Mishra, S.N., Patra, G.J., 2000. Effect of depth of planting and mulching on rainfed turmeric (*Curcuma longa*). Indian J. Agron. 45 (1), 210–213.

Mishra, R.K., Pandey, V.K., 1998. Intercropping of turmeric under different tree species and their planting pattern in agroforestry system. Range Manag. Agroforest. 19, 199–202.

Mishra, S., Mishra, S.S., 1982. Chemical weed control in turmeric Annual Conference of Indian Society of Weed Science. Orissa University of Agriculture and Technology, Bhubaneshwar, Orissa State, p. 33.

Mohanty, D.C., Sarma, Y.N., Panda, B.S., 1991. Effect of mulch materials and intercrops on the yield of turmeric cv. Suvarna under rainfed condition. Indian Cocoa Arecanut Spices J. 15 (1), 8–11.

Mudaliar, V.T.S., 1960. Turmeric. In: Viswanathan, S., Anandam, V. (Eds.), South Indian Field Crops Central Art Press, Madras (now Chennai), pp. 221–228.

Nair, P.C.S., 1982. Agronomy of Ginger and Turmeric. In: Nair, M.K., Premkumar, T., Tavindran, P.N., Sarma, Y.R. (Eds.), Agronomy of Ginger and Turmeric Central Plantation Crops Research Institute, Kasaragod, Kerala State, pp. 63–68.

Narayanpur, M.N., Sulikeri, G.S., 1996. Economics of companion cropping system in turmeric (Curcuma longa L.). Indian Cocoa Arecanut Spices J. 20 (3), 77–79.

Nel, A., 1988. Turmeric trials Important Findings, 188. Information Bulletin, Citrus and Subtropical Fruit Research Institute, South Africa, pp. 11–12.

Ngachan, S.V., Deka, V.C., 2008. Present status and future prospects of turmeric production in the north eastern states. In: Krishnamurthy, K.S., Prasath, D., Kandiannan, K., Suseela Bai, R., George Johnson, K., Parthasarathy, V.A. (Eds.), National Workshop on Zingiberaceous Spices—Meeting the Growing Demand Through Sustainable Production Indian Institute of Spices Research, Calicut, Kerala State, pp. 64–68.

Okoro, O.N.E., Olojede, A.O., Nwadili, C., 2007. Studies on the optimum minisett sizes for rapid multiplication of riza (Plectranthus esculentus), hausa potato (Selenosterum rotundifolius) and turmeric (Curcuma longa) in Nigeria. Niger Agric. J. 38, 24–30.

Om, H., Verma, V.K., Srivastava, R.P., 1978. Influence on the time of planting on the growth and yield of turmeric. Indian J. Hortic. 35 (2), 127–129.

Pal, S., Maiti, S., Mandal, N.N., 2000. Effect of N and K fertilizer on turmeric under sole and alley cropping system. J. Interacademicia 4, 274–283.

Panigrahi, U.C., Patro, G.K., Mohanty, G.C., 1987. Package of practices for turmeric cultivation in Orissa. Indian Farming 37 (4), 4–6.

Patil, B., Borse, C.D., 1980. Effect of planting material and transplanting of seedlings of different age on the yield of turmeric (Curcuma longa L.). Indian Cocoa, Arecanut Spices J. 4, 1–5.

Philip, J., 1983. Effect of different planting materials on growth, yield and quality of turmeric. Indian Cocoa Arecanut Spices J. 7 (1), 8–11.

Ponnuswamy, V., Muthuswami, S., 1981. Influence of spacings on yield and yield components of turmeric (Curcuma longa L.). South Indian Hortic. 29 (4), 229–230.

Pradhan, U.B., Maiti, S., Pal, S., 2003. Effect of frequency of pruning and tree spacing of Leucaena on the growth and productivity of turmeric when grown under alley cropping system with Leucaena leucocephala. J. Interacademicia 7, 11–20.

Prasad, R., Yadav, L.M., Yadav, R.C., 2004. Studies of suitable intercropping system for turmeric (Curcuma longa L.). J. Appl. Biol. 14 (2), 40–42.

Prasadji, J.K., Rao, D.M., Murthy, K.V.M.K., Pandu, S.R., Muralidharan, K., 2002. Effect of cultural practices on leaf blotch severity and yield in turmeric. Indian J. Plant Protection 30 (2), 115–119.

Rahman, A., Faruque, A.H.M.A., 1974. A study on the effect of number of eyes of a seed, spacing and depth of planting on the yield of turmeric. Bangladesh Hortic. 2 (1), 25–27.

Rajput, S.G., Patil, V.K., Warke, D.C., Ballal, A.L., Gunjkar, S.N., 1982. Effect of nitrogen and spacing on the yield of turmeric rhizomes. In: Nair, M.K., Premkumar, T., Ravindran, P.N., Sarma, Y.R. (Eds.), Ginger and Turmeric Central Plantation Crops Research Institute, Kasaragod, Kerala State, pp. 83–85.

Ram, P., Singh, T., 2007. Influence of types of rhizomes and plant geometry on growth, yield of turmeric(Curcuma longa). Progressive Agric. 7 (1/2), 110–112.

Randhawa, G.G., Mahey, R.K., Gill, S.R.S., Sidhu, B.S., 1984. Performance of turmeric (*Curcuma longa* L.) under different dates and methods of sowing. J. Res. 21 (4), 489–495. Punjab Agricultural University, Ludhiana, Punjab State.

Rao, A.M., Reddy, M.L., 1990. Population and fertilizer requirement of maize in turmeric + maize intercropping system. J. Plantation Crops 18 (1), 44–49.

Rao, A.M., Jagadeeshwar, P., Sivaraman, K., 2006. Turmeric. In: Ravindran, P.N., Nirmal Babu, K., Shiva, K.N., Johy, K.A. (Eds.), Advances in Spices Research History and Achievements of Spices Research in India Since Independence Agrobios (India) Jodhpur, Rajasthan State, pp. 433–491.

Rao, C.H., 1957. Profitable intercrops in coconut plantations of East Godavari district. Andhra Agric. J. 4 (3), 73–75.

Rao, M.R., Reddy, K.R.C., Subbarayudu, M., 1975. Promising turmeric types of Andhra Pradesh. Indian Spices 12 (2), 2–13.

Rao, T.S., 1978. Turmeric cultivation in Andhra Pradesh. Indian Arecanut Spices Cocoa J. 2 (2), 31–32.

Rashid, A., Khan, S., Hussain, F., Ayaz, Saljoqi, A.U.R., 1992. Weed flora of turmeric fields of Bannu district. Sarhad J. Agric. 8 (3), 289–299.

Rashid, A., Islam, M.S., Paul, T.K., Islam, M.M., 1996. Productivity and profitability of turmeric varieties as influenced by planting materials and spacing. Ann. Bangladesh Agric. 6 (2), 77–82.

Ray, A.K., Reddy, D.V.S., 2001. Performance of areca-based high-density multispecies cropping system under different levels of fertilizer. Tropical Agric. 78, 152–155.

Rethinam, P., Sankaran, N., Sankaran, S., 1977. Chemical weed control in turmeric (*Curcuma longa*) Weed Science Conference and Workshop in India. Andhra Pradesh Agricultural University, Hyderabad, Andhra Pradesh, pp. 218–219.

Rethinam, P., Selvarangaraju, G., Sankaran, N., Rathinam, S., Sankaran, S., 1984. Intercropping in turmeric. In: Ahmed Bavappa, K.V. (Ed.), Proceedings of the Fifth Annual Symposium on Plantation Crops on Economics, Marketing, Statistics (PLACROSYM V) Indian Society for Plantation Crops, Kasaragod, Kerala State, pp. 485–490.

Sahu, S.K., Mitra, G.N., 1992. Influence of physico-chemical properties of soil on yield of ginger and turmeric. Fertilizer News 37 (10), 59–63.

Sarma, U.J., Chowdhury, D., 2002. Synergistic effect of high density multiple cropping on soil productivity and yield of base crop (coconut). Indian Coconut J. 33, 18–22.

Shankaraiah, V., Prabhakar Reddy, J., Rao, R., 1987. Studies on inter cropping in turmeric with maize, chilly, caster and others. Indian Cocoa Arecanut Spices J. 11 (2), 50–52.

Shanmugam, P., Vijayakumar, M., Shanmugasuundaram, K.A., 2000. Standardization of planting material with reference to the stages of sprouting in turmeric (*Curcuma longa* L.). Spices and aromatic plants: challenges and opportunities in the new century. In: Ramana, K.V., Eapen, S.J., Babu, K.N., Krishnamurthy, K.S., Kumar, A. (Eds.), Centennial Conference on Spices and Aromatic Plants Indian Society for Spices, Calicut, Kerala State, pp. 113–116.

Shashidhar, T.R., Sulikeri, G.S., 1996. Effect of plant density and nitrogen levels on growth and yield of turmeric (*Curcuma longa* L.). Karnataka J. Agric. Sci. 9 (3), 483–488.

Silva, N.F., da Sonenberg, P.E., Borges, J.D., 2004. Growth and production of turmeric as a result of mineral fertilizer and planting density. Hortic. Brasil. 22 (1), 61–65.

Singh, A.K., Edison, S., 2003. Eco-friendly management of leaf spot of turmeric under partial shade. Indian Phytopathol. 56, 479–480.

Singh, J., Randhawa, G.S., 1988. Effect of intercropping and mulch on yield and quality of turmeric (*Curcuma longa* L.). Acta Hortic. 188A, 183–186.

Singh, J., Malik, Y.S., Nehra, B.K., Pratap, S., 2000. Effect of size of seed rhizomes and plant spacing on growth and yield of turmeric (*Curcuma longa* L.). Haryana J. Hortic. Sci. 29, 258–260.

Singh, R.V., Rai, M., 2003. Standardization of mango-based cropping system for sustainable production. J. Res. Birsa Agric. Univ. 15, 61–63.

Singh, U.R., Singh, N., Gupta, J.H., Singh, J.B., 1976. Effect of post-harvest fungicidal treatment on storage rot and viability of turmeric seed rhizomes. Punjab Hortic. J. 16, 75–76.

Singh, J., Mehta, C.P., Ram, M., Singh, I., 2006. Production potential and economics of intercropping various vegetable crops in turmeric (*Curcuma longa* L.). Haryana J. Agron. 22 (1), 49–51.

Sivaraman, K., Palaniappan, S.P., 1994. Turmeric–maize and onion intercropping systems. I. Yield and land use efficiency. J. Spices Aromat. Crops 3, 19–27.

Sreenivasan, K.G., Ipe, C.V., Haridasan, V., Mathew, M., 1987. Economics of intercropping in the first three years among new/replanted rubber. Rubb. Board Bull. 23, 13–17.

Sundarraj, D.D., Thulasidas, G., 1976. Botany of Field Crops. The Macmillan Company of India Ltd., New Delhi, p. 604.

Swain, S.C., Rath, S., Ray, D.P., 2007. Effect of NPK levels and mulching on growth, yield and economics of turmeric in rainfed uplands. Orissa J. Hortic. 35 (1), 58–60.

Tayde, G.S., Deshmukh, V.D., 1986. Yield of turmeric as influenced by planting material and nitrogen levels. PKV. Res. J. 10 (1), 63–65.

Umarani, N.K., Patil, R.B., Pawar, H.K., 1982a. Effect of planting materials and transplanting on yield of turmeric cultivar, Tekurpeta. In: Nair, M.K., Premkumar, T., Ravindran, P.N., Sarma, Y.R. (Eds.), Proceedings on National Seminar on Ginger and Turmeric Central Plantation Crops Research Institute, Kasaragod, Kerala State, pp. 79–82.

Umarani, N.K., Patil, R.B., Pawar, H.K., 1982b. Studies on time of harvesting ginger and turmeric. In: Nair, M.K., Premkaumar, T., Ravindran, P.N., Sarma, Y.R. (Eds.), Proceedings on National Seminar on Ginger and Turmeric Central Plantation Crops Research Institute, Kasaragod, Kerala State, pp. 90–92.

Valsala, P.A., Nair, G.S., Nybe, E.V., Abraham, K., Mathew, M.P., 1998. Developments in plantation crops research. optimum spacing for turmeric cultivation. In: Mathew, N.M., Jacob, C., Kuruvilla, C., Licy, J., Joseph., T., Meenattoor, J.R., Thomas, K.K. (Eds.), Proceedings of the 12th Symposium on Plantation Crops, PLACROSYM XII Allied Publishers Ltd, New Delhi, pp. 206–208.

Venkatesha, J., Vanamala, K.R., Khan, M.M., 1997. Effect of method of storage on the viability of seed rhizome in turmeric. Curr. Res. 26 (6–7), 114–115.

Viswanathan, R., 2008. Turmeric—harvesting, processing, and marketing. In: Krishanmurthy, K.S., Prasath, D., Kandiannan, K., Suseela Bhai, R., Johnson, K., George, Parthasarathy, V.A. (Eds.), National Workshop on Zingiberaceous Spices—Meeting the Growing Demand Through Sustainable Production Indian Institute of Spices Research, Calicut, Kerala State, pp. 89–96.

Wakhare, A.V., Bhatt, R.I., Kamble, B.M., Patil, Y.G., Bhende, S.N., 2007. Effect of planting methods, spacing and levels of nitrogen on growth and yield of turmeric (*Curcuma longa* L.). Adv. Plant Sci. 20 (1), 201–204.

Yamgar, V.T., Shirke, M.S., Kamble, B.M., 2006. Studies on the feasible intercropping in turmeric cv. Salem. Indian J. Arecanut Spices Med. Plants 8 (2), 44–47.

Yothasiri, A., Somwong, T., Tubngon, S., Kasirawat, T., 1997. Effect of types and sizes of seed rhizomes on growth and yield of turmeric (*Curcuma longa* L.). Kasetsart J. Nat. Sci. 31 (1), 10–19.

Zaman, M.M., Rahman, M.H., Rahim, M.A., Nazrul, M.I., 2004. Effect of seed rhizomes and fertilizers on the growth and yield of turmeric. J. Agric. Dev. Gazipur 2 (1), 73–78.

7 Nutrition and Nutrient Management in Turmeric

Agricultural systems differ from natural systems in one fundamental aspect: while there is a net outflow of nutrients by crop harvests from soils in the first, there is no such thing in the second (Sanchez, 1994). This is because nutrient losses due to physical effects of soil and water erosion are continually replenished by weathering of primary minerals or atmospheric deposition. Hence, the crucial element of sustainability of crop production is the nutrient factor. But, of all the factors, the nutrient factor is the least resilient (Fresco and Kroonenberg, 1992). The thrust of the "high-input technology," the hallmark of the so-called "green revolution," in retrospect, or the moderation of "low-input technology," the foundation stone of sustainable agriculture, in prospect, both dwell on this least resilient nutrient factor. If the pool of nutrients in the soil, both native and added, could be considered as the "capital," efficient management of this capital might be analogous to the "interest" accrued from this capital in such a way that there is no great danger to the erosion of the capital. Hence, sound prescriptive soil management should aim at understanding the actual link between this capital and the interest, so that meaningful management practices can be prescribed.

Soil Tests and Nutrient Availability

It is universal knowledge that soil tests are the basis for predicting nutrient "availability." There are perhaps as many soil tests for each nutrient, as there are nutrients. This chapter on turmeric nutrition has no scope to extensively dwell on this aspect. Suffice to say that turmeric nutrition still revolves around "conventional mode," that is, routine soil tests and fertilizer prescriptions based on these soil tests. The author has proposed a revolutionary soil testing program and fertilizer management, which is now globally known as "The Nutrient Buffer Power Concept" and an elaborate review is presented in Advances in Agronomy, the magnum opus of agricultural science, as an invitational chapter (Nair, 1996). While reviewing the currently available results on turmeric nutrition, an attempt will be made to show how investigations could have been modeled better, wherever relevant information is available.

Nutrient Requirement and Uptake

Dry matter production enhances correspondingly to the uptake of nutrients. Active vegetative growth coincides with the maximum uptake of nutrients. The nutrient

The Agronomy and Economy of Turmeric and Ginger. DOI: http://dx.doi.org/10.1016/B978-0-12-394801-4.00007-7

Table 7.1 Nutrients Removed by Turmeric (kg/ha) by Harvest Time

Location and Soil Type	Nutrient Removed					References
	N	P_2O_5	K_2O	Ca	Mg	
Kasaragod, Kerala State, India, Laterite	124	30	236	73	84	Nagarajan and Pillai (1979)
Vellanikkara, Kerala State, India, Laterite	72–115	14–17	141–233	–	–	Rethinam et al. (1994)
Bhavanisagar,Tamil Nadu, India, Sandy loam	166	37	285	–	–	Rethinam et al. (1994)
Coimbatore, Tamil Nadu, India, Clay loam	187	39	327	–	–	Rethinam et al. (1994)
Calicut, Kerala State, India, Laterite	86	31	194	–	–	Sadanandan et al. (1998)

status of the plant depends on its concentration in the third leaf from the top, identified as the "index leaf" and for the age of the plant to sample for the nutrient status, it was suggested that 120 DAP is the optimum time (Saifudeen, 1981). Uptake
increased up to the third, fourth, and fifth month after planting inasmuch as uptake
of K, N, and P, respectively were concerned, and thereafter uptake of these nutrients
decreased. Use of mother rhizomes enhances the uptake of N, P, and K, compared to
primary or secondary rhizomes. Uptake of nutrients by turmeric is in the following
order: K>N>Mg>Ca (Nagarajan and Pillai, 1979; Rethinam et al., 1994). Table 7.1
depicts the nutrient uptake pattern.

Average dry rhizome yield of 5.5 t/ha removes 91 kg N, 16.9 kg P_2O_5, 245 kg
K_2O (Sadanandan et al., 1998), which shows that the crop is a very heavy feeder
of potassium. In sandy loam soils, excess calcium was found to induce micronutrient deficiency (Subramanian et al., 2003). According to Kumar et al. (2003),
approximately 20% of the turmeric growing area in Tamil Nadu State, India, suf-
fers from severe mineral deficiency. These investigators worked out the optimum levels of 12 plant nutrients (N, P, K, Ca, Mg, Na, S, B, Zn, Cu, Fe, and Mn)
in turmeric leaves, using the Diagnosis and Recommendation Integrated System
(DRIS), Modified Diagnosis and Recommendation Integrated System (MDRIS),
and also Compositional Nutrient Diagnosis (CND) to attain optimum production. The reference norms for optimum concentration of these nutrients in leaves
of turmeric in India (Erode Turmeric variety, earlier referred to in this book) are
reported here with the following range values: 1.22–2.75% for N, 0.36–1.27%
for P, 3.66–6.6% for K, 0.18–0.33% for Ca, 0.61–1.25% for Mg, 0.16–0.31% for
Na, 0.13–0.29% for S, 14.3–26.3 mg/kg dry matter for B, 41.1–93.2 mg/kg dry
matter for Zn, 15.2–40.3 mg/kg dry matter for Cu, 143–1568 mg/kg dry matter for Fe, and 66–219 mg/kg dry matter for Mn. The order of nutrient imbalance was as follows: S>B>Mg>Cu>P>Na>Ca>K>Zn>N>Fe>Mn based on
DRIS, S>B>Cu>Ca>Na>Zn>Mg>P>Fe>Mn>K>N based on MDRIS, and
S>Ca>Na>Mg>Cu>Zn>K>B>Mn>P>N>Fe based on CND. Among the different

Table 7.2 Inorganic Fertilizer Prescriptions for Turmeric Growing in Various Agroclimatic Regions of India

Region	Recommended Rates			
	FYM (t/ha)	N (kg/ha)	P_2O_5 (kg/ha)	K_2O (kg/ha)
Andhra Pradesh	25	300	125	200
Assam	20	30	30	60
Bihar	NA	150	50	100
Kerala	40	30	30	60
Maharashtra	NA	120	60	60
Tamil Nadu	25	120	60	60

Source: Rethinam et al. (1994).
NA—Not available, FYM = farmyard manure (organic source).

techniques used, CND projected the nutrient imbalances better than DRIS. B deficiency in turmeric results in reduced accumulation of sugars, amino acids, and organic acids at all leaf positions. Translocation of the metabolites toward rhizome and roots and photoassimilate portioning to essential oil in leaf and to curcumin in rhizome decreases. Due to B deficiency, the overall rhizome yield and curcumin yield also decreases (Dixit et al., 2002).

Dixit and Srivastava (2000) also observed decrease in plant growth, fresh weight, rhizome size, photosynthetic rate, and chlorophyll content, but increased curcumin content in all the genotypes when there was Fe deficiency. N, Mn, and Zn contents of rhizome had a significant positive correlation with curcumin content, whereas P, Ca, Mg, S, Cu, and Fe contents had no correlation (Kumar et al., 2000). Total Zn concentration in the third leaf was considered for delineation of Zn deficiency which varied from 10 to 51.3 mg/kg dry matter content. Zn concentration was significantly and positively correlated with leaf weight and rhizome yield, and negatively correlated with Fe:Zn and P:Zn concentration ratios. P:Zn ratio in the absence of applied Zn proved a good indicator of efficiency of response to added Zn (Singh et al., 1986). Leaf P:Zn ratio had a positive correlation and soil P:Zn ratio had a negative correlation with rhizome yield (IISR Report, 2005).

Manuring of Turmeric

Inorganic Fertilization

Because turmeric is a very soil-exhausting crop, it needs adequate and ample supplemental nutrition by way of inorganic fertilizers. Location-specific fertilizer recommendations are listed in Table 7.2.

The overall results indicate that response to applied N varies from region to region, depending on the type of soil, variety cultivated, and irrigation schedule. Application of varying rates of N, P, and K significantly increased the growth attributes, such as tiller number per plant, leaf number per plant, plant height, total leaf area per plant, and total dry matter produced per plant. Principal growth parameters, such as leaf number, tiller number, leaf area index (LAI), weight of mother and finger rhizomes, and total curing percentage, except plant height, had significant and positive correlation with rhizome yield (Prasad et al., 2004). Fertilizer application increased turmeric yields ranging from 81% to 282%, compared to control treatments where no fertilizer was applied (Eyubov et al., 1984).

Field investigations in India indicated that turmeric responded well up to a rate of 300 kg N/ha. In Tamil Nadu, India, N rates up to 120 kg/ha significantly increased the yield of fresh rhizome up to 41 t/ha, the percentage increase being 62. Optimal N rate was 150 kg/ha for higher rhizome essential oil yield at Jabalpur, Madhya Pradesh, India, and Rajasthan State, India (Tiwari et al., 2003), while Rajput et al. (1982) suggested 100 kg/ha N for optimal yield.

Yield of fresh rhizome increased from N application starting from 0 to 130 kg/ha, and P application from 0 to 177 kg/ha, and split application of N, as is the common field practice, had no positive effect on rhizome yield (Silva et al., 2004).

In Punjab State, India, maximum rhizome yield was obtained by applying N in three splits, as compared to applying the whole quantity while planting, at a rate of 60 kg/ha. Leaf and rhizome contents of N increased with enhanced N rate application and also the number of splits (Gill et al., 2001). In Maharashtra State, India, enhanced response to N application was obtained up to a rate of 120 kg/ha in variety Krishna. The N X Cultivar interaction was found to be significant inasmuch as weight of fresh finger was concerned (Attarde et al., 2003). Niranjan et al. (2003) found no significant positive impact of N application on the quality components of rhizome, except noting that the total biomass produced was more.

The beneficial effects of P application are primarily on root proliferation, consequent to which there will be better water use efficiency in the plant. In conjunction with other nutrients, such as N and K, positive response to P application, up to a rate of 175 kg/ha has been noted. Application of rock phosphate (RP) and single super phosphate (SSP) in the ratio of 1:3, 1:1, 3:1 combinations also increased rhizome yield compared to no fertilizer application (control treatment). Among P sources, SSP and Farm Yard Manure (FYM)-incubated RP enhanced the rhizome P concentration. Significantly higher rhizome yield was obtained with *Gafsa Phos* followed by *Raj Phos* (both locally made phosphatic fertilizers), incubated with FYM. Phosphate sources did not have any discernible effect on curcumin content. Phosphate recovery, agronomic efficiency of use, and percentage increase in yield were observed in the former compared to the latter (Srinivasan et al., 2000). Venkatesh et al. (2003) recorded 50% higher rhizome yield due to the application of FYM alone. N and P uptake, curcumin content, and P use efficiency were increased with the application of a combination of RP + SSP + FYM in Meghalaya, India. Maximum yield was obtained with a combination of 90 kg N/ha and 60 kg P/ha in a poplar tree plantation intercropping system in sodic soils (Katiyar et al., 1999).

Turmeric is a heavy feeder of potassium and positive response has been found up to a rate of 180 kg K_2O/ha. Splitting K application in four doses, initial at planting, 40, 80, and 120 DAP, had a positive effect on rhizome yield (30 t/ha) up to a level of

90 kg K_2O/ha (Rethinam et al., 1994). Both N and P fertilizers significantly affected plant height and tiller number. K had a significant effect on curcumin content. A combination of N and K at 150 kg/ha increased yield and net profit (Indian Rupees 25,120/ha, equivalent to approximately US$55), showing net income of Indian Rupee 1.45 per Indian Rupee invested, which is equivalent to approximately US$10 (Singh and Mishra, 1995), the approximate Indian Rupee–US$ exchange rate being 1 US$=45 Indian Rupees approximately. The above calculations show that under Indian conditions, applying inorganic fertilizers to turmeric is indeed a profitable proposition. Application of 80 kg/ha of both N and K on the low hills of Nagaland, India, increased turmeric productivity and led to enhanced uptake of N, P, and K and led to more curcumin recovery (Singh et al., 1998). Under rainy conditions of Madhya Pradesh, India, the maximum yield of 16.3 t/ha was recorded with a combination of 120 kg N/ha and 60 kg K_2O/ha (Gupta and Sengar, 1998). In north Bihar State, India, both N and K application up to 150 kg/ha had enhanced rhizome yield (Yadav and Prasad, 2003). Meenakshi et al. (2001) obtained maximum curing percentage (22.07) and cured rhizome yield at the highest levels of P and K fertilizers (100:100 kg/ha P and K). Split application of K at 25 kg K_2O/ha, half at planting + half at "earthing up" (a postplanting operation involving raking up the soil mass on the ridges) resulted in higher rhizome yield (25.7 t/ha), as compared to a single application at planting in northeastern parts of Haryana State, India (Singh et al., 2006).

Maximum yield (35.3 t/ha) and investment–benefit ratio (one Rupee invested and 2.58 Rupees recovered) were recorded with the application of N, P, and K at rates 150:125:250 kg/ha, respectively, irrespective of the cultivars planted in the southern state of Karnataka, India (Venkatesha et al., 1998). Sadanandan and Hamza (1996) computed the following crop response–fertilizer application in turmeric:

N	$Y = 2346 + 54.8\,N - 0.4500\,N^2$	$r^2 = 0.981$[***]
P	$Y = 2346 + 65.8\,P - 0.5832\,P^2$	$r^2 = 0.981$[***]
K	$Y = 2346 + 27.4\,K - 0.1000\,K^2$	$r^2 = 0.981$[***]

***Significant at 0.1% confidence level.

The optimum level of N, P, and K was standardized at 60:50:120 kg/ha. This, once again, shows the heavy K requirement for optimal turmeric production. Rao and Swamy (1984) recommended N, P, and K rates at 187.6:62.5:125 kg/ha from their nutrient input–crop response functions. These rates were found optimal for two cultivars, namely, Mydukur and Gorakhpur. The fertilizer levels adopted by different investigators to maximize turmeric production in various agroclimatic regions of India are summarized in Table 7.3.

N:P:K at 200:100:100 kg/ha resulted in maximum plant height, green rhizome yield, and the highest net return (Indian Rupees 30,227/ha, which equates to approximately US$67) with a cost–benefit ratio of 1.62 in Maharashtra State, India (Yamgar et al., 2001). N was best applied in split doses. When N was applied in three split doses, maximum rhizome yield, coupled with maximum returns and maximum cost–benefit ratio were recorded. Application of NPK in the ratio of 125:62.5:62.5 kg/ha + 1 t/ha

Table 7.3 Optimum Rates of N, P, and K (kg/ha) for Turmeric in Different Agroclimatic Regions of India

Optimum Rates of NPK	Recommended Location	Reference
150:125:250	Karnataka State	Venkatesha et al. (1998)
120:60:120	Hilly zone, Karnataka State	Sheshagiri and Uthaiah (1994)
60:50:120	Calicut, Kerala State	Sadanandan and Hamza (1996)
130:90:70	Arunachal Pradesh	Dubey and Yadav (2001)
200:60:200	Alluvial Plains of West Bengal	Medda and Hore (2003)
75:60:150	Allahabad (Uttar Pradesh)	Thomas et al. (2002)
100:50:50	Simla Hills, Himachal Pradesh	Randhawa et al. (1973)
200:100:100	Maharashtra State	Yamgar et al. (2001)
100:100:100	Uttar Pradesh	Upadyay and Misra (1999)

"Farm Boon" (organic manure obtained from sugarcane harvest waste) resulted in highest rhizome yield. The "Farm Boon" rate can go up to 27.75 t/ha (Krishnamurthi et al., 2001). Babu and Muthuswami (1984) recorded the highest curcumin content in the mother, primary, and secondary rhizomes, when the N P_2O_5 K_2O rate applied was in the ratio of 120:60:60–90 kg/ha, which had only negligible effect on oleoresin and essential oil contents of rhizomes. N:P:K at 90:60:90 kg/ha gave the maximum rhizome yield (25.5 t/ha) in rainfed conditions (Patra, 1998). Application of 100% NPK fertilizer at recommended rates, coupled with humic acid as potassium humate, at 10 kg/ha supplemented with a 0.1% foliar spray of humic acid enhanced plant growth and yield attributes, leading to increased fresh and cured rhizome yield (Baskar and Sankaran, 2005). Swain et al. (2006) recorded the highest fresh yield of rhizome with 120:60:120 kg/ha of N P_2O_5 K_2O, respectively. But the highest cost–benefit ratio was obtained at 100:50:110 g/ha, respectively of N, P_2O_5, and K_2O. Poinkar et al. (2006) reported that application of NPK at 120:60:60 kg/ha followed by FYM at 10 t/ha + azotobacter + phosphate solubilizing bacteria (PSB) at 250 g/10 kg of seed, increased plant height, leaf number, size and surface area of leaf (LAI), girth of pseudostem, tiller number per plant, and rhizome fresh yield in Maharashtra State, where the C:B ratio was the maximum (2.17) in the treatment where FYM + azotobacter+PSB were used in conjunction. The best N:P:K combination of 60:60:120 kg/ha was recommended by Liu et al. (1974) for Taiwan.

Timing of fertilizer application has an important bearing on rhizome yield. Application in four splits, 30, 60, 90, and 120 DAP showed the maximum yield of 23.2 t/ha in Costa Rica (Silva et al., 2004). The highest dry rhizome yield was obtained where neem (*Azadirachta indica*) cake at a rate of 1.25 t/ha + FYM at 12.5 t/ha + the recommended NPK rates was applied, followed by FYM at 25 t/ha + the recommended NPK rates. Merely applying the recommended rates of NPK gave the least yield (Rao et al., 2005). Among the quality traits, curcumin content was higher with neem cake treatment at 1.25 t/ha + FYM at 12.5 t/ha along with recommended rates of NPK, whereas essential oil and oleoresin contents were higher with vermicompost treatment at 1 t/ha singly as compared to recommended rates of NPK. To maximize turmeric yield in brown hill soils of Chittagong hill tracts (Assam State, India),

150 kg/ha N and 100 kg/ha K along with the prescribed rates of other nutrients has been recommended by Haque et al. (2007).

Akamine et al. (2007) noted that combined application of N and K or N, P, and K (NPK) provided 4–6 times greater shoot biomass and 8–9 times more rhizome yield. Better leaf chlorophyll content and foliar nutrient composition were observed with N and K application at 150 and 160 kg/ha, respectively. However, the maximum rhizome yield and curcumin content were obtained through combined application of mulch, N at 120 kg/ha and K at 160 kg/ha (Sanyal and Dhar, 2008). Kao et al. (2007) did not find any significant difference among fertilizer treatments inasmuch as rhizome yield was concerned. Jagadeeswaran et al. (2007) found that fresh rhizome yield increased significantly by as much as 125%, when NPK fertilizers were applied as slow-release fertilizer tablets. Inclusion of S and Mg in the fertilizer schedule dramatically improved fresh rhizome yield in West Bengal, India. Maximum fresh rhizome yield of 26 t/ha was obtained with 44 kg/ha of S, and 22 kg/ha of Mg, along with adequate N, P, and K applied as per appropriate soil test. The yield of dry turmeric increased to 6 t/ha in West Bengal, where the average is 1.5 t/ha, without any significant effect on curcumin content (Bose et al., 2008). Nawalagatti et al. (2008) observed that application of S at the rate of 80 kg/ha significantly increased plant height, tiller number, total dry matter production, total chlorophyll content, fresh and cured rhizome yield. The same treatment also recorded significantly higher curing percentage and curcumin content, resulting in higher economic benefit to the farmer.

Sharma et al. (2003) observed that continuous application of inorganic fertilizers lead to reduction in rhizome yield over the years, unlike in the case of FYM and vermicompost application, which enhanced rhizome yield in turmeric var. Suma by as much as 7–10%. Application of 50% of the recommended rates of N, P, and K fertilizers (175 kg N, 60 kg P_2O_5) and 125 kg K_2O + 10 t/ha vermicompost improved soil porosity, reduced soil bulk density, and enhanced organic carbon content (from 0.44% to 0.72%), leading to highest rhizome yield, net return, and cost–benefit ratio of 3.35. The combined application of spent wash, biosulfur, and 50% of recommended rate of N and P produced high curcumin content and essential oil in rhizome, whereas spent wash application at 50 kl/ha along with 2.5 t/ha of biosulfur supplemented with 75% of recommended rates of N and P resulted in increased micronutrient status of the soil, as well as plant parts, such as tiller number and leaf number (Davamani and Lourdhuraj, 2006). Application of vermicompost at the rate of 5 t/ha + 100% of the recommended rates of N, P, and K fertilizers (125:60:60 kg of NPK/ha) along with humic acid at the rate of 0.2% foliar spray showed significant increase in fresh rhizome yield per plant, number, length, and girth and weight of mother, primary, and secondary rhizomes per plant in the cultivar Erode local in Tamil Nadu, India (Suchindra and Anburani, 2008).

Effect of Micronutrients on Turmeric Production

Many tropical soils, including Indian, are vulnerable to micronutrients' deficiency, principally Zinc (Zn). This section will review whatever data is available on this aspect with specific reference to Indian soils. In Tamil Nadu, application of $ZnSO_4$ at 15 kg/ha increased the rhizome yield by 15%. In iron-deficient soils, 24% increase in rhizome yield has been obtained with the application of $FeSO_4$ at 30 kg/ha. Combined application of $ZnSO_4$ and $FeSO_4$ at 50 kg/ha recorded a high yield of 21.4 t/ha. In addition to Zn and Fe, boron (B) and molybdenum (Mo) have also been found quite beneficial in turmeric production, especially where these micronutrients are deficient. Optimum dosage of Zn, B, and Mo have been found to be 5, 2, and 1 kg/ha, respectively, when applied along with 15 t of FYM plus recommended doses of N, P, and K fertilizers, and green mulching for cover (Sadanandan et al., 1998). Application of 120 kg/ha each of N and K along with 2 kg/ha of B and 10 kg/ha of Zn recorded the maximum economic yield in turmeric intercropped under partial shade of coconut trees (Meerabai et al., 2000). Foliar application of 0.25% $ZnSO_4$ twice resulted in higher rhizome yield and was found on par with 7.5 kg/ha of soil applied $ZnSO_4$. Investigations by Kumar et al. (2004) observed that there was a positive effect of FYM plus Zn solubilizing bacteria (ZSB), and soil and foliar application of Zn and Fe, and this positive effect was observed in constant increase in the uptake of all the major plant nutrients from early stage on, from planting time right up to harvest time. Compared to mere FYM application, FYM + ZSB application recorded 21.6% increase in rhizome yield. With recommended rates of fertilizers, 120 kg N + 60 kg P_2O_5 + 60 kg K_2O per hectare and 30 kg Zn/ha, cultivars Krishna, Suvarna, Salem, and Waigaon, showed increased plant height, tiller number, leaf number, LAI, total dry matter per plant, highest average, and total yield of fresh and dry rhizomes (Jadhao et al., 2005).

Lower Nutrient Imbalance Index (NII) in leaves was noted when Zn and Fe, singly or in combination, were applied to soil, as compared to their application as foliar spray (Kumar et al., 2006). They also observed significant and positive correlation (0.92**, significant at 1% confidence level) between Zn concentration in leaves and rhizome yield, pointing to the importance of nutrient indexing. Vishwakarma et al. (2006) reported that application of 16 kg/ha of Cu and 20 kg/ha of B increased growth and yield of turmeric in both open and shaded conditions. Foliar application 0.5% $FeSO_4$, 60 and 120 days after sprouting, was found to be very effective, followed by sprays of $MnSO_4$ at 0.4% concentration and that of $ZnSO_4$ at 0.5% concentration (Jirali et al., 2007). These treatments significantly increased photosynthetic rate, stomatal conductance, transpiration rate, nitrate reductase activity, and total chlorophyll content. These treatments were also effective in increasing both fresh and dry yield of rhizomes. Application of these micronutrients significantly increased curcumin content (Velmurugan et al., 2007). A combination of Zn (at 4.5 kg/ha) and B (at 3.0 kg/ha) was found to bring about the maximum increase in rhizome yield (Halder et al., 2007).

The Role of Organic Manures in Turmeric Production

There are many investigations, also farmers' experience, which show that using organics singly or in combination with inorganics, has a very beneficial effect on turmeric production. Among the organic, FYM is the most used. FYM or compost at a rate of 40 kg/ha is applied broadcast, and plowed at the time of land preparation for planting or as a basal dressing by spreading over the beds covering the seed sets after planting. The increase was over 37% in fresh rhizome yield. Kerala Agricultural University, India, recommends 40 t/ha compost or cattle manure as basal dressing for the soils of the state. In Odisha State, 15 kg per bed ($5 m^2$) in three split applications along with recommended rates of chemical (NPK) fertilizers were applied. Organic manure like groundnut cake, vermicompost, or neem cake can also be used. In such cases, the quantity of FYM used can be reduced. Application of both FYM and any other organic manure enhances the organic carbon content of soil, leading to enhancement of soil productivity. In addition, applying organics have been found to have a positive effect on cation exchange capacity, increase water retention, and solubilize more of soil P. These have beneficial effects reflected on plant growth, which ultimately will positively reflect on rhizome yield. FYM application has been found to enhance leaf content of N, P, and K coupled with increased rhizome yield (Gill et al., 2004). Application of 50 t/ha FYM increased fresh rhizome yield up to 11.72 t/ha in Mizoram, India (Saha, 1988).

In Meghalaya, FYM + 90 kg N/ha is optimum for turmeric production, with considerable N use efficiency (34.3%) and nutrient buildup in the soil (Majumdar et al., 2002). Gill et al. (1999) observed significant increase in rhizome yield and curcumin content with increased FYM application (up to 60 t/ha), coupled with wheat straw mulch. In turmeric, application of cow dung (cattle manure) at 50 t/ha followed by 90, 60, and 90 kg/ha of N, P_2O_5, and K_2O yielded the highest coupled with maximum profit in the prevalent conditions of Mizoram. On the other hand, in Waynad district, Kerala State, India, 100 kg/ha of N and FYM (15 t/ha) with green mulch (50 t/ha) produced the maximum yield. Application of groundnut cake (1.1 t/ha) significantly increased the dry rhizome yield and curcumin production (Table 7.4), and this treatment was found to be on par with neem cake application (2.5 t/ha) (Sadanandan and Hamza, 1998). No significant differences in dry yield and curcumin content of turmeric rhizomes were observed with application of a combination of chicken manure and inorganic fertilizers in China (Hu et al., 2003). Neem cake application seemed to control infestation by *Mimegralla* fly (Reddy and Reddy, 2000).

Organic Farming

In recent years, organic agriculture has gained a lot of momentum. This has happened mainly because of customer consciousness, because the so-called "green revolution," which saw the unbridled use of chemicals, led to many environment-related, and, more importantly, health-related problems, originating in the wrong soil management practices. For instance, in the district of Gurdaspur, Punjab State, India, the "cradle" of the green revolution, cancer incidence has been on the increase since the last two decades, commencing from the peaking of the green revolution; most of these cases have been traced to the excessive intake of harmful chemicals through food

Table 7.4 Effect of Organic and Inorganic Fertilizers on Yield and Quality of Turmeric

Treatment	Turmeric Dry Yield (kg/ha)	Curcumin Content (kg/ha)
Control	2374	169
FYM	2596	250
Neem cake	2602	287
Cotton cake	2640	284
Brassica cake	2784	243
Groundnut cake	2669	277
Gingily cake	2768	249
NPK fertilizer (60 kg N/ha N, 50 kg P/ha, and 120 kg K/ha)	2480	268
LSD (95% confidence level)	N.S.	32.2

Source: Sadanandan and Hamza (1998).
LSD = least significant difference, N.S. = not significant.

materials. Breastfeeding mothers' milk has been found to contain DDT and other toxic chemicals. Hence, the switch to organic farming has actually been initiated, on the question of sustainable agriculture, a switch from "high-input technology" to a "low-input" one. This, coupled with global warming, led to a "greening consciousness." Traditional agriculture in India has been organic farming for millennia, and it was in the early 1960s, with the introduction of dwarf wheat varieties, which were found to be very highly fertilizer responsive, that the farmers' craze for excessive use of chemical fertilizers and other chemicals, by way of insecticides and fungicides, that led to the current situation.

Organic farming favors lower input costs, conserves nonrenewable resources, promotes high-value markets, and, consequently, boosts farmers' income. Estimates in India show that the country has 41,000 ha (0.003% of the total arable area of the country) under organic farming with 5661 registered organic farms, producing agricultural crops like plantation crops (black pepper and cardamom), spices, pulses, vegetables, oilseeds, and fruits. Internationally, there is a shift from allopathic medicine to traditional/ethnic medicine. In India, this is best exemplified by the increasing focus on *Ayurveda*, the ancient Indian system of healing. This shift, naturally, has opened up a huge market for organically produced crops, some of the best examples being black pepper, turmeric, and ginger.

Turmeric is a best component crop in agri–horti systems (agriculture–horticulture crops combination) and silvi–horti systems (silviculture–horticulture crops combination). Recycling of farm waste can be effectively done when grown with coconut, arecanut, mango, *Leucaena*, and rubber. As a mixed crop, it can also be grown or rotated with green manure/legume crops or trap crops, enabling effective nutrient buildup and pest and disease control. For certified organic production, the crop should be under organic management for at least 18 months. Traditional varieties of crops adapted to the local soil and climatic conditions are promoted (Parthasarathy et al., 2008).

In the organic farming program, the most cumbersome and, often, most heart-breaking experience for the farmer is to get the proper accreditation by competent agencies. Unlike in Western countries, e.g., Germany, Europe, where organic farming has grown many-fold during the last two decades, where foolproof, corruption-free certification agencies exist to monitor carefully the preplanting and postharvest procedures, in India, things are more often than not fudgy. The farmers, on the whole, entertain the wrong notion that if a crop is grown without the use of chemical fertilizers, it would be entitled to be branded as "organic." And it is necessary to educate the farmer that use of inorganic fertilizers is only one aspect of the procedure of certification. Certification and labeling usually is done by an independent body to provide a guarantee that the production standards are adequately and scrupulously met. The government of India has taken adequate steps to put in place indigenous certification system to help small and marginal farmers involved in organic farming and issue valid certificates testifying the end products, through certifying agencies accredited by the central agency APEDA. This agency follows the codex alimentaries stipulations, which is required, because many of the organic products originating in the developing countries, including India, have a vast market both in Europe and in North America, including Japan, where the stipulations are extremely stringent. The Spices Board, headquartered in Kochi, Kerala State, India, has streamlined certification processes for turmeric, under the auspices of the National Program for Organic Production (NPOP).

Liming at 400 kg/ha for acidic soils, application of 40 t/ha of FYM, as a basal application at planting, 2 t/ha of neem cake and 5–8 t/ha of vermicompost and mulching with green leaves at 12–15 t/ha at 45-day intervals are recommended. In turmeric, combined application of different organic sources, such as FYM + pongamia (a green manure crop) oil cake + stera (a popular organic manure) meal + RP + wood ash yielded similar results as with recommended fertilizers, and the rhizomes were of high quality. It should be noted that when a farmer switches to organic farming away from inorganic fertilizers, in the initial 2–3 years there is a reduction in yield which can be as low as 15% and as much as 30–35%. These reductions make the farmer lose faith in organic farming. But, thereafter, yield stabilizes. This is mainly because the soil has to find a stable carbon index, which would have been lost due to unbridled use of chemicals. There are many instances where after the switch, with the first 3–5 years yield stabilize and the crop fetches good yields, even without any external addition of manure, including organic manure. This can happen in inherently fertile soils, which have been ravaged due to excessive and unbridled use of chemicals. Sadanandan and Hamza (1998) observed that application of FYM and organic cakes enhanced residual nutrient availability to the succeeding crop (after turmeric), and in these instances, fewer incidences of pests and disease were also observed. Turmeric quality parameters, measured by curcumin and starch contents, were found to be higher in organic farming (Srinivasan et al., 2008). Among organics, maximum fresh rhizome yield of 34.4 t/ha was recorded in combined application of vermicompost and coir pith compost at 5 t/ha each, followed by FYM, vermicompost, and coir pith compost treatment. Application of vermicompost at 10 t/ha increased rhizome yield from 6.7% to 25.5% (Vadiraj et al., 1998).

Turmeric showed better response to the application of organic manures and biofertilizers. Investigations by Hossain et al. (2002) reported that Manda 31 (fermented natural plant concentrate) can be efficiently used to improve the efficiency of FYM. Application of Manda compost to turmeric produced more numbers of leaves, tillers, dry weight of shoot, root, and rhizome yield, as compared to normal compost application. Combined application of FYM, azospirillum, and phosphobacteria and vesicular arbuscular mycorrhizae (VAM) recorded higher yield attributes, higher yield per plant, and higher estimated yield of 33.3 t/ha in turmeric cultivar BSR-2 (Velmurugan et al., 2007).

Biofertilizers

Turmeric shows good response to the application of biofertilizers as well. Integrated application of coir compost at 2.5 t/ha combined with FYM, biofertilizer (*Azospirillum*), and half the quantity of recommended NPK fertilizers, significantly increased yield and quality (Srinivasan et al., 2000). *Azospirillum* inoculation brought about a 10% increase in rhizome yield (Santhanakrishnan and Balashanmugam, 1993). Application of coir pith compost, *Azospirillum*, phosphobacteria, and VAM induced maximum IAA oxidase and peroxidase activity (Velmurugan et al., 2002). Application of FYM at 5 t/ha with 50% inorganic N and *Azospirillum* at 5 kg/ha produced higher yield of mother, primary, and secondary rhizomes resulting in a cost–benefit ratio of 1:2.28 (Subramanian et al., 2003). Up to 16% increase in rhizome yield by the application of 25 kg of *Azospirillum* with 50% of the recommended dose of inorganic N and 5 t/ha FYM over the recommended rate of NPK fertilizers was reported by Selvarajan and Chezhiyan (2001). This is, indeed, a substantial savings of up to 50% of inorganic N. Jena et al. (1999) noted that the rhizome yield and nutrient uptake of turmeric were significantly higher in both single and combined inoculation of *Azotobacter* and *Azospirillum* and the percentage increase in yield due to inoculation integrated with fertilizer N ranged from 15.2% to 30.5%, coupled with enhanced N use efficiency. The soil was left with a positive N-balance in the integrated treatments.

The effect of ZSB on increasing dry matter production in turmeric was found to be 14%, 14.3%, and 18.1% for $ZnSO_4$, Zn-enriched FYM, and Zn-enriched coir pith when applied with ZSB, respectively, than in treatments without ZSB. All the genotypes of turmeric were found to have effective mycorrhizal association by extra-, intra-, and intercellular hyphae of VAM fungi in their roots. Suguna, followed by Prabha and Sugandham, among cultivars, were seen to be most heavily colonized with mycorrhizal fungi with mycelium, arbuscules, vesicles, and spores of *Glomus*, *Gigaspora*, and *Sclerocystis* with domination of *Glomus* population (Reddy et al., 2003). The mycorrhizal inoculation is advantageous in improving plant growth compared to uninoculated plants (Kumar, 2004). The highest growth rate was found in plants inoculated with *Glomus fasciculatum*. Mohan et al. (2004) recorded a linear response of turmeric growth, yield, and quality of turmeric with inoculation of *Azospirillum* in combination with N levels compared to that of *Azotobacter*.

The Role of Growth Regulators in Turmeric Production

It has been observed that in turmeric, application of growth regulators has a significant impact on yield of cured produce, its curing percentage, and curcumin content. Growth regulators (thiourea, agallol (2-methoxyethylmercurychloride), and GA) improved the percentage of emergence; however, no significant differences in plant height and tiller number were observed (Randhawa and Mahey, 1984). NAA and KNO_3 applications increased the curing percentage, yield of cured produce, and curcumin content compared to that of kinetin sprays (Lynrah et al., 2002). Spraying Planofix (NAA)—a patented Indian growth regulator—at a concentration of 10 ppm once in 6 months after planting recorded the highest yield of fresh rhizomes (Balashanmugam and Vanangamudi, 1987) with the highest rhizome tuberization rate and rhizome yield at the 20 ppm application rate of NAA. Rao et al. (1989) recorded the highest peudostem height, per plant leaf number, and total leaf area per plant, with the application of 2,4-D at 20 ppm.

The combined application of two growth regulators, triacontanol and kinetin, in the form of foliar spray showed enhanced rhizome yield by as much as 25.7% and curcumin content by as much as 39% over their respective control treatments and may be adopted for improved productivity and quality (Masroor et al., 2006). The correlation analysis also revealed the significant positive influence on increased plant height ($r=0.985***$, significant at 0.1% confidence level), fresh weight of shoot ($r=0.997***$), number of leaves ($r=0.983***$), leaf N content ($r=0.999***$), total chlorophyll content ($r=0.997***$), on rhizome yield and leaf content of N ($r=0.992***$). Jirali et al. (2008) found that foliar application of CCC at 1000 ppm concentration (60 and 120 days after sprouting) was very effective, followed by Cytozyme (a patented Indian growth regulator) at 2000 ppm and Miraculan (a patented Indian growth regulator) at 2000 ppm, and these treatments significantly increased the photosynthetic rate, stomatal conductance, transpiration rate, and biochemical parameters, such as nitrate reductase activity, total chlorophyll content, fresh and dry rhizome yield, and curcumin content.

References

Akamine, H., Hossain, M.A., Ishimine, Y., Yogi, K., Hokama, K., Iraha, Y., et al., 2007. Effects of application of N, P and K alone or in combination on growth, yield and curcumin content of turmeric (*Curcuma longa* L.). Plant Prod. Sci. 10 (1), 151–154.

Attarde, S.K., Jadhao, B.J., Adpawar, R.M., Warade, A.D., 2003. Effect of nitrogen levels on growth and yield of turmeric. J. Spices Aromat. Crops 12 (1), 77–79.

Babu, M., Muthuswami, S., 1984. Influence of potassium on the quality of turmeric (*Curcuma longa* Linn.). South Indian Hortic. 32 (6), 343–346.

Balashanmugam, P.V., Vanangamudi, K., 1987. Foliar spraying of Planofix on growth and yield of turmeric. Madras Agric. J. 74 (3), 178–180.

Baskar, K., Sankaran, K., 2005. Effect of humic acid (potassium humate) on growth and yield of turmeric (*Curcuma longa* L.) in an alfisol. J. Spices Aromat. Crops 14 (1), 34–38.

Bose, P., Sanyal, D., Majumdar, K., 2008. Balancing sulfur and magnesium nutrition for turmeric and carrot grown on red lateritic soil. Better Crops 92 (1), 23–25.

Davamani, V., Lourdhuraj, C.A., 2006. Effect of spent wash, biocompost, biosuper and inorganic fertilizers on micronutrient content of turmeric (*Curcuma longa* L.) and soil. Crop Res. 32 (3), 568–572.

Dixit, D., Srivastava, N.K., 2000. Effect of iron deficiency stress on physiological and biochemical changes in turmeric (*Curcuma longa*) genotypes. J. Med. Aromat. Plant Sci. 22 (1B), 652–658.

Dixit, D., Srivastava, N.K., Sharma, S., 2002. Boron deficiency induced changes in translocation of $^{14}CO_2$-photosynthate into primary metabolites in relation to essential oil and curcumin accumulation in turmeric (*Curcuma longa* L.). Photosynthetica 40 (1), 109–113.

Dubey, A.K., Yadav, D.S., 2001. Response of turmeric (*Curcuma longa* L.) to NPK under foothill conditions of Arunachal Pradesh India. J. Hill Farming 14 (2), 144–146.

Eyubov, R.E., Isaeva, F.G., Ragimov, M.A., 1984. Effect of fertilizers on turmeric yield. Khimiya-v-Sel-skom-Khozyastve 22 (10), 18–20.

Fresco, L.O., Kroonenberg, S.B., 1992. Time and spatial scales in ecological sustainability. Land Use Policy July, 155–167.

Gill, B.S., Randhawa, R.S., Randhawa, G.S., Singh, J., 1999. Response of turmeric (*Curcuma longa*) to nitrogen in relation to application of farm yard manure and straw mulch. J. Spices Aromat. Crops 8 (2), 211–214.

Gill, B.S., Kroya, R., Sharma, K.N., Saini, S.S., 2001. Effect of rate and time of nitrogen application on growth and yield of turmeric (*Curcuma longa* L.). J. Spices Aromat. Crops 10 (2), 123–126.

Gill, B.S., Kaur, S., Saini, S.S., 2004. Influence of planting methods, spacing and farm yard manure on growth, yield and nutrient content of turmeric (*Curcuma longa* L.). J. Spices Aromat. Crops 13 (2), 117–120.

Gupta, C.R., Sengar, S.S., 1998. Effect of varying levels of nitrogen and potassium fertilization on growth and yield of turmeric under rainfed condition. In: Sadanandan, A.K., Krishnamurthy, K.S., Kandiannan, K., Korikanthimath, V.S. (Eds.), Proceedings Water and nutrient management for sustainable production and quality of spices. Indian Society of Spices, Calicut, Kerala State, India, pp. 47–51.

Halder, N.K., Shill, N.C., Siddiky, M.A., Sarkar, J., Gomez, R., 2007. Response of turmeric to zinc and boron fertilization. J. Biol. Sci. 7 (1), 182–187.

Haque, M.M., Rahman, K.M.M., Ahmed, M., Masud, M.M., Sarker, M.M.R., 2007. Effect of nitrogen and potassium on the yield and quality of turmeric in hill slope. Int. J. Sust. Crop Prod. 2 (6), 10–14.

Hossain, A., Matsuura, S., Doi, M., Ishimine, Y., 2002. Growth and yield of turmeric (*Curcuma* sp.) and sweet bell pepper (*Capsicum annum* L.) as influenced by Mandacompost. Sci. Bull. Fac. Agric. Univ. Ryukyus 49, 205–212.

Hu, M.-F., Tsai, S.J., Chang, I.-F., Liu, S.Y., 2003. Effects of combined chicken compost and chemical fertilizer application on the yield and quality of turmeric (*Curcuma longa* L.). J. Agri. Res. China 52 (4), 334–340.

IISR, 2005. Annual Report, Indian Institute of Spices Research (ICAR), Calicut, Kerala State, India, pp. 45–46.

Jadhao, B.J., Gonge, V.S., Panchbhai, D.M., Mohariya, A., Hussain, I.R., 2005. Performance of turmeric varieties (*Curcuma longa* L.) under varying levels of zinc and iron. Int. J. Agric. 1 (1), 94–98.

Jagadeeswaran, R., Murugappan, V., Govindswamy, M., Kumar, P.S.S., 2007. Influence of slow release fertilizers on soil nutrient availability under turmeric (*Curcuma longa* L.). Asian J. Agric. Res. 1 (3), 105–111.

Jena, M.K., Das, P.K., Pattnaik, A.K., 1999. Integrated effect of microbial inoculants and fertilizer nitrogen on N-use efficiency and rhizome yield of turmeric (*Curcuma longa* L.). Orissa J. Hortic. 27 (2), 10–16.

Jirali, D.I., Hiremath, S.M., Chetti, M.B., Patil, S.A., 2007. Biophysical, biochemical parameters, yield and quality attributes as affected by micronutrients in turmeric. Plant Arch. 7 (2), 827–830.

Kao, R.L., Yu, J.Z., Chen, C.L., Lee, Y.C., Chu, C.L., Hsieh, T.F., Hu, M.F., 2007. Effect of fertilizer treatments on rhizome yield and curcumin content of turmeric. J. Taiwan Agric. Res. 56 (3), 165–175.

Katiyar, R.S., Balak, R., Tewari, S.K., Singh, C.P., Ram, B., 1999. Response of turmeric to nitrogen and phosphorus application under intercropping system with poplar on sodic soils. J. Med. Aromat. Plant Sci. 21, 937–939.

Krishnamurthi, V.V., Kumar, K., Rajkannan, R.K., 2001. Efficacy of farm yard manure on yield of turmeric. Madras Agric. J. 87 (1/3), 182.

Kumar, G.S., 2004. Association of VAM fungi in some spices and condiments. J. Ecobiol. 16 (2), 113–118.

Kumar, P.S.S., Geetha, S.A., Savithri, P., Mahendran, P.P., 2006. Effect of enriched organic manures and zinc solubilizing bacteria in the alleviation of Zn deficiency in turmeric and its critical evaluation using nutrient imbalance index (NII) of DRIS approach. Indian J. Agric. Res. 40 (1), 1–9.

Kumar, G.V.V., Khan, M.A.A., Begum, H., 2000. Influence of mineral nutrient composition of turmeric rhizome on curcumin content of different cultivars. Int. J. Trop. Agric. 18 (3), 265–269.

Kumar, D., Singh, R., Gadekar, H., Patnaik, U.S., 2003. Effect of different mulches on moisture conservation and productivity of rainfed turmeric. Indian J. Soil Conserv. 31 (1), 41–44.

Kumar, P.S.S., Geetha, S.A., Rajarajan, A., Savithri, P., Raj, S.A., 2004. Evaluation of Zn use efficiency of Zn enriched organic manures and zinc solubilizing bacteria in turmeric crop and distribution of applied Zn among various Zn fractions of the soil using Zn^{65} radiotracer technique. Res. Crops 5 (1), 118–125.

Liu, L.S., Yang, Y.Y., Chu, Y.T., 1974. Effects of nitrogen, phosphorus and potassium on the yield and curcumin content of *Curcuma longa* Linn. J. Taiwan Agric. Res. 23 (4), 284–291.

Lynrah, P.G., Chakraborty, B.K., Chandra, K., 2002. Effect of CCC, kinetin, NAA and KNO_3 on yield and curcumin content of turmeric. Indian J. Plant Physiol. 7 (1), 94–95.

Majumdar, B., Venkatesh, M.S., Kumar, K., 2002. Effect of nitrogen and farmyard manure on yield and nutrient uptake of turmeric (*Curcuma longa*) and different forms of inorganic N build-up in an acidic Alfisol of Meghalaya. Indian J. Agric. 72 (9), 528–531.

Masroor, M., Khan, A., Singh, M., Naeem, M., Nasir, S., 2006. Preharvest combined application of triacontanol and kinetin could ameliorate the growth, yield and curcumin content of turmeric (*Curcuma longa* L.). Planta Med. 72. (no pages available please).

Meenakshi, N., Sulikeri, G.S., Hegde, R.V., 2001. Effect of planting material and P & K nutrition on yield and quality of turmeric. Karnataka J. Agric. 14 (1), 197–198.

Meerabai, M., Jayachandran, B.K., Asha, K.R., Geetha, V., 2000. Boosting spice production under coconut gardens of Kerala: maximizing yield of turmeric with balanced fertilization better. Crops Intl. 14 (2), 10–12.

Medda, P.S., Hore, J.K., 2003. Effects of N and K on the growth and yield of turmeric in alluvial plains of West Bengal. Indian J. Hortic. 60 (1), 84–88.

Mohan, E., Melanta, K.R., Guruprasad, T.R., Herle, P.S., Gowda, N.A.J., Naik, C.M., 2004. Effects of graded levels of nitrogen and biofertilizers on growth, yield and quality in turmeric (*Curcuma domestica* Val.) cv. DK local. Environ. Ecol. 22 (3), 715–719.

Nagarajan, M., Pillai, N.G., 1979. A note on the nutrient removal by ginger and turmeric rhizomes. Madras Agric. J. 66, 56–59.

Nair, K.P.P., 1996. Buffering of plant nutrients and effects on nutrient availability. Adv. Agron. 57, 237–287.

Nawalagatti, C.M., Surendra, P., Chetti, M.B., Hiremath, S.M., 2008. Influence of sulphur on morpho-physiological traits, yield and quality attributes in turmeric. Int. J. Plant Sci. Muzaffarnagar 3 (1), 78–81.

Niranjan, A., Prakash, D., Tewari, S.K., Pande, A., Pushpangadan, P., 2003. Chemistry of *Curcuma* species cultivated on sodic soil. J. Med. Aromat. Plant Sci. 25 (1), 69–75.

Parthasarathy, V.A., Kandiannan, K., Srinivasan, V., 2008. Organic Spices. New India Publishing Agency, New Delhi, p. 694.

Patra, S.K., 1998. Fertilizer management in turmeric (*Curcuma longa* L.) under rainfed farming system. Environ. Ecol. 16 (2), 480–482.

Poinkar, M.S., Shembekar, R.Z., Neha, C., Nisha, B., Archana, K., Kishore, D., 2006. Effect of organic manure and biofertilizers on growth and yield of turmeric. J. Soils Crops 16 (2), 417–420.

Prasad, R., Yadav, L.M., Yadav, R.C., 2004. Studies of suitable intercropping system for turmeric (*Curcuma longa* L.). J. Appl. Biol. 14 (2), 40–42.

Randhawa, G.S., Mahey, R.K., 1984. Effect of chemicals on emergence of turmeric. J. Res. 21, 470–471. Punjab Agricultural University, Ludhiana, Punjab State.

Randhawa, K.S., Nandpuri, K.S., Bajwa, M.S., 1973. Response of turmeric (*Curcuma longa*) to NPK fertilization. J. Res. 1, 45–48. Punjab Agricultural University, Ludhiana, Punjab State.

Rao, A.M., Rao, P.V., Reddy, Y.N., Reddy, M.S.N., 2005. Effect of organic and inorganic manurial combinations on growth, yield and quality of turmeric (*Curcuma longa* L). J. Plantation Crops 33 (3), 198–205.

Rao, D.V.R., Swamy, G.S., 1984. Studies on the effect of N, P and K on growth, yield and quality of turmeric. South Indian Hortic. 32 (5), 288–291.

Rao, D.V.R., Sreehari, D., Reddy, N.T., Reddy, K.S., 1989. Influence of certain plant growth regulators on growth, tuberization and rhizome yield of turmeric (*Curcuma longa* L.)—a note. Prog. Hortic. 21 (3–4), 194–197.

Reddy, M.N., Devi, M.C., Sridevi, N.V., 2003. Evaluation of turmeric for VAM colonization. Indian Phytopathol. 56 (4), 465–466.

Reddy, M.R.S., Reddy, P.V.R.M., 2000. Preliminary observations on neem-cake, against rhizome fly of turmeric (*Curcuma longa*). Insect Environ. 6 (2), 62.

Rethinam, P., Sivaraman, K., Sushama, P.K., 1994. Nutrition of turmeric. In: Chadha, K.L., Rethinam, P. (Eds.), Advances in Horticulture. Vol 9.Plantation and Spice Crops Part 1 Malhotra Publishing House, New Delhi, pp. 477–489.

Rajput, S.G., Patil, V.K., Warke, D.C., Ballal, A.L., Gunjkar, S.N., 1982. Effect of nitrogen and spacing on the yield of turmeric rhizomes. In: Nair, M.K., Premkumar, T., Ravindran, P.N., Sarma, Y.R. (Eds.), Ginger and Turmeric Central Plantation Crops Research Institute, Kasaragod, Kerala State, pp. 83–85.

Sadanandan, A.K., Hamza, S., 1996. Response of four turmeric (*Curcuma longa* L.) varieties to nutrients in an oxisol on yield and curcumin content. J. Plantation Crops 24 (Suppl.), 120–125.

Sadanandan, A.K., Hamza, S., 1998. Effect of organic manures on nutrient uptake, yield and quality of turmeric (*Curcuma longa* L.). In: Mathew, N.M., Kuruvilla, J. (Eds.), Developments in Plantation Crops Research. Proceedings Plantation Crops Symposium

XII (PLACROSYM XII) Indian Rubber Research Institute, Kottayam, Kerala State, pp. 175–181.

Sadanandan, A.K., Kandiannan, K., Hamza, S., 1998. Soil nutrients and water management for sustainable spices production. In: Sadanandan, A.K., Krishnamurthy, K.S., Kandiannan, K., Korikanthimath, V.S. (Eds.), Proceedings Water and Nutrient Management for Sustainable Production and Quality of Spices Indian Society for Spices, Calicut, Kerala State, pp. 12–20.

Saha, A.K, 1988. Note on response of turmeric to manure and source of N and P under terrace conditions of mid-altitude Mizoram. Indian J. Hortic. 45 (1–2), 139–140.

Saifudeen, N., 1981. Foliar Diagnosis, Yield and Quality of Turmeric (*Curcuma longa* L.) in Relation to Nitrogen, Phosphorus and Potassium. M.Sc. (Ag) Thesis, Kerala Agricultural University, Thrissur, Kerala State.

Sanchez P.A. Tropical soil fertility research: towards the second paradigm. In: Proc. XV ISSS Symposium, Acapulco, Mexico; July 10–16, 1994. Vol 1. pp. 89–104.

Santhanakrishnan, P., Balashanmugam, S., 1993. Paper presented in All India Tuber Crops Meet at Tamil Nadu Agricultural University, Coimbatore, Tamil Nadu, February 1993, India.

Sanyal, D., Dhar, P.P., 2008. Effects of mulching, nitrogen and potassium levels on growth, yield and quality of turmeric grown in red lateritic soil. Acta Hortic. 769.

Selvarajan, M., Chezhiyan, N., 2001. Studies on the influence of Azospirillum and different levels of nitrogen on growth and yield of turmeric (*Curcuma longa* L.). South Indian Hortic. 49, 140–141.

Sharma, D.P., Sharma, T.R., Agrawal, S.B., Rawat, A., 2003. Differential response of turmeric to organic and inorganic fertilizers. JNKVV Res. 37 (2), 17–19.

Sheshagiri, K.S., Uthaiah, B.C., 1994. Effect of nitrogen, phosphorus and potassium levels on growth and yield of turmeric (*Curcuma longa*) in the hill zone of Karnataka. J. Spices Aromat. Crops 3 (1), 28–32.

Silva, N.F. da., Sonenberg, P.E., Borges, J.D., 2004. Growth and production of turmeric as a result of mineral fertilizer and planting density. Hortic. Brasil. 22 (1), 61–65.

Singh, B.P., Mishra, H., 1995. Economics of nitrogen, phosphorus and potassium fertilization on turmeric in Andaman. Spice India 8, 12–13.

Singh, B.P., Maurya, K.R., Sinha, M.K., Singh, R.A., 1986. Reaction of genotypes of turmeric (*Curcuma longa* L.) to Zn stress in a calcareous soil. Proc. Indian Natl. Sci. Acad. Biol. Sci. 52 (3), 385–390.

Singh, V.B., Singh, N.P., Swer, B., 1998. Effect of potassium and nitrogen on yield and quality of turmeric (*Curcuma longa*). J. Potassium Res. 14 (1), 89–92.

Singh, J., Mehta, C.P., Ram, M., Singh, I., 2006. Production potential and economics of intercropping various vegetable crops in turmeric (*Curcuma longa* L.). Haryana J. Agron. 22 (1), 49–51.

Srinivasan, V., Sadanandan, A.K., Hamza, S., 2000. Efficiency of rock phosphate sources on ginger and turmeric in an Ustic Humitropept. J. Indian Soc. Soil Sci. 48 (3), 532–536.

Srinivasan, V., Sangeeth, K.P., Samsudeen, M., Thankamani, C.K., Hamza, S., Zachariah, T.J., et al., 2008. Evaluation of organic management system for sustainable production of turmeric, In: Programme and abstracts, national conference on organic farming in horticultural crops with special reference to plantation crops 15–18 october. Central Plantation Crops Research Institute, Kasaragod, Kerala State, India, pp. 67–68.

Subramanian, S., Rajeswari, E., Chezhiyan, N., Shiva, K.N., 2003. Effect of Azospirillum and graded levels of nitrogenous fertilizers on growth and yield of turmeric (*Curcuma longa* L.). Proceedings National Seminar on New Perspectives in Spices Medicinal and

Aromatic Plants, 27–29 November, Indian Society for Spices, Calicut, Kerala State, pp. 158–160.

Suchindra, R., Anburani, A., 2008. Influence of graded levels of N and K and biostimulants on yield parameters in turmeric (*Curcuma longa* L.) cultivar Erode local. Plant Arch. 8 (2), 923–926.

Swain, S.C., Rath, S., Ray, D.P., 2006. Effect of NPK levels and mulching on yield of turmeric in rainfed uplands. J. Res. Birsa Agric. Univ. 18 (2), 247–250.

Thomas, A., Barche, S., Singh, D.B., 2002. Influence of different levels of nitrogen and potassium on growth and yield of turmeric (*Curcuma longa.* L). J. Spices Aromat. Crops 11 (1), 74–77.

Tiwari, G., Shah, P., Agrawal, V.K., Harinkhede, D.K., 2003. Influence of nitrogen application on the growth, biomass productivity and leaf essential oil yield in turmeric. JNKVV Res. J. 37 (2), 90–91.

Upadyay, D.C., Misra, R.S., 1999. Nutritional study of turmeric (*Curcuma longa* Linn.) cv. Roma under agroclimatic conditions of eastern Uttar Pradesh. Prog. Hortic. 31 (3/4), 214–218.

Vadiraj, B.A., Siddagangaiah, Poti, N., 1998. Effect of vermicompost on the growth and yield of turmeric. South Indian Hortic. 46 (3/6), 176–179.

Velmurugan, M., Chezhiyan, N., Jawaharlal, M., 2002. Effect of biofertilizers on physiological and biochemical parameters of turmeric cv. BSR 2. Proceedings National Seminar on Strategies for Increasing Production and Export of Spices, 24–26 October, Indian Society for Spices, Calicut, Kerala State, pp. 138–145.

Velmurugan, M., Chezhiyan, N., Jawaharlal, M., Anand, M., 2007. Micronutrient studies in turmeric. Asian J. Hortic. 2 (2), 291–293.

Venkatesh, M.S., Majumdar, B., Kumar, K, 2003. Effect of rock phosphate, single super phosphate and their mixtures with FYM on yield, quality and nutrient uptake by turmeric (*Curcuma longa* L.) in acid alfisol of Meghalaya. J. Spices and Aromatic Crops 12 (1), 47–51.

Venkatesha, J., Khan, M.M., Farooqi, A.A., Sadanandan, A.K., 1998. Effect of major nutrients (NPK) on growth, yield and quality of turmeric (*Curcuma domestica* Val.) cultivars. In: Sadanandan, A.K., Krishnamurthy, K.S., Kandiannan, K., Korikanthimath, V.S. (Eds.), Proceedings on Water and Nutrient Management for Sustainable Production and Quality of Spices Indian Society of Spices, Calicut, Kerala State, pp. 52–58.

Vishwakarma, S.K., Kumar, A., Prakash, S., 2006. The effect of micro-nutrient on the growth and yield of turmeric under different shade conditions in mango orchard. Intl. J. Agric. 2 (1), 241–243.

Yadav, L.M., Prasad, R., 2003. Effect of nitrogen and potash on yield of turmeric (*Curcuma longa* L.). Appl. Biol. 13 (1), 69–72.

Yamgar, V.T., Kathmale, D.K., Belhekar, P.S., Patil, R.C., Patil, P.S., 2001. Effect of different levels of nitrogen, phosphorus and potassium and split application of N on growth and yield of turmeric (*Curcuma longa*). Indian J. Agron. 46 (2), 372–374.

8 Turmeric Entomology

In all, 59 species of insects are known to infest *Curcuma* spp., in the world, among which the shoot borer (*Conogethes punctiferalis* Guen.) and rhizome scale (*Aspidiella hartii* Sign.) are major insect pests, especially on turmeric (*C. longa* L.) in India. The following discussion pertains to the major and minor insect pests and their field control.

Major Insect Pests

Shoot borer (*Conogethes punctiferalis* Guen.).

Distribution

Among the insect pests infesting turmeric, the shoot borer, like in many other field crops, is the most serious one, and was recorded as early as 1914 (Ravindran et al., 2007). However, published information on the distribution of the pest in various turmeric-growing areas in the country is limited, except for the reports of Kotikal and Kulkarni (2000) in the northern districts of Karnataka. The shoot borer is widely distributed throughout Asia, Africa, America, and Australia on many crops, though records of the pest on *Curcuma* spp. are lacking. In Australia and Southeast Asia, *C. punctiferalis* has been suggested to be a complex of more than one species (Honda, 1986). Table 8.1 lists insect pests recorded on *Curcuma* spp. while table 8.2 lists insect pests recorded on dry turmeric during storage.

Nature of Damage

The larvae of the shoot borer initially scrape and feed on the margins of the unopened or newly opened leaf and subsequently bore into the pseudostem and feed on the central growing shoot, resulting in yellowing and drying of the central shoot. The symptoms of pest infestation include the presence of the bore-holes on the pseudostem through which frass is extruded and the presence of withered central shoot.

Life History

The adults are medium-sized moths with a wingspan of 18–24mm; the wings and body are pale straw yellow with minute black spots. Descriptions of adults have been provided by Hampson (1856). The eggs are elliptical, pitted on the surface, and creamy white when freshly laid. The larval stage comprises of five instars; fully grown larvae measure 16–26mm in length and are light brown in color with sparse hairs. However, the size of adults and larvae may vary depending on the host, in which they

The Agronomy and Economy of Turmeric and Ginger. DOI: http://dx.doi.org/10.1016/B978-0-12-394801-4.00008-9

Table 8.1 Pests Infesting the *Curcuma* spp.

Host and Insect Pest	Order, Family	Distribution
C. longa L.		
Letana inflata Bru.	Orthoptera, Tettigonidae	India
Phenoroptera gracillus Burm.	Orthoptera, Tettigonidae	India
Orthacris simulans B.	Orthoptera, Acrididae	India
Cyrtacanthacris ranacea Stoll.	Orthoptera, Acrididae	India
Oxyrachis tarandus Fab.	Hemiptera, Membracidae	India
Tricentrus bicolor Dist.	Hemiptera, Membracidae	India
Tettigoniella ferruginea Wlk.	Hemiptera, Cicadellidae	India
Pentalonia nigronervosa Coq.	Hemiptera, Aphididae	India, China
Uroleucon compositae (Theobald)	Hemiptera, Aphididae	India
Paracocus marginatus Williams & Granara de Willink	Hemiptera, Pseudococcidae	India
Planococcus sp.	Hemiptera, Pseudococcidae	India
Aspidiella hartii Sign.	Hemiptera, Coccidae	India, West Africa, West Indies
Aspidiotus curcumae Gr.	Hemiptera, Coccidae	India
Cocus hesperidium (Ckll.)	Hemiptera, Coccidae	Fiji, Papua New Guinea
Hemiberlesia palmae L.	Hemiptera, Diaspididae	Fiji, Papua New Guinea
Stephanitis typicus Dist.	Hemiptera, Tingidae	India
Cletus rubidiventris West.	Hemiptera, Coreidae	India
C. bipunctatus West	Hemiptera, Coccidae	India
Riptortus pedestris Fab.	Hemiptera, Coreidae	India
Coptosoma cibraria Fab.	Hemiptera, Pentatonide	India
Nezara viridula (L.)	Hemiptera, Pentatonide	India
Anaphothrips sudanensis (Trybom)	Hemiptera, Pentatonide	India
Asprothrips indicus Bagn.	Thysanoptera, Thripidae	India
Panchaetothips indicus Bagn.	Thysanoptera, Thripidae	India
Sciothrips cardamom Ramk.	Thysanoptera, Thripidae	India
Haplothrips sp.	Thysanoptera, Phlaeothripdae	India
Holotrichia serrota Fab.	Coleoptera, Scarabaeidae	India
Epilachna sparsa (Hbst.)	Coleoptera, Coccinellidae	India
Chirida bipunctata Fab.	Coleoptera, Chrysomelidae	India
Ceratobasis nair (Loc.)	Coleoptera, Chrysomelidae	India
Colasposoma splendidum (Fab.)	Coleoptera, Chrysomelidae	India
Cryptocephalus rajah Jac.	Coleoptera, Chrysomelidae	India
C. schestedti Fab.	Coleoptera, Chrysomelidae	India
Lema praeusta Fab.	Coleoptera, Chrysomelidae	India
L. fulvicornis Jac.	Coleoptera, Chrysomelidae	Sri Lanka
L. lacordairei Baly	Coleoptera, Chrysomelidae	India
L. signatipennis Jac.	Coleoptera, Chrysomelidae	India
L. semiregularis Jac.	Coleoptera, Chrysomelidae	India
Monolepta signata (Olivier)	Coleoptera, Chrysomelidae	India
Raphidopalpa abdominalis (Fab.)	Coleoptera, Chrysomelidae	India

(Continued)

Table 8.1 (Continued)

Host and Insect Pest	Order, Family	Distribution
Hedychorus rufomaculatus M.	Coleoptera, Curculionidae	India
Myllocerus discolor Boheman	Coleoptera, Curculionidae	India
M. udecimpunctatus Faust	Coleoptera, Curculionidae	India
M. viridinus Fab.	Coleoptera, Curculionidae	India
Libnotes punctipennis Meij	Diptera, Tipulidae	India
Mimegralla coeruleifrons Macq.	Diptera, Micropezidae	India
M. albimana (Doleschall)	Diptera, Micropezidae	India
Eumerus albifrons Wlk.	Diptera, Syrphidae	India
E. pulcherrimus Bru.	Diptera, Syrphidae	India
Conogethes punctiferalis Guen.	Lepidoptera, Pyralidae	India, Sri Lanka
Udaspes folus Cram.	Lepidoptera, Hesperidae	India
Notocrypta curvifascia C & R Felder	Lepidoptera, Hesperidae	India
Bombotelia nugatrix Gr.	Lepidoptera, Noctuidae	India
Spodoptera litura (Fab.)	Lepidoptera, Noctuidae	India
Amata passalis Fab.	Lepidoptera, Arctiidae	India
Creatonotos gangis (L.)	Lepidoptera, Arctiidae	India
Spilarctia oblique Wlk.	Lepidoptera, Arctiidae	India
Catopsilia pomona Fab.	Lepidoptera, Pieriidae	India
Euproctis latifolia Hamp.	Lepidoptera, Lymantridae	India
C. amada Roxb.		
C. punctiferalis Guen.	Lepidoptera, Pyralidae	India
C. aromatica Salisb.		
C. punctiferalis Guen.	Lepidoptera, Pyralidae	India
C. pseudomontana Grah.		
Sciothrips cardamomi Ramk.	Thysanoptera, Thripidae	India
C. zeodaria Rosc.		
S. cardamomi Ramk.	Thysanoptera, Thripidae	India

Table 8.2 Pests Infesting Stored Dry Turmeric (*C. longa* L.)

Genus/Species	Order/Family	Regional Distribution
Lasioderma serricorne Fab.	Coleoptera, Anobiidae	Bangladesh, India
Stegobium paniceum L.	Coleoptera, Anobiidae	India
Tribolium castaneum (Hbst.)	Coleoptera, Bostrychidae	India
Oryzaephilus surinamensis L.	Coleoptera, Sylvanidae	India
Rhizopertha dominica (F.)	Coleoptera, Sylvanidae	India
Tenebroides mauritanicus (L.)	Coleoptera, Tenebroinidae	India
Araecerus fasciculatus DeG.	Coleoptera, Anthribiidae	India
Setomorpha rutella Zell	Lepidoptera, Tineidae	India
Ephestia sp.	Lepidoptera, Pyralidae	India

breed. The morphometrics of various stages of the shoot borer when reared on turmeric have been shown by Jacob (1981). A method for determination of sex in the shoot borer (infesting cardamom) has also been suggested (Thyagaraj et al., 2001).

The life history of the shoot borer in turmeric has been investigated in Kasaragod district (Kerala State, India) in controlled laboratory conditions. The preoviposition and egg-laying periods lasted 4–7 days and 3–4 days, respectively. The larval instars comprising five stages lasted 3–4, 3–7, 3–8, and 7–14 days, respectively, and the pre-pupal and pupal stages, which lasted 3–4 and 9–10 days, respectively. Adult females laid 30–60 eggs during their life span, and 6–7 generations were completed during a crop season in the field. Variations were also observed in the duration of the life cycle during various seasons (Jacob, 1981).

Seasonal Incidence of Insect Pests

There is no published scientific information on the incidence of the seasonal population dynamics of the turmeric shoot borer. In Kerala, the shoot borer is generally found to manifest itself in the crop between June and December.

Host Plants of Turmeric

The turmeric shoot borer is highly polyphagous and has been recorded on 65 host plants belonging to 30 families, many of which are economically important (Devasahayam, 2004). Various parts of these plants, such as shoots, buds, flowers, and fruits, are infested by the shoot borer. In India, apart from *C. longa*, the shoot borer also infests zedoary (*C. aromatica* Salis.) and mango borer-ginger (*C. amada* Roxb.)

Resistance

Information on the reaction of various turmeric types to the shoot borer is available mainly from investigations conducted in Kerala and Karnataka States in India (Sheila et al., 1980). Dindigam Ca-69, an aromatic type of turmeric variety and another Manuthy Local, the former from Tamil Nadu, India and the latter from Kerala, India, were identified as tolerant varieties to the pest in the Kerala State, India. Incidence of the shoot borer was recorded in 489 accessions belonging to 21 morphotypes, at Vellanikkara (Kerala State, India), where the State's only agricultural university is located, and the incidence of the pest was found to be the lowest in morphotype II (Velayudhan and Liji, 2003).

Natural Enemies

A rich biodiversity of natural enemies has been recorded on the shot borer infesting various crops, especially in Sri Lanka, China, Japan, and India (Jacob, 1981). These include viruses, mermithid nematodes, spiders, dermapterans, asilid flies, and hymenopterous parasitoids. The details are given in Table 8.3.

Table 8.3 The Natural Enemies of *Conogethes puciferalis* Guen.

Genus/Species	Order/Family
Dichocrocis punctiferalis NPV	Baculoviridae
Steinemema glaseri Steiner 1929	Rhabditida, Steinernematidae
Hexamermis sp.	Dorylamida, Mermithidae
Araneus sp.	Araenea, Araneidae
Micarisa sp.	Araenea, Gnaphosidae
Thyene sp.	Araenea, Salticidae
Euborellia stali Dohm	Demaptera, Carcinophoridae
Philodicus sp.	Diptera, Asilidae
Heligmoneura sp.	Diptera, Asilidae
Palexorista parachrysops Bazzi	Diptera, Tachinidae
Agrypon sp.	Hymenoptera, Ichneumonidae
Apechthis copulifera	Hymenoptera, Ichneumonidae
Eriborus trochanteratus (Mori.)	Hymenoptera, Ichneumonidae
Friona sp.	Hymenoptera, Ichneumonidae
Gotra sp.	Hymenoptera, Ichneumonidae
Nythobia sp.	Hymenoptera, Ichneumonidae
Sambus persimilis	Hymenoptera, Ichneumonidae
Temelucha sp.	Hymenoptera, Ichneumonidae
Theronia inareolata	Hymenoptera, Ichneumonidae
Trathala flavoorbitalis (Cam.)	Hymenoptera, Ichneumonidae
Xanthopimpla sp.	Hymenoptera, Ichneumonidae
X. australis Kr.	Hymenoptera, Ichneumonidae
X. kandiensis Cram.	Hymenoptera, Ichneumonidae
Apanteles sp.	Hymenoptera, Braconidae
A. taragamme	Hymenoptera, Braconidae
Bracon brevicornis West.	Hymenoptera, Braconidae
Microbracon hebator Say	Hymenoptera, Braconidae
Myosoma sp.	Hymenoptera, Braconidae
Phanerotomma hendecasisella Cam.	Hymenoptera, Braconidae
Angitia trochanterata Morl.	Hymenoptera, Chalcidae
Brachymeria sp.nr. *australis* kr.	Hymenoptera, Chalcidae
B. brevicornis West.	Hymenoptera, Chalcidae
B. euploeae West.	Hymenoptera, Chalcidae
B. lasus West.	Hymenoptera, Chalcidae
B. nosatoi Habu	Hymenoptera, Chalcidae
B. obscurata Wlk.	Hymenoptera, Chalcidae
Synopiensis sp.	Hymenoptera, Sphegidae
Dolichurus sp.	Hymenoptera, Sphegidae

Managing Insect Control in Turmeric

There are different ways in which insect pests of turmeric could be controlled in the field. A brief description of each follows.

Chemical Control

Despite the very serious damage the turmeric shoot borer causes to the crop, and the consequent yield loss, no systematic investigations have been made on its chemical control and no results have been documented. In general, malathion (a patented Indian insecticide, and one quite popular among farmers) at a concentration of 0.1% has been recommended to control the pest (IISR Report, 2001).

Biopesticidal Control

Owing to the several environmental damages caused by the chemically driven high-input technology of the green revolution, more and more farmers are turning to biopesticidal control in many crops. Turmeric is no exception. Evaluation of two commercially available products, namely Bioasp and Dipel, derived from *Bacillus thuringiensis* (Bt), have been tried in turmeric fields attached to the Indian Institute of Spices Research (IISR), Calicut, under the administrative control of the Indian Council of Agricultural Research (ICAR). Spraying Dipel at a concentration of 0.3% during the months from July to October was found effective (Devasahayam, 2002). Entomopathogenic nematodes, such as *Steinernema* sp. and *Heterorhabditis* sp., were found promising in the control of *C. punctiferalis* (infesting chestnut) in a laboratory bioassay conducted in Korea (Choo et al., 1995, 2001).

Plant-Based Products

There have been extensive investigations on the efficacy of neem oil from neem (*Azadirachta indica*) seed on insect control, in general. However, consistent results were not obtained when plant-based products such as above (neem oil at 1% concentration) or commercial neem-based products, also at 1% concentration, were used to control shoot borer infestation (IISR Report, 2002).

Sex Pheromones

Many investigators have demonstrated the presence of sex pheromones in the shoot borer infesting other crops in China, Japan, and Korea (Jung et al., 2000; Konno et al., 1980). In India, field investigations on the efficacy of sex pheromones in cardamom (*Elettaria cardamomum* Maton.) fields indicated that male moths were attracted and caught in traps, baited with (10)-10-hexadecenal and (Z)-10-hexadecenal in 9:1 ratio (Chakravarthy and Thyagaraj, 1997, 1998).

Rhizome Scale (*Aspidiella hartii* Ckll.)

Distribution

The rhizome scale is distributed mainly in the tropical regions of Asia, Africa, Central America, and the Caribbean Islands. However, authentic records of pest infestation on turmeric are scarce to come by in different parts of the world, including India. Apart from India, the pest has also been recorded to infest turmeric in West Africa and West Indies (Hill, 1983).

Crop Damage

Rhizome scale infests turmeric rhizomes, both in the field and in storage. In the field, the pest normally manifests later in the crop growth stage, and where infestation is severe, the plants completely wither and dry up. In storage, the pest infestation results in the shriveling of buds and rhizomes, when infestation is severe, and it adversely affects sprouting of rhizome in the field.

Life History of the Pest

The adult females of rhizome scale are circular and light brown to gray in color, and measure about 1.5 mm in diameter. Females are ovoviviparous and also reproduce parthenogenetically. Only limited scientific information is available on the life history of this pest on turmeric. A single female lays about 100 eggs and the life cycle from egg to adult is completed in about a month (Jacob, 1982).

Host Plants

Rhizome scale also infests a few other rhizomatous plants, such as tuber crops. In India, the pest has been recorded on ginger (*Zingiber officinale* Rosc.), elephant foot yam (*Amorphophallus paeoniifolius* (Dennst.) Nicolson), yams (*Dioscorea alata* L., *Dioscorea esculenta* (Lour.) Burkill, and *Dioscorea rotundata* Poir), tannia (*Xanthosoma sagittifolium* (L.) Schott), and taro (*Colacasia esculenta* (L.) Schott) (Ayyar, 1940). In other countries, such as West Indies, Ivory Coast, Ghana, and Nigeria, rhizome scale has been found to infest yams and tannia. It also infests sweet potato in Africa (Ballou, 1916).

Turmeric Resistance

The incidence of rhizome scale was investigated in 191 turmeric types in Coimbatore district, Tamil Nadu, India and it was found that 87 accessions were free from infestation (Regupathy et al., 1976). Velayudhan and Liji (2003) recorded the incidence of rhizome scale in 489 accessions belonging to 21 morphotypes at Vellanikkara, Kerala State, in the Kerala Agricultural University fields, and it was found that 80 were free from infestation and the lowest scale incidence was observed in morphotype VI.

Natural Enemies

A few natural enemies have been found to infest rhizome scale, which control it, including *Physcus comperei* Hayat (Aphelinidae), *Adelencyrtus moderatus* Howard (Encyrtidae), and two species of mites in Kasaragod, Kerala State, India and *Cocobius* sp., a predatory beetle and ant at the experimental farm at Peruvannamuchi of the ISSR, Kerala State, India (Sika et al., 2005).

Management of Rhizome Scale

The management schedule of this pest in ginger is also applicable in the case of turmeric, which involves discarding (by uprooting infested plants and burying the same in soil far away from the field where the main crop stands) and also by dipping the rhizome in quinalphos at 0.075% concentration after harvest and before storage (IISR Report, 2001).

Other Insect Pests

Grasshoppers

The grasshoppers *Orthacris simulans* B. and *Cyrtacanthacris ranacea* Stoll. cause higher damage (about 10%) in the northern districts of Karnataka State, India, whereas *Letana inflata* Bru. and *Phenoroptera gracillus* Burm. cause only minor damage to the crop in the region (Kotikal and Kulkarni, 2000).

Scale Insects and Mealy Bugs

The scale insects recorded on turmeric apart from *A. hartii*, include *Aspidiotus curcumae* Gr. from India (Nair, 1975) and *Coccus hesperidium* (Ckll.) and *Hemiberlesia palmae* L. from Fiji and Papua New Guinea (Ecoport, 2009). The mealy bugs recorded on turmeric rhizome in the field include *Planococcus* sp. from Waynad district, Kerala State, India (Devasahayam, 2006) and *Planococcus marginatus* Williams & Granara de Willink from Erode district, Tamil Nadu, India (Suresh, unpublished).

Aphids

The aphids recorded on turmeric include *Pentalonia nigronervosa* Coq. from China and India and *Uroleucon compositae* (Theobald) from India (Kotikal and Kulkarni, 2000; Zhou et al., 1995).

Hemipteran Bugs

The first report on *Stephanitis typicus* Dist. infestation on turmeric leading to the drying up of leaves was that of Fletcher (1914). The incidence of *S. typica* in different districts of northern Karnataka, India, has been extensively investigated (Kotikal and Kulkarni, 2000). In Maharashtra State, India, the population of *S. indica* appeared in the field during the month of September and peaked during the month of November (Zhou et al., 1995). Spraying of dimethoate or phosphamidon (0.05%) has been recommended to control the pest (Thangavelu et al., 1977). Phosphamidon has been in use in India and many other countries for many years. The other sap feeders recorded on turmeric are *Nezara viridula* (L.), *Croptosoma cribraria* Fab., *Tettigoniella ferruginea* Wlk., *Oxyrachis tarandus* Fab., *Tricentrus bicolor* Dist., *Cletus rubidiventris* West., *C. bipunctatus* West., and *Riptortus pedestris* Fab. in northern Karnataka State, India (Kotikal and Kulkarni, 2000).

Thrips

Pachaetothrips indicus Bagn. infesting turmeric was included in the list of economically important Thysanoptera from India by Ayyar (1920). The infested leaves roll up, turn pale, and gradually dry; the development of rhizomes is reduced when the pest infestation is severe. The incidence of *P. indicus* in the northern districts of Karnataka has been investigated (Kotikal and Kulkarni, 2000). Spraying dimethoate 0.06%, fenpropathrin 0.02%, benidiocarb 0.08%, or methyl dedmeton 0.05%, has been suggested for the control of the pest (Balasubramanian, 1982). The other species of the thrips recorded in northern Karnataka include *Asprothrips indicus* Bagn., *Anaphothrips sudanensis* (Trybom), *Haplothrips* sp. (Kotikal and Kulkarni, 2000). The cardamom thrips *Sciothrips cardamomi* Ramk., which is a major insect pest of cardamom (*Elettaria cardamomum* Maton.) also breeds *C. longa*, *C. pseudomontana* Grah., and *C. zeodaria*.

Leaf-Feeding Beetles

Lema praeusta Fab., *L. signatipennis* Jac., and *L. semiregularis* Jac., the main leaf-eating beetles of turmeric, have been recorded in Odisha State, India (Sengupta and Behura, 1955), whereas *L. fulvicornis* Jac. has been recorded in Sri Lanka (Hutson, 1936). The life history of *L. praeusta* and *L. semiregularis* has been described by Sengupta and Behura (1956, 1957). The other leaf-feeding beetles recorded on turmeric in India, include *Pseudocophora* sp., *Colasposoma splendidum* (Fab.), *Cryptocephalus rajaj* Jac., *Epilachna sparsa* (Hbst.), *Ceratobasis nair* (Loc.), *Myllocerus viridinus* Fab., *L. lacordairei* Baly., *Hedychorus rufomaculatus* M., *Chirida bipunctata* Fab., *Cryptocephalus schestedti* Fab., *Monolepta signata* (Olivier), *Raphidopalpa abdominalis* (Fab.), *Myllocerus discolor* Boheman, and *M. undecimpustulatus* Faust (Kotikal and Kulkarni, 2000; Nair, 1975).

Rhizome Borers

Holotrichia serrota Fab., the white grub, was found to infest rhizomes and roots of turmeric in the western districts of Maharashtra State, India (Patil et al., 1988). An unidentified species *Holotrichia* was also found to infest turmeric roots in northern Karnataka State, India (Kotikal and Kulkarni, 2000).

Rhizome Maggots

Dipteran maggots, which belong to various species, bore into the rhizome and root of turmeric and are generally seen in plants affected by the rhizome rot disease, thus proving a cohabitation of disease and insect pest. The maggots recorded on turmeric include *Mimegralla coeruleifrons* Macq., *M. albimana* (Doleschall), *Eumerus pulcherrimus* Bru., *E. albifrons* Wlk., *Libnotes puctipennis* Meij., and *Tipula* sp. (Fletcher, 1914; Kotikal and Kulkarni, 2000). Ghorpade et al. (1983) reported that *M. coeruleifrons* was endemic in the Sangli and Satara districts of Maharashtra State, India, which resulted in 25.4% reduction in turmeric yield. The incidence, life history, seasonal population, and natural enemies of *M. coeruleifrons* on turmeric were investigated in Maharashtra and Karnataka States, India (Ghorpade et al., 1983, 1988; Kotikal and Kulkarni, 2000). Several varieties of turmeric were screened to observe resistance against rhizome fly, in Maharashtra State, India and it was found that *Sughandham* and *Duggirala* varieties were found resistant (Jhadav et al., 1982). Eight genotypes of turmeric were screened for resistance against "rhizome fly in Belgaum district, Karnataka State, India, and all of them were found susceptible (Kotikal and Kulkarni, 2001). *M. albimana* (*C. albimana* Macq.) was considered a major pest on turmeric in Andhra Pradesh, India, and was more common in the crop raised in ill-drained soils causing 6.7–13.8% loss in yield. Investigations on the life history and seasonal population incidence of the pest were also carried out (Rao and Reddy, 1990). In the State of Meghalaya, India, the larvae of the "Crane Fly," *L. punctipennis*, were observed to bore into the rhizome, and the life cycle of the fly was completed in 40–55 days in *in vitro* conditions. Damage to the rhizome was very low among local cultivars, such as Lakadong, VK-77, RCT-1, and PTS-11 (Shylesha et al., 1998).

Application of Phorate (a patented Indian insecticide) at 0.75 kg a.i/ha at 30 days interval or spraying (in all six sprays) of Parathion (another patented Indian insecticide) along with soil application of Diazinon, Chloradane, or Aldrin (all patented Indian insecticides) at the rate of 0.75 kg/acre, twice during the crop cycle, was found to be very effective in controlling the insect in Maharashtra State, India (Dhoble et al., 1978, 1881). It is worth noting that many of these insecticides are now banned from use in India. These insecticides belong to the "Red Triangle," the most toxic of all in use and this is the outcome of an increasingly environmentally conscious situation in the country. However, clandestine use of such products, often found to control the pest instantaneously but adversely impacts the environment, is resorted to by the farmers, where such practices go unnoticed. Soaking seed rhizomes in dimethoate (again a patented insecticide in use in India) at a concentration of 1.15 m/l resulted effectively controlled emergence of the adult fly. And this practice also ensured maximum germination of the turmeric seedlings (Kotikal and Kulkarni, 1999). In Andhra Pradesh, India, turmeric fields where neem (*Azadirachta indica*) cake was applied were found to be free of the insect infestation (Reddy and Reddy, 2000).

There are yet no critical and systematic investigations carried out on the precise role of the rhizome fly on the incidence of rhizome rot disease. Such investigations, though, have been conducted in ginger. Infestation of the pest is light, in well-drained and light-textured soils (where the incidence of the rhizome rot disease also was found to be of lower intensity) in both Maharashtra State and Andhra Pradesh, India (Ghorpade et al., 1983; Rao and Reddy, 1990). In northern Telengana, currently within Andhra Pradesh, but now (at the time of writing this book) under a strong political upsurge for secession into a separate state in India, cultivars such as PCT-8, PCT-10, Suguna, and Sudarshana which were free from the infestation of rhizome rot disease, were also found to be free from the insect infestation (Rao et al., 1994). This strongly suggests that health of the turmeric plant, free from pathogenic infestation, predetermines the insect attack. In the northern districts of Karnataka as well, the insect was generally observed in the rhizome infested with the rhizome rot disease (Kotikal and Kulkarni, 2000). In Kerala State, the rhizome fly rarely infests rhizomes, where the incidence of rhizome rot is negligible, whereas the situation is just the opposite case with ginger—the insect infestation being intense where the ginger rhizome rot was also prevalent. These observations also suggest that rhizome fly infestation in turmeric is of a secondary and less important nature.

Leaf-Feeding Caterpillar

Among the leaf-feeding caterpillars, *Udaspes folus* Cram., the turmeric skipper, is the most serious one, especially in India, where it was recorded as early as 1909 (Maxwell-Lefroy and Howlett, 1909). Investigations on the life history and seasonal population fluctuation of the caterpillar have been conducted in the Godavari delta of Andhra Pradesh, India (Sujatha et al., 1992), Kasaragod district, Kerala State, India (Abraham et al., 1975), and Maharashtra State, India (Patil et al., 1988). Investigations at IISR on the natural enemies of the pest and its alternate hosts have been detailed by Koya et al. (1991).

Severe leaf defoliation caused by heavy infestation of *Creatonotos gangis* (L.), another leaf-eating caterpillar on turmeric has been reported by Ramakrishna and Raghunath (1982) in Andhra Pradesh, India. The other leaf-feeding caterpillars recorded on turmeric in India include *Bombotelia nugatrix* Gr., *Catopsilia pomona* Fab., *Spilarctia obliqua* Walk., *Notocrypta curvifascia* C & R Felder, *Amata passalis* Fab., *Euproctis latifolia* Hamp., and *Spodoptera litura* (Fab.) (Kotikal and Kulkarni, 2000; Nair, 1975).

Storage Pests of Turmeric

Different insects have been reported to infest stored turmeric. A detailed account is given in Table 8.4. These pests are the cigarette beetle (*L. serricorne*), drugstore beetle (*L. paniceum*), and coffee bean weevil (*A. fasciculatus*).

Distribution of the Pests

Many storage pests have been found to infest turmeric in the tropical regions of Asia and Africa. In the temperate regions, they are found mostly in heated stores. In the State of Kerala, India, 30–60% of the market samples were infested by the cigarette beetle and coffee bean beetle (Abraham et al., 1975). Srinath and Prasad (1975) collected market samples of stored turmeric across India, and 88 of the 115 samples collected were found to be infested by the cigarette beetle. In Udaipur, Rajasthan State, India, 67.7% of the market samples were found to be infested by the cigarette beetle (Kavadia et al., 1978). In Dhaka, Bangladesh, the cigarette beetle was considered as the most serious pest of turmeric (Rezaur et al., 1982).

Damage

The larvae of cigarette beetle and drugstore beetle tunnel into the produce and also contaminate it with frass. The adult cigarette beetles do not feed, but bore through the produce while leaving the pupal cocoon. Both adults and larvae of coffee bean weevil feed on dry turmeric rhizomes, leaving only the outer covering intact. Studies conducted at Udaipur indicated that the cigarette beetle caused 39.8% loss of weight

Table 8.4 Storage Pests of Turmeric (*C. longa* L.)

Genus/Species	Order/Family
Blattisocius tarsalis (Berl.)	Mesostigmata, Ascidae
B. keegan Fox	Mesostigmata, Ascidae
Acaropsis docta (Berl.)	Prostigmata, Cheyletidae
Acaropsellina solers (Kuzin)	Prostigmata, Cheyletidae
Cheyletus sp.	Prostigmata, Cheyletidae
Tydeus sp.	Prostigmata, Cheyletidae
Tyrophagus putrescentiae (Sch.)	Prostigmata, Pyemotidae
Pyemotus tritici (Lagreze-Fossat & Montane)	Prostigmata, Pyemotidae
Chortoglyphus gracilipes Banks	Astigmata, Chortoglyphidae
Peregrinator biannulipes (Montr. & Sign.)	Hemiptera, Reduviidae
Alloeocranum biannulipes (Montr. & Sign.)	Hemiptera, Reduviidae
Xylocoris flavipes (Reuter)	Hemiptera, Anthocoridae
Termatophyllum insigne (Reuter)	Hemiptera, Miridae
Tribolium castaneum (Hbst.)	Coleoptera, Bostrychidae
Thaneroclerus buqueti (Lefevre)	Coleoptera, Cleridae
Tilloidea notata (Klug)	Coleoptera, Cleridae
Alphitobius diaperinus (Panzer)	Coleoptera, Tenebroinidae
Tenebroides mauritanicus (L.)	Coleoptera, Tenebroinidae
Anisopteromalus calandrae (Howard)	Hymenoptera, Pteromalidae
Lariophagus distinguendus (Forst.)	Hymenoptera, Pteromalidae
Pteromalus cerealellae (Ashmead)	Hymenoptera, Pteromalidae
Cephalonomica gallicola (Ashmead)	Hymenoptera, Bethylidae
Israelius carthami Richards	Hymenoptera, Bethylidae
Perisierola gestroi Keifer	Hymenoptera, Bethylidae

to the produce (Kavadia et al., 1978). Infestation by the cigarette beetle and drug-store beetle resulted in significant reduction in protein, fat, and ash contents and an increase in uric acid levels in the dry turmeric rhizomes after 3–6 months of infestation (Gunasekaran et al., 2003).

Life History of the Insect

The cigarette beetle has a smooth elytra and measures 3–4 mm in length, with a serrated antennae. The larvae are gray-white with dense hairs. There are 4–6 larval instars and the later instars are scarabaeiform. In the State of Kerala, India, the incubation period lasts 9–14 days and the larval and pupal periods last 17–29 days and 2–8 days, respectively (Abraham et al., 1975). The drugstore beetle resembles the cigarette beetle but is smaller in size with striated elytra, and the distal segments of the antenna are clubbed. The larvae are pale white in color, with abdomen terminating in two dark, horny points in fully grown specimens. The eggs hatch in 6 days and the larval and pupal periods last 10–20 days and 8–12 days, respectively in Kerala State, India (Abraham et al., 1975). In Rajasthan State, India, the life cycle of the pest is completed in about 6 weeks (Srivastava, 1959). The coffee bean weevil is a gray-colored beetle with pale marks on the elytra and measures 3–5 mm in length, with long-clubbed antennae. The life cycle lasts 21–38 days in Kerala State, India (Abraham et al., 1975).

A wide range of produce, including cocoa and coffee beans, cereals, pulses, dried fruits, oil seeds, confectionery products, processed food stuffs, and even animal products, are infested by these storage pests. The olfactory response of adults of *L. serricorne* and *S. paniceum* to various spices, including turmeric, has been investigated. In *L. serricorne*, the highest attraction value of 42.6% was observed in turmeric; however, in *S. paniceum*, the attraction value was only 1.1% (Jha and Yadav, 1991).

Management of the Insect Pest

Various management strategies, including storage in suitable containers, fumigation, and radiation, have been suggested for the management of the storage pests. Polypropylene (2 μm in size, a commercially available low-density plastic baggage), low-density black polyethylene (250 μm), aluminum foil laminated with LDPE (low-density polyethylene 50 μm), and printed polypropylene (50 μm) were found resistant to biting by the insects and have been suggested for packing spice products, including turmeric (Jha and Yadav, 1991). Fumigation with a mixture of ethylene dibromide and carbon tetrachloride, methyl bromide, and phosphine has been suggested to manage the infestation by storage pests (Abraham et al., 1975; Kavadia et al., 1978). The concentration, time of exposure, and residual effects of various fumigants for controlling insect infestations in spices, including turmeric, have been furnished by Muthu and Majumdar (1974). However, new regulations on pesticide residues in many importing countries render many of these fumigants unsafe for application. Exposure of dry turmeric produce to gamma radiation has been suggested by some investigators for the management of storage pests (Rezaur et al., 1982). Sex pheromones have been identified in *L. serricorne* and *S. paniceum* and have been used to monitor population of these two species in stores (Kuwahara et al., 1975). Aggregation pheromones have also been identified in *A. fasciculatus* (Singh, 1993).

Other Insect Pests

The other insect pests recorded in stored produce include *Tenebroides mauritanicus* (L.), *Ephestia* sp., *Oryzaephilus surinamensis* L., *Setomorpha rutella* Zell., *Rhizopertha dominica* (Fab.), and *Tribolium castaneum* (Hbst.) (Abraham et al., 1975).

References

Abraham, V.A., Pillai, G.B., Nair, C.P.R., 1975. Biology of *Udaspes folus* Cram. (Lepidoptera: Hesperidae), the leaf roller pest of turmeric and ginger. J. Plantation Crops 3, 83–85.

Ayyar, T.V.R., 1920. A note on the present knowledge of Indian Thysanoptera and their economic importance. Report, Proceedings of the Third Entomological Meeting, February 1919, Pusa, Bihar State, India, pp. 618–622.

Ayyar, T.V.R., 1940. Handbook of Economic Entomology for South India. Government Press, Madras (now Chennai), Tamil Nadu, p. 565.

Balasubramanian, M., 1982. Chemical control of turmeric thrips *Panchaetothrips indicus* Bagnall. South Indian Hort 30, 54–55.

Ballou, H.A., 1916. Report of the prevalence of some pests ad diseases in the West Indies during 1915. Part I. Insect pests. West Indian Bull. Barbados 16 (1), 1–30.

Chakravarthy, A.K., Thyagaraj, N.E., 1997. Response of the cardamom (*Elettaria cardamomum* Maton) shoot and fruit borer (*Conogethes punctiferalis* Guenee Lepidoptera: Pyralidae) to different pheromone compounds. Insect Environ. 2, 127–128.

Chakravarthy, A.K., Thyagaraj, N.E., 1998. Evaluation of selected synthetic sex pheromones of the cardamom shoot and fruit borer, *Conogetheses punctiferalis* Guenee (Lepidoptera: Pyralidae) in Karnataka. Pest Manag. Trop. Ecosyst. 4, 78–82.

Choo, H.Y., Lee, S.M., Chung, B.K., Park, Y.D., Kim, H.H., 1995. Pathogenicity of Korean entomopathogenic nematodes (Steinernematidae and Heterorhabditidae) against local agricultural and forest insect pests. Korean J. Appl. Entomol. 34, 314–320.

Choo, H.Y., Kim, H.H., Lee, S.M., Park, S.H., Choo, Y.M., Kim, J.K., 2001. Practical utilization of entomopathogenic nematodes, *Steinernema carpocapsae* Pocheon strain and *Heterorhabditis bacteriophora* Hamyang strain for control of chestnut insect pests. Korean J. Appl. Entomol. 40, 69–76.

Devasahayam, S., 2002. In: Rethinam, P., Khan, H.H., Reddy, V.M., Mandal, P.K., Suresh, K. (Eds.), Evaluation of biopesticides for the management of shoot borer (*Conogethes punctiferalis* Guen.) on turmeric. Plantation crops research and development in the new millennium. Coconut Development Board, Kochi, Kerala State, India, pp. 489–490.

Devasahayam, S., 2004. Insect pests of ginger. In: Ravindran, P.N, Nirmal Babu, K (Eds.), Ginger, the Genus *Zingiber* CRC Press, Washington, DC, pp. 367–389.

Devasahayam, S., 2006. Bio-ecology and integrated management of root mealy bug (*Planococcus* sp.) infesting black pepper. Final report of ad-hoc research scheme. Indian Institute of Spices Research, Calicut, Kerala State, p. 24.

Dhoble, S.Y., Kadam, M.V., Dethe, M.D., 1978. Control of turmeric rhizome fly by granular systemic insecticides. J. Maharashtra Agric. Uni. 3, 209–210.

Dhoble, S.Y., Kadam, M.V., Dethe, M.D., 1981. Chemical control of turmeric rhizome fly, *Mimegralla coeruleifrons* Macquart. Indian J. Entomol. 43, 207–210.

Ecoport (Homepage on the internet) EcoPort Foundation Inc (updated 17 November 2006; cited 30 June 2009) Available from <http://www.ecoport.org>.

Fletcher, T.B., 1914. Some South Indian Insects and Other Animals of Importance Considered Especially from an Economic Point of View. Government Press, Madras (now Chennai), Tamil Nadu, p. 565.

Ghorpade, S.A., Jhadav, S.S., Ajri, D., 1983. Survey of rhizome fly on turmeric and ginger in Maharashtra. J. Maharashtra Agric. Univ. 8, 292–293.

Ghorpade, S.A., Jhadav, S.S., Ajri, D., 1988. Biology of rhizome fly, *Mimegralla coeruleifrons* Macquart (Micropezidae: Diptera) in India, a pest of turmeric and ginger crops. Trop. Pest. Manag. 34, 48–51.

Gunasekaran, N., Bhaskaran, V., Rajedran, S., 2003. Effect of insect infestation on proximate composition of selected stored spice products. J. Food Sci. Tech. Mysore 40, 239–242.

Hampson, G.F., 1856. Fauna of British India Including Ceylon and Burma. Moths Vol. 1. Taylor and Francis, London, p. 594.

Hill, D.S., 1983. Agricultural Insect Pests of the Tropics and Their Control, Second ed. Cambridge University Press, Cambridge, p. 746.

Honda, H., 1986. EAG responses of the fruit- and Pinaceae-feeding type of yellow peach moth, *Conogethes punctiferalis* (Guenee) (Lepidoptera: Pyralidae) to monoterpene compounds. Appl. Entomol. Zool. 21, 399–404.

Hutson, J.C., 1936. Report on the work of the entomological division. Administrative Report, Division of Agriculture, Ceylon (Sri Lanka now), pp. D22–D28.

IISR (Indian Institute of Spices Research) Report. 2001. Turmeric (Extension pamphlet). Indian Institute of Spices Research, Calicut, Kerala State, India, p. 8.

IISR (Indian Institute of Spices Research) Report. 2002. Indian Institute of Spices Research, Calicut, Kerala State, India, p. 131.

Jacob, S.A., 1981. Biology of *Dichocrocis punctiferalis* Guen. on turmeric. J. Plantation Crops 9, 118–123.

Jacob, S.A., 1982. Biology and bionomics of ginger and turmeric scale *Aspidiotus hartii* Green. In: Nair, M.K., Premkumar, T., Ravindran, P.N, Sarma, Y.S (Eds.), Proceedings of National Seminar on Ginger and Turmeric Central Plantation Crops Research Institute, Kasaragod, Kerala State, pp. 131–132.

Jha, A.N., Yadav, T.D., 1991. Olfactory response of *Lasioderma serricorne* Fab. and *Stegobium paniceum* (Linn.) to different spices. Indian J. Entomol. 53, 396–400.

Jhadav, S.S., Ghorpade, S.A., Ajri, D., 1982. Field screening of some turmeric varieties against rhizome fly. J. Maharashtra Agric. Univ. 7, 260.

Jung, J.K., Han, K.S., Choi, K.S., Boo, K.S., 2000. Sex pheromone composition for field trapping of *Dichocrocis punctiferalis* (Lepidoptera: Pyralidae) males. Korean J. Appl. Entomol. 39, 105–110.

Kavadia, V.S., Pareek, B.L., Sharma, K.P., 1978. Control of *Lasioderma serricorne* Fab. infestation of turmeric by phosphine fumigation. Entomon 3, 57–58.

Konno, Y., Honda, H., Matsumato, Y., 1980. Observations on the mating behavior and bioassay for the sex pheromone of the yellow peach moth, *Conogethes punctiferalis* (Guenee) (Lepidoptera: Pyralidae). Appl. Entomol. Zool. 15, 321–327.

Kotikal, Y.K., Kulkarni, K.A., 1999. Management of rhizome fly, *Mimegralla coeruleifrons* Macquart (Micropezidae: Diptera), a serious pest of turmeric in northern Karnataka. Pest Manag. Hortic. Ecosyst. 5, 62–66.

Kotikal, Y.K., Kulkarni, K.A., 2000. Incidence of insect pests of turmeric (*Curcuma longa*) in northern Karnataka, India. J. Spices Aromat. Crops 9, 51–54.

Kotikal, Y.K., Kulkarni, K.A., 2001. Reaction of selected turmeric genotypes to rhizome fly and shoot borer. Karnataka J. Agric. Sci. 14, 373–377.

Koya, K.M.A., Devasahayam, S., Premkumar, T., 1991. Insect pests of ginger (*Zingiber officinale* Rosc.) and turmeric (*Curcuma longa* Linn.) in India. J. Plantation Crops 19, 1–13.

Kuwahara, Y., Fukami, H., Ishil, S., Matsumura, F., Burkholderm, W.E., 1975. Studies of the isolation and bioassay of the sex pheromone of the drugstore beetle, *Stegobium paniceum* (Coleoptera: Anobiidae). J. Chem. Ecol. 1, 413–422.

Maxwell-Lefroy, H., Howlett, F.M., 1909. Indian Insect-Life—A Manual of the Insects of the Plains (Tropical India). Government Press, Calcutta (Kolkata now), p. 396.

Muthu, M., Majumdar, S.K., 1974. Insect control of spices. Proceedings of Symposium on Development and Prospects of Spice Industry in India, Association of Food Science and Technology, Mysore, Karnataka State, p. 35.

Nair, M.R.G.K., 1975. Insects and Mites of Crops in India. Indian Council of Agricultural Research, New Delhi, p. 404.

Patil, A.P., Thakur, S.G., Mohalkar, P.R., 1988. Incidence of pests of turmeric and ginger in western Maharashtra. Indian Cocoa Arecanut Spices J. 12, 8–9.

Ramakrishna, T.A.V.S., Raghunath, T.A.V.S., 1982. New record of *Creatonotos gangis* (Linneaus) on turmeric in India. Indian Cocoa Arecanut Spices J. 5, 63.

Rao, P.S., Krishna, M.R., Srinivas, C., Meenakumari, K., Rao, A.M., 1994. Short duration, disease resistant turmerics for northern Telengana. Indian Hortic. 39, 55–56.

Rao, S.V., Reddy, P.S., 1990. The rhizome fly, *Calobata albimana* Macq, a major pest of turmeric. Indian Cocoa Arecanut Spices J. 14, 67–69.

Ravindran, P.N., Babu, K.N., Shiva, K.N., 2007. In: Ravindran, P.N., Babu, K.N., Sivaraman, K. (Eds.), Turmeric-the genus curcuma botany and crop improvement of turmeric. CRC Press, Boca Raton, pp. 15–70.

Reddy, M.R.S., Reddy, P.V.R.M., 2000. Preliminary observations on neem cake against rhizome fly of turmeric (*Curcuma longa*). Insect Environ. 6 (2), 62.

Regupathy, A., Santhanam, G., Balasubramanian, M., Arumugam, R., 1976. Occurrence of the scale *Aspidiotus hartii* C. (Diaspididae: Hemiptera) on different types of turmeric *Curcuma longa* Lin. J. Plantation Crops 4, 80.

Rezaur, R., Ahmed, M., Hossain, M., Nazar, G., 1982. A preliminary report on the problems of dried spice pests and their control. Bangladesh J. Zool. 10, 141–144.

Sengupta, G.C., Behura, B.K., 1955. Some new records of crop pests from India. Indian J. Entomol. 17, 283–285.

Sengupta, G.C., Behura, B.K., 1956. Note on the life history of *Lema semiregularis* Jac. (Coleoptera: Chrysomelidae). J. Bombay Nat. Hist. Soc. 53, 484–485.

Sengupta, G.C., Behura, B.K., 1957. On the biology of *Lema praeusta* Fab. J. Econ. Entomol. 50, 471–474.

Sheila, M.K., Abraham, C.C., Nair, P.C.S., 1980. Incidence of shoot borer (*Dichocrocis punctiferalis*) Guen. (Lepidoptera: Pyraustidae) on different types of turmeric. Indian Cocoa Arecanut Spices J. 3, 59–60.

Shylesha, A.N., Azad-Thakur, N.S., Ramachandra, S., 1998. Occurrence of *Libnotes punctipennis* Meij (Tipulidae: Diptera) as rhizome borer of turmeric in Meghalaya. In: Mathew, N.M., Jacob, C.K. (Eds.), Developments in Plantation Crops Research. Proceedings of 12th Symposium on Plantation Crops. PLACROSYM—XII, 27–29 November 1996, Kottayam, Kerala State, pp. 276–278.

Sika, N.A., Yao, N.A., Foua-Bi, K., 2005. Action of the parasitic Hymenoptera: *Adelencyrtus femoralis* (Hymenoptera, Encyrtidae) against the yam's brown yam cochineal: *Aspidiella hartii* Cockerell, Homoptera, Diaspididae. Agron. Africaine 17 (2), 153–162.

Singh, K., 1993. Evidence of male and female of coffee bean weevil, *Araecerus fasciculatus* (DeG) (Coleoptera: Anthriibidae) emitted aggregation and sex pheromones. J. Crop Res. Hissar. 6, 97–101.

Srinath, D., Prasad, C., 1975. *Lasioderma serricorne* F. as a major pest of stored turmeric. Bull. Grain Technol. 13, 170–171.

Srivastava, B.K., 1959. *Stegobium paniceum* as a pest of stored turmeric in Rajasthan, India, and its control by fumigation. FAO Plant Protect. Bull. 7, 113–114.

Sujatha, A., Zaherudeen, S.M., Reddy, R.V.S.K., 1992. Turmeric leaf roller, *Udaspes folus* Cram. and its parasitoids in Godavari Delta. Indian Cocoa Arecanut Spices J. 15, 118–119.

Thangavelu, P., Sivakumar, C.V., Kareem, A.A., 1977. Control of lace wing bug (*Stephantis typica* Distant) (Tingidae: Homoptera) in turmeric. South Indian Hortic. 25, 170–172.

Thyagaraj, N.E., Singh, P.K., Chakravarthy, A.K., 2001. Sex determination of cardamom shoot and fruit borer, *Conogethes punctiferalis* (Guenee) pupae. Insect Environ. 7, 93.

Velayudhan, K.C., Liji, R.S., 2003. Preliminary screening of indigenous collections of turmeric against shoot borer (*Conogethes punctiferalis* Guen.) and scale insect (*Aspidiella hartii* Sign.). J. Spices Aromat. Crops 12, 72–76.

Zhou, Z.J., Lin, Q.Y., Xi, L.H., Zheng, G.Z., Huang, Z.H., 1995. Studies on banana bunchy top. III. Occurrence of its vector *Pentalonia nigronervosa*. J. Fujian Agric. Univ. 24, 32–38.

9 Turmeric Nematology

Nematode Pests of Turmeric

A number of plant parasitic nematodes have been reported to be associated with turmeric in India and the list is given in Table 9.1.

Among the species identified, *Meloidogyne* sp., *R. similis*, and *P. coffeae* are the ones which are economically important (Ayyar, 1926). The root knot nematode *Heterodera radicicola* (Greef) Muller was first reported on turmeric by Ayyar (1926). Two species of root knot nematodes, namely *M. incognita* (Kofoid & White, 1919) Chitwood 1949 and *M. javanica* (Treub, 1885) Chitwood, 1949, have been reported to be prevalent in India. *R. reniformis* Linford & Oliveira, 1940 and *M. incognita* were the most dominant and frequently recorded nematode species in Chittoor and Cuddapah districts, Andhra Pradesh, India (Mani and Prakash, 1992) and Bihar State, India (Haidar et al., 1998). *R. reniformis* caused significantly higher growth reduction at lower population levels, and hence was reported to be more harmful to turmeric than *M. incognita* in Bihar State, India (Haidar et al., 1998). Records of nematode infestation on turmeric in other countries include *M. incognita* from China (Chen et al., 1986) and *M. javanica*, *R. similis*, *R. reniformis*, *Macroposthonia onoensis* (Lue, 1959) De Grisse & Loof, 1965, *Helicotylenchus indicus* Siddiqi, 1963, *Hemicriconemoides cocophilus* (Loos, 1949) Chitwood & Birchfield, 1957, and *Pratylenchus brachyurus* (Godfrey, 1929) Flipjev & S. Stekhoven, 1941 from Fiji (Williams, 1980).

Symptoms of Nematode Infestation

Yellowing, marginal and tip necrosis, reduced tillering, culminating in stunted growth, coupled with root galling and rotting are the principal symptoms of the infestation by *M. incognita* on turmeric. The infested rhizomes also tend to lose their characteristic bright yellow color (Mani et al., 1987). Population density of *M. incognita* on turmeric increased with age of the crop and decreased crop resistance (Poornima and Sivagami, 1999). Roots of turmeric damaged by *R. similis* become rotten and most of these decayed roots retain only the epidermis and are devoid of the cortex and stellar portions. The infested plants show a tendency to age and dry faster than healthy ones. The infested rhizomes are yolk yellow in color compared with the golden yellow color of healthy rhizomes and have shallow water-soaked brownish areas on the surface (Sosamma et al., 1979). *P. coffeae* has been reported to be associated with discoloration and rotting of mature rhizomes of *C. aromatica*. The rhizomes turn deep red to dark brown in color, less turgid, and wrinkled with dry rot symptoms.

The Agronomy and Economy of Turmeric and Ginger. DOI: http://dx.doi.org/10.1016/B978-0-12-394801-4.00009-0

The finger rhizomes are more severely affected than the mother rhizomes and the affected rhizomes show dark brown necrotic patches internally (Sarma et al., 1974).

Economic Consequences of Nematode Infestation

When turmeric plants were inoculated with *M. incognita* and *R. reniformis*, significant reductions in both growth and yield were observed individually at the rate of 1000 juveniles per plant (Haidar et al., 1998). Poornima and Sivagami (1998) reported that an initial level of >5000 *M. incognita* larvae per plant was highly pathogenic to turmeric. The yield loss under field conditions was 45.3% due to *M. incognita* infestation but was only 33.3% in a mixed infestation of *M. incognita* and *R. reniformis* (Bai et al., 1995). Pathogenicity investigations using *R. similis* showed that an inoculum level of 10 nematodes per plant led to 35% reduction in rhizome weight after 4 months of infestation and 46% reduction at the end of the crop season, i.e., about 8 months. With 100,000 nematodes, the extent of rhizome yield reduction was 65% and 76%, respectively after 4 and 8 months of infestation (Sosamma et al., 1979).

Field Management of Nematode Infestation

Nematode control is basically centered around prevention. This is because once a plant is parasitized, it is extremely difficult to eradicate the nematode without destroying the host plant. There is seldom a single method available, to alleviate nematode problems in the field, in entirety, and a sustainable approach integrating several tools and strategies, such as crop rotation, soil solarization, application of nematicides, biological control, and cultivating resistant varieties, are all needed in a comprehensive manner for effective management.

Preventive Measures

Heat treatments, such as steam, hot water, and sunlight, are very effective means to control nematode infestation. Disinfestation of turmeric rhizomes was achieved by hot water of 45°C continuously for 3 h in China (Chen et al., 1986). Soil solarization is another technique that is effective in controlling soil-borne diseases caused by nematode attack. But these techniques are unreliable as an effective economic means for the control of nematodes, especially when used in isolation. This is why integrated pest management is preferred, where several parameters of pest control are brought under one umbrella.

Planting Nematode-Resistant Lines

Several investigators have screened a number of turmeric lines against infestation by root knot nematodes and reported a few lines as resistant, especially to

Table 9.1 Nematode Pests of Turmeric (*C. longa* L.)

Nematode Species	Order/Family	Geographic Region
Heterodera radicicola (Greef) Muller	Tylenchida, Heteroderidae	India
Tylenchus sp.	Tylenchida, Tylenchidae	India
Bitylenchus sp.	Tylenchida, Dolichodoridae	India
Criconemella sp.	Tylenchida, Criconematidae	India
Hemicriconemoides sp.	Tylenchida, Criconematidae	India
H. cocophilus (Loos, 1949) Chitwood & Birchfield, 1957	Tylenchida, Criconematidae	Fiji
Macroposthonia sp.	Tylenchida, Criconematidae	India
M. onoensis (Lue, 1959), De Grisse & Loof, 1965	Tylenchida, Criconematidae	Fiji
Meloidogyne sp.	Tylenchida, Criconematidae	India
M. incognita (Kofoid & White, 1919) Chitwood, 1949	Tylenchida, Meloidogynidae	Fiji, India
M. javanica (Treub, 1885), Chitwood, 1949	Tylenchida, Meloidogynidae	China, Fiji, India
Radopholus similis (Cobb. 1893), Thorne, 1949	Tylenchida, Meloidogynidae	Fiji, India
Tylenchorhynchus sp.	Tylenchida, Pratylenchidae	India
Basirolaimus sp.	Tylenchida, Dolichodoridae	India
Hoplolaimus sp.	Tylenchida, Hoplolaimidae	India
Helicotylenchus sp.	Tylenchida, Hoplolaimidae	India
H. indicus Siddiqi, 1963	Tylenchida, Hoplolaimidae	Fiji
Pratylenchus sp.	Tylenchida, Hoplolaimidae	India
P. brachyurus (Godfrey, 1929), Filipjev & S. Stekhoven, 1941	Tylenchida, Pratylenchidae	Fiji
Pratylenchus coffeae (Zimmerman, 1898) Filipjev & S. Stekhoven, 1941	Tylenchida, Pratylenchidae	India
Rotylenchulus sp.	Tylenchida, Hoplolaimidae	India
R. reniformis Linford & Oliveira, 1940	Tylenchida, Rotylenchulidae	Fiji, India
Hemycycliophora sp.	Tylenchida, Hemycycliophoridae	India
Caloosia sp.	Tylenchida, Caloosiidae	India
Paratrichodorus sp.	Triplonchida, Trichidoridae	India
Aphelenchus sp.	Aphelenchida, Aphelenchidae	India
Longidorus sp.	Dorylaimida, Longidoridae	India
Xiphinema sp.	Dorylaimida, Xiphinematidae	India

M. incognita (Mani et al., 1987; Rao et al., 1994). They are 5379-1-2, 5363-6-3, Kodur, Cheyapuspa, 5335-1-7, 5335-27, Ca-17/1, Cli-124/6, Cli-339, Armoor, Duggirala, Guntur-1, Guntur-9, Rajampet, Sugandham, and Appalapadu. In China, *C. zedoaria* was found to be more resistant to *M. incognita* than *C. domestica* (Chen et al., 1986). The high-yielding varieties PCT-8, PCT-10, Suguna, and Sudharshana were found to be free from the infestation of *M. incognita* in Andhra Pradesh, India (Rao et al., 1994). Eight turmeric accessions were identified to be resistant to

M. incognita among the germplasm collection at IISR, Calicut, Kerala State, India (Eapen et al., 1999).

Chemical Control of Nematodes

In turmeric, chemical control of nematodes offers immense possibilities. Significant increases in rhizome yield have been reported following the use of nematicides. The following are some important examples. Aldicarb and Carbofuran (both patented products in use in India, though both have now been taken off the list of prescribed nematicides in the State of Kerala, India, as they fall under the "Red Triangle" category, the most toxic group of chemicals) used at 1 kg a.i/ha led to yield increases of 71% and 68%, respectively, over untreated control treatments in Tamil Nadu, India. Aldicarb is now banned in many other countries. Carbofuran at 4 kg a.i/ha applied to a 4-month-old turmeric crop resulted in 81.6% reduction in root knot population, in Andhra Pradesh, India (Mani et al., 1987). Application of phorate or carbofuran at 1 kg a.i./ha was effective in reducing the population of root knot nematodes in the field (Haidar et al., 1998).

Biological Control of Turmeric Nematodes

Various antagonistic fungi have been evaluated at the Experimental Station of IISR, Calicut, Kerala State, India, to control the spread of root knot nematode in turmeric, and among them *Fusarium oxysporum* Schlecht emend. Synd. & Hans. (nonpathogenic isolate) and *Paecilomyces lilacinus* (Thom) Samson were found to be quite promising, compared to others (Eapen et al., 2008). Several other rhizobacteria have also been evaluated for their efficacy in controlling root knot nematodes of turmeric in the field of the Experimental Farm at Peruvannamuzhi, Kerala State, affiliated with the IISR.

References

Ayyar, P.N.K.A., 1926. A preliminary note on the root gall nematode, *Heterodera radicicola* and its economic importance in South India. Madras Agric. J. 14, 113–118.

Bai, H., Sheela, M.S., Jiji, T., 1995. Nemic association and avoidable yield loss in turmeric, *Curcuma longa* L. Pest Manag. Hortic. Ecosyst. 1, 105–110.

Chen, C.M., Li, H.Y., Li, D.Y., 1986. The study on root knot nematodes of common turmeric (*Curcuma domestica* Valet.). Herald Agric. Sci. 1, 16–22.

Eapen, S.J., Ramana, K.V., Sasikumar, B., Johnson, K.G., 1999. Screening ginger and turmeric germplasm for resistance against root knot nematodes. In: Dhawan, S.C (Ed.), Proceedings of National Symposium on Rational Approaches in Nematode Management for Sustainable Agriculture Nematological Society of India, New Delhi.

Eapen, S.J., Beena, B., Ramana, K.V., 2008. Evaluation of fungal bioagents for management of root-knot nematodes in ginger and turmeric fields. J. Spices Aromat. Crops 17, 122–127.

Haidar, M.G., Jha, R.N., Nath, R.P., 1998. Efficacy of some nematicides and some organic soil amendments on nematode population and growth characteristics of turmeric (*Curcuma longa* L.). J. Res. Bisra. Agric. Univ. 10, 211–213.

Mani, A., Prakash, K.S., 1992. Plant parasitic nematodes associated with turmeric in Andhra Pradesh. Curr. Nematol. 3, 103–104.

Mani, A., Naidu, P.H., Madhavachari, S., 1987. Occurrence and control of *Meloidogyne incognita* on turmeric in Andhra Pradesh. India Int. Nematol. Network Newslet. 4, 13–18.

Poornima, K., Sivagami, V., 1998. Pathogenicity of *Meloidogyne incognita* of turmeric (*Curcuma longa* L.). In: Mehta, U.K (Ed.), Nematology: Challenges and Opportunities in 21st Century. Proceedings of the Third International Symposium of Afro-Asian Society of Nematologists (TISAASN) Afro-Asian Society of Nematologists, Luton.

Poornima, K., Sivagami, V., 1999. Occurrence and seasonal population behavior of phytonematodes in turmeric (*Curcuma longa* L.). Pest Manage Hort. Ecosyst. 5, 42–45.

Rao, P.S., Krishna, M.R., Srinivas, C., Meenakumari, K., Rao, A.M., 1994. Short-duration, disease resistant turmerics for northern Teleganna. Indian Hortic. 39, 55–56.

Sarma, Y.R., Nambiar, K.K.N., Nair, C.P.R., 1974. Brown rot of turmeric. J. Plantation Crops 2, 33–34.

Sosamma, V.K., Sudarraju, P., Koshy, P.K., 1979. Effect of *Radopholus similis* on turmeric. Indian J. Nematol. 9, 27–31.

Williams, K.J.O., 1980. Plant parasitic nematodes of the Pacific. Technical report (UNDP/FAO-SPEC Survey of agricultural pests and diseases in the south pacific) v.8. Commonwealth Institute of Helminthology, Herts, UK.

10 Diseases of Turmeric

Turmeric is attacked by many diseases and is listed in Table 10.1.

Major Diseases of Turmeric

Rhizome Rot

This disease is characterized by the rotting of underground stem and was first reported by Park (1934) in Sri Lanka (then Ceylon). In India, the disease was reported by Ramakrishnan and Sowmini (1954) from Krishnagiri, Tiruchirapally, and Coimbatore in Tamil Nadu, India, and from Kasaragod in Kerala State, India (Anon, 1974, Joshi and Sharma, 1980), and also, from Assam State, India (Rathaiah, 1982a). Rhizome rot pertains to the economic aspects of turmeric cultivation, because of its severe damage on the consumable part of the crop, leading to much pecuniary distress for the farmer. It is the most damaging of all the diseases affecting turmeric and is reported as a major soil-borne problem in all turmeric-growing tracts.

Symptoms

The initial symptom of the disease is the appearance of water-soaked lesions at the base of the pseudostem, leading to yellowing of the leaves. Infection on roots is manifested as browning and rotting of the roots, the symptoms later advancing to the rhizomes, which turn soft and putrefied and the color changes from bright orange to varying shades of brown. The infection gradually spreads to all the fingers of the rhizome and the mother rhizome as well, and eventually the plant dies. When the affected rhizomes are split open, brown to dark brown fibrovascular tissues are seen. The infected plants show gradual drying up of leaves along the margins and later the entire leaf dries (Rao, 1995). The appearance of symptoms depends on the time of infection in relation to crop phenology and extent of damage to the rhizome. If it is confined to just the roots, or to the secondary or tertiary fingers (cormlets), hardly any symptoms can be observed.

Crop Loss

Crop loss can be as much as 50%, as reported in the Telengana region of Andhra Pradesh, India (Rao and Rao, 1988), but can vary in extent from region to region. Sankaraiah et al. (1991) also reported high incidence of the disease in Andhra Pradesh, where the turmeric crop is one of the important economic mainstays.

The Agronomy and Economy of Turmeric and Ginger. DOI: http://dx.doi.org/10.1016/B978-0-12-394801-4.00010-7

Table 10.1 Diseases of Turmeric and Their Causal Organisms

S. No.	Disease	Causal Organism(s)	References
1	Rhizome rot	*Pythium aphanidermatum*	Park (1934)
	Leaf blotch	*P. graminicolum*	Middleton (1943)
			Ramakrishnan and
			Sowmini (1954)
		P. myriotylum	Rathaiah (1982a)
		Pythium sp. and Fusarium solani	Dohroo (1988)
2		*Taphrina maculans*	Butler (1918), Pavgi
			and Upadhyay (1964)
3	Leaf spots	*Colletotrichum capsici*	
		C. curcumae	
		C. gloeosporioides	
		Phaeodactylium alpiniae	
		Thirumalacharia curcumae	
		Phyllosticta zingiberii	
		Phaeorobillarda curcumae	
		Cercospora curcumae-longae	
		Myrothecium roridum	
		Pyricularia curcumae	
		Pestalotiopsis sp.	
4	Leaf blast	*Pyricularia curcumae*	Rathaiah (1980)
		Alternaria alternate	Kumar and Roy (1990)
5	Leaf blight	*Corticum sasaki*	Saikia and Roy
			(1975), Mallikarjun
			and Kulkarni (1998a),
			Mallikarjun and
			Kulkarni (1998b)
6	Nematode	*M. incognita, R. similis*	Nadakkal and Thomas
	infestations		(1964), Nirula and
			Kumar (1963), Mani
			and Prakash (1992)
		Pratylenchus sp.	Haidar et al. (1995)
7	Root rot	*F. solani*	Dohroo (1988)
8	*Pseudomonas* leaf rot	*Pseudomonas cichorii*	Maringoni et al. (2003)
9	Storage Rots	*Macrophomina phaseolina*	Sarma et al. (1994)
		A. flavus, A. niger	Sharma and Roy
			(1984)
		Cladosporium cladosporioides	Kumar and Roy (1990)
		Drechsleza rostrata	Kumar and Roy (1990)
		F. moniliforme, F. oxysporum	Kumar and Roy (1990)
		P. aphanidermatum, R. solani	Kumar and Roy (1990)
		Sclerotium rolfsii	Kumar and Roy (1990)
		Fusarium sp.	
10	Brown rot	*Pratylenchus sp.*	

Epidemiology

The etiology of rhizome rot has been a subject of much scientific speculation. Several fungi are said to be involved, as well as the involvement of dipteran maggots, at times attributing a complex fungal–insecticidal interrelationship and chemical and physiological interaction in the causation of rhizome rot. The disease was reported to be caused by different species of *Pythium*, namely *Pythium myriotylum, Pythium graminicolum*, and *Pythium aphanidermatum*. The disease reported to be caused by *P. aphanidermatum* as early as 1934 (Park, 1934). Subsequently, Ramakrishnan and Sowmini (1954) reported that *P. graminicolum* as the causal agent, while Rathaiah (1982a) attributed the causative agent as *P. myriotylum* in Assam State, India. In 1982, Ajiri et al. (1982) observed the disease to be associated with the maggots of *Mimigrella coerulifrons* found in diseased rhizomes.

Investigations on the rhizome rot complex on turmeric in Nizamabad district of Andhra Pradesh, India during the period 1979–1988 showed that *Pythium* infestation was the primary cause for the disease incidence, and the incidence of the maggots was secondary and consequential. However, rotting of the rhizome would be accelerated with the entry of the insects into the rhizome (Sankaraiah et al., 1991). Reports also exist in scientific literature attributing *Aspergillus niger* and *Aspergillus flavus* as causative organisms in rhizome rot (Sharma and Roy, 1987). In Telengana region of Andhra Pradesh, India, the causative agents were *Pythium* and *Fusarium* sp., namely, *Fusarium solani* (Anon, 1996–1997). Though various fungal species are reported to be associated with turmeric rhizome rot, it is irrefutably proved that the *Pythium* species is the predominant causal organism. Other species reported are only of secondary nature. Premkumar et al. (1982) and Koya (1990), based on systematic investigations conducted under controlled conditions, have concluded that soft rhizome rot is predominantly caused by *Pythium* sp. and dipteran maggots play only a secondary role, in putrefying the rotten tissues. These authors also argue that the dipterans do not have the required mouth parts to penetrate the intact tissues of the rhizome or roots. They found the association of the maggots with only rotten rhizomes, and stated that they do not play any significant role in causing the disease. Careful examination of the disease at various stages proves that the maggots are not the primary agent but only accelerate the putrefying process. Initial infection by Pythiaceous fungi are soon followed by saprotrophs and often lead to erroneous conclusions (Erwin and Ribeiro, 1996). Observations at various stages of infection reveal the secondary role of the maggots in aggravating infection. In the later stages of growth of the plant, infection is restricted to the roots and/or secondary/tertiary fingers. In partially infected rhizomes, the presence of the maggots and pupae could be seen besides restriction of lesions.

A detailed study of the etiology of rhizome rot was undertaken at IISR, Calicut, Kerala State, India, during 2005–2009. Systematic field survey was conducted in different turmeric-growing regions of the country, in the four southern states, namely, Andhra Pradesh, Tamil Nadu, Kerala, and Karnataka. During the survey, 118 samples were collected from 38 locations. It became clear that *Pythium* was the principal causative organism to cause rhizome rot, from this study as well. This was to the extent of 72.9%. This trend was followed by the causative fungus *Rhizoctonia* sp. which accounted for 30.5% infestation. *Fusarium* sp. caused the least infestation at 27.1%. Pathogenic tests revealed that 75% of the pathogenic isolates were *P. aphaniderma-tum* (Anon, 2006–2007). Ramakrishnan (1954) studied the biology of the pathogen *P. graminicolum* and showed that the fungus grows over a wide range of pH, from 3 to 9, and the best growth was at a pH of 7–8. Oospores, the perennating structure, are produced to the maximum at pH ranging from 6 to 9, and the fungus can tolerate a high concentration of phosphates (125–250 kg/ha) in the culture medium, while it is quite sensitive to the presence of urea in the same (Ramakrishnan and Sowmini, 1954).

The manifestation of rhizome rot in India is principally observed during the main SW monsoon period, especially when the crop is in the active growing stage. The plant is primarily disposed to infestation when there is waterlogging, as is bound to happen in this period, when there could be incessant rains for days on end. Sankaraiah et al. (1991) suggested that there is a positive relationship between continuous rains and the incidence of rhizome rot. These authors noted that when excessive amounts of tank silt are applied to the soil, the disease erupts, as excessive tank silt causes drainage problems.

The Management of Turmeric Rhizome Rot

Chemical Control

Turmeric diseases are both soil and seed borne, which harbor maggots of *M. coeru-lifrons* in partly diseased rhizomes. Dipping rhizomes in a combination of solution containing both fungicides and pesticides preempt disease incidence. This operation is carried out just before storage or sowing. The rhizomes so treated may be dried under shade before storage. The treatment of rhizome with a combination of Mancozeb, at a concentration of 0.25%, and Quinalphos at a concentration of 0.75% for half an hour (the former a patented fungicide and the latter a patented insecticide), has been found to be the best practice to contain the incidence of the disease. According to Sharma and Dohroo (1982), drenching of the soil with Metalaxyl man-cozeb (a patented fungicide) or simply Manocozeb (a patented fungicide), twice at 15–20 days interval, coinciding with the first appearance of the disease symptoms, has been found to be quite effective in containing the disease.

A very effective method to contain the disease is by first digging out all the infected plants followed by soil drenching with chestnut compound or cere-san wet (patented fungicide) at a concentration of 1% (Joshi and Sharma, 1980). Concurrently, treating the seed sets with Agallol (another patented fungicide) at a concentration of 0.25% for half an hour prior to storage, and also prior to sow-ing, is also found to prevent the disease onset. Crop rotation is also recommended to reduce the incidence of the disease. When the disease is noticed in the field, the plant beds are drenched with chestnut compound at 0.3% concentration and Agallol at 0.1% concentration, though these recommendations are now obsolete (Ikisan, 2000). Evaluation of fungicides to control the incidence of *A. niger* infestation showed that Carbendazim and Benomyl (both patented fungicides) are effective even at a low concentration of 100 ppm in both *in vitro* and *in vivo* conditions (Sharma and Roy, 1987). At Rudrur in Andhra Pradesh, India, rhizome rot incidence was found to be best when turmeric plants were spaced 30 cm × 15 cm with an intercrop of maize, sown in a separate row after two rows of turmeric (Sankaraiah et al., 1991). The same authors, based on their investigations in Nizamabad, Andhra Pradesh, India, formulated an effective field package of practice to contain the disease. The package advocates, most importantly, the use of disease-free seed sets obtained from a disease-free region, presowing treatment with Carbendazim, and drenching the soil with 1% Bordeaux mixture, when the disease symptoms appear. It was also recommended to intercrop with maize, as above, with a spacing of 30 cm × 15 cm for the main crop and 60 cm × 30 cm for maize. The cultivation of highly susceptible varieties like Armoor in the disease-prone areas was strongly discouraged (Sankaraiah et al., 1991).

Investigations on rhizome rot in the Cuddapah district of Andhra Pradesh, India, proved that among the fungicides, it was Ridomil (a patented fungicide) that was most effective, which was followed by Mancozeb, Aliette, and Thride, in this order (all patented fungicides) in their effectiveness (Anandam et al., 1996). Dipping the seed sets in Ridomil and soil drenching with the same fungicide at the first appearance of the disease symptoms effectively controlled the incidence of the disease and increased rhizome yield. Treatment of seed rhizome with Dithane M-45 at 0.25% concentration or Bavistin at 1% concentration (both patented fungicides) for half an hour prior to storage and also at the time of planting and drenching the infected plant beds with Blitox (also a patented fungicide) at 0.3% concentration or spraying the ubiquitous Bordeaux mixture (1% concentration) or copper oxychloride at 0.25% concentration on affected plants effectively contained the spread of the disease, especially when the causative fungus is *Panicum graminicolum*.

Investigations at IISR, Calicut, Kerala State, India, conclusively showed that preplanting treatment of seed rhizomes with Metalaxyl mancozeb at a concentration of 0.125% for half an hour or using copper oxychloride at 0.25% for the same purpose, followed by soil drenching with these fungicides at the time of planting, and repeating the treatment at, 30, 60, and 90 days after planting effectively controlled the disease in Annoor, a severely disease-prone region in Coimbatore district, Tamil Nadu, India (Anon, 2007). Almost 33–45% disease reduction was obtained by the above treatment. Mercury- and copper-based fungicides were recommended earlier to control rhizome rot, but on account of their heavy metal concentration, which is toxic, they are no longer recommended. These heavy metals get adsorbed to soil particles (especially clay particles, through physicochemical processes in soils) and create a grave health hazard when they pollute the ground water and also seep into the plant parts through root absorption. The increasing environmental consciousness in India has also focused far greater attention on the use of such toxic fungicides, and they are to a large extent avoided by the farmers, though some farmers do resort to the use of such fungicides to obtain immediate control of the disease. Such practices are detrimental to soil health, which in turn, adversely affect human health when plant parts grown in such situations are consumed by unwary consumers.

Biological Control of Rhizome Rot

A number of attempts have been made using antagonistic microflora for the control of rhizome rot. Most prominent among them is *Trichoderma*, the widely known antagonistic fungus against many pathogens. Application of *Trichoderma* at the time of planting can check the incidence of rhizome rot (Ravikumar, 2002). A fermented mixture of 100 kg cow dung/gobar gas slurry (gobar gas is a fermented gas obtained from cow dung in rural Indian homes, used for cooking) with 10 l of "Panchakavya" (a liquid organic manure prepared through knowledge of ancient Indian texts on agriculture) and 200 l of water, when incorporated with irrigation water to turmeric at fortnightly intervals, was recommended as effective in controlling rhizome rot and nematode infections, as part of organic farming practices (Ravikumar, 2002). In Andhra Pradesh, the disease suppressive role of *Trichoderma harzianum, Trichoderma virens* (*Gliocladium virens*), and VAM fungus, along with fungicides, to control rhizome rot was tested in both pot culture and field trials (Reddy et al., 2003). VAM-treated plants in infested soil showed reduced disease incidence compared to plants exposed to *F. solani* alone. *Trichoderma* and *Gliocladium* treatments resulted in reduced disease intensity when compared to VAM-treated plants. Among the three biological control agents tested, *T. harzianum* was found to be the most effective in disease control. Plants treated with biological control agents were healthy resulting in very high biomass production, coupled with rhizome yield. The combination of *Trichoderma* sp. and Ridomil effectively controlled the disease in the field (Reddy et al., 2003). Ushamalini et al. (2008) developed a biomanure, containing MG 6 strain of *Trichoderma viride*, effective in inhibiting the growth of *P. aphanidermatum*, the principal causal agent of rhizome rot in turmeric in glass house conditions. Rhizome rot incidence was found to be reduced in plants applied with farmyard manure (FYM)-based organic manure. Investigations conducted at IISR, Calicut, Kerala State, India, on biological control of rhizome rot using different isolates of *Trichoderma* also showed that certain isolates are very effective in both the disease incidence and the disease control.

Disease Resistance

Rhizome rot disease is endemic to Andhra Pradesh, India, where turmeric is extensively cultivated. In these endemic situations, varieties such as PCT 8, PCT 10, PCT 13, PCT 14, Suguna, and Sudharshana were found tolerant to the disease (Rao et al., 1992, 1994). The local high-yielding varieties like Armoor, Duggirala, and Mydukur were found to be highly susceptible, whereas in Assam State, India, the cultivars like Mydukur, Tekurpet, and Duggirala were found susceptible, while cultivars Ca 69 and Shillong were found to be resistant (Rathaiah, 1982b).

The variety CLI 370 was found moderately tolerant to rhizome rot (Sankaraiah et al., 1991). Screening germplasm for disease resistance under artificial inoculation showed that CA 17/1, CA 146/A, and Suvarna were free of disease incidence. In field conditions, rotting was higher among long-duration genotypes, followed by medium duration and then short-duration ones, sequentially (Anandam et al., 1996). Turmeric variety TCP-2, a clonal selection from the State of West Bengal, India, and developed by Uttar Banga Krishi Viswavidyalaya (UBKVV) in Pundibari, Cooch Behar district, West Bengal, India, showed moderate resistance to rhizome rot (Ravindran et al., 2002).

Host range studies of the rhizome rot pathogen, *P. aphanidermatum*, conducted at IISR, Calicut, Kerala State, India, showed that among the five turmeric species, *C. longa*, *C. zedoaria*, *C. amada*, *C. aromatica*, and *C. caesia*, *C. amada* and *C. caesia* were found resistant to pathogenic infection (Anon, 2007–2008).

Leaf Blotch

Ascomycetous fungus causes leaf blotch, which is one of the major foliar diseases of turmeric. The disease affects the turmeric plant at various stages of its growth, and in susceptible varieties, it causes severe economic loss to the farmer. Certain varieties fall prey to the onslaught of the disease toward the end of their growing cycle, when leaf senescence commences.

Symptoms

The disease is characterized by the appearance of small, scattered, oily-looking translucent spots on the lower side of the leaves when the turmeric plant is in the 3–4 leaf stage. The leaf spots subsequently turn dirty yellow, then deepen in color to that of gold, and sometimes to bay shade. The adjacent individual leaf spots of 1–2 mm in diameter coalesce forming reddish brown blotches leading to varying degrees of leaf blight. The lower leaves are more prone to infection than the upper ones, which are relatively younger in age. In severe cases of infection, hundreds of spots appear and they coalesce on both sides of the leaf, which severely reduces the leaf's photosynthetic activity, as almost all green parts turn color from green to gold.

Crop Loss

There will be considerable reduction in rhizome yield due to foliar infection if it occurs especially in the early part of plant growth, as noted by Butler (1918). A crop loss assessment study conducted at Bidhan Chandra Krishi Vishwa Vidyalaya (BCKVV), in North Bengal, during 1996–1998, showed that highly susceptible genotypes exhibited not only a loss of biomass by as much as 21.6–32.5% but also a loss of fresh rhizome yield by as much as 24.5–32.0%. In certain genotypes, a higher rate of dry matter loss has been reported (Panja et al., 2000).

Epidemiology

The leaf blotch disease is caused by *Taphrina maculans*, an ascomycetous fungus. The fungus is reported to be active during moist cloudy weather which is very common during the SW monsoon in India, especially during the months of August and September. A temperature range of 25–30°C is reported to predispose the plant to infection (Upadhyay and Pavgi, 1967a). The primary source of inoculum is soil borne which survives in dried trash leaves of the host in the field and starts to affect first the lower leaves. Subsequent spread of the inoculum is through the air, which can be severe in intensity during October and November, when ambient temperature falls to 21–23°C, with relative humidity of 80% (Ahmad and Kulkarni, 1968). Secondary infection is by ascospores discharged from successively maturing asci, which grow into octosporus microcolonies and infect fresh leaves without any dormancy. The primary infections are less harmful than the secondary ones, inciting profuse spots covering the entire foliage (Upadhyay and Pavgi,1966). The disease perpetuates from one season to another through ascospores and blastospores ejected from mature asci during the crop season and oversummering in the soil and leaf trash (Upadhyay and Pavgi, 1967b). *C. amada* (Pavgi and Upadhyay, 1964), *C. angustifolia, Zingiber cassumnar, Z. zerumbet,* and *Hedychium* sp. (Butler, 1918) are reported to serve as alternate hosts of *T. maculans.*

The Management of the Disease

Disease-management strategies to control leaf blotch include besides others, crop rotation, phytosanitation, and, most important of all, use of chemical fungicides, in addition to the cultivation of resistant varieties. Because the infected plant material serves as the disease-perpetuating base and medium for the fungus, crop rotation is an important component of the integrated disease management practice. Of all the management practices, phytosanitation assumes the most crucial place in the disease-management structure. When infection is seen in earlier stages of plant growth, removal of the infected plant and burning it away from the field is the most important practice in containing the disease. Removal and burning should not be restricted only to the infected leaves, but to the entire plant so as to assure a foolproof disease spread control technique. Growing of alternate hosts, such as *C. amada, C. angustifolia, Zingiber cassumnar,* and *Hedychium* sp., near turmeric plots, must be avoided. Inasmuch as chemical control is concerned, foliar spray containing a mixture of various formulations of Mancozeb, copper, and Carbendazim are reported to be effective in controlling the disease (Srivastava and Gupta, 1977). Sprays of Metalaxyl at a concentration of 500 ppm or Thiophanate methyl at a concentration of 0.1% (both patented fungicides in India) at monthly intervals, from initiation of the symptoms is also recommended (Singh et al., 2000).

Disease Resistance

All short-duration cultivars are reported to be resistant to *Taphrina maculans.* The varieties, such as CLL 324, Amalapuram, Mydukur, Karhadi local, CLL 326, Ochira 24, and Alleppey, among *C. longa* group and Ca 88, Ca 67, Dahgi, and Kasturi, among *C. aromatica* types were reported to be resistant to the fungal infection (Nambiar et al., 1977). At Raigarh in Chattisgarh State, India, 34 cultivars of turmeric were evaluated for resistance against leaf blotch, which showed that Roma, Rashmi, Suguna, Suroma, Sudharshana, TCP 56, Pratibha, TCP 11, were resistant to the fungal infection (Singh, 2007). Panja et al. (2001) evaluated 15 genotypes of turmeric to identify the best suited for high-yielding and resistance characteristics to leaf blotch for the Tarai region of the fertile Indo-Gangetic Plains in West Bengal State, India, and found that PTS-62, ACC-360, Roma, BSR-1, and Kasturi, among the genotypes tested, as most resistant to fungal infection.

Leaf Spot

The fungus *Colletotrichum* sp. causes the leaf spot disease. In severity, it is next to leaf blotch, and the disease is prevalent in almost all of the turmeric-growing regions, and often it can occur in epidemic form. Several investigators have extensively studied the disease, in various species of turmeric, such as *C. capsici*, *C. curcumae* (Kandaswamy, 1958), and *C. gloeosporioides* (Patel et al., 2005).

Disease Symptoms

The disease appears in the form of elliptic or oblong spots of variable size. In the initial stages of infection, the spots are small and measure only few centimeters in length and are 2–4 cm in breadth, which later may increase in size. In advanced stages of infection, two or more leaf spots coalesce and develop into irregular patches which occupy a major portion of the leaf, which eventually dries up. The center of the spot is grayish white and thin with numerous black dots called acervuli, springing up on both upper and lower surfaces, and arranged in concentric rings. The grayish white portion is surrounded by a brown region with a yellow halo around it. The spots are markedly visible on the upper surface of new leaves. The central region of the spot may become papery and easily tear. The infection is usually confined to lamina and may occasionally extend to leaf sheaths.

The infection is also seen on turmeric flowers as spots. The pale brownish hyphae accumulate inside the epidermal cells and form the basis of stomatal development. The stomata are made up of light to dark brown pseudoparenchymatous cells. The outer wall of the epidermis is ruptured and the setae and conidiophores are exposed. The pathogen survives on infected leaf debris for at least 1 year, which forms the primary source for succeeding infection.

Crop Loss

The primary causative mechanism in which yield losses occur due to leaf spot infection is through the loss of photosynthetic green surface. These losses depend on the time of initiation of the disease. During the periods of heavy infection, yield loss could be as much as 50% (Ramakrishnan, 1954). Nair and Ramakrishnan (1973) observed yield loss up to 62.7% due to foliar infection by *Colletotrichum capsici*.

Epidemiology

The disease appears usually during the SW monsoon period, in the month of September, when there is high and continuous humidity in the atmosphere. The pathogen attacks mostly the leaves. When the infection is intense, most of the leaves dry up and the turmeric field can present a completely scorched appearance (Joshi and Sharma, 1980). The epidemiology has been systematically investigated by Ghogre and Kodmelwar (1986). Palarpawar and Ghurde (1992, 1995) investigated the role of the debris of infected rhizomes in the soil on the survival of *C. curcumae* in the Vidharbha region of Maharashtra State, India. They observed that the infected rhizomes were not the primary source of the inoculum, but the infected leaf debris, which remain buried up to a depth of 15 cm, forming a viable source of inoculum even up to 8 months. They also observed leaf spots on plants grown in infested soil containing bits of leaf petioles. The disease has been found to adversely affect rhizome yield, more when the crop is cultivated in the open, as compared to the one growing in full or partial shade (Singh and Edison, 2003).

The Management of Leaf Spot

Both cultural and chemical practices can be effectively employed to contain the disease. The primary source of inoculum is from the previous crop's residues, in the form of leaf debris and other crop residues. Hence, to effectively control the onset of the disease, good phytosanitation is a must (Ali et al., 2002). The importance of spraying Bordeaux (1% concentration) to contain the disease was stressed as early as 1951 by Govinda Rao (1951). Ramakrishnan (1954) and Subbaraja (1981) reported that application of chemical fertilizers (N at 120 kg/ha, K at 70–120 kg/ha) was effective in reducing disease incidence, obviously, due to the increased plant vigor obtained through the mineral nutrients from these chemical fertilizers. Chemical control of the disease also includes spraying of fungicides, such as Mancozeb and Difolatan (both patented fungicides in use in India). Palarpawar and Ghurde (1989) experimented with 14 fungicides against the onslaught of the fungus *C. curcumae* and found that Carbendazim, among the ones used, was most effective, followed by Thiophanate methyl, edifenphos, and dinocap, in that order. However, maximum rhizome yield was recorded after treatment with edifenphos, and this fungicide was recommended as the most effective against *C. curcumae* in turmeric. Thamburaj (1991) reported that two sprays of Carbendazim (0.2% concentration), at the time of the disease onset, twice more at 15 days interval offered the most effective control of leaf spot. Four sprays of Captan (0.2% concentration) or Mancozeb at monthly intervals or six sprays of Mancozeb (0.25% concentration) at 15 days interval throughout the crop cycle was also found effective in controlling the disease (Dakshinamurthy et al., 1966). Gorawar et al. (2006) evaluated nine plant extracts for their effectiveness against *C. capsici* and found Duranta and Parthenium leaf extracts at 10% concentration quite effective against the fungus, but *in vitro* culture. Because leaf spot disease is a major threat to turmeric production in almost all of the turmeric-growing regions, the possibilities for exploring the use of these plant extracts as foliar sprays to contain the disease merits further examination. It is interesting that Parthenium, which was imported into India, along with the shipment of PL 480 (Public Loan) wheat imports from the United States, turned out to be a perennial uncontrollable weed in the wheat crop and grows extensively in turmeric fields, and its antifungal nature is worth further scientific evaluation. There is growing environmental consciousness in India, against the backdrop of the so-called "green revolution," where indiscriminate use of fertilizers and pesticides ruined the soil resources and polluted the ground water and has been noted as the prime cause for the spread of cancer in Gurdaspur district of Punjab State, India, the "cradle" of the green revolution. It is in this backdrop that more and more of environmentally friendly products deserve to be used in farmers' fields. However, the spread of these practices is very slow, as there is the entrenched chemical lobby peddling fertilizers and chemicals, which more often than not mounts strong opposition to the use of the "green products." But, it is also a healthy development among some chemical companies involved in the manufacture of insecticides and fungicides to wake up to this ground reality, and they are coming out with newer "green products." Rallis India Limited, an arm of the multi-billion Tata group, is one such company.

Disease Resistance

The reaction of turmeric cultivars, both short- and long-term, to the leaf spot disease has been investigated by some authors (Reddy et al., 1963; Sarma and Dakshinamurthy, 1962). The varieties, namely Nallakatla, Sugandham, Duvvur, and Gandikota, were found resistant to leaf spot. Subbaraja (1981) screened 150 turmeric types for their resistance to leaf spot and found Sugandham as the best. Roma, Rashmi, Suguna, Suroma, Sudharshana, Pratibha, TCP 11, and TCP 56 were also found resistant to the disease.

Minor Diseases of Turmeric

Turmeric is also susceptible to a number of minor diseases, which are classified as leaf spots, leaf blights, blasts, and root rot. The discussion on these are as follows.

Leaf Blast

The disease is caused by *Pyricularia curcumae* and *Alternaria alternata* (Rathaiah, 1980), especially on older leaves, leaf sheaths, and flower petals. A yellow halo develops around the spots showing infection, and this is visible when the infected leaf is held against sunlight. When the infection is severe, the leaves turn yellow and then dry up.

Leaf Spots

Several fungi, namely *Phaeodactylium alpiniae*, *Thirumalacharia curcumae*, *Phyllosticta zingiberii*, *Phaeorobillarda curcumae*, *Cercospora curcumae-longae* sp. nov., *Myrothecium roridum*, *Pyricularia curcumae*, and *Pestalotiopsis* sp. were found to cause the leaf spot symptoms in turmeric (Rathaiah, 1980). The disease symptoms include water-soaked spots of varying shapes and sizes, which can be seen on both the surfaces of the leaf, which gradually increase in size. The spots caused by *Thirumalacharia curcumae* cause oval to oblong, 4 cm–9 cm × 3 cm–5 cm gray brown leaf spots on *Curcuma* sp.

Leaf Blight

Corticum sasaki causes leaf blight and the disease is characterized by water-soaked spots of varying shapes and sizes, gradually increasing in size, which appear on lower leaves (Saikia and Roy, 1975). The blighted leaf area is divided or banded into various sectors, which are characteristic symptoms for early diagnosis.

Damage Caused by Nematodes

Meloidogyne incognita, *Radopholus similis*, and *Pratylenchus* sp. are the three principal nematode species causing great damage to the turmeric crop. Plants infested by *M. incognita* show stunted growth, yellowing, marginal and tip drying of leaves resulting in reduced tillering, coupled with galling and rotting of roots. Roots of turmeric infected by *R. similis* become rotten and most of these decayed roots retain only the epidermis devoid of cortex and stele portions. The infested plants show a tendency to age quickly and dry faster than healthy ones. Root rot due to *F. solani* has also been reported.

Brown Rot

Fusarium sp. and *Pratylenchus* sp. are reported to cause brown rot of rhizomes. The rhizome becomes deep gray to dark brown, less turgid, wrinkled, and exhibit dry rot symptoms. On cutting open the affected rhizome, dark brown necrotic lesions starting from the margin into the internal tissues are seen.

Bacterial Diseases of Turmeric

Turmeric is comparatively less susceptible to bacterial infection. This may be due to its antimicrobial properties, which also has come into much use as an antiseptic. However, the bacterial ginger strain *Ralstonia solanacearum* was found to infect turmeric on artificial inoculation. Similarly, occurrence of a new pathogen *Pseudomonas cichorii* was reported on turmeric plants, grown for bulb multiplication, in experimental fields in Sao Paulo, Brazil (Maringoni et al., 2003). The strain was identified using Micro Log2 System, which proved its pathogenicity when infected after inoculation, on leaves.

The occurrence of these diseases is sporadic in nature and noticed only in few turmeric-growing regions and do not warrant much concern, as they are of no economic consequence to the farmer.

Storage Diseases of Turmeric

Because turmeric is propagated vegetatively, the rhizomes have to be stored at least 4–5 months before they are ready for planting in the field. When stored in summer, preceding planting, the rhizomes have to be protected against dessication because of the summer heat, and they are to be protected from other infestations, from pests and from diseases. In Andhra Pradesh, India, where turmeric is grown extensively, large-scale storage of rhizomes is done in the open under trees and covering them with dried leaves. In Odisha State, India, seed rhizomes are stored under thatched sheds made of coconut fronds, as large heaps, covered with dried leaves or straw and plastered with a protective layer of mud. At times, unusual rains cause seepage of moisture into the heaps and produce dampness, which is conducive for the onslaught of disease-causing fungi. This leads to the start of diseases and insect pests also attack the rhizomes. Improper storage and heaping harvested rhizomes under the sun leads to rotting of the rhizomes, besides loss of their moisture content, which will subsequently render them unviable in the field, as they will not germinate. Rotting is also reported in processed and stored samples that were not properly dried.

Crop Loss During Storage

The losses occurring due to improper storage could be as high as 60% under adverse conditions, and it appears to be a very serious problem in turmeric production. This is particularly so in some states like Andhra Pradesh, India (Sharma et al., 1987). Rhizome rot in storage was found to be maximum in the month of September, and minimum in May in the Delhi, India, turmeric market (Kumar and Roy, 1990). It is the seed rhizome that is most adversely affected by storage rot, as was observed in Barua Sagar, Jhansi, Uttar Pradesh, India.

Many fungal species are found associated with rotting rhizomes. Sharma and Roy (1984) investigated the conditions for storage and found that favorable incubation temperature and relative humidity around 60% leads to maximum spoilage. But no rotting occurs at 15°C even when the relative humidity varied from 30% to 90%. Maximum rotting due to the infection by *Aspergillus* was observed at 35°C and relative humidity of 60–90%. The most virulent and widespread among the fungi attacking stored rhizomes and causing storage rot of *C. longa* are *Macrophomina phaseolina*, *Cladosporium cladosporioides*, *Drechsleza rostrata*, and other species of *Aspergillus*, such as *A. flavus*, *A. niger*, *C. cladosporioides*, *D. rostrata*, *F. moniliforme*, *F. oxysporum*, *P. aphanidermatum*, *R. solani*, and *Sclerotium rolfsii* (Kumar and Roy, 1990; Rao, 1995).

Management of Storage Losses

To obtain healthy seed material, the seed rhizome has to be properly stored under optimum conditions of air temperature and relative humidity in the storage space to preempt any fungal or bacterial infection, including insect pest attack. To ensure seed safety, the normal course of action is to treat seed rhizome with appropriate fungicides. Following fungicidal treatment, seed rhizomes should be dried thoroughly under shade before storage. Sharma et al. (1987) reported that hot water (50°C) for half an hour can be resorted to eradicate any fungal infection, without adversely affecting germination. Currently, the prevalent practice is to dip seed rhizomes in a mixture of Mancozeb (0.2% concentration) and quinalphos (0.075% concentration) solution for half an hour and to dry under shade before storage to eradicate storage infection due to fungi and also contain any insect pest attack in storage. Storage rot due to infection by *A. flavus* was effectively controlled by pre- and postinoculation treatment with Carbendazim and Benomyl at a concentration of 10 ppm (Sharma and Roy, 1984).

Table 10.2 Disease–Varietal Responses in Turmeric

Variety	Rhizome Rot	Leaf Blotch	Leaf Spot
Kesari Type	NA	Susceptible	NA
Duggirala	Susceptible	NA	Susceptible
Tekurpet	Susceptible	NA	NA
Mydkur	Highly susceptible	NA	Highly susceptible
Armoor	Susceptible	NA	Susceptible
Sugandham	Susceptible	Susceptible	NA
CI-316	NA	Tolerant	NA
CII-320	NA	Tolerant	NA
CII-324	NA	Susceptible	Tolerant
CII-325	Susceptible	Tolerant	Susceptible
CII-326	Susceptible	Tolerant	Susceptible
CII-327	NA	Tolerant	Tolerant
Shillong	Tolerant	Tolerant	Resistant
Tallkashi	Tolerant	Susceptible	Tolerant
Rajapuri	Susceptible	Susceptible	Moderate resistance
Suvarna (PCT 8)	Tolerant	Tolerant	Tolerant
Suguna (PCT 13)	Tolerant	Tolerant	Tolerant
Sudharshana (PCT 14)	Tolerant	Tolerant	Tolerant
Suvarna	Tolerant	Tolerant	Tolerant
Manuthy local	NA	Tolerant	Tolerant
Tekurpet	NA	Susceptible	Tolerant
Roma (PTS 10)	NA	Tolerant	NA
C. aromatica (Ca 69)	Resistant	NA	Resistant
Kodur	NA	NA	Tolerant
Vontimitta	NA	NA	Susceptible
GL Puram 2	NA	Susceptible	NA
GL Puram	NA	Susceptible	NA
Amalapuram	NA	Tolerant	NA
Ethamukkali	NA	Tolerant	NA
Thanuku	NA	Tolerant	NA
Nallakatla	NA	NA	Resistant
Duvvur	NA	NA	Resistant
Gandikota	NA	NA	Resistant
Clone DKHT-6	NA	Resistant	Resistant
Roma	NA	Resistant	Resistant
Rashmi	NA	Resistant	Resistant
Suguna	NA	Resistant	Resistant
Suroma	NA	Resistant	Resistant
Sudharshana	NA	Resistant	Resistant
TCP 56	NA	Resistant	Resistant
Pratibha	NA	Resistant	Resistant
TCP 11	NA	Resistant	Resistant
Nagaland local	NA	Highly susceptible	NA
Tall clone Assam	NA	NA	NA

(Continued)

Table 10.2 (Continued)

Variety	Rhizome Rot	Leaf Blotch	Leaf Spot
Sonajuli local	NA	Highly susceptible	NA
Sugandham	NA	Highly susceptible	NA
Meghalaya local	NA	Highly susceptible	NA
RH-5	NA	Medium susceptible	NA
Rajendra Sonia	NA	Medium susceptible	NA
PTS-62	NA	Highly resistant	NA
ACC-360	NA	Highly resistant	NA
ACC 361	NA	Highly resistant	NA
BSR-1	NA	Highly resistant	NA
Kasturi	NA	Highly resistant	NA

NA: Information not available.

Host Resistance

In the integrated disease management program, concentration on development of host resistance, apart from other management practices directed at containing diseases, forms an important component of the overall disease-management approach. In this context, a large number of turmeric varieties have been developed, by both individual researchers and also collectively at research institutions spread across the length and breadth of India, to identify resistant hosts to specific diseases, under widely varying agroclimatic conditions prevalent in the country, and integrate them in the overall breeding program. To begin with, one has to have a clear understanding of the different turmeric varietal responses to an array of principal turmeric diseases. A summary of this information is given in Table 10.2. The overall observations indicate that, in general, long-duration types are more susceptible to foliar diseases, such as leaf blotch and leaf spot, than short-duration types. Varying reactions are shown by varieties growing in different locations to the infection by the same pathogen. This is but expected, as the emergence of a disease symptom in a plant type is an intimate resultant interaction between the specific variety, with its inherent traits, and the environment that it grows in. It must also be understood that the screening methodology put in place in different locations, which may not be uniform, can also induce a degree of variation, that is more due to the experimental technique in question than that resulting from an interaction between the type in question and the environment per se.

References

Ahmad, L., Kulkarni, N.B., 1968. Studies on *Taphrina maculans* butler inciting leaf spot of turmeric (*Curcuma longa* L.). Isolation of the pathogen. Mycopathol. Mycol. Appl. 34, 40–46.

Ajiri, D.A., Ghorpade, S.A., Jhadav, S.S., 1982. Research on Rhizome Fly *Mimigrella coerulifrons* on Turmeric and Ginger in Maharashtra. Indian Council of Agriculture Research, New Delhi, (India Report).

Ali, M.A., Fakir, G.A., Sarkar, A.K., 2002. Research on seed borne fungal diseases of spices in Bangladesh Agricultural University. Bangladesh J. Train. dev. 15, 245–250.

Anandam, R.J., Rao, A.S., Babu, K.V., 1996. Studies on rhizome rot of turmeric (Curcuma longa L.). Indian Cocoa Arecanut and Spices J. 20 (1), 17–20.

Anon, 1974. Annual Report of Central Plantation Crops Research Institute, Kasaragod, Kerala State, India.

Anon, 1996–1997. Annual Report of All India Coordinated Project on Spices, Indian Council of Agricultural Research, New Delhi, India, p. 46.

Anon, 2006–2007. Annual Report of Indian Institute of Spices Research, Calicut, Kerala State, India, pp. 57–58.

Anon, 2007. Annual Report, Indian Institute of Spices Research, Calicut, Kerala State, India, p. 71.

Butler, E.J., 1918. Fungi and Diseases in Plants. Thacker Spink & Co., Calcutta (now Kolkata).

Dakshinamurthy, V., Reddy, G.S., Rao, D.K., Rao, P.G., 1966. Fungicidal control of turmeric leaf spot caused by Colletotrichum capsici. Andhra Agric. J. 13, 69–72.

Dohroo, P., 1988. Fusarium solani on curcuma longa. Indian Phytopath. 41, 504.

Erwin D.C., Ribeiro O.K. (Eds.), 1996. Phytopthora Diseases Worldwide. ISBN 0 89054 212 0 APS Press, St Paul MN, USA, p. 592.

Ghogre, S.B., Kodmelwar, R.V., 1986. Influence of temperature, humidity, pH and light on mycelial growth and sporulation of Colletotrichum curcumae (Syd.) and Bisby. PKV Res. J. 10, 77–79.

Gorawar, M.M., Hegde, Y.R., Kulkarni, S., 2006. Biology and management of leaf spot of turmeric caused by Colletotrichum capsici. J. Plant Dis. Sci. 1, 156–158.

Govinda Rao, P., 1951. Control of plant diseases in the northern region of the Madras (now Chennai) State in the year 1949–50. Plant Protect. Bull. 3, 21–23.

Haidar, M.G., Jha, R.N., Nath, R.P., 1995. Studies on the nematodes of spices Nemic association of turmeric in Bihar and reaction of certain lines to some of the dominating nematodes. Indian J. Nematol. 25, 212–213.

Ikisan. com., 2000. Disease Management, p. 5.

Joshi L.K., Sharma N.D. Diseases of ginger and turmeric. In: Nair, M.K., Premkumar T., Ravindran P.N., Sarma Y.R. (Eds.), Proceedings of the National Seminar on Ginger and Turmeric. 1980, Calicut April 8–9 CPCRI Kasaragod Kerala State,India, pp. 104–119.

Kandaswamy, T.K., 1958. Control of leaf spot disease of turmeric caused by Colletotrichum capsici. Madras Agric. J. 45, 55–60.

Koya, K.M.A., 1990. Role of rhizome maggot Mimegralla coerulifrons Macquart (Diptera:Micropezidae) associated with ginger. Entomon 15, 75–77.

Kumar, H., Roy, A.N., 1990. Occurrence of fungal rot of turmeric (Curcuma longa L.) rhizome in Delhi market. Indian J. Agric. Sci. 60, 189–191.

Mallikarjun, G., Kulkarni, S., 1998a. Physiological studies on Alternaria alternata (Fr.) Keissler – a causal agent of leaf blight of turmeric. Karnataka J. Agric. Sci. 11 (3), 684–686.

Mallikarjun, G., Kulkarni, S., 1998b. Survey for the incidence of turmeric leaf blight caused by Alternaria alternata (Fr.) Keissler in Karnataka. Karnataka J. Agric. Sci. 11 (4), 1096–1097.

Mani, A., Prakash, K.S., 1992. Plant parasitic nematodes associated with turmeric in Andhra Pradesh. Curr. Nematol. 3, 103–104.

Maringoni, A.C., Theodoro, G.F., Ming, L.C., Cardoso, J.C., Kurozawa, C., 2003. First report of Pseudomonas cichorii on turmeric (Curcuma longa) in Brazil. Plant Pathol. 52 (6), p. 794.

Middleton, J.T., 1943. The taxonomy, host range and geographic distribution of the genus. Pythium Mem. Torrey Bot. Cl. 20, 1–140.

Nadakkal, A.M., Thomas, N., 1964. Studies on plant parasitic nematodes of Kerala. Curr. Sci. 33, 247–248.

Nair, M.C., Ramakrishnan, K., 1973. Effect of *Colletotrichum* leaf spot disease of turmeric (*Curcuma longa* L.) on the yield and quality of rhizomes. Curr. Sci. 42, 549–550.

Nambiar, K.K.N., Sarma, Y.R., Brahma, R.N., 1977. Field reaction of turmeric types to leaf blotch disease. J. Plantation Crops 5, 124–125.

Nirula, K.K., Kumar, R., 1963. Collateral host plants root-knot nematodes. Curr. Sci. 32, 221–222.

Palarpawar, M.Y., Ghurde, V.R., 1989. Perpetuation and host range of *Colletotrichum curcumae* causing leaf spot disease of turmeric. Indian Phytopathol. 42, 576–578.

Palarpawar, M.Y., Ghurde, V.R., 1992. Survival of *Colletotrichum curcumae* causing leaf spot of turmeric. Indian Phytopathol. 45, 255–256.

Palarpawar, M.Y., Ghurde, V.R., 1995. Variability of *Colletotrichum curcumae*, the incitant of leaf spot of turmeric (*Curcuma longa* L.) in Vidharbha, Maharashtra State, India. J. Spices Aromat. Crops 4, 74–77.

Panja, B.N., De, D.K., Majumdar, D., 2000. Assessment of yield losses in turmeric genotypes due to leaf blotch diseases (C.O) *Taphrina maculans* (Bult.) from Tarai region of West Bengal. Plant Protect. Bull. 52 (3–4), 13–15.

Panja, B.N., De, D.K., Majumdar, D., 2001. Evaluation of turmeric (*Curcuma longa* L.) genotypes for yield and leaf blotch disease (C.O) *Taphrina maculans* Butler for tarai region of West Bengal. Environ. Ecol. 19 (1), 125–129.

Park, M.,1934. Report on the work of the mycology division. Adm. Rept. Dir. Agric. Ceylon (now Sri Lanka), p. 126.

Patel, R.V., Joshi, K.R., Solanky, K.U., Sabalpara, A.N., 2005. *Colletotrichum gloesporioides*, a new leaf spot pathogen of turmeric in Gujarat. Indian Phytopathol. 58, 125.

Pavgi, M.S., Upadhyay, R., 1964. Artificial culture and pathogenicity of *Taphrina maculans* Butler. Sci. Cult. 30, 558–559.

Premkumar, T., Sarma, Y.R., Goutam, S.S., 1982. Association of Dipeteran maggots in rhizome rot of ginger. In: Nair, M.K., Premkumar, T., Ravindran, P.N., Sarma, Y.R. (Eds.), Proceedings of National Seminar on Ginger and Turmeric Central Plantation Crops Research Institute, Kasaragod, Kerala State. 8–9 April 1980, Calicut, Kerala State, India.

Ramakrishnan, T.S., 1954. Leaf spot disease of turmeric caused by *Colletotrichum capsici*. Indian Phytopathol. 7, 111–117.

Ramakrishnan, T.S., Sowmini, C.K., 1954. Rhizome rot and root rot of turmeric caused by *Pythium graminicolum* Sub. Indian Phytopathol. 7, 152–159.

Rao, P.S., Rao, T.G.N., 1988. Diseases of turmeric in Andhra Pradesh. In: Proceedings of National Seminar on Chillies, Ginger and Turmeric, Hyderabad, India, pp. 162–267.

Rao, P.S., Reddy, M.L.N., Rao, T.G.N., Krishna, M.R., Rao, A.M., 1992. Reaction of turmeric cultivars to *Colletotrichum* leaf spot, *Taphrina* leaf blotch and rhizome rot. J. Plantation Crops 20 (20), 131–134.

Rao, P.S., Krishna, M.R., Srinivas, C., Meenakumari, K., Rao, A.M., 1994. Short duration, disease resistant turmeric for Northern Telengana. Indian Hortic. 39 (3), 55–56.

Rao, T.G.N., 1995. Diseases of turmeric (*Curcuma longa* L.) and their management. J. Spices Aromat. Crops 4, 49–56.

Rathaiah, Y., 1980. Leaf blast of turmeric. Plant Dis. 64 (1), 104–105.

Rathaiah, Y., 1982a. Rhizome rot of turmeric. Indian Phytopathol. 35, 415–417.

Rathaiah, Y., 1982b. Ridomil for control of rhizome rot of turmeric. Indian Phytopathol. 35, 297–299.

Ravikumar, P., 2002. Production technology for organic turmeric. Spice India 15 (2), 2–6.

Ravindran, P.N., Kallupurackal, J.A., Shiva, K.N., 2002. New promising spice varieties. Indian Spices 39 (1), 20.

Reddy, M.N., Charita Devi, M., Sreedevi, N.V., 2003. Biological control of rhizome rot of turmeric (*Curcuma longa* L.) caused by *Fusarium solani*. J. Biol. Contr. 17 (2), 193–195.

Reddy, G.S., Dakshinamurthy, V., Sarma, S.S., 1963. Notes on varietal resistance against leaf spot diseases in turmeric. Andhra Agric. J. 10, 146–148.

Saikia, U.N., Roy, A.K., 1975. Leaf blight of turmeric caused by *Corticum sasakii*. Indian Phytopathol. 28, 519–520.

Sankaraiah, V., Zaheeeruddin, S.M., Reddy, L.K., Vijaya, 1991. Rhizome rot complex on turmeric crop in Nizamabad District in Andhra Pradesh. Indian Cocoa Arecanut Spices J. 14, 104–106.

Sarma, Y.R., Anandaraj, M., Venugopal, M.N., 1994. Diseases of spice crops. In: Chadha, K.L., Rethinam, P. (Eds.), Advances in Horticulture. Vol 10, Plantation and Spice Crops Part 2, Malhotra Publishing House, New Delhi India, pp. 1015–1057.

Sarma, S.S., Dakshinamurthy, D., 1962. Varietal resistance against leaf spot disease of turmeric. Andhra Agric. J. 9, 61–64.

Sharma, M.P., Roy, A.N., 1984. Storage rot in seed rhizomes of turmeric (*Curcuma longa* L.) and its control. Pesticides 18, 26–28.

Sharma, M.P., Roy, A.N., 1987. Prevention and control of rhizome rot of turmeric caused by *Aspergillus niger*. Arab. J. Plant Protect. 5, 35–38.

Sharma, M.P., Gupta, M., Roy, A.N., 1987. Physicochemical control of rhizome rot of turmeric. Pesticides April, 33–38.

Sharma, S.L., Dohroo, N.P., 1982. Efficacy of chemicals in controlling rhizome rot of ginger (*Zingiber officinale* Rosc.). In: Proceedings on National Seminar on Ginger and Turmeric, 120–122.

Singh, A.K., 2007. Screening of turmeric cultivars against *Taphrina* and *Colletotrichum* leaf spot disease. Indian J. Plant Prot. 35 (1), 142–143.

Singh, A.K., Edison, S., 2003. Ecofriendly management of leaf spot of turmeric under partial shade. Indian Phytopathol. 56 (4), 479–480.

Singh, A.K., Edison, S., Shashank, S., Singh, S., 2000. Evaluation of fungicides for the management of *Taphrina* leaf blotch of turmeric (*Curcuma longa* L.). J. Spices Aromat. Crops 9 (1), 69–71.

Srivastava, V.P., Gupta, J.H., 1977. Fungicidal control of turmeric leaf spot incited by *Taphrina maculans*. Indian J. Mycol. Plant Pathol. 1, 76–77.

Subbaraja, K.T., 1981. Studies on Turmeric Leaf Spot Caused by *Colletotrichum capsici* (Syd.) Butler and Bisby. Ph.D. Thesis, Tamil Nadu Agricultural University, Coimbatore, Tamil Nadu, India.

Thamburaj, S., 1991. Research in spice crops at TNGDNAU. Spice India 4, 17–18.

Upadhyay, R., Pavgi, M.S., 1966. Secondary host infection by *Taphrina maculans* Butler, the incitant of turmeric leaf spot diseases. Phytoath. Z. 56, 151–154.

Upadhyay, R., Pavgi, M.S., 1967a. Varietal resistance in turmeric to leaf spot diseases. Indian Phytopathol. 20, 29–31.

Upadhyay, R., Pavgi, M.S., 1967b. Some factors affecting incidence of leaf spot of turmeric by *Taphrina maculans* Butler. Ann. Phytopath. Soc. Japan 33, 176–180.

Ushamalini, C., Nakkeran, S., Marimuthu, T., 2008. Development of biomanure for the management of turmeric rhizome rot. Arch. Phytopathol. Plant Protect. 41 (5), 365–376.

11 Harvesting and Postharvest Management of Turmeric

Harvesting of Turmeric

It normally takes 7–9 months for turmeric to be ready for harvest, and this depends on the variety that is cultivated and the time the crop is sown in the field. On maturity, the leaves turn dry attaining a light brown and yellowish color. Normally, the harvest takes place between the months of January and March in India. The land is plowed using a wooden bullock-drawn plow (as tractor-mounted ones will damage the fresh rhizomes underground), and the uprooted rhizomes are hand-picked or the clumps are carefully lifted from the soil using a spade, without damaging the rhizomes. Thereafter, the harvested rhizomes are cleaned from the mud and any other extraneous matter sticking to them manually, using hands.

Tamil Nadu Agricultural University (TNAU), Coimbatore, India, has developed a "Power Tiller" which is a mechanical turmeric harvester with a capacity to harvest 0.6 ha of turmeric field crop per day. The mechanical unit consists of a blade with three bar points for easy penetration into the soil. From the gear box, the power is transmitted to the shaft of the turmeric digger unit through the V-belt transmission. The pneumatic wheels are replaced with a pair of special cage wheels to accommodate the height of the ridges on which the seed sets are planted. This causes only minimal damage to the rhizomes, which on average is about 0.5%, as compared to 4.2% damage in manual harvesting. The undug rhizomes left in the field are 0.8% as compared to 4.8% in manual harvesting. There is 60% saving in the cost incurred and 90% saving in time.

A tractor-drawn turmeric harvester mounted on a 35–45 horsepower (hp) tractor with a capacity to harvest turmeric from a field of 1.6 ha/day has been developed by the TNAU. The unit is 120 cm wide and consists of a blade with five bar points for easy penetration into the soil. To the rear end of the blade, seven lift rods of 250 mm length are provided. For digging, the bar points with the blade penetrate into the soil and lift the turmeric rhizomes from the soil. The soil slips back to the ground and the dug-out rhizomes are deposited at the center of the unit. The extent of damage caused to the rhizomes is much less (2.83%). There is 70% saving in cost and 90% saving in time, when compared to manual digging out of the rhizomes.

Washing the Rhizomes

The dug-out rhizomes will be covered by soil particles and soil microflora, which stick to the surface of the rhizomes or enter into the small pores of the raw rhizome. Generally, washing is done manually in the field, immediately after harvest.

The Agronomy and Economy of Turmeric and Ginger. DOI: http://dx.doi.org/10.1016/B978-0-12-394801-4.00011-9

According to FAO guidelines, adequacy of any washing system of horticultural produce can be gauged by the total surface microbial load of the produce, before and after washing. Ideally, there should be a sixfold reduction in microbial load after washing, which is equivalent to 80% microbial washing efficiency (Arora et al., 2007). Orissa Agricultural University (OUAT) in Bhubaneswar, Odisha State, India, has developed a hand-operated washer, which consists of a perforated drum (60 cm × 40 cm) which rests on a stand (Pal et al., 2008). Below the drum there is a water tank (73 cm × 20 cm × 23 cm) which permits the bottom of the drum to remain submerged inside the water in the tank. Turmeric is loaded in the perforated drum through the door provided on the surface of the drum. When the drum is rotated, the adhering earth on the surface is washed away and collected in the water tank. The capacity of the washer is 200 kg/ha. Test results showed that there was 12% loss in rhizome weight due to washing and the total water consumed was 150 l/ha.

A power-operated turmeric washer has been developed by OUAT, which consists of a vertical container 25 cm high, having a rotating base of 77 cm in diameter. Water is sprayed through a perforated pipe fitted inside the container. The chamber is rotated by a motor of 1 hp at a speed of 240 rpm. Unwashed turmeric is loaded at the top opening and the washed turmeric is discharged at the side opening of the cylindrical casing. The capacity of the washer is 300 kg/ha with 16% weight loss during washing at the base speed of 120 rpm. Water consumption for the washing process is 150 l/ha.

Punjab Agricultural University (PAU) in Ludhiana, Punjab State, India, has developed a power-operated rotary drum mechanical washer-cum-polisher used for turmeric polishing (Arora et al., 2007). The unit consists of a stainless steel drum 62 cm in diameter and 61 cm long with 6 mm wide holes on the surface and operated by an electric motor of 1 hp. The parameters optimized for the turmeric washing machine were 2.2 min washing time at a rotational speed of 47.2 rpm, and the corresponding microbial washing efficiency was found to be 87.2% without any mechanical injury to the rhizomes. Microbial washing efficiency and "bruising percentage" (mechanical injury to the rhizome) are the main criteria deciding washing efficiency of the washer, which doubtless will be affected by the capacity of the washer, the rotational speed, and the time consumed.

Processing of the Rhizomes

The harvested and cleaned rhizomes have to attain stability before they enter the market as a dependable and saleable commodity. This involves a thorough post-harvest treatment including boiling, drying, polishing, and coloring. Curing of the rhizomes is begun 3–4 days after harvest. The rhizome fingers and bulbs or mother rhizomes are separated and cured separately, and not whole, as the latter will take more time to cook (Sasikumar, 2003). The weight of the resultant product turns out to be manyfold more than what the raw material was before it entered the curing process. Curing percentages of different turmeric varieties vary widely, ranging from 14% to 26% (Purseglove et al., 1981). The curing quality of turmeric and the proportion of cured and dried produce to the green produce vary with the variety in question. Mother rhizomes normally record a higher curing percentage than the fingers (Pruthi, 1998). Higher initial moisture content in the raw rhizomes has been found to be associated with lower curing percentage, resulting in poor quality end product. The variation in curing percentage among different varieties can be attributed to the genetic traits of the variety in question (Jalgaonkar and Rajput, 1991).

Boiling

Among the different postharvest operations, the first one is boiling. This is done in the field itself, and essentially involves cooking fresh rhizomes in water until they turn soft for drying. Boiling destroys the vitality of the rhizomes, avoids raw odor, reduces drying time, and yields a uniformly colored end product. In the traditional method, a vessel made of galvanized iron sheet is used for turmeric boiling. In certain parts of India, e.g., Tamil Nadu in southern India and Rajasthan in northern India, cow dung slurry is used in open-mouthed vessels for curing. This leads to a lower price of the end product. It also results in loss of energy, loss of time, and higher labor cost (Kachru and Srivastava, 1991). However, this practice has now been discontinued. Cooking turmeric in salt solution results in lower curcumin levels, compared to cooking in plain water (Sampathu et al., 1988). A systematic investigation at the Central Food Technology Research Institute (CFTRI) in Mysore, India, showed that the final end product obtained through cooking in water, alkaline water (0.1% sodium bicarbonate solution), and traditional processing resulted in identical appearance and color of the end product. There was no significant difference in drying time, yield, moisture content, coloring matter, volatile oil, and oleoresin contents in the finger rhizomes as a consequence of the various treatments (Sampathu et al., 1988). Boiling of turmeric is carried out until froth forms at the surface of the material in the boiling pan and white fumes emanate from the pan in which the rhizomes are boiled, with a typical odor emanating from the pan. When a brown stick is pressed into the rhizomes with a slight pressure, it can be decided whether boiling is complete because that gives a special feel. The other indication which shows that boiling is complete is when some rhizomes are removed and pressed between the fingers, with the rhizomes breaking and having a yellow interior, instead of an earlier red one. Sometimes a few turmeric leaves are also put in the cooking pan (Kachru and Srivastava, 1991). If the boiling and cooking process is not done thoroughly, the rhizomes will most likely be attacked by insect pests. The length of cooking can vary from half an hour to 6 h, depending on the location where the crop is cultivated (Kachru and Srivastava, 1991). The Spices Board of India, which deals with several aspects of production technology of different spices in the country, recommends boiling for 45–90 min, for the process to be effective. There are many different mechanical boilers available in the Indian market, which source different forms of energy, including simple firewood, steam, electricity, or biogas.

A mechanical boiler utilizing steam as energy has been devised by TNAU, Coimbatore, India (Viswanathan et al., 2002). The device consists of a trough, inner perforated drums, and a lid. The outer drum is made of 18 SWG (Indian specifications) thick mild steel to a size of 122 cm × 122 cm × 55 cm. A lid is provided with hooks to lift it easily, and it also has an inspection door. To facilitate draining of used water and cleaning, an outlet is placed at the bottom of the drum. Four inner drums of 48 cm × 48 cm × 45 cm size are provided within the outer drum. The inner drums are provided with a stand of 10 cm high for support. This prevents rhizomes from coming in contact with water filled to a depth of about 6–8 cm in the outer drum. The outer drum is placed with more than half of its depth below the ground level by digging a pit which serves as a furnace. This furnace is provided with openings, one for feeding of the fuel, and the other for removal of ash from burnt fuel (especially when wood/other wooden materials are used) and unburnt fuel.

Following the placement of the turmeric boiler on the furnace, about 75 l of water is added to reach a depth of 6–8 cm. Following this, about 50–75 kg of well-washed rhizome is taken in each inner drum and placed in the boiler and the lid is placed in position. Using the available agricultural waste materials, mostly dried turmeric leaves, the fire is lit in the furnace. During the boiling process, it takes about 25 min to produce enough steam to start the boiling process of the initial batch of rhizomes placed in the drum, and about 10–15 min for the subsequent batches. Through the inspection door, the stage of boiling of the rhizome is assessed by pressing the rhizomes with a hard pin or needle. Using a long pole, the lid is removed and the inner drums are lifted, one by one. For the next batch, about 20 l of water is added to the outer drum, depending on the water lost by evaporation. The next batch of rhizomes is loaded in all the drums and heating is resumed. At the end of the boiling process, all the drums ought to be cleaned free of sticking mud and soil to avoid damage, which ensures the lengthening of the life span of the gadget. The capacity of the boiler is about 200–300 kg/batch and 40 quintal (q) during a day's working of 8 h. Fuel requirement is about 70–75 kg of agricultural waste materials.

The heat utilization factor for curing turmeric rhizomes in open steam and pressure boiling at varying pressures (0.5, 0.75, and 1 kg/cm^2) for varying durations (5, 10, and 15 min) was compared (Athmaselvi and Varadharaju, 2003). The rhizomes boiled in open steam for 15 min showed the highest heat utilization factor (0.046) and the rhizomes boiled at 0.5 kg/cm^2 for 5 min showed the highest percentage of curcumin, oleoresin, and essential oil contents.

The optimum curing time for the Deshala variety of turmeric was found to be 40 min at 100°C when the amount of water used to cook was equal to that of the quantity of turmeric used. The final moisture content and hardness at the optimum curing conditions were 76.21% and 3.4 kg force, respectively (Sharma et al., 2008). The curcumin loss by heat treatment of rhizome by different methods, such as boiling for 10 min, boiling for 20 min, and pressure cooking for 10 min, showed that curcumin loss varied from 27% to 53%, with maximum loss in pressure cooking for 10 min. The results indicated diminished availability of spice active principles from cooked food, when food ingredients have been subjected to either boiling or pressure cooking for a few minutes (Suresh et al., 2007).

Viswanathan (2008) has reported on the use of a large-scale steamer to boil large quantities of turmeric. A typical large-scale steamer consists of a furnace, water tank, steam-production unit, barrel, stand, and steam line. The steam-production unit is fabricated using a mild steel sheet 4 mm thick to a size of 1500 mm × 1000 mm × 1500 mm. This unit is placed on a suitable furnace. Water is brought from a nearby source in the water tank, which is let into the steam-production unit by gravity through a pipeline before starting the steam production. The water in the steam-producing unit is heated by burning turmeric crop waste from after harvest, such as leaves and leaf sheath (Spices Board, Home Page Internet).

Turmeric for boiling is taken in two barrels, each of 550 mm in diameter and 880 mm high, made of mild steel sheet of 3 mm thickness. Steam from the steam-producing unit is conveyed through a pipeline of 55 mm diameter provided with a control valve. The barrels are mounted on a stand made of mild steel angle and channel sections. A vent pipe from the steam-production unit is provided to act as a pressure-release mechanism for safety, in case excessive pressure of steam develops. In the steam-production unit, it takes about half an hour to produce steam to a pressure of 2 kg/cm^2. Turmeric rhizomes are manually loaded into the barrels. Each barrel holds about 135 kg of fresh turmeric rhizomes. The time for boiling rhizomes is about half an hour. Completion of boiling is ensured by the characteristic turmeric flavor and also by the softness of the boiled rhizome using a sharp piece of wood. After ensuring the complete boiling of the rhizomes, the shutter provided at the bottom of the barrel is opened for unloading. The boiled rhizomes are then collected in the trolley provided with rubber wheels. The fuel required for steam production is about 18–20 kg of crop residues (dried leaves, leaf sheaths, or other crop wastes) per batch. After discharging the boiled rhizome from the first barrel, steam to the next barrel is opened. It is observed that about 10–12 l of water evaporates for the steam to form during each boiling process (per batch).

Drying

The cooked fingers are dried in the sun by spreading 5–7 cm thick layers on bamboo mat or on the drying floor. A thinner layer is not desirable, as the color of the dried product may be adversely affected. During the night, the material should be heaped and properly covered with a polythene sheet so that rains do not soak and spoil the produce or other field animals like rodents attack the produce. It takes 10–15 days for the rhizomes to become completely dry. The spheres and fingers are dried separately, where the former takes more time to dry (Spices Board, Home Page on Internet). Turmeric should be dried on clean surfaces to ensure that the product does not get contaminated by extraneous matter such as soil, other plant litter, and insect droppings. Care should be taken to prevent the onset and growth of mold on the produce, as this can happen where dampness is present. The rhizomes should be turned over intermittently to ensure safety of drying. A satisfactorily dried product is obtained when drying is done by hot cross-circulating air (60°C) (Prasad, 1980). Solar driers can also be used for drying turmeric (Balakrishnan, 2007). Yield of the dry product varies from 20% to 30% depending on the variety dried and the agroclimate zone where the crop is grown. The starch gelatinized during boiling shrink and during the drying process causes intercellular spaces to expand. This enhances water diffusion from the produce and quickens drying, reducing drying time. In a comparative investigation on drying, it was observed that the drying time of cured produce was 18 h, whereas for noncured produce it was 108 h at 60°C drying (Praditdoung et al., 1996).

A systematic investigation on the effect of solar drying, as compared to sun drying of turmeric variety Erode, showed that the key biochemical constituents indicated that boiling and drying enhanced the curcumin content of cured turmeric (Gunasekar et al., 2006). However, volatile essential oil, oleoresin, and total protein contents progressively diminished as the moisture content of the rhizomes decreased. The results also revealed that solar drying was more effective than sun drying, as it achieved the desired moisture and quality of the produce in a minimum of 64 h compared to sun drying in 96 h, thus saving considerable time (32 h) in drying time.

Prasad et al. (2006) used a direct natural convection solar-cum-biomass drier to dry boiled (cooked for 2 min in boiling water) and unboiled sliced turmeric rhizomes 50–70 mm long and 8 mm thick. The rhizomes dried in the solar biomass drier by two different treatments were similar in quality with respect to physical appearance, such as color and texture, but there was significant difference in volatile oil content. The volatile oil content in the treatment with sliced rhizomes was 2.89%, while in the unsliced rhizome treatment it was 3.35%, and in open sun drying treatment it was just 1.75%. The quantitative analysis showed that the traditional open sun drying had taken 11 days for the rhizomes to dry, while it took just a day and a half to dry in the solar biomass drier, and the end product was of better quality. The efficiency of the entire unit was 28.57%.

Drying in hot air is an alternative to the traditional solar-drying process. Investigations on the drying kinetics at different air flow rates (0.2, 0.5, 0.7, 1.2, 2.1, 2.6, 3, and 4 m/s) on the blanched and unblanched rhizomes at different temperatures (60°C, 70°C, 80°C, 90°C, and 100°C) revealed that the diffusion model provided valuable information on the phenomenon of water removal (Blasco et al., 2006). Blanching previous to drying increased the drying rate at all temperatures tested, but the effect was reduced when the air-drying temperature was increased. The effect of air flow rate on external resistance was observed, and the air velocity transition zone between the external and internal resistance control zone was identified as 1–2 m/s.

Model drying kinetics of rhizomes of C. longa in monolayer at different temperatures (60°C, 70°C, 80°C, 90°C, and 100°C) for cured and uncured rhizomes showed that mathematical models based on Fick's first law, $F = -D(dC/dx)$, where F = the flux, dC/dx = concentration gradient across a particular section, and D = the diffusion coefficient could be used to describe water removal, though in this equation, concentration of specific ions was considered (Nair, 1996). Effective moisture diffusivities identified from modeling presented an Arrhenius-type relationship, and the mass transfer coefficient (k) was identified to be 9.7×10^{-5} kg water/m^2/s.

Polishing the Rhizomes

A rough, dull outer surface with root bits and scales characterizes the dried rhizomes, presenting a poor appearance. This appearance can be improved by smoothing and polishing. This can be done by manual or mechanical rubbing. Polishing is done until the recommended polish of 7–8% is achieved (Shankaracharya and Natarajan, 1999). Usually 5–8% of the weight of turmeric is "polishing wastage" during the full operation of polishing, but this is reduced to 2–8% in half polishing (Purseglove et al., 1981). Polishing dried turmeric also helps in removing wrinkles (Kachru and Srivastava, 1991). Manual polishing consists of rubbing the dried turmeric fingers on a hard surface, or trampling them under feet wrapped in a gunny sack. Manual polishing provides a rough appearance and dull color to the dried rhizomes. Often, unscrupulous farmers or middlemen add undesirable and sometimes dangerous chemicals to the dried rhizomes which might impart a glowing yellow color, but lead to toxicity when consumed. This cannot be easily detected, and Indian markets are notorious for such malpractices. Such practices are also seen in chile (red paprika) polishing where the chemical used is the dangerous mercuric oxide. All of this causes adulteration of food stuffs. In the improved polishing method, the use of a hand-operated barrel or drum mounted on a central axis, the sides of which are made of expanded metal screen, is used. When the drum filled with turmeric rhizomes is rotated, polishing is done by abrasion of the surface against each other as they roll inside the drum. The turmeric rhizomes are also polished in power-operated drums (Pruthi, 1998).

Multipurpose rotary washing-cum-polishing machines without the use of water and with certain modifications could be used for polishing turmeric rhizomes on the farm itself (Arora et al., 2007). The central shaft of the washer is removed and three detachable perforated screens of galvanized iron sheets are attached along the inner periphery of the drum with an abrasive surface on the inner side to increase friction. The optimum parameters for polishing a batch of 50 kg capacity at 7–8% polish were found to be 40 rpm for half an hour. The color of the polished rhizomes changed from dark brown to olive yellow, with their surfaces becoming smoother.

Agricultural engineers at the Andhra Pradesh Agricultural University (APAU), Rajendranagar, India, developed a mechanical polisher for turmeric (Sukumaran and Satyanarayana, 1999). The polisher consists of a mild steel drum, 88 cm in diameter, with meshes wrapped one above the other, mounted on ball bearings at two ends of a rectangular stand. The polisher is operated by a 2 hp electric motor. The speed of the drum rotation is maintained at 30–32 rpm and the polisher has a capacity to polish 600–700 kg/ha of rhizomes. Pal et al. (2008) at OUAT, Bhubaneswar, India, developed a pedal-operated hexagonal drum having six polishing plates of 30 cm × 60 cm size. The drum is made of inner expanded wire mesh (2.5 cm × 2.5 cm) and outer oven wire mesh (0.5 cm × 0.5 cm) resting on ball bearings at the two ends of the stand. The polishing drum is rotated by pedal through a chain-and-sprocket arrangement. Due to rotation of the drum, turmeric is rubbed against the expanded wire mesh surface and polishing is achieved. The outer skin rubbed off by polishing fell through the perforations of the drum. The capacity of the polisher was 100 kg/ha and 6% polishing was achieved. Large-scale polishing units with the capacity to polish 500–1000 kg/batch are used for polishing turmeric rhizomes on the farm. It takes about 4560 min/batch, and about 4% is wasted as dust. The yield after polishing was found to be about 15–25% on the basis of fresh weight of the rhizomes (Viswanathan, 2008).

Coloring the Rhizomes

In India, some turmeric exporters take recourse to the practice of coloring the end product, while exporting to make it look attractive. Yellow color is used for this purpose. Coloring is done by two methods to the half-polished rhizomes. One is the dry-coloring method, and the other is the wet-coloring method. In the former, turmeric powder is added to the polishing drum during the last 10 min of the polishing process (Balakrishnan, 2007). In the wet-coloring process, turmeric powder suspension in water is sprayed on half-polished and cured rhizomes. The product is mixed thoroughly and left to dry for about a week. Subsequently these rhizomes are bagged for export. Wet coloring imparts a better finish than the dry-coloring process (Viswanathan, 2008).

In some parts of India, a special treatment is imparted to enhance the appearance of the product by soaking it in an aqueous extract of tamarind (a tropical tree whose dried fruit provides a spongy pulp which is widely used in Indian cuisine) for 10 min to which a paste of turmeric is added. To give a bright color, the boiled, dried, and half-polished fingers are put in baskets which are shaken continuously and the emulsion is poured in. When the fingers are uniformly coated with this emulsion, they are dried in the sun. The composition of the emulsion required for coating 100 kg of half-boiled turmeric is: alum 0.04 kg, turmeric powder 2 kg, castor seed oil 0.14 kg, sodium bisulfite 30 g, and concentrated HCl 30 ml. This is a harmless process of coloring, dispensing with chemicals which are injurious to health (Viswanathan, 2008).

Grading of Turmeric

Although Indian turmeric is considered to be the best in the world, about 90% of the total produce is consumed internally. And this leads to the next fact that only a small percentage is exported because of huge internal demand (Balakrishnan, 2007). Some of the popular Indian varieties are: Alleppey, Erode, Duggirala, Rajapuri, Nizamabad, and Cuddapah (Sasikumar, 2003). India supplies most of the world's requirements of turmeric. Research centers devoted to the development of the crop, scattered in India, have developed more than 20 improved varieties, which suit the widely varying agroclimatic conditions of the country and its industrial needs (Parthasarathy et al., 2003). Turmeric of commercial importance is described in three ways (Balakrishnan, 2007). The description of these are as follows:

1. *Fingers*: These are the lateral branches or secondary "daughter" rhizomes, which are detached from the central rhizome before curing. Fingers usually range in size from 2.5 to 7.5 cm in length and may be over 1 cm in thickness. Broken and very small fingers are combined and marketed separately from whole fingers.
2. *Bulbs*: These are central "mother" rhizomes, which are ovate in shape and are shorter in length, which are larger in thickness than the finger rhizomes.
3. *Splits*: These are the bulbs that have been split into halves or quarters to facilitate curing and subsequent drying.

To ensure quality of all the spices exported from the country, the Government of India (GOI) introduced the scheme of compulsory quality control and preshipment inspection of the produce in 1963. Spices are graded, based on the standards stipulated for this purpose. These grades are popularly known as the "Agmark Grades." This is an internal legal arrangement of the GOI, unlike the "Codex Alimentaris," which is an international one, again referring to quality specifications of all products of plant or animal origin. The "Agmark Grading" takes into consideration the type of the rhizome (finger, bulb, or splits, as the case may be), its color, shape, extraneous matter, and presence of light pieces. The various grades of Indian turmeric under the Agmark system and their specifications (Spices Board, 2001) are given in Tables 11.1 and 11.2.

Storage of Turmeric Rhizomes

In view of the wide gap between the time of harvest and consumption, there is an important need for turmeric to be properly stored. In high-lying areas of India, cured turmeric rhizomes are usually stored in pits $5\,m \times 3\,m \times 2\,m$. Following the digging of a pit, its sides are left to dry for 2 days, the bottom and sides of the pit are thickly lined with grass, and mats made of locally available materials are spread over it. Pits are covered with earth. About 200 bags are stored in each pit (Kachru and Srivastava, 1991). Polyethylene bags are found to be the best packaging material, though tin containers can also be used to store powdered turmeric (Kachru and Srivastava, 1991).

Goyal and Korla (1993) investigated changes in the quality of cured and uncured turmeric rhizomes of four turmeric cultivars, namely, PCT-2, EM-321, ST-85, and ST-323 Y, during storage in unsealed polyethylene bags at room temperature (10–35°C) and relative humidity (23–95%), during one year. A gradual decrease in curcumin content was observed as the storage period increased up to 10 months. Thereafter, either there was no reduction in curcumin content or when there was a reduction, it was negligible. However, the essential oil and oleoresin contents decreased continuously throughout the storage period. The maximum losses in curcumin, essential oils, and oleoresin contents were 23.4%, 27.5%, and 24.2%, respectively, after one year in cured rhizomes of cultivar EM-321.

General Composition

The general composition of turmeric is given in Table 11.3. On average, turmeric contains 6.3% protein, 5.1% fat, 3.5% minerals, 69.4% carbohydrates, and 13.1% moisture. The rhizomes contain 2.5–6% curcuminoids, which impart the yellow color. The essential oil obtained by steam distillation from rhizomes is about 5.8%.

Quality Specifications of Turmeric

Quality specifications for turmeric can be classified under three categories: (i) cleanliness specifications, (ii) health requirements, and (iii) commercial requirements. The

Table 11.1 Agmark Grade Designations of Quality of Turmeric "Fingers" Produced in India

Grade Designation	Flexibility	Pieces[a], Percentage by Weight, Minimum	Foreign Matter, Percentage by Weight, Maximum	Chura and Defective Bulbs, Percentage by Weight, Maximum	Percentage of Bulbs by Weight, Maximum	General Characteristics
Special	Should be hard to touch and break with metallic twang	2	1.0	0.5	2.0	(A) Turmeric fingers shall be secondary rhizomes (B) They shall be well set and closely gained and be free from bulbs (primary rhizomes) and ill developed porous fingers
Good	Same as above	3	1.5	1.0	3.0	
Fair	Should be hard	5	2.0	1.5	5.0	
Nonspecified		–	–	–	–	This is not a grade in strict sense but provided for the produce not covered by other grades

Source: Spices Board, 2001.
For varieties other than Alleppey variety.
[a]Pieces are fingers, broken or whole, 15 mm or less in length.

Table 11.2 Agmark Grade Designations of the Quality of Turmeric Bulbs Produced in India

Grade Designation	Foreign Matter, Percentage by Weight	Chura and Defective Bulbs, Percentage by Weight, Maximum	General Characteristics
Special	1.0	1.0	(A) The turmeric bulbs shall be primary rhizomes of the plant *Curcuma longa* (B) They shall be well developed, smooth, round, soft, and free from rootlets (C) Be free from damage caused by weevils, moisture over boiling or fungus attack except that 0.1% and 0.2% by weight of rhizomes damaged by moisture and over boiling shall be allowed in grades, Good and Fair, respectively (D) have not been artificially colored with chemicals or dyes
Good	1.5	3.0	
Fair	2.0	5.0	
Nonspecified	–	–	This is not a grade in its strict sense but provided for the produce not covered by the other grades

Source: Spices Board (2001).
For varieties other than Alleppey variety.

quality specifications for turmeric are to ensure that turmeric as unprocessed agricultural commodity have been properly handled and stored properly, before being further processed into acceptable finished products for consumption, at consumer, food service, and industrial levels (Sivadasan and MadhusudanaKurup, 2002).

Cleanliness Specifications

A major concern for the importing countries is cleanliness. This, by and large, holds true for almost all imports from the developing world to the developed world, especially with regard to biological material, both of plant and of animal origin. Trade in spices is governed by numerous national as well as regional regulations. For instance, dried turmeric exported to the United States must confirm to the specifications laid out by the American Spice Trade Association (ASTA). The specifications of ASTA for cleanliness of turmeric imported into the United States are given in Table 11.4 (Sivadasan and MadhusudanaKurup, 2002). Similar specifications have

Table 11.3 Chemical Composition (in %) of Turmeric Samples Traded

Source	Moisture	Starch	Protein	Fiber	Ash	Fixed Oil	Volatile Oil	Alcohol Extractives
China	9.0	48.7	10.8	4.4	6.7	8.8	2.0	9.2
Pulenea	9.1	50.1	6.1	5.8	8.5	7.6	4.4	7.3
Alleppey	8.1	50.4	9.7	2.6	6.0	7.5	3.2	4.4
Indian	13.1	69.4	6.3	4.0	3.5	5.1	5.8	–
Alleppey (Fingers)	11.0	30.8	–	4.6	–	–	3.4	24.2
Alleppey (Bulbs)	12.0	26.3	–	3.7	–	–	3.4	16.2
Kadur	19.0	32.1	–	1.8	–	–	4.5	16.3
Duggirala	11.0	32.8	–	–	–	–	2.9	13.9

Source: Govindarajan (1980).

Table 11.4 ASTA Cleanliness Specifications for Turmeric

Parameter Considered	Maximum Permissible Limit
Whole insects, dead (by count)	3
Mammalian excreta (mg/lb)	5
Other excreta (mg/lb)	5
Mold (% by weight)	3
Defiled/Infested insect (% by weight)	2.5
Extraneous/Foreign matter (% by weight)	0.5

Source: Sivadasan and MadhusudanaKurup (2002).

Table 11.5 Microbiological Specification of Spices Under German Law

Parameter Considered	Standard Value (per gram)	Toxic Level (per gram)
Total aerobic bacteria	10^5	10^6
Escherichia coli	Absent	Absent
Bacillus cereus	10^4	10^5
Staphylococcus aureus	100	1000
Salmonella	Absent in 25 g	Absent in 25 g
Sulfite-reducing *Clostrides*	10^4	10^5

Source: Balakrishnan (2007).

also been laid out by other importing countries on the European continent such as Germany, France, the Netherlands, and the United Kingdom, as well.

In case of defective spices, the US Food and Drug Administration (USFDA) has laid out some defect action levels (DAL). If the defects in spices exceed the DAL, the import consignment at the port of disembarkation is retained there until cleaned to remove the defect. In case some specific defects cannot be eliminated and/or destroyed completely by reconditioning, the consignment will be destroyed at the site or returned to the shipper. This is the general specification. However, for turmeric no DAL has been prescribed (Balakrishnan, 2007).

Health Requirements

The presence of microorganisms in food stuff is critical to human health, and hence it must be ensured that all these products are free of contaminating organisms. Inadequate and unhygienic drying and storage leads to accumulation of microbial load on the spice. Strict regulations have been specified by importing countries for the limits of microbial load in spices. The microbiological specifications for turmeric under German law are given in Table 11.5.

Uncontrolled application of chemical pesticides at various stages of plant growth results in accumulation of their residues in spices, sometimes to a level beyond acceptable limits. With the growing concern about the carcinogenic properties of various pesticide residues, the importing countries are tightening the tolerance

Table 11.6 Tolerance Levels Against Aflatoxins in Spice Products

	Aflatoxins	Maximum Permissible Limit (ppb)
Country under consideration	$B_1 + B_2 = G_1 + G_2$	10
European Spice Association	B_1	5

Source: Balakrishnan (2007).

limits. Pesticide residue continues to be a major concern in all of the spices currently exported from India.

Another major issue in the spices is the presence of aflatoxins. Aflatoxins are a group of secondary metabolites of fungi, *Aspergillus flavus* and *Aspergillus parasticus*, and are rated as potent carcinogens. Inadequate and unhygienic drying process leads to the growth of these fungi on the spice products. Aflatoxins in spices are generally classified into four groups: B_1, B_2, G_1, and G_2, where B_1 and B_2 are produced by *Aspergillus flavus*, while G_1 and G_2 are produced by *Aspergillus parasticus*. Of these, B_1 is the most virulent carcinogen, which can lead even to human death when infected agricultural products are consumed inadvertently. Also, this group has received the most scientific attention, both from agricultural scientists and from medical practitioners. The tolerance limits for aflatoxins under the European Spice Association (ESA) is given in Table 11.6.

Commercial Requirements

Turmeric must comply with ISO-1980 specifications as per the requirements of the International Standard Specifications.

Value Additions in Turmeric

Turmeric Powder

Value addition is the means to enhance marketability of an agricultural product. Turmeric is no exception. Ground turmeric is mostly used in the retail market, primarily by food processors. Rhizomes are ground to approximately 60–80 mesh particle size. Since curcuminoids, the color constituents of turmeric, deteriorate with incident light and to a lesser extent by incident heat and oxidative conditions, it is very important that ground turmeric is packed in UV-protective packaging and appropriately stored under cool and dampness-proof conditions. Turmeric powder is the ubiquitous product used globally in "Indian Curry." It comes as powder, paste, and also in combination with other ground processed spices such as black pepper. In the food industry, turmeric powder is mostly used as a coloring and as a flavoring agent (Kachru and Srivastava, 1991).

In olden days (it also continues to be a practice even now), turmeric was traditionally pound in a wooden container, known locally as "Chakki." It is a wooden mortar with a wooden pestle, and it is hand operated. The use of the commercial burr mill to produce turmeric powder was reported by Kachru and Gupta in 1993. The material to be powdered is fed into the top hopper, which runs through a controlled chute opening and goes to the center of the grinding stones. The rotating stone crushes the particles against the stationary stone. The clearance between the stones could be adjusted by the mechanism provided in the machine depending upon the material to be milled and the product size required. But the problem encountered normally is the accumulation of rhizomes in the feeding section. This problem has been overcome by providing spiral splines with a pitch of 144 mm, with a 45° pressure angle, augmenting the feeding process of rhizomes toward the grinding zone. But of late, a hammer mill is used for grinding. Turmeric powder obtained should be so fine-particled that it passes through a 300 μm sieve, and there is nothing left over in the sieve (Kachru and Srivastava, 1991).

The spice to be powdered in the mill will most likely heat up and the volatiles contained might be lost, depending on the type of the mill used and the speed of crushing. In the case of turmeric, heat and oxygen during the process of milling may contribute to degradation of curcumin. Cryogenic milling in liquid nitrogen prevents oxidation and volatile loss, but it is an expensive process and is not in widespread use in the food industry. Ground spices are size-stored through screens and larger particles are further ground (Balakrishnan, 2007). The specifications of turmeric powder under the Indian Agmark specifications are given in Table 11.7.

Storage studies conducted on turmeric powder using various packing materials revealed that aluminum foil laminate offers the maximum protection against loss of volatile oil and ingress of moisture. Polyethylene pouches alone are inadequate to provide desired protection against loss of volatile oils because as much as 60% of the volatile oils are lost within 135 days (Balasubramanyam et al., 1979). Polyethylene film pouches of 200 and 300 gauges are also found inadequate to offer desired protection against volatile oil losses during storage, because nearly 60–70% of volatile oils were found to be lost within a period of 150 days (Balasubramanyam et al., 1980).

According to the specifications of the Bureau of Indian Standards (BIS), whole turmeric shall comply with the requirements specified by IS: 3576–1994, and turmeric powder with the requirements specified by IS:2446–1980.

Turmeric Oleoresin

Solvent extraction of the pulverized or powdered rhizomes is used to obtain turmeric extractives, or oleoresins. The process yields about 12% of an orange- or red-colored viscous liquid, depending on the type of solvent used and turmeric cultivar. This viscous liquid contains various proportions of the coloring matter, namely the curcuminoids, volatile oils (the latter imparting the flavor), in addition to nonvolatile fatty and resinous materials. The compounds of commercial interest in turmeric are the curcuminoids, which make up about 40–55% and volatile oils, which make up about 15–20%. The curcuminoids, which consist mainly of curcumin, can be further purified to a crystalline material, which is used preferably in the manufacture of products where the turmeric flavor is not desired (FAO, 2009). As per BIS, turmeric oleoresin shall comply with the specifications of IS: 10925–1984.

Table 11.7 Grade Designations and Definitions of the Quality of Turmeric Powder (Fine and Coarse Ground)

Grade Designation	Moisture (Percentage by Weight, Maximum)	Total Ash (Percentage ash by Weight, Maximum)	Acid Insoluble (Percentage by Weight, Maximum)	Lead (Parts Per Million, Maximum)	Starch (Percentage, by Weight)	Chromate Test	Fineness
Fine ground							
Standard	10	7.0	1.5	2.5	60.0	Negative	Powder should pass through a 300 µm sieve
Coarse ground							
Standard	10	9.0	1.5	2.5	60.0	Negative	Powder should pass through a 500 µm sieve

Source: Spices Board (2001).

Essential Oils

There is negligible academic or commercial interest in turmeric essential oil, especially inasmuch as far as the food industry is concerned. This is in sharp contrast to the importance of oleoresin in food industry. However, there is now an emerging body of scientific evidence pointing to the medicinal value of essential oils in turmeric. The medicinal properties of the essential oils in turmeric are attributable to the specific chemical compounds present in them. Turmeric essential oils are obtained by steam distillation or by supercritical fluid extraction of the powdered rhizomes. It is also a byproduct of the purification of curcuminoids from oleoresins. The major compounds found in turmeric essential oil are the sesquiterpene ketones, β-, and ar-termerones, which together make up 50–60%. The sesquiterpene, zingeberene, and ar-curcumene do not exceed 25% and 35%, respectively (Kachru and Srivastava, 1991), and generally are not reported in scientific literature.

Curry Powder

A major component of the curry powder available in commercial markets in packed form meant for culinary purposes is turmeric powder. Curry powder is a mixture of different spices, which are used for seasoning dishes, such as meat, fish, poultry, or vegetables, primarily in the tropical countries. In fact, the ubiquitous "Indian curry" has turmeric as an important base. About 40–50% of the curry powder is made up of turmeric powder. The curry powder is now increasingly used in the United States, Europe, and Japan. There also it is used for seasoning dishes. India is the principal exporter of curry powder, and the major importers are the United Kingdom, Australia, Fiji, and others. Turmeric powder provides the required attractive color and background aroma to the curry powder (Sasikumar, 2003).

Newer Methods of Value Addition to Turmeric

Spray-Dried Turmeric Oleoresin

Turmeric oleoresin is highly sensitive to light, heat, oxygen, and alkaline conditions, which can be contained through the process of microencapsulation of oleoresin by spray drying. Turmeric oleoresin may be spray dried on a sugar matrix, such as maltodextrin, to a powder form, which can then be used as a colorant in dry cereals or beverages. The advantage of spray-dried turmeric oleoresin powder is that it is devoid of starch, proteins, and fibers (Balakrishnan, 2007).

Supercritical CO_2 Extraction of Curcuminoids and Turmerones

Supercritical fluid extraction is a separation process where the substances are dissolved in a fluid, which is able to modify its dissolving power under specific conditions, above their critical temperature and pressure. CO_2 has become the ideal supercritical fluid in the food industry, due to its unique characteristics, which are: (i) the critical temperature which is 31.06°C; (ii) the critical pressure which is 73.83 bar; and (ii) the critical density which is 0.46 g/cm^3. The influence of the drying temperature of fresh turmeric on the extraction of curcuminoids with supercritical CO_2 showed that a higher concentration of curcuminoids in oleoresin was obtained by supercritical fluid extraction from turmeric rhizomes dried at 343 K at a pressure of 30 MPa. Separation and purification of turmerones from the turmeric essential oil extracted by supercritical CO_2 showed that the extraction at a combination of 320 K and 26 MPa gives an optimum production of turmeric oil which has a purity of 71% of turmerones. Consequent purification leads to the production of ar-turmerone which is 86% pure and 81% pure α- + β-turmerone, both of which have been separated from one another.

Microwave-Assisted Extraction Technique for Curcuminoids

For rapid and selective extraction of curcuminoids into organic solvents, a novel technique was developed, now known as microwave-assisted extraction (MAE) (Dandekar and Gaikar, 2002). The extraction process was optimized employing acetone at 20% microwave power level, which leads to 60% extraction of curcuminoids of 75% purity within a time span of just 1 min. The dielectric heating of cellular matrix resulted in vaporization of volatile components, increase in internal pressure of the cell, leading to remarkable swelling and finally rupture of the cells. The degradation of the cellulosic cell wall at higher temperatures, when subjected to microwaves, also enhanced permeability of solvents into the biomatrix.

References

Arora, M., Sehgal, V.K., Sharma, S.R., 2007. Quality evaluation of mechanically washed and polished turmeric rhizomes. J. Agric. Eng. 44 (2), 39–43.

Athmaselvi, K.A., Varadharaju, N., 2003. Heat utilization in different methods of turmeric boiling. Madras Agric. J. 90 (4/6), 332–335.

Balakrishnan, K.V., 2007. Post harvest technology and processing of turmeric. In: Ravindran, P.N., Nirmal Babu, K., Sivaraman, K. (Eds.), Turmeric: The Genus *Curcuma* CRC Press, Boca Raton, FL, pp. 193–256.

Balasubramanyam, N., Kumar, R.K., Anandaswamy, B., 1979. Packaging and storage studies on ground turmeric (*Curcuma longa* L.) in flexible consumer packages. India Spices 16 (6), 10–13.

Balasubramanyam, N., Baldevraj Indiramma, A.R., Anandaswamy, B., 1980. Evaluation of polyprolene and other flexible materials for packaging of ground spices. India Spices 17 (2), 15–20.

Blasco, M., Garcia-Perez, J.V., Bon, J., Carreres, J.E., Mulet, A., 2006. Effect of blanching and air flow rate on turmeric drying. Food Sci. Technol. Int. 12 (4), 315–323.

Dandekar, D.V., Gaikar, V.G., 2002. Microwave assisted extraction of curcuminoids from *Curcuma longa*. Sep. Sci. Technol. 37 (11), 2669–2690.

Food and Agriculture Organization of the United Nations, 2009. Post-Production Management for Improved Market Access for Herbs and Spices—Turmeric (cited May 14). Available from <http://www.fao.org/inpho/content/compend/text/ch29/ch29.htm/>.

Govindarajan, V.S., 1980. Turmeric—chemistry, technology and quality. Crit. Rev. Food Sci. Nutr. 12 (3), 200–2134.

Goyal, R.K., Korla, 1993. Changes in the quality of turmeric rhizomes during storage. J. Food Sci. Technol. 30 (5), 362–364.

Gunasekar, J.J., Doraiswamy, P., Kaleemullah, S., Kamaraj, S., 2006. Evaluation of solar drying for post harvest curing of turmeric (Curcuma longa L.). Agric. Mech. Asia Afr. Lat. Am. 37 (1), 9–13.

Jalgaonkar, R., Rajput, J.C., 1991. Possibility of growing turmeric in heavy rainfall area of Maharashtra. Spice India 4 (8), 9–11.

Kachru, R.P., Gupta, R.K., 1993. Modified commercial burr mill for production of turmeric powder (Curcuma longa L.) powder. Ind. J. Agric. Eng. 3 (3–4), 114–117.

Kachru, R.P., Srivastava, P.K., 1991. Processing of turmeric. Spice India 4 (9), 2–5.

Nair, K.P.P., 1996. The buffering power of plant nutrients and effects on availability. Adv. in Agron. 57, 237–287.

Pal, U.S., Khan, K., Sahoo, N.R., Sahoo, G., 2008. Development and evaluation of farm level turmeric processing equipment. Agric. Mech. Asia Afr. Lat. Am. 39 (4), 46–50.

Parthasarathy, V.A., John Zachariah, T., Chempakam, B., 2003. Spices products—global and Indian scenarios. In: Chadha, K.L., Singh, A.K., Patel, V.B. (Eds.), Recent Initiatives in Horticulture Horticultural Society of India, New Delhi, pp. 589–611.

Praditdoung, S., Kaeumanae, P., Ganjanagoochorn, W., Wanichgamjanakul, K., 1996. Effect of curing on turmeric tissue and drying time. Kasetart J. Nat. Sci. 30 (4), 485–492.

Prasad, J., 1980. Processing of turmeric. Fiji Agric. J. 42 (2), 23–25.

Prasad, J., Vijay, V.K., Tiwari, G.N., Sorayan, V.P.S., 2006. Study on performance evaluation of hybrid drier for turmeric (Curcuma longa L.) drying at village scale. J. Food Eng. 75 (4), 497–502.

Pruthi, J.S., 1998. Major Spices of India–Crop Management and Post Harvest Technology. Indian Council of Agricultural Research, New Delhi, pp. 289–314.

Purseglove, J.W., Brown, E.G., Green, C.L., Robin, S.R.J., 1981. Turmeric Spices, vol. II. Longman, New York, NY, pp. 532–580.

Sampathu, S.R., Krishnamurthy, N., Sowbhagya, H.B., Shankarnarayana, M.L., 1988. Studies on quality of turmeric (Curcuma longa) in relation to curing methods. J. Food Sci. Technol. 25 (3), 152–155.

Sasikumar, B., 2003. Turmeric In: Peter, K.V. (Ed.), Handbook of Herbs and Spices, vol. 1 Woodhead Publishing Ltd, England, pp. 297–310.

Shankaracharya, N.B., Natarajan, C.P., 1999. Turmeric—chemistry, technology and uses. India Spices 10, 7–11.

Sharma, P.D., Sinha, M.K., Kumar, V., 2008. Curing characteristics and quality evaluation of turmeric finger. J. Agric. Eng. 45 (2), 58–61.

Sivadasan, C.R., MadhusudanaKurup, P. (Eds.), 2002. Quality Requirements of Spices for Export Spices Board, Cochin, Kerala State.

Spices Board Agmark Grade Specifications for Spices. 2001. Spices Board, Cochin, Kerala State, India.

Sukumaran, C.R., Satyanarayana, V.V., 1999. In: Souvenir cum Proceedings of the National Seminar on Food Processing: Challenges and Opportunities. Gujarat Agricultural University, Gujarat State.

Suresh, D., Manjunatha, H., Srinivasan, K., 2007. Effect of heat processing of spices on the concentrations of their bioactive principles: turmeric (Curcuma longa), red pepper (Capsicum annum) and black pepper (Piper nigrum). J. Food Comp. Anal. 20 (3–4), 346–351.

Viswanathan, R., 2008. Turmeric—harvesting, processing and marketing. Paper presented at National Workshop on Zingiberaceous Spices–Meeting the Growing Demand Through Sustainable Production. 19–20 March 2008, IISR, Calicut, pp. 89–96.

Viswanathan, R., Devdas, C.T., Sreenarayanan, V.V., 2002. Farm level seam boiling of turmeric rhizomes. Spice India 15 (7), 2–3.

12 Neutraceutical Properties of Turmeric

The following review on the neutraceutical properties of turmeric presents, in some detail, its clinical effects established through scientific investigations such as anti-inflammatory activity, wound healing, antioxidative, chemopreventive, antimutagenic, anticarcinogenic, antimicrobial, antidiabetic, antiangiogenic, antithrombotic, and hepatoprotective properties. Turmeric also has therapeutic properties against Alzheimer disease (AD). It is also useful against gastrointestinal and respiratory disorders and arthritis. The role of turmeric in cosmetology is also documented. Turmeric is an important spice among rice consumers of Asia, specially India, Southeast Asia, and Indonesia, and is the main ingredient in the ubiquitous "Indian curry" powder. Turmeric is also a well-known remedy used in ancient Indian traditional medicine and cosmetics, serving as a multipurpose herbal remedy for practitioners of *Ayurveda*, the ancient system of Indian medicine, and the more recent *Siddha* and *Unani*, and also practitioners of Chinese medicine (Sakarkar et al., 2006).

Turmeric has several traditional uses, starting from the *Ayurveda*. The common use of turmeric is as an antiseptic. Johnson & Johnson, the well-known US drug company, makes turmeric bandages for the Indian market (MacGregor, 2006). In Northern India, women are given a tonic of fresh turmeric paste with dried ginger powder mixed with honey in a glass of milk to drink twice a day immediately after childbirth, which is thought to rejuvenate the postnatal condition of the mother. Powdered turmeric is taken with boiled milk to cure cough and related respiratory ailments (Sundaram, 2005). This ancient treatment also combats dental problems and digestive disorders such as indigestion dyspepsia, acidity, indigestion, flatulence, and ulcers. It also alleviates adverse effects of hallucination due to consumption of *hashish* (cannabis) or other psychotropic drugs. Turmeric is used extensively now in the food and cosmetic industries, as a natural coloring agent and as an additive to flavor curries. The medicinal properties of turmeric originate in curcumin, a powerful antioxidant and antiinflammatory ingredient.

Indian natives used turmeric more extensively than what the current populace uses. Aryan culture gave turmeric a preeminent position in religious ceremonies. The belief then was because turmeric has a golden yellow color depicting sunlight, it must possess protective properties. Even today turmeric finds its place of importance in Hindu rituals. Traditionally, turmeric was also used to dye clothing used in marriage ceremonies in Hindu society. The belief was that it protects the wearer from any infection or fever. Use of turmeric in *Ayurveda* is legendary. It is called by 46 different synonyms, including "pitta" (yellow), "gauri" (brilliant), and all words that indicate "night". In *Ayurveda*, turmeric is believed to maintain a balance between all the three states of disease–"vata," "pitta," and "kapha"—the three "doshas" (states of

The Agronomy and Economy of Turmeric and Ginger. DOI: http://dx.doi.org/10.1016/B978-0-12-394801-4.00012-0

disease). There are many Ayurvedic formulas which utilize turmeric. To relieve muscular sprain and inflammation, a paste made of turmeric, lime, and salt was applied.
Smoke made by sprinkling turmeric over burned charcoal was used to relieve pain
from scorpion stings, to provide relief from indigestion, and combat hysteric fits
(Nadkarni and Nadkarni, 1975). It has also been used as an insect repellent. The
ancient Indian system of medicine, *Charaka Samhita*, chronicles turmeric's use as an
important remedy against snake bites, chicken and small pox, ulcers, conjunctivitis,
skin blemishes, malaria, etc.

The active ingredient of turmeric is curcumin (2–5%; Sundaram, 2005). Curcumin
is the substance that imparts to turmeric most of its therapeutic properties. The rhizome is 70% carbohydrates, 7% protein, 4% minerals, and a minimum of 4% essential oils, and about 1% resin. It also contains vitamins and other alkaloids. Clinical
and laboratory research indicates that diets which include turmeric or curcumin stabilize and protect biomolecules in the body at the molecular level, which is shown in its
antioxidant, antimutagenic, and anticarcinogenic action (Sundaram, 2005). Turmeric
has powerful antioxidant properties, which protect the person taking it regularly
against cancer, in particular, colon and breast cancer. Current research shows that
about 98% of all diseases are controlled by a molecule called NF-Kappa B, a powerful protein which promotes abnormal inflammatory response in the body. Excess
of NF-Kappa B can lead to cancer, arthritis, and a wide range of other diseases, and
studies show that the curcumin subdues NF-Kappa B, thus preventing nearly all
diseases afflicting the human body (Sundaram, 2005). Its principal drawback is its
low bioavailability, about 60–66%. Based on the most recent reports on nanodrug
delivery, curcumin encased in nanoparticle and made of silk fiber rich in keratin and
chitosan has been found to be highly effective against cancer (Gupta et al., 2009).
Turmeric enhances the flavor of food, aids digestion (in particular, that of protein),
promotes absorption of the digested material, and in turn, regulates body metabolism.
It helps to regulate intestinal microflora and is particularly recommended to be taken
after a course of antibiotics, thereby reducing the risk of gastritis and ulcer formation
in the intestines. In diabetics, turmeric lowers blood sugar (Arun and Nalini, 2002).
Because it contains natural cyclooxygenase inhibitors, it shows good antiinflammatory action. Turmeric extract, volatile oil from turmeric, and curcuminoids have been
found to be quite effective against arthritis, because of their antiinflammatory properties (Arora et al., 1971). The antiinflammatory property is due to the capacity of
turmeric to lower histamine levels, while possibly increasing production of natural
cortisone by adrenal glands. It also helps both liver and gall bladder function. It has
been shown to be helpful in the treatment of arthritis, rheumatoid arthritis, osteoarthritis, injuries, trauma, and stiffness both under normal activity and hyperactivity
(Deodhar et al., 1980). It shows that natural cortisone production by adrenal glands
is enhanced. Its most important antiinflammatory mechanism is based on its positive effects on the prostaglandins (PGs), a large family of potent lipids produced by
the body (Majeed et al., 1995). PG1 and PG3 calm the body while PG2 inflames it.
Turmeric is a potent inhibitor of cyclooxygenase, 5-lipooxygenase, and 5-HETE production in neutrophils, which reduces PG2 levels, resulting in reduced body pain and
inflammation.

The neutraceutical properties of turmeric are now widely accepted in modern medicine. The Indian traditional medical systems use turmeric for wound healing, rheumatic disorders, gastrointestinal symptoms, deworming, rhinitis, and as a cosmetic. Studies in India have explored its antiinflammatory, cholekinetic, and antioxidant potentials, its preventive effect on precarcinogenic, antiinflammatory, and antiatherosclerotic effects in biological systems both under *in vivo* and *in vitro* conditions in animals and humans. Both turmeric and curcumin, the major bioactive principle in lending color to turmeric, were found to enhance detoxifying enzymes, prevent DNA damage, improve DNA repair, decrease mutations and tumor formation, and exhibit antioxidative potential in animals. Limited clinical studies suggest that turmeric can significantly impact excretion of mutagens in urine among smokers and help regress precancerous palatal lesions. It reduces DNA adducts and micronuclei in oral epithelial cells. It prevents formation of nitrous compounds both *in vivo* and *in vitro*. It delays induced cataract in diabetics and reduces hyperlipidermia in obese rats. Recently, several molecular targets have been identified turmeric for therapeutic/preventive effects (Krishnaswamy, 2008). Turmeric is also used to treat asthma, dysmenorrhea, psoriasis, eczema, arthritis, and hepatic and digestive disorders and to prevent and treat cardiovascular diseases (Sakarkar et al., 2006).

In the *Unani* system of medicine (Indian), turmeric is used to treat liver obstruction, dropsy, jaundice, ulcers, and inflammation. In the Himalayan system of medicine, turmeric is used as a contraceptive and skin tonic, and to treat swelling, insect stings, wounds, whooping cough, inflammation, internal injuries, pimples, and also external injuries (Sakarkar et al., 2006).

Curcumin has a great biological significance. Because of this, a number of investigations have centered on curcumin. According to these investigations, curcuminexhibitsantiinflammatory(Chainani-Wu,2003),antioxidant(Masudaetal.,1993),antiviral (Suai et al., 1993), antimicrobial (Mahady et al., 2002), anticarcinogenic (Frank et al., 2003), and antimutagenic properties protecting the body from mutagens such as smoke and other pollutants. Recent investigations suggest that curcuminoids are active in the external treatment of certain cancerous conditions, and this is presumably related to the cytotoxicity of these substances, which has been demonstrated on cell cultures, including tumor cells (Jayaprakasha et al., 2005). Besides these, curcumin has a variety of potentially therapeutic properties such as antineoplastic, antiapoptotic, antiangiogenic, cytotoxic, immunomodulatory (Strimpakos and Sharma, 2008), antithrombotic, wound healing, antidiabetogenic, antistressor, and antilithogenic actions (Chainani-Wu, 2003). Turmeric has been used in traditional Indian medicine for the treatment of jaundice and other ailments related to the liver, ulcers, parasitic infections, various skin diseases, sprains, inflammation of the joints, cold and influenza symptoms, and for preserving food stuff because of its antimicrobial properties.

The Principal Compounds in Turmeric

Turmeric rhizomes contain both volatiles and nonvolatiles. Chemical constituents of the volatile oil are identified using gas chromatography (GC) and gas chromatography–mass spectrometry (GC–MS), and these constituents are ar-turmerone, turmerone, zingiberene, and curlone. The nonvolatile compounds of turmeric are coloring agents, and principal bioactive components of turmeric, named curcuminoids. These curcuminoids are a rich source of phenolic compounds, namely, curcumin, demethoxycurcumin, and bis-demethoxycurcumin. Turmeric consists of 3–5% curcuminoids, which include mainly curcumin I (diferuloyl methane), curcumin II (demethoxycurcumin), and curcumin III (bisdemethoxy curcumin) (Chainani-Wu, 2003). The melting point of curcumin, $C_2H_2OO_6$, is 184 °C; it is soluble in ethanol and acetone but insoluble in water (Joe et al., 2004). Curcumin exists in solution as keto-enol tautomers (Payton et al., 2007). Curcuminoids can be isolated by a variety of analytical methods, of which solvent extraction is the most effective. Curcuminoids are estimated by absorption, flourimetric, and HPLC methods. HPLC methods are suitable for the determination of individual curcuminoids (Jayaprakasha et al., 2005).

Curcumin is lost due to heat processing of turmeric to an extent of 27–53%, with maximum loss occurring from pressure cooking (Suresh et al., 2007). In the presence of tamarind, the loss of curcumin was only 12–30%. The samples of curcuminoids, with varying composition of curcumin I, curcumin II, and curcumin III, decomposed very rapidly (more than 90% within 12 h) in the absence of serum and were unstable in the presence of serum in physiological media (Pfeiffer et al., 2003). Curcumin I was found to be the least stable, while curcumin III was found to be the most stable. Several degradation products of curcumin I were detected, most of which are yet to be identified. Ferulic acid and vanillin were products of minor consequence.

Medicinal Properties of Turmeric

Antiinflammatory Activity

Several commonly used plant products, including those from curcumin, have anti-inflammatory properties. Through *in vitro* test systems, it has been established that water-soluble extracts were not cytotoxic and did not exhibit biological activity. Organic extracts of turmeric were cytotoxic only at concentrations above 50 µg/ml. Crude organic extracts of turmeric were capable of inhibiting lipopolysaccharide (LPS)-induced tumor necrosis factor (TNF) and prostaglandin E2 (PGE_2) production. Purified curcumin was more active than either demethoxy or bis-demethoxycurcumin (Lantz et al., 2005).

Within the developing world, the most extensive use of turmeric in traditional medicine has been from India and China, especially for its immune-modulatory properties. Dendritic cells (DCs) are antigen-presenting cells specialized to initiate and regulate immunity. The ability of DCs to initiate immunity is linked to their activation status. The hydroethanolic extracts (HEE), and not the lipophilic fraction, of turmeric inhibits the activation of human DCs in response to inflammatory cytokines. Treatment of DCs with HEE also inhibits the ability of DCs to stimulate the mixed lymphocyte reaction (MLR). More importantly, the lipophilic fraction does not synergize with the hydroethanolic fraction for the ability of inhibiting DC maturation. Rather, culturing of DCs with the combination of HEE and SCE leads to partial abrogation of the effects of HEE on the MLR initiated by DCs. These studies provide a mechanism for the antiinflammatory properties of turmeric. However, they suggest that these extracts are not synergistic and may contain components with mutually antagonistic effects on human DCs. Harnessing the immune effects of turmeric may benefit from specifically targeting the active fractions.

Wound Healing

A central consideration in common medical practice has been the healing of irradiated wounds, because radiation disrupts normal response to injury, leading to a protracted recovery period. Pretreatment of irradiated wounds with curcumin was reported to significantly enhance the rate of wound contraction, decrease mean wound healing time, increase synthesis of collagen (the principal protein of connective tissues in animals), hexosamine, DNA, and nitric oxide, and improve fibroblast and vascular densities, indicating that curcumin pretreatment has a conducive effect on the irradiated wound and could be a substantial therapeutic strategy in initiating and supporting the cascade of the tissue repair process (Jagetia and Rajanikant, 2004).

Arthritis

Turmeric has been used for centuries in *Ayurvedic* medicine, the traditional Indian system of medicine, as a treatment for inflammatory disorders including arthritis. On the basis of this ancient knowledge, dietary supplements containing turmeric rhizome and turmeric extract are also being increasingly recommended for use in the West, especially in Europe and the United States as a curative and preventive process to contain arthritis, which is now increasingly affecting the elderly in these regions. The *in vivo* antiarthritic efficacy of an essential oil-depleted turmeric fraction has been attributed to the three major curcuminoids, while the remaining compounds in the crude turmeric extract may inhibit this antiarthritic effect (Funk et al., 2006).

Turmeric, a close relative of ginger, is a promising disease-preventive agent as well, probably due largely to its antiinflammatory properties. At least one new study suggests that it can be used effectively for the treatment of arthritis. This research, carried out in Italy, was a three-month trial involving 50 patients, diagnosed through X-ray having osteoarthritis of the knee. The Italian team was investigating the effect on arthritic symptoms of a special formulation of turmeric designed to improve its absorption by the body. Half the participating patients took the turmeric formulation, in addition to standard medical treatment; those in the second group continued following their physicians' recommendations.

After 90 days, the researchers found a 58% decrease in overall reported pain and stiffness in the knee joint, as well as improvement in physical functioning among the group of patients on turmeric formulation, compared to the controls. These changes were documented with a standard medical scoring method used to assess symptoms of knee and hip osteoarthritis. In addition, another scoring method showed a 300% improvement in the emotional well-being of the patients on the turmeric formulation, compared to the others. And blood tests showed a 16-fold decline in C-reactive protein, a marker for inflammation. Patients in the turmeric group were able to reduce their use of nonsteroidal antiinflammatory drugs by as much as 63%, a remarkable improvement, compared with the other group.

Results of this investigation are good news for the millions of people worldwide who suffer from osteoarthritis and have not been adequately helped by currently available treatments for the disease. The dose of the turmeric formulation used in the study was 1 g/day. It is now commercially available in the United States and Europe.

Turmeric may also be useful for prevention of the symptoms of rheumatoid arthritis, but this evidence comes from animal studies, not human trials. The research also suggests that turmeric may prevent changes which lead to the dreaded AD, causing severe loss of memory and making life miserable, and investigations using animals have shown that turmeric may be effective in the prevention or treatment of colon, breast, and prostate cancer.

Inflammatory Bowel Disease

Inflammatory bowel disease (IBD) is characterized by oxidative and nitrosative stress, leukocyte infiltration, and upregulation of proinflammatory cytokines. The protective effects of curcumin, an antiinflammatory and antioxidant food derivative, on 2,4,6-trinitrobenzene sulfonic acid-induced colitis in mice was demonstrated by the significant reduction in the degree of neutrophil infiltration (measured as decrease in myeloperoxidase activity), lipid peroxidation (measured as decease in malondialdehyde activity) in the inflamed colon, and the decreased serine protease activity in curcumin pretreated mice. Curcumin also reduced the levels of nitric oxide (NO) and O_2 associated with the favorable expression of cytokines and inducible NO synthase. Thus, curcumin or diferuloylmethane exerts beneficial effects in experimental colitis and may, therefore, be useful in the treatment of IBD (Ukil et al., 2003).

Antioxidant Property

For an aerobic organism, oxygen is a double-edged sword, because it is essential for aerobic processes, and 5% of the inhaled oxygen is converted to reactive oxygen species (ROS). Cellular antioxidant enzymes and free-radical scavengers normally protect the cells from the damaging effects of ROS, but when the dynamic equilibrium is upset, pathological conditions result from the oxidative damage to the cellular macromolecules: DNA, lipids, proteins, and carbohydrates. Free-radical-mediated peroxidation has been implicated in multiple diseases. A major oxidation by-product of this deleterious biochemical process is 4-hydroxy-2-nonenal (HNE), which is cytotoxic, mutagenic, and genotoxic, and is involved in disease pathogenesis. Using an ELISA test that employed HNE-modification of solid phase antigen, following immobilization, it was found that curcumin solubilized in dilute alkali-inhibited HNE-protein modification (Kurien et al., 2007). Turmeric also inhibited HNE-protein modification similarly, but at a much lower alkali level. Alkali by itself was found to enhance HNE-modification substantially. Curcumin/turmeric has to inhibit this alkali-enhanced HNE-modification prior to inhibiting the normal HNE-protein modification induced by HNE. Thus, inhibition of HNE-modification could be a mechanism by which curcumin exerts its antioxidant effects. The pH at which the inhibition of HNE-modification of substrate was observed was close to the physiological pH, making this formulation of curcumin potentially useful in a very practical sense.

The aqueous extract of fresh rhizomes showed higher antioxidant properties as compared to the extracts from dry rhizomes, while the commercially available dry turmeric powder, commonly sold and used as a spice in Indian market and used as a very common culinary ingredient in Indian cooking, had the least antioxidant properties (Vankar, 2008). There was substantial loss of antioxidant properties when turmeric rhizomes are turned into dry marketable powder. The order of reactivity is curcuma long (dry) < dry spice turmeric powder < curcuma long (wet) < curcuma short (dry) < curcuma short (wet).

The loss of antioxidant properties during the dry spice preparation thus signifies that its beneficial pharmacological activities were definitely reduced. It is, therefore, recommended that fresh turmeric rhizome be preferred for consumption, than dry powder, to harness the most health-giving benefits.

It is important to note that significant alterations can occur in the combinations of phytochemicals of turmeric rhizomes, based on the plant genotype, conditions in which they are grown, and postharvest processing (Cousins et al., 2007). Tissue drying negatively affected the ability of extracts to scavenge the DPPH radical in all the accessions tested, whereas the effect of tissue drying on ferrous iron chelating ability of extracts was cultivar-specific. Fresh tissue extracts were more potent than extracts from commercially available turmeric powder in all the cases for both assays. The iron chelation assay revealed that extracts from recently dried tissue were significantly more potent than extracts from aged commercially available turmeric powder. DPPH scavenging capacity of the dried tissue was usually of similar intensity to the commercially available powder, with only one clone showing a significant difference in potency. Commercial drying methods may have negative effects on the antioxidants present in the turmeric rhizome, thus adversely affecting their health-giving benefits. Genotypic selection minimized this adverse effect, as superior ones would end up losing the quantum of antioxidants only to a lesser extent compared to inferior ones.

Curcumin can be removed from turmeric oleoresin, and extracted with hexane to get turmeric oil, which can be fractionated using silica gel column chromatography to obtain three fractions: turmeric oil containing aromatic turmerone (31.32%), turmerone (15.08%), and curlone (9.7%), whereas fraction III has aromatic turmerone (44.5%), curlone (19.22%), and turmerone (10.8%) as major compounds. Also, oxygenated compounds were enriched in fraction III. Turmeric oil and its fractions were tested for antioxidant activity, using the β-carotene–lineolate model system and the phosphomolybdenum method. Faction III showed maximum antioxidant capacity. All the fractions and turmeric oil exhibited a marked antimutagenicity by means of the Ames test, but fraction III was the most effective. The antioxidant effects of turmeric oil and its fractions may provide an explanation for their antimutagenic action (Jayaprakasha et al., 2002). The protective effect of aqueous extracted turmeric antioxidant protein on H_2O_2-induced red blood cells, in which lipid peroxidation and hemolysis was inhibited by 70% and 80%, respectively, has been reported. This was more effective as an antioxidant than were α-tocopherol and curcumin (Lalitha and Selvam, 1999). The antioxidant activity of *Curcuma longa* free phenolics (CLFP) was attributed to curcumin and that of *Curcuma longa* bound phenolics (CLBP) to ferulic acid and *p*-coumaric acid (Kumar et al., 2006). These extracts exhibited significant protection to DNA against oxidative damage. Studies have also established that turmeric, silymarin, ginkgo, alone and in combination, might be useful herbal remedies to suppress oxidative damage caused by iron overload and emphasize the additive effect of the dietary antioxidants (Sarhan et al., 2007).

The effects of natural dietary antioxidants for their effectiveness in protecting cells from free radical damage in the cell, to DNA, protein, and lipids induced by antitumor agents or radiation which leads to the generation of free radical in normal cells *in vivo* and *in vitro* are the areas of extensive scientific scrutiny. Curcumin is a natural antioxidant known to possess therapeutic properties and has been reported to scavenge free radicals and to inhibit clastogenesis in mammalian cells. However, curcumin has been reported to act as prooxidant and induce a significant increase in the frequency of chromosomal aberrations in Chinese hamster ovary (CHO) cells mediated by hydroxyl radical generation (Araujo et al., 2001). Curcumin can generate ROS as a prooxidant in the presence of transition metals in cells, resulting in DNA injuries and apoptotic cell death. The prooxidant action of curcumin may be related to the conjugated β-diketone structure of this compound (Yoshino et al., 2004). Curcumin in the presence of Cu (II) causes strand cleavage in the DNA through generation of ROS, in particular, the hydroxyl radical. Of curcumin, and its two naturally occurring derivatives, namely, demethoxycurcumin (dmC) and bidemethoxycurcumin (bdmC), curcumin, which is considerably more active as an antioxidant, was also found to be the most effective in the DNA cleavage reaction and a reducer of Cu (II) followed by dmC and bdmC. The DNA cleavage activity is the consequence of binding of Cu (II) to the phenolic and methoxy groups on the two benzene rings of curcumin and the third site is due to the presence of 1,3-diketone system between the rings. Furthermore, both the antioxidant and the prooxidant effects of curcuminoids are determined by the same structural moieties (Ahsan et al., 1999).

Chemopreventive and Bioprotectant Properties

It is important to note that the effects of turmeric as a preventive agent against malignancy and cardiac problems have been extensively investigated. Researchers believe that curcumin, which gives the rhizome its bright yellow color, has inhibitive properties against tumor-promoting enzymes and interferes with the growth of cancerous tumors. Because turmeric is an antioxidant, curcumin neutralizes free radicals that accumulate in the cell, which raises the risk of cancer and cardiac diseases.

The antimutagenic and antioxidant properties of turmeric have been extensively investigated both in humans and in animals. Oral cancer is common in India. Turmeric or curcumin administered in the diet or applied as a paint in the buccal cavity is most likely to have a chemopreventive effect on pancreatic lesions in experimental model of 7,1'2-dimethylbenzanthracene-induced buccal pouch tumors in Syrian golden hamsters (Krishnaswamy et al., 1998). Turmeric and/or its main coloring component, curcumin (diferuloylmethane), have been shown to inhibit benzo(a) pyrene (B(a)P)-induced forestomach papillomas in mice. However, the mechanisms of turmeric-mediated chemoprevention are not well understood. Deshpande et al. (2001) established that administration of turmeric through diet significantly inhibited, in forestomach (target organ), liver, and lungs, the activity of isozymes of cytochrome P-450, namely, ethoxyresorufin-O-deethylase and methoxyresorufin-O-demthylase, which predominantly are involved in the metabolism of B(a)P, by monitoring the formation of resorufin. *In vitro* investigations employing curcumin, demethoxycurcumin, and bis-demethoxycurcumin suggest that curcumins are the inhibitors in turmeric. Inhibition of B(a)P metabolizing phase I enzymes may be at least, in part, one of the possible modes of chemopreventive action of turmeric/curcumin.

To elucidate the possible interaction of turmeric and curcumin, with conjugation reactions, which in many cases are involved in the activation of procarcinogens, their effects in the conjugation of 1-naphthol in Caco-2 cells, a human colon carcinoma cell line, within a 24 h period was measured (Naganuma et al., 2006). Turmeric exhibits inhibitory activity toward both sulfo-and glucuronosyl conjugations of 1-naphthol at approximately the same levels. Curcumin inhibits sulfo-conjugation at lower concentrations, and only showed weak inhibition toward glucuronosyl conjugation of 1-naphthol in Caco-2 cells. In addition, turmeric and curcumin were found to strongly inhibit *in vitro* phenol sulfotransferase (SULT) activity and demonstrate moderate inhibitory properties against UDP-glucuronosyl transferase (UGT) activity in Caco-2 cells. Moreover, and in contrast to the moderate inhibition of UGT activity by turmeric and curcumin, both induce the expression of UGT1 A1 and UGT 1 A6 genes. These findings are indicative of a possible interaction of both turmeric and curcumin with conjugation reactions in the human intestinal tract and colon. This, in turn, may affect the bioavailability of therapeutic drugs and toxicity levels of environmental chemicals, in particular, procarcinogens.

Both dietary turmeric and catechin (the principal phenolic constituent of *Acacia catechu*) in drinking water significantly inhibited the tumor burden and tumor incidence in B(a)P-induced forestomach tumors in Swiss mice and methyl-(acetoxymethyl I)-nitrosomine-induced oral mucosal tumors in Syrian golden hamsters and delayed induction of oral tumors in golden hamsters. Adjuvant chemoprevention utilizing both catechin and dietary turmeric inhibited both the gross tumor yield and the burden more effectively than when compared with individual components in both tumor models. Thus, catechin and turmeric are effective chemopreventive agents in mice and golden hamsters, when regularly consumed (Azuine and Bhide, 1994).

Antimutagenic and Anticarcinogenic Properties

Curcumin is recognized as an anticarcinogenic agent because of its propensity to induce cell suicide or apoptosis *in vivo* and *in vitro*. The human diet contains several substances capable of inhibiting chemical carcinogenesis. It is known that such inhibitors may either act directly by scavenging the reactive substances or indirectly by promoting mechanisms which enhance detoxification. Turmeric is an active anti-mutagen both *in vivo* and *in vitro*. The effects of turmeric on xenobiotic metaboliz-ing enzymes in hepatic tissue of rats fed turmeric indicated that UDP glucuronosyl transferase and glutathione-*S*-transferase (GST) were significantly elevated, suggest-ing that turmeric may increase detoxification systems in addition to its antioxidant properties. Curcumin perhaps is the active principle in turmeric. It is suggested that turmeric used widely as a spice may mitigate the effects of several dietary carcino-gens (Goud et al., 1993). Earlier, it was shown that curcumin protects cells against oligonucleosomal DNA fragmentation and induces a novel apoptosis-like pathway in Jurkat cells (Piwocka et al., 1999). Curcumin has the ability to induce cell death in all tested cells that can be classified as apoptosis-like, and only in HL-60 cells can it be recognized as classical apoptosis (Bielak-Zmijewska et al., 2000).

P-glycoprotein is an ATP-dependent drug efflux pump linked to the development of multidrug resistance (MDR) in cancer cells. Curcumin I, it is proved, is the most effec-tive MDR modulator among the three curcuminoids, and may be used in combination with conventional chemotherapeutic drugs to reverse MDR in cancer cells (Chearwae et al., 2004). Turmeric has been shown to prevent B(a)P or dimethylbenz(a)anthracene (DMBA)-induced forestomach, skin, and mammary tumors in mice and/or in rats. In an investigation on nitrosodiethylamine (NDEA)-induced heptocarcinogenesis in rats, NDEA-treated rats receiving turmeric before, during, and after carcinogen exposure showed significant decrease in the number of gamma-glutamyl transpeptidase (GGT) positive foci and decrease in the incidence of NDEA-induced focal dysplasia (FD) and heptocellular carcinomas. Decrease in the number of GGT positive foci was also observed in NDEA-treated rats, although no decrease in tumor incidence was noted. On the other hand, different levels of turmeric treatment after exposure to NDEA did not show any protective effects (Thapliyal et al., 2003).

The ethanol turmeric extract was found stimulatory for murine lymphocytes and inhibitory for ascitic fibrosarcoma cells. Viability of lymphocytes was better with turmeric. DNA synthesis increased and the majority of the lymphocytes were driven toward mitotic stage. However, turmeric caused a significant level of death in ascetic fibrosarcoma cells *in vitro* and possibly arrested the cell cycle at S-phase, it induced programmed cell death (Chakravarty et al., 2004).

The mechanism of turmeric-mediated chemoprevention in DMBA-induced ham-ster buccal pouch (HBP) carcinogenesis was attributed to the augmentation of apop-tosis of the initiated cells and decreasing cell proliferation in DMBA-treated animals, which in turn is reflected in decreased tumor burden, multiplicity, and enhanced latency period. Some of these biomarkers are likely to be helpful in monitoring clini-cal trials and evaluating drug effect measurements (Garg et al., 2008).

Curcumin inhibits proliferation and induction of apoptosis in cancer cells, but the sequence of events leading to the cell death is poorly understood and is yet poorly defined. The molecular mechanisms by which multidomain proapoptotic Bcl-2 family members Bax and Bak regulate curcumin-induced apoptosis using mouse embryonic fibroblasts (MEFs) deficient in Bax or Bak or both genes was elucidated by Shankar and Srivastava (2007). Curcumin treatment resulted in an increase in the protein levels of both Bax and Bak, and mitochondrial translocation and activation of Bax in MEFs to trigger drop in mitochondrial membrane potential, cytosolic release of apoptogenic molecules (cytochrome c and second mitochondria—derived activator of caspases (Smac)/direct inhibitor of apoptosis protein-binding protein with low isoelectric point), activation of caspase-9 and caspase-3, and ultimately apoptosis. Furthermore, MEFs derived from Bax and Bak double-knockout (DKO) mice exhibited even greater protection against curcumin-induced release of cytochrome c and Smac, activation of caspase-3 and caspase-9, and induction of apoptosis compared with wild-type MEFs or single-knockout Bax/Bak/MEFs. Interestingly, curcumin treatment also caused an increase in the protein level of apoptosis protease-activating factor-1 in wild-type MEFs. Smac N7 peptide enhanced curcumin-induced apoptosis, whereas Smac siRNA inhibited the effects of curcumin on apoptosis. The mature form of Smac sensitized Bax and Bak DKO MEFs to undergo apoptosis by acting downstream of mitochondria. The study demonstrates that the role of Bax and Bak as a critical regulator of curcumin-induced apoptosis and overexpression of Smac as interventional approaches to deal with Bax- and/or Bak-deficient chemoresistant cancers for curcumin-based therapy. Curcumin, when mixed with cells from head and neck cancer affected patients, stopped their proliferation and induced apoptosis in malignant cells (Seppa, 2004). Curcumin had no effect on healthy cells. Previous research suggested that curcumin stops proliferation of prostate cancer cells. It also kills human breast and liver cancer cells in laboratory cultures. Cancer researchers have taken a keen interest in curcumin because many countries with curry-rich cuisines, such as India, Sri Lanka, and Pakistan, and other Southeast Asian countries on the subcontinent, have lower incidence of cancer than in Western countries, where use of turmeric is rather limited.

Turmeric extract in ethanol plays a diabolically opposite role on murine lymphocytes and on Ehlrich ascitic carcinoma cells. The variability of the cells, blastogenesis, DNA synthesis, and SEM studies established that turmeric is a conducive agent for lymphocytes and inhibitory as well as apoptotic for tumor cells (Chakravarty and Yasmin, 2005). The unilateral ureteral obstruction (UUO) model of renal injury in rat is characterized by nuclear factor kB (NF-kB) activation and tumor necrosis factor-α (TNF-α) production, which induces apoptosis via activation of caspase, resulting in cell death. Curcumin has been reported to provide protection against fibrosis and apoptosis elicited by UUO. A turmeric-based diet can delay apoptosis without modulating NF-kB, so as not to sensitize the mesangial cells to the apoptotic stimuli (Reem et al., 2008).

Antimicrobial Property

The essential oil of turmeric rhizomes show toxicity to seven fungi involved in deterioration of stored agricultural commodities. *Aspergillus flavus, Fusarium semitectum, Colletotrichum gloeosporioides*, and *C. musae* were most sensitive with growth inhibition of more than 70%. The bioautography of the oil produced only one antifungal band representing 40% of the total oil. The fractionation of this band by reverse phase preparative HPLC yielded two peaks, in the proportion of 57.9% and 42.1%. The larger peak had only one compound, which was identified as ar-turmerone by

MS and nuclear magnetic resonance (NMR) spectra. The smaller peak contained two compounds, in the proportion of 31% and 69%, which were identified as β-turmerone and ar-turmerone, respectively. Thus, ar-turmerone constituted 87% of the fungitoxic component to the oil. The purified ar-turmerone showed antifungal activity similar to the crude oil (Dhingra et al., 2007). The antibacterial activity of the aqueous heated extracts was greater than that of the unheated extracts of turmeric (T), ginger (G), and mango ginger (M), alone or as mixtures, against the bacterial strains of *Escherichia coli*, *Bacillus subtilis*, and *Staphylococcus aureus*. Of the organic acid extracts, the antibacterial activity of 1,4-dioxan extracts of T, GT, and TM showed the highest activity against *E. coli* (Chandarana et al., 2005).

An important role in preventing infection in wound covering and coating for surgical implants is a biodegradable polymeric bioadhesive emulsion, which is nontoxic and contains antimicrobial agents. Therefore, a bioadhesive polymer was synthesized by a semiinterpenetrating network process using a blend of shellac, casein, and polyvinyl alcohol and maleic anhydride as reactive compatibilizer by Jagannath and Radhika (2006). The synthesized polymer was mixed with neem and turmeric extract and homogenized using an emulsifier. Antimicrobial properties were satisfactory for biomedical use for the human pathogenic organisms *E. coli*, *S. aureus*, *Bacillus cereus*, and *Salmonella typhimurium* using well-diffusion assay, as were the stability, miscibility, and antimicrobial properties of the bioadhesive.

The crude ethanol and aqueous extracts of garlic, ginger, turmeric, and neem had antibacterial effects against *S. aureus*, *Salmonella typhi*, and *E. coli in vitro* (Neogi et al., 2007). The inhibition was more effective in combinations than in single extracts. The highest inhibition was observed with synergistic combinations of garlic and turmeric (ethanolic) extracts of *S. aureus*, garlic, and turmeric (aqueous), garlic and turmeric (70% ethanolic), ginger and turmeric (70% ethanolic) on *S. typhi*. *E. coli* was more resistant than the other two organisms. The maximum inhibition was observed with garlic and turmeric (aqueous). Results of this kind herald great potential for designing active antibacterial synergized agents of plant origin.

Effectiveness of turmeric oil and curcumin, against several isolates of *Trichophyton rubrum*-induced dermatophytosis, pathogenic molds, and yeasts in guinea pigs was tested. All the isolates of dermatophytes and pathogenic fungi isolates were inhibited by turmeric oil and none by curcumin. All yeast isolates proved to be insensitive to both turmeric oil and curcumin. In the experimental animals, turmeric oil as dermal application, showed quicker improvement in controlling lesions and rapid disappearance of the lesions was observed (Apisariyakul et al., 1995).

The addition of turmeric powder (TMP) to aflatoxin B_1 (AFB_1) diet significantly improved the weight gain in chicks. The addition of TMP or an adsorbent-hydrated sodium–calcium aluminosilicate (HSCAS) and TMP with HSCAS ameliorated the adverse effects of AFB_1 on serum chemistry parameters (total protein, albumin, cholesterol, and calcium). It decreased antioxidant functions, such as those of peroxides, superoxide dismutase activity, and total antioxidant concentration in liver homogenate due to AFB_1, which were alleviated by the inclusion of either TMP or HSCAS, singly, or both together. The reduction in the severity of hepatic microscopic lesions due to supplementation of the AFB_1 diet with TMP and HSCAS, demonstrated the protective action of the antioxidant and adsorbent (Gowda et al., 2008). Turmeric root powder and mannanoligosaccharides are found to be very satisfactory alternatives to antibiotics in broiler feeds. Both possessed antimicrobial effect *in vivo*. Turmeric may also depress fat deposition in broilers (Samarasinghe et al., 2003). Turmeric provided the maximum protection to *Labeo rohita* fingerlings challenged with opportunistic pathogen of *Aeromonas hydrophila* (Sahu et al., 2008).

The antifungal properties of several chemical and herbal compounds were tested *in vitro* on *Aspergillus parasiticus* and the study indicated that all the selected chemical and herbal compounds reduced fungal growth (i.e., fungal spore count) and aflatoxin production. Among the herbal compounds tested in feeds, clove oil was the best, followed by turmeric, garlic, and onion; among the chemical compounds, propionic acid, sodium propionate, benzoic acid, and ammonia were the best antifungal compounds, followed by urea and citric acid (Gowda et al., 2008).

Antidiabetic Property

Diabetes (diabetes mellitus) is increasingly seen as a global health problem, more so in India. Per current medical projections, in next one decade, India will be the "global capital of diabetes." Among Type I and Type II diabetes, the latter is vastly spread among Indian population. Among the spices, turmeric rhizomes have recently been used in herbal preparations to combat diabetes. In alloxan-induced diabetic rats, the lower rate of weight gain in diabetic rats, compared with nondiabetic rats, was normalized by oral administration of an aqueous extract of turmeric or curcumin. Treatment of diabetic rats with either turmeric or curcumin drastically reduced glucose-6-phosphatase and elevated the activity of liver and serum hexokinase, lactate dehydrogenase, and glucose-6-phosphatase and elevated the activity of liver and serum hexokinase, lactate dehydrogenase, and glucose-6-phosphate dehydrogenase, indicating a decrease in the cellular leakage of acid phosphatase, alkaline phosphatase, and lactate dehydrogenase into the serum of diabetic animals. Curcumin appeared to be more effective in attenuating diabetes mellitus than turmeric (Narayanasamy et al., 2002).

One of the important consequences of diabetes is that chronic hyperglycemia leads to the overproduction of free radicals in the cells and medical evidence is mounting that these contribute to the development of diabetic nephropathy. Streptozotocin (STZ)-induced diabetic rats showed significant increase in blood glucose, polyuria, and a decrease in body weight. After 6 weeks, renal dysfunction, as evidenced by reduced creatine and urea clearance and proteinuria was noticed, as also a marked increase in oxidative stress, as determined by lipid peroxidation and activities of key antioxidant enzymes. Chronic treatment with curcumin significantly reduced both renal dysfunction and oxidative stress in diabetic rats due to the possible antioxidative mechanism of curcumin (Sharma et al., 2006). Changes in glycoconjugate metabolism during the development of diabetic complications and their modulation by feeding bitter gourd and spent turmeric as fiber-rich source of STZ-induced diabetic rats has been reported by Vijayalakshmi et al. (2009). The total sugar content decreased in liver, spleen, and the brain, while an increase was observed in heart and lungs.

Uronic acid content in liver, spleen, and the brain decreased, and a marginal increase was observed in testis. Amino sugar content decreased in liver, spleen, lungs, and heart during diabetes. Decrease in sulfation of glycoconjugates was observed in liver, spleen, lungs, and heart during diabetes, and this condition was significantly ameliorated by bitter gourd and spent turmeric, but this positive action was absent in the brain. Protein content decreased in the liver, while an increase was observed in the brain. These results clearly demonstrate alteration of glycoconjugate metabolism during diabetes and amelioration to varying extents by feeding bitter gourd and turmeric. Improvement is due to slow release of glucose by fiber in the gastrointestinal tract and short-chain fatty acid production from fiber by colon microbes.

The beneficial effect of feeding fenugreek seed mucilage and spent turmeric on STZ-induced diabetic rats was investigated by Kumar et al. (2005a). Diabetic rats lost weight, but body weights improved more by feeding spent turmeric than fenugreek seed mucilage. In diabetic rats, a 30% decrease in urine sugar and urine volume profiles was observed with feeding fenugreek seed mucilage and spent turmeric. Fasting blood glucose showed a 26% and 18% reduction with fenugreek seed mucilage and spent turmeric feeding, respectively, in diabetic rats. Fenugreek seed mucilage compared with spent turmeric was more effective in ameliorating diabetic state in the disease-affected rats. These results were explained further by the same authors (Kumar et al., 2005b) by elucidating the effect of feeding fenugreek seed mucilage and spent turmeric (10%) on the activity of disaccharidases, and the specific activity of intestinal and renal disaccharidases, namely sucrase, maltase, and lactase in STZ-induced diabetic rats. The specific activities of intestinal disaccharidases were increased significantly during diabetes and amelioration of these activities during diabetes was clearly visible by supplementing fenugreek seed mucilage and spent turmeric in the diet of the affected rats. However, during diabetes, the activity of renal disaccharidases was significantly lower than that in the control (placebo) treatment. Fenugreek seed mucilage and spent turmeric supplements were beneficial in alleviating the reduction in maltase activity during diabetes, though not perceptibly in the activity of sucrase and lactase. This positive influence of feeding fenugreek seed mucilage and spent turmeric on intestinal and renal disaccharidases clearly indicates their beneficial role in the management of diabetes in rats, a result which probably can be extended to the treatment of human diabetes as well.

In *Ayurveda*, the traditional Indian system of medicine, several spices and herbs are thought to possess medicinal properties. Administration of turmeric or curcumin to diabetic rats reduced significantly their blood glucose, Hb, and glycosylated hemoglobin levels in blood. Turmeric and curcumin supplements also reduced the oxidative stress encountered by diabetic rats. This was demonstrated by the lower levels of thiobarbituric acid reactive substances (TBARS), which may have been due to the reduced influx of glucose into the polyol pathway leading to an increased NADPH/NADP ratio and elevated activity of the potent antioxidant enzyme glutathione peroxidase (GPx). Moreover, the activity of sorbitol dehydrogenase, which catalyzes the conversion of sorbitol to fructose, was significantly lowered on treatment with turmeric or curcumin. These results also corroborate the observation that curcumin rather than turmeric is more effective in reducing diabetes mellitus-related symptoms in rats affected by the disease (Arun and Nalini, 2002).

There are wide ranging positive effects of turmeric on human health. Turmeric oleoresin, an extract of turmeric, is commonly used to flavor and color food stuff. Curcuminoids and turmeric essential oil are both contained in turmeric oleoresin, and both these fractions have hypoglycemic (reducing blood glucose) effects. The effect of turmeric oleoresin on hepatic gene expression in obese diabetic KK-Ay mice, using DNA microarray analysis, and quantitative real-time polymerase chain reaction, was assessed (Honda et al., 2006). The analysis suggested that curcuminoids regulated turmeric oleoresin ingestion-induced expression of glycolysis-related genes and also that curcuminoids and turmeric essential oil acted synergistically to regulate the peroxisomal β-oxidation-related gene expression induced by turmeric oleoresin ingestion. These changes in gene expression were considered to be the mechanism by which turmeric oleoresin affected the control of both blood glucose and excessive abdominal adipose tissue mass. All of these results suggest that the use of whole turmeric oleoresin is more effective in diabetic control and the secondary effects of diabetes, rather than using them singly.

While curcumin and turmeric did not prevent STZ-induced hyperglycemia, as quantified by blood glucose and insulin level, slit lamp microscope observations indicated that these supplements delayed the progression and maturation of cataract. Curcumin and turmeric treatment seem to counter onset of hyperglycemia-induced oxidative stress because there was a reversal of changes with respect to lipid peroxidation, reduced glutathione, protein carbonyl content, and the activity of antioxidant enzymes significantly. Also, treatment with turmeric or curcumin appears to have minimized osmotic stress, as assessed by polyol pathway enzymes. Most importantly, aggregation and insolubilization of lens proteins due to hyperglycemia was prevented by turmeric and curcumin. Turmeric was more effective than its corresponding levels of curcumin, and its efficacy as an anticataractogenic agent that prevents or one that delays the onset of cataracts must be explored further (Palla et al., 2005).

Supplements containing a compound found in curry spice may help prevent diabetes in people at high risk of the disease, quotes a recent Thai study.

Researchers, whose results were published recently in the journal *Diabetes Care*, found that over nine months, a daily dose of curcumin seemed to prevent new cases of diabetes among people with so-called prediabetes—abnormally high blood sugar levels—that may progress to full-blown Type II diabetes.

Curcumin is a compound in turmeric. Previous laboratory research has suggested that it can fight inflammation and so-called oxidative damage to body cells. These two processes are thought to feed a range of diseases, including Type II diabetes.

"Because of its benefits and safety, we propose that curcumin extract may be used for an intervention therapy for pre-diabetes population," wrote research leader Dr. Somlak Chuengsamarn of Srinakharinwirot University of Nakomnayok, Thailand. The investigation included 240 Thai adults with prediabetes, who were randomly assigned to take either curcumin capsules or a placebo. The ones taking curcumin took six supplement capsules a day, each of which contained 250mg of "curcuminoids."

After nine months, 19 of the 116 placebo patients had developed Type II diabetes, compared with none of 119 patients taking curcumin.

The researchers found that the supplement seemed to improve the function of beta cells, which are cells in the pancreas which release the blood sugar regulating hormone insulin. They speculate that antiinflammatory effects of curcumin help protect beta cells from damage.

Despite the above-described positive effects of curcumin controlling Type II diabetes, a diabetes expert (not involved in the investigation) said that it is still too early for people with Type II diabetes to head to the health food store for curcumin supplements. "This looks promising, but, there are still a lot of questions," said Constance Brown-Riggs, a certified diabetes educator and spokesman for the Academy of Nutrition and Dietetics.

The investigation lasted only nine months, and it is already known from longer-lasting, larger trials that lifestyle changes, including calorie cutting and exercise can prevent Type II diabetes in people predisposed to prediabetes.

Antiangiogenic and Antithrombotic Effects

Cardiac-related diseases are on the increase in the world, and in India, even among the relatively young population. In a human tissue-based angiogenic assay, curcumin, isolated from turmeric rhizome, was found to show antiangiogenic properties. As a liposoluble compound, curcumin can be extracted from turmeric rhizome with organic extractants (solvents), such as ethanol or acetone. Curcumin in its pure form has poor water solubility, potentially limiting its medicinal use in humans when it is taken orally or injected. Liu et al. (2008) have investigated the possibility of improving curcumin's low solubility to help maintain antiangiogenic properties with improved water solubility. Pure curcumin ($85\,\mu M$ 1% ethanol v/v) totally suppressed angiogenic responses; in contrast, a curcumin concentration of $18.5\,\mu M$ (as $100\,\mu g/ml$ turmeric extract) achieved the same level of total inhibition of angiogenesis. This, nearly a five-fold gap, reflected the unaccounted involvement of other antiangiogenic compounds, including curcumin derivatives, and/or enhancement of curcumin by nonantiangiogenic compounds in the extract.

The protection of endothelial integrity by elimination of certain risk factors has proven to be effective in maintaining hemostasis and in slowing the progress of the cardiovascular disease. Indigenous drugs are the natural source of protection against these disorders, which can be used more effectively by the knowledge of their active ingredients as well as by the mechanism of action. Most prominent among these drugs are garlic and turmeric, which are very commonly used culinary spices. A notable restoration of arterial blood pressure was seen in animals when fed with garlic and turmeric-supplemented diet (Zahid et al., 2005). Animals on a supplemented diet showed significantly enhanced vascorelaxant response to adenosine, acetylcholine, isoproterenol, and contractile effect of 5-hydroxytryptamine as significantly attenuated. Inhibition of these responses by L-NMMA (L-N^{G}-monomethyl arginine) was smaller in tissues from herbal-treated animals. Incubation of tissues with L-arginine resulted in a significant reversal of L-NMMA induced inhibition of endothelium-mediated relaxation, which appeared to be pronounced in rings from animals supplemented with herbs as compared to hypercholesterolemic animals. Addition of indomethacin augmented the relaxation in all groups of animals. The study demonstrated that garlic and turmeric are potent vascorelaxants apart from their capacity in reducing the atherogenic properties of cholesterol.

Hepatoprotective Effects

There was a significant cellular recovery among carbon tetrachloride-induced cellular hepatic damage hepatocytes, a significant restoration of lymphocyte viability and CD25, CD71, and Con A receptor expression in both immature thymocytes and splenic helper T lymphocytes with treatment of curcuminoid extract at different doses. Turmeric crude extract, at both high and low doses, was found to be more efficient than the purified extract in its action (Abu-Rizq et al., 2008).

Despite extensive investigations carried out on the medicinal properties of curcumin in various animal models, no systematic investigation has been carried out to examine whether crude extract of turmeric can reduce hepatotoxicity and oxidative stress in rats, induced by chronic feeding of p-dimethyl aminoazobenzene and phenobarbital, two known carcinogens of liver. In carcinogen-intoxicated rats, there was an increase in activities of acid and alkaline phosphatases, alanine and aspartate amino transferases, gamma-glutamyl transferase, lipid peroxidation, and in serum triglyceride level, cholesterol, creatine, urea, bilirubin, blood urea nitrogen and a decrease in reduced glutathione content, catalase, glucose-6-phosphate dehydrogenase and blood glucose, high-density lipoprotein (HDL) cholesterol, albumin, and hemoglobin contents (Khuda-Buksh et al., 2008). Most of these changes were reversed by the administration of crude extract of turmeric, indicating its hepatoprotective potential and ability to reduce oxidative stress in experimental rats. Further, from the analysis of expression of matrix metalloproteinases, p53 and Bcl-2 proteins in liver at 90 and 120 days (post-tumor development), its antitumorigenic activity is also evident. The results would thus validate its traditional use in various ailments, in particular against liver disorders.

The hepatoprotective effect of turmeric with its sesquiterpenes and curcuminoid fractions were examined on D-galactosamine-induced liver injury in rats. All the diets individually containing the turmeric extract, the curcuminoid fraction, and the sesquiterpenes fraction suppressed the increase of LDH, ALT, and AST levels caused by D-GalN treatment. Since only few antioxidative activities are expected in the sesquiterpenes fraction, it is presumed that hepatoprotective mechanism of sesquiterpenes in turmeric is different from that of curcuminoids (Miyakoshi et al., 2004).

The effects of lead acetate in the diet (positive control) on reduced GSH activity of phase II metabolizing enzyme GST, lipid peroxidation in liver homogenate and bone marrow chromosomes of mice simultaneously supplemented with powdered turmeric and myrrh revealed a significant decrease in the amount of GSH in all treated groups compared to negative control (El-Ashmawy et al., 2006). Also, the activity of GSH S-transferase was significantly decreased in positive control compared with other groups. However, coadministration of the protective plants resulted in a significant increase in the activity of GST, compared with both positive and negative control groups. Furthermore, lipid peroxidation was significantly enhanced in the positive control group alone, while cotreatment with the protective plants resulted in reduction in the level of liquid peroxidation receiving turmeric powder. Lead genotoxicity was confirmed through significant reduction in the number of dividing cells, increased total number of aberrant cells, and increased frequency of chromosomal aberrations. Simultaneous treatment with the plant significantly reduced the genotoxicity induced by lead administration and powerful protection was observed with powdered turmeric. Thus, turmeric is a useful herbal remedy, especially for controlling oxidative damages and genotoxicity induced by lead acetate intoxication.

Effect on AD

One of the most distressful diseases of the contemporary times globally is the widely spreading AD, which leads to severe dementia. Curcumin has been used to treat dementia and traumatic brain injury. It has also shown potential in the prevention and treatment of AD. Curcumin, with its antioxidant, antiinflammatory, and lipophilic action, improves the cognitive functions in patients with AD. A growing body of evidence indicates that oxidative stress, free radicals, β-amyloid, cerebral deregulation caused by biometal toxicity, and abnormal inflammatory reactions contribute to the key event in AD pathology. Owing to the observed effects of curcumin, such as decreased β-amyloid plaques, delayed degradation of neurons, metal chelation, antiinflammatory, antioxidant, and decreased microglia formation properties, the overall memory in patients with AD will definitely improve (Ishra and Palanivelu, 2008). Ahmed and Gilani (2009) report that curcuminoids and all individual components except curcumin possess pronounced acetylcholinesterase (AChE) inhibitory activity. Curcumin was relatively weak in the *in vitro* assay and without effect in the *ex vivo* AChE model, while equally effective in memory-enhancing effect, suggestive of additional mechanism(s) involved. Thus, curcuminoid mixture might possess a better therapeutic profile than curcumin for its medicinal use in AD.

Dye (2007) has reported the ability of curcumin to prevent the development of β-amyloid plaques in humans. When macrophages, taken from the blood of patients suffering from AD, were exposed to curcumin *in vitro*, bis-demethoxycurcumin removed 50% of β-amyloid in patients suffering from AD.

Epidemiological studies show a reduced risk of AD among patients using nonsteroidal inflammatory drugs (NSAIDs), indicating the role inflammation plays in the onset and spread of AD. Studies have shown a chronic CNS inflammatory response associated with increased accumulation of amyloid peptide and activated microglia in AD. Previous studies have shown that interaction of $A\beta_{1-40}$ or fibrilar $A\beta_{1-42}$, caused activation of nuclear transcription factor, early growth response-1 (Egr-1), which resulted in increased expression of cytokines (TNF-α and IL-1β) and chemokines (MIP-1β, MCP-1, and IL-8) in monocytes. Giri et al. (2004) have reported that curcumin suppressed the activation of Egr-1 DNA-binding activity in THP-1 monocyclic cells. Curcumin abrogated $A\beta_{1-40}$-induced expression of cytokines (TNF-α and IL-1β) and chemokines (MIP-1β, MCP-1, and IL-8) in both peripheral blood monocytes and THP-1 cells; curcumin inhibited $A\beta_{1-40}$-induced MAP kinase activation and the phosphorylation of ERK-1/2 and its downstream target Elk-1; curcumin inhibited $A\beta_{1-40}$-induced expression of CCR5, but not of CCR2b in THP-1 cells. This involved abrogation of Egr-1 DNA binding in the promoter of CCR5 by curcumin. Finally, curcumin inhibited chemotaxis of THP-1 monocytes in response to chemoattractant. The inhibition of Egr-1 by curcumin may represent a potential therapeutic approach to ameliorate the inflammation of progression of AD.

Additional Therapeutic Properties of Turmeric

Tardive Dyskinesia

Tardive dyskinesia (TD) is a motor disorder of the orofacial region resulting from chronic neuroleptic treatment. The high incidence and irreversibility of this hyperkinetic disorder has been considered a major clinical issue in the treatment of schizophrenia. The molecular mechanism related to the pathophysiology of TD is not completely known. Various animal studies have demonstrated an enhanced oxidative stress and increased glutamatergic transmission as well as inhibition in the glutamate uptake after the chronic administration of haloperidol. Bishnoi et al. (2008) have investigated the effect of curcumin, an antioxidant, in haloperidol-induced TD by using different behavioral (orofacial dyskinetic movements, stereotypy, locomotor activity, percentage retention), biochemical (lipid peroxidation, reduced glutathione levels, antioxidant enzyme levels (SOD), and catalase), and neurochemical (neurotransmitter levels) parameters. Chronic administration of haloperidol significantly increased vacuous chewing movements (VCMs), tongue protrusions, facial jerking in rats, and decrease in turnover of dopamine, serotonin, and norepinephrine in both cortical and subcortical regions was inhibited by curcumin in a dose-dependent manner; oxidative damage to all major regions of the brain was attenuated by curcumin, especially in the subcortical region containing striatum. Chronic administration of haloperidol also resulted in increased dopamine receptor sensitivity, as evidenced by increased locomotor activity and stereotypy, and also decreased retention time on elevated plus maze paradigm. Pretreatment with curcumin reversed these behavioral changes.

Gastrointestinal and Respiratory Disorders

Gilani et al. (2005) reported that the crude extract of turmeric (C1.Cr) relaxed the spontaneous and K^+-induced contractions in isolated rabbit jejunum, as well as shifted the $CaCl_2$ concentration–response curves. In rabbit tracheal preparation, C1.Cr inhibited carbachol and K^+-induced contractions. In anesthetized rats, C1.Cr produced variable responses on blood pressure with a mixture of weak hypertensive and hypotensive actions. In rabbit aorta, C1.Cr caused a weak vasoconstrictor and vasodilator effect on K^+ and phenylephrine-induced contractions. In guinea pig atria, C1.Cr inhibited spontaneous rate and force of contractions at 14 to 24 times higher concentrations. Activity directed fractionation revealed that the vasodilator and vasoconstrictor activities are widely distributed in the plant with no clear separation into the polar or nonpolar fractions. When used for comparison, both curcumin and verapamil caused similar inhibitory effects in all smooth muscle preparations with relatively higher effect against K^+- induced contractions and that both were devoid of any vasoconstrictor effect, and curcumin had no effect on atria. These data suggest that the inhibitory effects of C1.Cr are mediated primarily through calcium channel blockade, though additional mechanisms cannot be ruled out, and this study forms the basis for the traditional use of turmeric in hyperactive states of the gut and airways. Furthermore, curcumin, the main active principle, does not share all effects of turmeric. Treatment with tablets of turmeric was reported to help reduce irritable bowel syndrome symptomology (Bundy et al., 2004).

The Protective Effect on Iron Overload

In humans, iron overload causes many adverse effects, namely a significant increase in all serum iron profiles, tissue iron deposition and tissue lipid peroxidation, and marked liver and kidney damage. A significant decrease in GPx activity was seen, while GST activity and glutathione levels (GSH) were significantly increased in all tissues. Administration of turmeric singly induced a significant decrease in serum and tissue iron profile. The powerful antioxidant effect of turmeric resulted in the marked increase of GPx and GST activities and reduced GSH level in all examined tissues, except in the liver, where the activity of GPx and reduced glutathione levels were significantly decreased. Although administration of turmeric for a continued period induced mild liver damage, its administration decreased the above-mentioned toxic effects, induced by iron overload in rats (El-Bahr et al., 2007).

Both red chili and spices like black pepper, ginger, and turmeric are very commonly used and widespread in Indian culinary. Because they are rich in phenolic compounds, they would be expected to bind iron Fe^{3+} in the intestine and inhibit Fe absorption in humans. Tuntipopipat et al. (2006) have reported that the addition of freeze-dried chili in the diet reduced iron absorption by as much as 38%. Turmeric did not inhibit iron absorption. A possible effect of chili on gastric acid secretion was indirectly assessed by comparing Fe^{2+} absorption from acid soluble ^{57}Fe ferric pyrophosphate relative to water-soluble ^{58}Fe ferrous sulfate from the same meal in the presence and absence of chili. Chili did not enhance gastric acid excretion. Relation iron bioavailability of ferric pyrophosphate was 5.4% in the presence of chili and 6.4% in the absence of chili. Despite the much higher amount of phenolics in the turmeric meal, it did not affect iron absorption. It is hence concluded that both phenol quality and quantity determine the inhibitory effect of phenolic compounds on iron absorption.

Effect on Hyperoxaluria

High oxalate intake resulting from consuming supplemental doses of cinnamon and turmeric may increase the risk of hyperoxaluria, a significant risk factor for urolithiasis. Compared with the cinnamon and control treatment, turmeric ingestion led to a significant enhancement in urinary oxalate excretion during oxalate load tests. There were no significant changes in fasting plasma glucose or lipids in conjunction with either cinnamon or turmeric supplements. Thus, the percentage of oxalate that was water soluble differed markedly between cinnamon and turmeric, which appeared to be the primary cause for the higher urinary oxalate excretion/oxalate absorption from turmeric. The consumption of supplemental doses of turmeric, but not cinnamon, can significantly increase urinary oxalate levels, thereby increasing the risk of kidney stone formation in susceptible individuals (Tang et al., 2008).

Use of Turmeric in Cosmetology

The cosmetic industry is a global multibillion endeavor. Men and women, becoming more and more conscious of the fact and how they look, take recourse and use many products, mostly synthetic. But of late there is a growing awareness about the use of naturally occurring herbal products to enhance one's external appearance/beauty. For millennia, turmeric has been used for this purpose in India. Even now, there are "bridal baths" where the main product used is turmeric paste. Ancient Indian medical texts, especially *Ayurveda*, have touted turmeric as a herb with the ability to provide glow and luster to one's skin. It also is supposed to provide vigor and vitality to the entire human body. As described in the introductory section of this chapter on turmeric, the spice is also referred to as "Haridara" in *Sanskrit*, ancient Indian language. The word "Haridara" has two parts, first "Hari," referring to *Vishnu*, the ominipresent and omnipotent *Hindu* God, and "Dara" referring to what one wears. Hence, turmeric is supposed to be "worn" (or used) by *Vishnu*. Even now, in many temples scattered all over India, turmeric is given as "Prasad" (a benedictory material) to the devotees. Most interestingly, on the Malabar coast of Kerala State, India, the "Theyyam" (a human depicted as a particular God, in special attire) distributes turmeric powder to the devotees. Turmeric is deeply ingrained in the Hindu psyche as a material (herb) of manifold uses. Turmeric has found its use even in dentistry (Chaturvedi, 2009). Contrary to common claims, a study by Shaffrathul et al. (2007) has shown that turmeric does not arrest hair loss and has no influence on acne.

Additional Curative Properties of Turmeric

Recently, researchers have found that Indian "curries" which contain curcumin through the addition of turmeric, blocks tendon inflammation in the joints, a finding which may pave the way for a remedy for a painful condition popularly known as "tennis elbow" (*Business Line Newspaper*, August 11, 2011).

The researchers have shown that curcumin, which gives turmeric the trademark bright yellow color, can be used to suppress the biological mechanisms that spark inflammation in tendon diseases (originally reported in the *Daily Mail*, London). For their study, the researchers at the University of Nottingham in London and Ludwig-Maximilians University in Germany have described laboratory experiments that show the ingredient can switch off the inflammatory cell cycle involved. Dr. Ali Mobasheri, who coled the research, said, "Our research is not suggesting that curry, turmeric or curcumin are cures for inflammatory conditions, such as, tendinitis and arthritis. However, we believe that it could offer scientists an important new lead in the treatment of these painful conditions through nutrition."

In the laboratory, researchers used a culture model of human tendon inflammation to study the antiinflammatory effects of curcumin on tendon cells. The results showed that introducing curcumin in the culture system inhibits NF-kB and prevents it from switching on and promoting more inflammation.

Dr. Mobasheri added, "Further research into curcumin, and chemically modified versions of it, should be the subject of future investigations and complementary therapies aimed at reducing the use of nonsteroidal antiinflammatory drugs, the only drugs currently available for the treatment of tendinitis and various forms of arthritis" (*Business Line Newspaper*, August 11, 2011).

References

Abu-Rizq, H.A., Mansour, M.H., Safer, A,M., Afzal, M., 2008. Cyto-protective and immunomodulating effect of *Curcuma longa* in Wistar rats subjected to carbon tetrachloride-induced oxidative stress. Inflammopharmacology 16 (2), 87–95.

Ahmed, T., Gilani, A.H., 2009. Inhibitory effect of curcuminoids on acetylcholinesterase activity and attenuation of scopolamine-induced amnesia may explain medicinal use of turmeric in Alzheimer's disease. Pharmacol. Biochem. Behav. 91 (4), 554–559.

Ahsan, H., Parveen, N., Khan, N.U., Hadi, S.M., 1999. Pro-oxidant, anti-oxidant and cleavage activities on DNA of curcumin and its derivatives demethoxycurcumin and bisdemethoxycurcumin. Chem. Biol. Interact. 121 (2), 161–175.

Apisariyakul, A., Vanittanakom, N., Buddhasukh, D., 1995. Antifungal activity of turmeric oil extracted from *Curcuma longa* (Zingiberaceae). J. Ethnopharmacol. 49 (3), 163–169.

Araujo, M.C.P., Antunes, L.M.G., Takahashi, C.S., 2001. Protective effect of thiourrea, a hydroxyl-radical scavenger, on curcumin-induced chromosomal aberrations in an *in vitro* mammalian cell system. Teratog. Carcinog. Mutagen. 21 (2), 175–180.

Arora, R.B., Basu, N., Kapoor, V., Jain, A.P., 1971. Anti-inflammatory studies on *Curcuma longa* L. Ind. Med. Res. 59, 1289.

Arun, N., Nalini, N., 2002. Efficacy of turmeric on blood sugar and polyol pathway in diabetic albino rats. Plant Foods Human Nutr. 57 (1), 41–52.

Azuine, M.A., Bhide, S.V., 1994. Adjuvant chemoprevention of experimental cancer: catechin and dietary turmeric in fore stomach and oral cancer models. J. Ethnopharmacol. 44 (3), 211–217.

Bielak-Zmijewska, A., Koronkiewicz, M., Skierski, J., Piwocka, K., Radziszewska, E., Sikora, E., 2000. Effect of curcumin on the apoptosis of rodent and human non proliferating and proliferating lymphoid ells. Nutr. Cancer 38 (1), 131–138.

Bishnoi, M., Chopra, K., Kulkarni, S.K., 2008. Protective effect of curcumin, the active principle of turmeric (*Curcuma longa*) in haloperidol-induced orofacial dyskinesia and associated behavioural, biochemical and neurochemical changes in rat brain. Pharmacol. Biochem. Behav. 88 (4), 511–522.

Bundy, R., Walker, A.F., Middelton, R.W., Booth, J., 2004. Turmeric extract may improve irritable bowel syndrome symptomology in otherwise healthy adults: a pilot study. J. Altern. Complement. Med. 10 (6), 1015–1018.

Business Line Newspaper, August 11, 2011.

Chainani-Wu, N., 2003. Safety and anti-inflammatory activity of curcumin: a component of turmeric (*Curcuma longa*). J. Altern. Complement. Med. 9, 161–168.

Chakravarty, A., Yasmin, H., Das, S., 2004. Two-way efficacy of alcoholic turmeric extract: stimulatory for murine lymphocytes and inhibitory for fibrosarcoma cells. Pharm. Biol. 42 (3), 217–224.

Chakravarty, A.K., Yasmin, H., 2005. Alcoholic turmeric extract simultaneously activating murine lymphocytes and inducing apoptosis of Ehlrich ascetic carcinoma cells. Int. Immunopharmacol. 5 (10), 1574–1581.

Chandarana, H., Baluja, S., Chanda, S.V., 2005. Comparison of antibacterial activities of selected species of zingiberaceae family and some synthetic compounds. Turk. J. Biol. 29 (2), 83–97.

Chearwae, W., Anuchapreeda, S., Nandigama, K., Ambudkar, S.V., Limtrakul, P., 2004. Biochemical mechanism of modulation of human P-glycoprotein (ABCB1) by curcumin I, II and III purified from turmeric powder. Biochem. Pharmacol. 68 (10), 2043–2052.

Chaturvedi, T.P., 2009. Uses of turmeric in dentistry: An update. Indian J. Dent. Res. 20 (1), 107–109.

Cousins, M., Adelberg, J., Chen, F., Rieck, J., 2007. Antioxidant capacity of fresh and dried rhizomes from four clones of turmeric (*Curcuma longa* L.) grown *in vitro*. Indian Crops Prod. 25 (2), 129–135.

Deodhar, S.D., Sethi, R., Srimal, R.C., 1980. Preliminary study on anti-rheumatic activity of curcumin (diferuloyl methane). Indian J. Med. Res. 71, 632–634.

Deshpande, S., Maru, G.B., Thapliyal, R., 2001. Effects of turmeric on the activities of benzo(a)pyrene-induced cytochromeP-450 isozymes. J. Environ. Pathol. Toxicol. Oncol. 20 (1), 59.

Dhingra, O.D., Jham, G.N., Barcelos, R.C., Mendonca, F.A., Ghiviriga, I., 2007. Isolation and identification of the principal fungitoxic component of turmeric essential oil. J. Essent. Oil Res. 19 (4), 387–391.

Dye, D., 2007. Turmeric compound helps immune system clear Alzheimer's plaques. Life Extension 13 (11), 16.

El-Ashmawy, I.M., Ashry, K.M., El-Nahas, A.F., Salama, O.M., 2006. Protection by turmeric and myrrh against liver oxidative damage and genotoxicity induced by lead acetate in mice. Basic Clin. Pharmacol. Toxicol. 98 (1), 32–37.

El-Bahr, S.M., Korshom, M.A., Mandour, A.E.A., El-Bessomy, A.A., Lebdah, M.A., 2007. The protective effect of turmeric on iron overload in albino rats. Egyptian J. Biochem. Mol. Biol. 25 (2), 94–113.

Frank, N., Knauft, J., Amelung, F., Nair, J., Wesch, H., Bartch, H., 2003. No prevention of liver and kidney tumors in long-evans cinnamon rats by dietary curcumin, but inhibition at other sites and of metastases. Mutat. Res./Fundam. Mol. Mech. Mutagen. 523–524, 127–135.

Funk, J.L., Oyarzo, J.N., Frye, J.B., Chen, G.J., Lantz, R.C., Jolad, S.D., et al., 2006. Turmeric extracts containing curcuminoids prevent experimental rheumatoid arthritis. J. Nat. Prod. 69 (3), 351–355.

Garg, R., Ingle, A., Maru, G.B., 2008. Dietary turmeric modulates DMBA-induced p21^{ras}, MAP kinases and AP-1/NF-eB pathway to alter cellular responses during hamster buccal pouch carcinogenesis. Toxicol. Appl. Pharmacol. 232 (3), 428–439.

Gilani, A.H., Shah, A.J., Ghayur, M.N., Majeed, K., 2005. Pharmacological basis for the use of turmeric in gastrointestinal and respiratory disorders. Life Sci. 76 (26), 3089–3105.

Giri, R.K., Rajagopal, V., Kalra, V.K., 2004. Curcumin, the active constituent of turmeric, inhibits amyloid peptide-induced cytochemokine gene expression and CCR5-mediated chemotaxis of THP-1 monocytes by modulating early growth response - 1 transcription factor. J. Neurochem. 9 (5), 1199–1210.

Goud, V.K., Polasa, K., Krishnaswamy, K., 1993. Effect of turmeric on xenobiotic metabolizing enzymes. Plant Foods Hum. Nutr. 44 (1), 87–92.

Gowda, N.K.S., Ledoux, D.R., Rottinghaus, G.E., Bermudez, A.J., Chent, Y.C., 2008. Efficacy of turmeric (*Curcuma longa*), containing a known level of curcumin, and a hydrated sodium calcium aluminosilicate to ameliorate the adverse effects of aflatoxin in broiler chicks. Poult. Sci. 87 (6), 1125–1130.

Gupta, V., Aseh, A., Rios, C.N., Aggarwal B.B., Mathur, A.B., 2009. Fabrication and characterization of silk fibrin derived curcumin nanoparticles for cancer therapy. Paper presented at Nanotech CXancer Conference, 2–3 May 2009, Houston.

Honda, S., Aoki, F., Tanaka, H., Kishida, H., Nishiyama, T., Okada, S., et al., 2006. Effects of ingested turmeric oleoresin on glucose and lipid metabolisms in obese diabetic mice: a DNA microarray study. J. Agric. Food Chem. 54 (24), 9055–9062.

Ishra, S., Palanivelu, K., 2008. The effect of curcumin (turmeric) on Alzheimer's disease: an overview. Ann. Indian Acad. Neurol. 11 (1), 13–19.

Jagannath, J.H., Radhika, M., 2006. Antimicrobial emulsion (coating) based on biopolymer containing neem (*Melia azardichta*)and turmeric (*Curcuma longa*) extract for wound covering. Biomed. Mater. Eng. 16 (5), 329–336.

Jagetia, G.C., Rajanikant, G.K., 2004. Role of curcumin, a naturally occurring phenolic compound of turmeric in accelerating the repair of excision wound, in mice whole-body exposed to various doses of α radiation. J. Surg. Res. 120 (1), 127–138.

Jayaprakasha, G.K., Jena, B.S., Negi, P.S., Sakariah, K.K., 2002. Evaluation of antioxidant activities and anti-mutagenicity of turmeric oil: a by-product from turmeric production. Z. Naturforsch. C. 57 (9/10), 828–835.

Jayaprakasha, G.K., Jagan Mohan Rao, L., Sakariah, K.K., 2005. Chemistry and biological activities of *C. longa*. Trends Food Sci. Technol. 16 (12), 533–548.

Joe, B., Vijaykumar, M., Lokesh, B.R., 2004. Biological properties of curcumin -cellular and molecular mechanisms of action. Crit. Rev. Food Sci. Nutr. 44, 97–111.

Khuda-Buksh, A.R., Banerjee, A., Biswas, R., Pathak, S., Boujedaini, N.B., 2008. Crude extract of turmeric reduces hepato-toxicity and oxidative stress in rats chronically fed carcinogens. J. Complement. Integr. Med. 5 (1), 1–36.

Krishnaswamy, K., 2008. Traditional Indian spices and their health significance. Asia Pacific J. Clin. Nutr. 17S (1), 265–268.

Krishnaswamy, K., Goud, V.K., Sesikeran, B., Mukundan, M.A., Krishna, T.P., 1998. Retardation of experimental tumorigenesis and reduction in DNA adducts by turmeric and curcumin. Nutr. Cancer 30 (2), 163–166.

Kumar, G., Suresh, N., Harish, D., Shylaja, M., Salimath, P.V., 2006. Free and bound phenolic antioxidants in amla (*Emblica officinalis*) and turmeric (*Curcuma longa*). J. Food Compos. Anal. 19 (5), 446–452.

Kumar, G.S., Shetty, K., Sambaiah, K., Salimath, P.V., 2005a. Antidiabetic property of fenugreek seed mucilage and spent turmeric in streptozotocin-induced diabetic rats. Nutr. Res. 25 (11), 1021–1028.

Kumar, G.S., Shetty, A.K., Salimath, P.V., 2005b. Modulatory effect of fenugreek seed mucilage and spent turmeric on intestinal and renal disaccharidases in streptozotocin-induced diabetic rats. Plant Foods Hum. Nutr. 60 (2), 87–91.

Kurien, B., Scofield, T., Hal, R., 2007. Curcumin/turmeric solubilized in sodium hydroxide inhibits HNE protein modification. An *in vitro* study. J. Ethnopharmacol. 110 (2), 368–373.

Lalitha, S., Selvam, R., 1999. Prevention of H_2O_2-induced red blood cell lipid peroxidation and hemolysis by aqueous extracted turmeric. Asia Pacific J. Clin. Nutr. 8 (2), 113–114.

Lantz, R.C., Chen, G.J., Solyom, A.M., Jolad, S.D., Timmermann, B.N., 2005. The effect of turmeric extracts on inflammatory mediator production. Phytomedicine 12, 445–452.

Liu, D., Schwimer, J., Zhijun, L., Woltering, E.A., Greenway, F., 2008. Antiangiogenic effect of curcumin in pure versus in extract forms. Pharm. Biol. 46 (10/11), 677–682.

Mac Gregor, H.E., 2006. http://www.latimes.com/features/health/la-he-turmeric February 2006, 2713647.

Mahady, G.B., Pendland, S.L., Yun, G., Lu, Z.Z., 2002. Turmeric (Curcuma longa) and curcumin inhibit the growth of Helicobacter pylori, a group of 1 carcinogen. Anticancer Res. 22, 4179–4181.

Majeed, M., Vladimir, B., Shivkumar, U., Rajendran, R., 1995. Curcuminoids, Antioxidant Phytonutrients. NutriScience Publishers Inc.

Masuda, T., Jitoe, A., Isobe, J., Nakatani, N., Yonemori, S., 1993. Anti-oxidative and anti-inflammatory curcumin-related phenolics from rhizomes of Curcuma domestica. Phytochemistry 32, 1557–1560.

Miyakoshi, M., Yamaguchi, Y., Takagaki, R., Mizutani, K., Kambara, T., Ikeda, T., et al., 2004. Hepatoprotective effect of sesquiterpernes in turmeric. Biofactors 21 (1–4), 67–170.

Nadkarni, K.M., Nadkarni, A.K (Eds.), 1975. Indian Material Medica Popular Prakashan, Bombay.

Naganuma, M., Saruwatari, A., Okamura, S., Tamura, H., 2006. Turmeric and curcumin modulate the conjugation of 1-naphthol in Caco-2 cells. Biol. Pharm. Bull. 29 (7), 1476–1479.

Narayanasamy, S., Nalini, N., Kavitha, R., 2002. Effect of turmeric on the enzymes of glucose metabolism in diabetic rats. J. Herbs Spices Med. Plants 10 (1), 75–84.

Neogi, U., Saumya, R., Irum, B., 2007. In vitro combinational effect of bio-active plant extracts on common food borne pathogens. Res. J. Microbiol. 2 (5), 500–503.

Palla, S., Megha, S., Tiruvalluru, M., Krishna, T.P., Kamala, K., Reddy, G.B., 2005. Curcumin and turmeric delay streptozotocin-induced diabetic cataract in rats. Invest. Opthalmol. Vis. Sci. 46 (6), 2092–2099.

Payton, F., Sandusky, P., Alworth, W.L., 2007. NMR study of the solution structure of curcumin. J. Nat. Prod. 70, 143–146.

Pfeiffer, E., Hohle, S., Solyom, A.M., Metzler, M., 2003. Studies on the stability of turmeric constituents. J. Food Eng. 56 (2/3), 257.

Piwocka, K., Zablocki, K., Wieckowski, M.R., Skierski, J., Feiga, I., Szopa, J., et al., 1999. Novel apoptosis-like pathway, independent of mitochondria and caspases, induced by curcumin in human lymphoblastoid T(Jurkat) cells. Exp. Cell. Res. 249, 299–307.

Reem, M.H., Soliman, H.M., Shaapan, S.F., 2008. Turmeric-based diet can delay apoptosis without modulating NF-kB in unilateral ureteral obstruction in rats. J. Pharm. Pharmacol. 60 (1), 83–89.

Sahu, S., Das, B.K., Mishra, B.K., Pradhan, J., Samal, S.K., Sarangi, N., 2008. Effect of dietary Curcuma longa on enzymatic and immunological profiles of rohu, Labeo rohita (Ham.), infected with Aeromonas hydrophila. Aquacul. Res. 39 (16), 1720–1730.

Sakarkar, D.M., Gonsalvis, L., Fariha, F., Khandelwal, L.K., Jaiswal, B., Pardeshi, M.D., 2006. Turmeric: an excellent traditional herb. Plant Archives. 6 (2), 451–458.

Samarasinghe, K., Wenk, C., Silva, K.F.S.T., Gunasekera, J.M.D.M., 2003. Turmeric (Curcuma longa) root powder and mannanoligosaccharides as alternatives to antibiotics in broiler chicken and diets Asian-Australasian. J. Anim. Sci. 16 (10), 1495–1500.

Sarhan, R., Abd El-Azim, S.A., Motawi, T.M.K., Hamdy, M.A., 2007. Protective effect of tur-
    meric, *Ginkgo biloba*, silymarin separately or in combination, on iron-induced oxidative
    stress and lipid peroxidation in rats. Int. J. Pharmacol. 3 (5), 375–384.
Seppa, N., 2004. Turmeric component kills cancer cells. Sci. News 166 (15), 238.
Shaffrathul, J.H., Karthick, P.S., Reena, R., Srinivas, C.R., 2007. Turmeric: role in hypertri-
    chosis and acne. Indian J. Dermatol. 52 (2), 116.
Shankar, S., Srivastava, R.K., 2007. Bax and Bak genes are essential for maximum apoptotic
    response by curcumin, a polyphenolic compound and cancer chemopreventive agent
    derived from turmeric, *Curcuma longa*. Carcinogenesis 28 (6), 1277.
Sharma, S., Kulkarni, S.K., Chopra, K., 2006. Curcumin, the active principle of turmeric
    (*Curcuma longa*), ameliorates diabetic nephropathy in rats. Clin. Exp. Pharmacol.
    Physiol. 33 (10), 940–945.
Strimpakos, A.S., Sharma, R.A., 2008. Curcumin: preventive and therapeutic properties in lab-
    oratory studies and clinical trials. Antioxid. Redox Signal. 10, 511–546.
Suai, Z., Salto, R., Li, J., Craik, C., Montellano, P.R.C., 1993. Inhibition of the HIV-1 and
    HIV-2 proteases by curcumin and curcumin boron complexes. Bioorg. Med. Chem.
    1, 415–422.
Sundaram, V., 2005. Don't Go Easy on Turmeric. It Prevents and Cures Cancer. India-West.
Suresh, D., Manjunatha, H., Srinivasan, K., 2007. Effect of heat processing of spices on the con-
    centrations of their bioactive principles: turmeric (*Curcuma longa*), red pepper (*Capsicum
    annum*) and black pepper (*Piper nigrum*). J. Food Compos. Anal. 20 (3/4), 346–351.
Tang, M.H., Larson-Meyer, D.E., Lieman, M., 2008. Effect of cinnamon and turmeric on
    urinary oxalate excretion, plasma lipids, and plasma glucose in healthy subjects. Am. J.
    Clin. Nutr. 87 (5), 1262–1267.
Thapliyal, R., Naresh, K.N., Rao, K.V.K., Maru, G.B., 2003. Inhibition of nitrosodiethylamine-
    induced hepatocarcinogenesis by dietary turmeric in rats. Toxicol. Lett. 139 (1), 45.
Tuntipopipat, S., Judprasong, K., Zeder, C., Wasantntwisut, E., Winichagoon, P., Charoenkiatkul,
    S., et al., 2006. Chilli, but not turmeric, inhibits iron absorption in young women from an
    iron-fortified composite meal. J. Nutr. 136 (12), 2874–2970.
Ukil, A., Maity, S., Karmakar, S., Datta, N., Vedasiromoni, J.R., Pijush, K.D., 2003. Curcumin
    the major component of food flavor turmeric, reduces mucosal injury in trinitrobenzene
    sulphonic acid-induced colitis. Br. J. Pharmacol. 139 (2), 209–218.
Vankar, P.S., 2008. Effectiveness of antioxidant properties of fresh and dry rhizomes of *Curcuma
    longa* (long and short varieties) with dry turmeric spice. Int. J. Food Eng. 4 (8), 1–8.
Vijayalakshmi, B., Suresh Kumar, G., Salimath, P.V., 2009. Effect of bitter gourd and
    spent turmeric on glycoconjugate metabolism in streptozotocin-induced diabetic rats.
    J. Diabetes Complications 23 (1), 71–76.
Yoshino, M., Haneda, M., Naruse, M., Htay, H.H., Tsubouchi, R., Qiao, S.L., et al., 2004.
    Prooxidant activity of curcumin: copper-dependent formation of 8-hydroxy-2^2-
    deoxyguanosine in DNA and induction of apoptotic cell death. Toxicol. In Vitro 18 (6),
    783–789.
Zahid, A.M., Hussain, M.E., Fahim, M., 2005. Antiatherosclerotic effects of dietary supple-
    mentations of garlic and turmeric: restoration of endothelial function in rats. Life Sci.
    77 (8), 837–857.

13 The Ornamental *Curcuma*

The cut-flower industry is growing in several parts of the world and there is a huge market for the same in India, as well as in overseas countries. For Indians engaged in this enterprise, it is a money spinner. The boom has occurred especially during the last decade. Many cut flowers are sent from India to overseas countries, the Netherlands being a very important destination. Turmeric also falls in this category and ornamental *Curcuma* falls in this category of exportable tropical cut flowers. The beautiful and attractive flowers of the inflorescence, arising from the apical buds of the underground rhizomes are the ones cut and exported or locally consumed. The turmeric cut-flower group is a new entrant into the cut-flower (floriculture) industry in India. The cut flowers from India occupy the ninth position among those traded in the Dutch market auctions. The flowers are popularly cultivated in Southeast Asian countries, for both the domestic and export markets. Some of the popular ornamental *Curcuma* species include *Curcuma alismatifolia*, *Curcuma zedoaria*, *Curcuma amada*, and *Curcuma angustifolia*. The crop is amenable for hybridization and a large number of promising hybrids have been developed. Studies are also being conducted across the world on various aspects of cultivation, such as the use of growth regulators, shade levels, photoperiodic responses, storage of rhizomes, postharvest physiology, and so forth. Since some of the species are native to India, there is potential to develop new varieties and technologies for commercial cultivation.

Gingers form an important group of tropical flowers of which *Curcuma* are known for their beauty and elegance. Having originated in the Indo-Malayan region, the genus is widely distributed in the tropics of Asia to Africa, including Australia (Sasikumar, 2005). The maximum diversity is concentrated in India and Thailand, with at least 40 species in each region, followed by Myanmar (Burma), Bangladesh, Indonesia, and Vietnam.

Curcuma occupies the ninth position in the cut-flower trade in the Dutch auction market. World over, the market for tropical flowers is small, representing approximately 4% of all cut flowers traded. Germany, Italy, and the United States are the chief importers of tropical flowers. In Dutch auctions, the change in turnover for *Curcuma* is 2.8% in a period of 4 years (Table 13.1). The trade of these flowers is characterized by standard products, large volumes of one or a few chosen varieties, efficiently produced against the background of minimal costs.

Description of Species

Curcuma alismatifolia

This species is native to northern Thailand and Cambodia and is popular as "Siam Tulip" or "Summer Tulip," although it is not related to the tulip as such. The flowers

The Agronomy and Economy of Turmeric and Ginger. DOI: http://dx.doi.org/10.1016/B978-0-12-394801-4.00013-2

Table 13.1 Important Tropical Flowers Sold in the Dutch Auctions

Product	Turnover × 1000				Supply × 1000 stems			
	2002	2004	2006	Change (%)	2002	2004	2006	Change (%)
Cymbidium	66,216	65,217	72,034	2.1	33,233	32,725	33,111	-0.1
Anthurium	41,566	39,631	46,447	2.8	63,827	81,531	84,356	7.2
Strelitzia	6,450	6,858	7,046	2.2	4,905	5,330	5,505	2.9
Leucadendron	4,584	4,836	5,160	3.0	29,486	31,698	33,540	3.3
Heliconia	2,761	3,256	3,651	7.2	2,550	2,897	3,176	5.6
Phalaenopsis	2,217	2,097	3,348	10.9	4,542	5,968	7,185	12.1
Protea	3,231	4,597	3,339	0.8	2,969	4,106	3,074	0.9
Leucospermum	2,857	3,386	3,321	3.8	6,645	7,193	7,727	3.8
Curcuma	1,821	1,810	2,033	2.8	3,473	3,244	3,864	2.7
Paphiopedilum	553	454	455	-4.8	748	616	877	4.1

Source: Federation of Dutch Flower Auctions (VBN) (2007).

and foliage of this are similar to the tulip in appearance. Plant height varies from 20 to 30 cm depending on the variety. The plant spread is 25–35 cm. It can grow as an indoor plant and is commercially cultivated for the cut-flower industry. The plant prefers high humidity, partial sun to partial shade, and sandy loam soil, rich in organic matter. In cool climates, it is planted outside but is dug up and stored in winter. It is sensitive to frost damage.

Curcuma amada

This is a stout underground rhizome. Foliage dies down in late autumn and the rhizomes remain dormant in winter. The inflorescence appears in spring from the base of the rhizome. The peduncle grows to about a height of 20–25 cm. Leaves appear after the emergence of the flowers. When in full growth, the plants can reach a height of about 90 cm. Leaves are broad and very decorative. Good cut flowers have a vase life of about 10 days, when the stem is fresh. This species is also popularly known as "mango ginger" in the countryside, as the rhizomes have an aroma similar to that of green mango. Fresh and dried rhizomes are used to flavor curry.

Curcuma angustifolia

This species is found in the eastern Himalayas and inhabits bright open hillsides and woods. Leaves grow to about 60 cm in length and wither in autumn. Deciduous rhizomes remain dormant underground in water. In early spring, the inflorescence is produced before the emergence of the leaves. Very colorful brachts make this a showy species. The shape and color of the brachts are very variable. The inflorescence lasts in full bloom on the plants for about three weeks or more. The species is good for making cut flowers and has a vase life of about 10 days or more, especially when the cut blooms are fresh.

Curcuma aromatica

This species is found in the eastern Himalayas and inhabits warm forest areas. It grows quickly and vigorously during the summer monsoon months. This is a robust ginger with stout underground rhizomes. Foliage dies down in late autumn and the rhizomes remain dormant in winter. The inflorescence appears in early spring from the base of the rhizome. The peduncle length is about 20–25 cm. Leaves appear after the emergence of the inflorescence. Fully grown, the plants reach a height of about 3 ft. Leaves are broad and very decorative. Good for cut-flower use with a vase life of about 10 days, especially when the cut blooms and stem are fresh.

Curcuma zeodaria

In the Western world, this species is also known as Zedoary. The habitat is in the eastern Himalayas. Flowers appear in early spring, when the weather warms up. Individual inflorescence grows up to 20–25 cm high. Sheaths covering the flower stem are brownish green in color. Floral brachts are light green and the apical flower brachts are deep pink flushed with green. Color of brachts is variable within the red/pink shades. The whole inflorescence remains in a good condition on the plant for about a month or more. It makes a good cut flower with a vase life of more than 10 days, especially if the cut stem is fresh. Leaves are decorative with a deep purple/brown line along the center. Height of the plant is about 90 cm, when fully grown. Leaves dry off in the months of November/December. Rhizomes remain underground and dormant in winter.

There are many other species and varieties of Thai *Curcuma*, in Thailand, Australia, and other countries with respect to germplasm collection, conservation, evaluation, sustainability, as cut flower and potted plants, production, propagation, postharvest management, and storage aspects. The scientific literature has been reviewed and presented below.

Genetic Diversity

Evaluation studies conducted in *Curcuma* showed 16 accessions of *Curcuma* germplasm which includes *C. alismatifolia*, "Chiang Ma Pink" (named after the late great Chinese leader) and "Lady Di" (named after the famous late Lady Diana), and *C. thorelii*, "Chiang Mai Snow," and *C. alismatifolia* "Pink," *C. parviflora* "White Angel," and *Curcuma* sp. "CMU Pride," all of which were evaluated for use as cut flowers and potted plants. All the cultivars of *C. alismatifolia* and *C. thorelii* "Chiang Mai Snow" were considered suitable for cut flower and potted plants. Similar morphological characteristics between *C. thorelii* "Chiang Mai Snow" and *C. parviflora* "White Angel" emphasized that the identification of *Curcuma* accessions using DNA-markers is required for future investigations (Roh et al., 2006). Sujatha and Sujatha (2006) evaluated four species of *Curcuma* for cut-flower use. *C. zedoaria*, on average, produced 2–3 inflorescences per plant with a stalk length of 15–17 cm and flower rachis length of 20 cm. *C. aromatica* produced flowers with a stalk length of 8–10 cm and rachis length of 12–15 cm. *C. amada* flowers had a stalk length of 10–12 cm with a rachis length of 10–15 cm. Optimum harvesting stage of flower stalks was when three-fourths of the flower on the inflorescence was open. *C. zedoaria* had greater potential for use as cut flower as compared to other species.

Genetic Diversity Based on Isozyme Investigations

Investigations were carried out on isozyme analyses to identify genetic diversity in *Curcuma* species, among natural populations compared with cultivated ones. Of the seven enzyme systems analyzed in these investigations, five (ADH, GDH-1, LAP-1, GPI-2, and PGM) showed reproducible and consistent bands. The GPI-2 loci showed the most variable (6 allozymes and 10 zymogram) patterns, followed by GDH-1 with 5 and 5, ADH with 3 and 5, PGM with 4 and 4, and LAP-1 with three allozymes and four zymogram patterns. Cultivated populations from Japan (cJ) and Thailand (cT) had the lowest percentage of polymorphic loci (P = 40–60%), alleles per locus (A_1 = 1.8), alleles per polymorphic locus (A_p = 2.33–3.00), and gene diversity (H_s = 0.216–0.304), compared to lowland populations (L1, L2) with P = 100%, A_1 = 3.0–3.2, A_p = 3.0–3.2, H_s = 0.465–0.496; and six highland populations (H1–H6) with P = 80–100%, A_1 = 2.4–3.8, A_p = 2.4–3.8, H_s = 0.342–0.659.

Breeding

Breeding work on *Curcuma*, which resulted in 16,000 hybrid seeds, demonstrates that conventional breeding can be a practical approach toward developing new varieties. In recent years, a large number of new *Curcuma* hybrids have been developed in Thailand, both for cut flowers and potted plants, which comprise the majority of *Curcuma* commercially marketed (exported) to Japan and Europe.

Cultural Investigations in *Curcuma*

Tuberous (t) Root Number on Flowering Date and Yield

In *C. alismatifolia*, studies on the effect of tuberous root number on flowering date and yield of inflorescences showed that in both varieties, Jezreel Valley and Jordan Valley, propagules bearing fewer than 2 t-roots flowered later than those with 2 or more t-roots. There were no significant differences in the number of days from planting to flowering among plants originating from propagules with 2–5 t-roots. Both the scape length and number of inflorescences per plant were correlated with the number of t-roots on the propagule. However, in Jordan Valley, plants started flowering about 40 days earlier than those in Jezreel Valley, possibly due to higher prevailing temperatures (Hagiladi et al., 1997).

Studies on off-season flowering of *C. alismatifolia* Gagnep. in Chiang Mai Province of Thailand, where the weather in winter is cool, with temperatures between 16°C and 30°C, relative humidity (RH) is 65–70%, and 10 h of daylight, showed that plant growth and flower qualities were similar with and without the night break treatment in August planting, while September to October planting dates required night break treatments to promote flowering and maintain flower qualities (Ruamrungsri et al., 2007).

Mulching

These are interesting studies. Mulching studies in *C. alismatifolia* showed that forcing rhizomes, which were treated at 30°C for 30 days prior to planting, planted in a pipe-house, and mulched with straw/polyethylene sheet or straw/polyethylene tunne, advanced the harvesting date and increased the yield of the cut flower by 1.8–2.2 times, as compared to mere outdoor culture (Hsu, 2001).

Growth Regulators

Curcuma spp. "Precious Petuma," *C. parviflora* "White Angel," and *C. alismatifolia* "Chiang Mai Pink" produced marketable flowering potted plants requiring no provision of shade. "Precious Petuma" and "White Angel" could be grown in shade levels up to 60%, without the use of growth retardants and no deleterious effect on plant quality was observed. "Chiang Mai Pink" could be grown in the open, in full sunlight, and an application of either 10 mg active ingredient (ai) per pot of uniconazole or over 20 mg ai per pot of paclobutrazol for production of a quality flowering plant. These ornamental gingers had excellent postproduction longevity of up to 40 days (Kuehny et al., 2002).

Gibberllic acid (GA) did not increase the number of inflorescence of *C. alismatifolia* but delayed the emergence of shoots and later flowering. GA normally stimulates stem lengthening. However, at a high rate of 600 ppm, it did shorten the inflorescence height in *C. alismatifolia*. Soaking rhizomes in a GA solution at concentrations of 200, 300, 400, and 600 ppm did not inhibit the sprouting of shoots but delayed the process of sprouting. Thus, GA could be used to prolong storage time of ornamental ginger rhizomes prior to planting. However, GA should not be used to promote or increase flowering (Keuhny, 2002).

Sarmiento and Kuehny (2003) investigated the efficacy of paclobutrazol and GA_{4+7} on growth and flowering of three *Curcuma* species. Rhizomes of *C. alismatifolia* "Chiang Mai Pink," *C. gracillima* "Violet," and *C. thorelii* were soaked in GA_{4+7} at 0, 200, 400, and 600 ppm and grown in a greenhouse at 30°C during the day and 23°C at night. When shoot height was 10 cm, the plants were drenched with 118 ml paclobutrazol at 0, 2, 3, or 4 mg ai per container measuring 15.2 cm in diameter. GA_{4+7} delayed shoot emergence and flowering but did not affect flower number. Paclobutrazol rates were not effective in controlling the height of *C. alismatifolia* "Chiang Mai Pink," averaging 85 cm, *C. gracillima* "Violet" averaging 25 cm, or *C. thorelii* averaging 17 cm. *C. alismatifolia* "Chiang Mai Pink," *C. gracillima* "Violet," and *C. thorelii* had postproduction longevities of 4.6, 2.6, and 3.8 weeks, respectively, making these three species of *Curcuma* excellent candidates for use as flowering potted plants.

The effect of growth retardants on "Pink" Thai tulip (*C. alismatifolia*) production was evaluated in controlled growing conditions in Brazil. The treatments were as follows:

1. Absolute control where only water was applied.
2. Paclobutrazol at 20, 25, 30, 35 mg ai/pot, applied as a single drench.
3. A solution mixture of Daminozide sprayed two, three, and five times at a concentration of 2.125 g/l and Chlormequat (0.2 mg/l) sprayed two, three, and five times.

Rhizomes were planted in a 11 plastic pot filled with commercial medium and fertilized (with N, P, and K, secondary and micronutrients) daily. Paclobutrazol (35 mg ai per pot) significantly reduced plant and foliage height and flower stem length, without adversely affecting inflorescence length and prolonging production cycle. However, plants were not compact enough to meet the demands of the market in terms of quality. Foliage height and flower stem length were not adversely affected in a significant manner by the growth retardants, as compared to the absolute control treatment, where only water was applied (Pinto et al., 2006).

Sujatha and Sujatha (2009) conducted studies to improve the quality of *Curcuma* flowers through GA$_3$ application, at flower initiation. *C. amada* and *C. zedoaria* plants were sprayed with GA$_3$ at concentrations of 0, 50, 100, and 200 ppm, and a second spray was applied a fortnight later. It was observed that only in the case of *C. zedoaria*, GA$_3$ spray at a concentration of 200 ppm had a significant effect on stalk length (16.21 and 35.46 cm after the first and second spray, respectively) as compared to control treatment. Similarly, rachis length was also significantly affected by the same treatment (21.23 and 24.23 cm after the first and second spray, respectively) compared to the control treatment. GA$_3$ spray did not have any significant influence on stalk length, as well as rachis length, on *C. amada*. Also, flower yield in both *C. amada* and *C. zedoaria* was not influenced by GA$_3$ application.

Photoperiodic Studies

Photoperiod affects the height of almost all species of ginger. The plants grown under 20 or 16 h photoperiod treatments were taller than those grown under 12 or 8 h photoperiod treatments. Photoperiod affected the number of unfolded leaves of all plants except *C. thorelii*. The 16 and 20 h photoperiod treatments increased the number of unfolded leaves compared to 12 and 8 h photoperiod treatments. Effect of photoperiod on number of rhizomes and number of t-roots was dependent on the species of ginger. *C. petiolata*, *C. cordata*, and *C. alismatifolia* "Siam Tulip White" produced more t-roots when grown under 8 or 12 h photoperiod treatments than when grown under 20 or 16 h photoperiod treatments. The exception was *C. thorelii*, where more t-roots were produced in the plants growing at 20 or 16 h photoperiods than those growing in 12 or 8 h photoperiods. *C. petiolata*, *C. condata*, and *C. alismatifolia* "Siam Tulip White" produced more rhizomes under longer photoperiods 16 or 20 h than those growing under shorter photoperiods (8 or 12 h). The best production strategy for growers of gingers as flowering potted plants is to plant in the spring (April), grow the plants throughout summer, and terminating production in the fall. If production in the winter is desired, a 16 h photoperiod must be employed (Kuehny et al., 2002).

Sarmiento and Kuehny (2004) studied the responses of ornamental gingers to photoperiod variations. In this study, rhizomes of *C. alismatifolia* "Chiang Mai Pink" and tissue-cultured plants of *C. cordata*, *C. petiolata* "Emperor," and *C. thorelii* were grown in a greenhouse under 8–12, 16, and 20 h photoperiods. The results suggest that vegetative growth of gingers, except for *C. thorelii*, was maintained and increased at photoperiods of 16 and 20 h. Photoperiods of 8 and 12 h induced dormancy and t-root production.

Hagiladi et al. (1997) investigated the possibility of extending flower production from autumn to winter, at minimal heating expenses, in *C. alismatifolia*. Initially all the plants were grown 11 weeks in a phytotron, at 29/21°C day/night and natural daylength (11–12 h). Thereafter, the plants bearing five leaves were allocated to four

groups and transferred to grow under a factorial combination of day/night temperatures (23/15°C and 26/18°C) and daylength (10h sunlight, SD, and 10h sunlight +
10 hours supplemental incandescent light, LD). The growth of both above and underground parts was greater at higher temperatures. Yields of inflorescences and propagules were higher under LD than under SD, and there was significant difference
between temperatures under LD.

   Curcuma "Laddawan" is a hybrid cross between *C. cordata* and *C. alismatifolia*
developed by horticulturists of DBIRD, Australia. Over the dry season, all of the
"Laddawan" plants went dormant with the first flush of shoots appearing around
mid-October. Inflorescences first appeared at the beginning of December with the
first stems harvested from plants grown in full sunlight around mid-December and a
week later from plants grown in the shaded area. The total number of inflorescences
harvested from plants grown in full sunlight ranged from 10 to 15 stems/plant, and
15 to 20 stems/plant from those grown in the shaded area. There was only a slight
increase in stem length with inflorescences in plants growing in the shade compared
with those growing in full sunlight, and no significant difference in head length or
diameter could be found. The intensity of flower color was found to be better in the
plants grown in shaded areas. An extended flowering season, almost by four weeks,
was observed in plants grown in shade, as compared to those grown in the sun. To
obtain improved vase life, "Laddwan" should be picked relatively young, when day
flowers are open only at the lower half of the inflorescence (Marcsik et al., 2003).

Nutrition of Ornamental *Curcuma*

Ruamrungsri et al. (2005) found that the optimum levels of N and K for *C. alismatifolia* grown in soilless medium (sand + rice husk + charcoal in the ratio of 1:1 (v/v))
was 200mg/l, three times in a week.

   Ohtake et al. (2006) found that when *Curcuma* (*C. alismatifolia* cv.Gagnep.) was
cultivated in a pot with vermiculite and supplied with different levels of nitrogen (N),
with a high level of N supply, flower number increased and continuous rhizome formation was promoted, but storage root growth was depressed. The N supply to plants
increased N concentration both in the rhizomes and in the storage roots. The predominant nitrogenous compounds related to total N increase related to proteins in
the rhizomes.

Tissue Culture in Ornamental *Curcuma*

Toppoonyanont et al. (2005) conducted an investigation to develop an efficient and
repeatable micropropagation scheme for *C. alismatifolia*, starting from initiation up to
multiplication and rooting stage. During the initiation stage, *C. alismatifolia* coflorescences were removed from each pouch of the inflorescence and used as starting material. These coflorescences were cultured in a modified MS medium, supplemented with
10mg benzyladenine and 0.1 mg IAA/l. After one month in the culture, they developed

and reverted to vegetative shoots located at the same positions and arranged spirally within the brachteole, similar to those *in vivo*. The shoots directly emerged from the flower organs, not via callus formation. During the multiplication stage, factors such as different concentrations of imazalil (IMA) at 0, 2, and 4 mg/l play a crucial role. It was observed that 0.5 mg TDZ combined with 4 mg IMA/l increased emergence of new shoots, by as many as 30–40, from one explants with retarded shoot morphology. Prior to transferring to the greenhouse, these retarded shoots elongated and rooted in a medium containing 0.3 mg benzyladenine and 0.1 mg IAA/l.

Wannakrairoj (1997) investigated cloning methodology to produce plants for export and to conserve germplasm in *C. alismatifolia* Gagnep. Young inflorescences and rhizomes were found to be suitable as sources of lateral bud explants. The rhizomes were air-dried for one week before use. Pretreatment in water at 52°C for 5–10 min greatly reduced bacterial contamination. Plantlets were multiplied on modified MS medium with 0, 6.67, 13.32, 19.98, and 26.64 µmol benzyladenine (BA) or 0.19, 0.56, 1.67, and 5.0 µmol of kinetin/l. The maximum multiplication rate (4.83-fold) was obtained when a longitudinally divided rhizome was cultured in a medium supplemented with 13.32 µmol BA/l. The germplasms collected from wild and selected clones responded in a similar manner to the medium.

Udomdee et al. (2003) could successfully tissue culture rhizome explants of *C. alismatifolia* "Lotus Pink" on 2 mg/l BA-supplemented MS medium. Factors such as sucrose concentration, substrate culture base, and incubation temperature affected germination efficiency of pollen, estimated by *in vivo* and *in vitro* methods, while viability was tested by Aniline blue staining (98–99%). Pollen could not be stored for longer than 24 h, with a subsequent exponential decrease in viability, and cryopreservation or other methods, such as low temperature storage or the use of organic solvents which did not improve the viability of pollen following storage. Flow cytometric analyses showed genetic stability, i.e., no polysomaty, in material derived from any plant part or from different cultivars. A hypothetical model is suggested to explain the development of inflorescences and florets which develop in a predictable sequence of events.

Storage of Ornamental *Curcuma* Rhizome

Investigations on storage conducted in *C. alismatifolia* showed that days to emergence and days to flower were hastened when rhizomes were stored at 25°C for at least 10 weeks (Paz et al., 2005).

Postharvest Physiology of Ornamental *Curcuma*

Chanasut (2005) conducted a preliminary vase life experiment in *C. alismatifolia* "Chiang Mai Pink" flowers. The growth regulators used were GA_3 and BA, which delayed the withering and collapsing of the plant stem. A pulsing treatment with 1 mM silver thiosulfate (STS) for less than 1 h increased the number of opened flowers but had no significant influence on the longevity of the inflorescence.

Cut flowers of Patumma (*C. alismatifolia* cv. Chiang Mai) were held in 50–200 ppm GA$_3$ or distilled water (control treatment) at 25°C with 75–80% RH. GA$_3$ at 100 ppm increased vase life by 4 days compared to the control treatment. Increasing the concentration to 150–200 ppm did not correspondingly extend the vase life despite the improvement in weight retention, absorption capacity, and water conductivity of stem tissues. The vase life of the flower held in 150–200 ppm GA$_3$ was similar to that with 100 ppm GA$_3$. GA$_3$ did not generally affect respiration rate but significantly reduced ethylene production. At a lower concentration of 50 ppm GA$_3$, responses of flowers were comparable to that of the control treatment (Kjonboon and Kanlayanarat, 2005).

Postharvest physiological investigations showed that *C. alismatifolia* cut flower has a considerable length of vase life (usually more than two weeks, in freshly harvested stems). The flowers are chilling sensitive, and cannot be stored dry, but can be stored in water at 7°C for about six days. Because vase life is rather long, it is also possible to store the flowers in water for a few days at ambient temperature (Bunya Atichart et al., 2004). Sujatha and Sujatha (2006) found that *C. zedoaria* recorded a maximum vase life of 10 days, followed by 8 days in the case of *C. longa*, and six days in the case of *C. amada*.

References

Bunya Atichart, K., Ketsa, S., Doorn, W.G., Van, 2004. Post harvest physiology of *Curcuma alismatifolia* flowers. Postharvest Biol. Technol. 34 (2), 219–226.

Chanasut, U., 2005. Treatments to maintain the post harvest quality of "*Patumma*" (*Curcuma alismatifolia* "*Chinag Mai Pink*") flowers. Acta Hortic. 682 (2), 1097–1101.

Hagiladi, A., Umiel, N., Yang, X.H., Gilad, Z., 1997. *Curcuma alismatifolia* I. Plant morphology and the effect of tuberous root number on flowering date and yield of inflorescences. Acta Hortic. 430, 747–753.

Hsu, Y.-M., 2001. Forcing and soil temperature maintenance improved the cut flower productivity of *Curcuma alismatifolia*. J. Chin. Soc. Hortic. Sci. 47 (2), 137–146.

Kjonboon, T., Kanlayanarat, S., 2005. Effects of gibberllic acid on the vase life of cut *Patumma* (*Curcuma alismatifolia* Gagnep) "Chaing Mai" flowers. Acta Hortic. 673 (2), 525–529.

Kuehny, J.S., Sarmiento, M.J., Branch, P.C., 2002. Cultural studies in ornamental ginger. In: Janick, J., Whipkey, A. (Eds.), Trends in New Crops and New Uses ASHS Press, Alexandria, VA.

Marcsik, D., Hoult, M., Connelley, M., Ford, C. 2003. New Heliconia and Ginger Varieties for the Ornamental Industry. <http://www.nt.gov.au/d/content/File/Flower/TAR_ORNAMENTALS. pdf/>.

Ohtake, N., Ruamrungsri, S., Ito, S., Sueyoshi, K., Ohyama, T., Apavatjrut, P., 2006. Effect of nitrogen supply on nitrogen and carbohydrate constituent accumulation in rhizomes and storage roots of *Curcuma alismatifolia* Gagnep. Soil Sci. Plant Nutr. 52 (6), 711–716.

Paz, M., Kuehny, J.S., McClure, G., Graham, C., Criley, R., 2005. Effect of rhizome storage duration and temperature on carbohydrate content, respiration, growth, and flowering of ornamental ginger. Acta Hortic. 673 (2), 737–744.

Pinto, A.C.R., Graziano, T.T., Barbosa, J.C., Lasmar, F.B., 2006. Growth retardants on production of flowering potted. Thai tulip Bragantia 65 (3), 369–380.

Roh, M.S., Lawson, R., Lee, J.S., Suh, J.K., Criley, R.A., Apavatjrut, P., 2006. Evaluation of *Curcuma* as potted plants and cut flowers. J. Hortic. Sci. Biotechnol. 81 (1), 63–71.

Ruamrungsri, S., Suwanthada, C., Apavatjrut, P., Ohtake, N., Sueyoshi, K., Ohyama, T., 2005. Effect of nitrogen and potassium on growth and development of *Curcuma alismatifolia* Gagnep. Acta Hortic. 673 (2), 443–448.

Ruamrungsri, S., Uthai Butra, X., Wichailiux, O., Apavatjrut, P., 2007. Planting date and night break treatment affected off-season flowering in *Curcuma alismatifolia* Gagnep. Gard. Bull. Singapore 59 (1/2), 173–182.

Sarmiento, M.J., Kuehny, J.S., 2004. Growth and development responses of ornamental gingers to photoperiod. Hort. Technol. 14 (1), 78–83.

Sarmiento, M.J., Kuehny, J.S., 2003. Efficacy of paclobutrazol and gibberellin 4+7 on growth and flowering of three Curcuma species. Hort. Technol. 13 (3), 493–496.

Sasikumar, B., 2005. Genetic resources of *Curcuma*: diversity, characterization and utilization. Plant Genet. Resour. 3 (2), 230–251.

Sujatha, K., Sujatha, A.N., 2006. Sangama. *Curcuma*—a potential tropical cut flower. National Symposium on Ornamental Bulbous Crops, 5–6 December 2006, Meerut, Uttar Pradesh, India.

Sujatha, K., Sujatha, A.N., 2009. Improving quality of *Curcuma* flowers through GA application. National Conference on Floriculture for Livelihood and Profitability, 16–19 March, 2009, New Delhi, India.

Toppoonyanont, N., Chongsang, S., Chujan, S., Somsueb, S., Nuamjaroen, P., 2005. Micropropagation scheme of *Curcuma alismatifolia* Gagnep. Acta Hortic. 673 (2), 705–712.

Udomdee, W., Fukai, S., Petpradap, L., Teixeira da Silva, J.A., 2003. Curcuma: studies on tissue culture, pollen germination and viability, histology and flow cytometry. Propag. Ornamental Plants 3 (1), 34–41.

Wannakrairoj, S., 1997. Clonal micropropagation of *Patumma* (*Curcuma alismatifolia* Gagnep). Kasetart J. Nat. Sci. 31 (3), 353–356.

14 Turmeric in *Ayurveda*

The use of turmeric in Indian tradition can be traced to *Vedic* (Hindu scriptural times, derived from the word *Veda*, pertaining to very ancient Indian —Hindu—scripture) times. Turmeric is used to color skin patches in the *Atharva Veda* and *Taittiriya Brahmana*. There are also references to the intake of turmeric powder with honey to enhance memory and with ghee (melted butter) to counteract snake bite. In patients suffering from cardiac diseases and jaundice, rice mixed with turmeric is taken internally and also applied all over the body. In the earliest writings of *Ayurveda*, dating several centuries prior to the Common Era (Christian era), the medicinal properties of turmeric have been elucidated in great detail. The *Charaka Samhita, Susruta Samhita, Bhela Samhita, and Kasyapa Samhita* make mention of turmeric as an essential dietary ingredient and medicine as well.

Nomenclature of Turmeric

Turmeric is known by a variety of synonyms in *Sanskrit*, and these synonyms sketch its main characteristics. Apart from *Haridara*, the name *Ranjani* gave to turmeric, means that which is used to dye clothes. *Pita* indicates the bright yellow color of the rhizome. The names *Varavarnini* points to color and its auspiciousness. *Lomasamulika* means that the rhizome is hairy and *Pindaharidra* indicates that the rhizome assumes an entangled shape. The yellow color of turmeric is likened to a golden hue, hence the name *Kancani.*

Turmeric has another very interesting *Sanskrit* name—*Hattavilasini*, which means one that shines in the market. This simply emphasizes the tremendous commercial value of the plant. *Yoshitpriya* indicates that turmeric was popular among women for anointing the body. Names such as Mangalya, Lakshmi, and Pavitra, all of which pertain to purity and prosperity in Hindu culture, indicate the auspiciousness associated with turmeric. Other names of turmeric offer clues to its medicinal properties. Turmeric has antihelmintic and antimicrobial properties, which the name *Krmighna* (*Krimi* in Sanskrit means microbes) suggest. So it is also an antidote against poison because it carries the name *Visagnhi* in Sanskrit. The term has a root *Visa* meaning poison, and *Agnhi*, which means fire. Thus, it is a fire that extinguishes poison. Turmeric also carries the Sanskrit name "Varnavilasini," ("varna" in Sanskrit meaning color, "vilasini" meaning one which makes an object glitter/shine) meaning it imparts excellent complexion to the human body (skin), and, therefore, the person who wears it "shines" in lustrous appearance. It finds special application in the management of diabetes and is therefore known as "Mehaghni."

Interestingly enough, turmeric is also known as *Nisha* and *Rajani*, both meaning night. The connection between turmeric and night is not explicitly clear in *Ayurveda*,

The Agronomy and Economy of Turmeric and Ginger. DOI: http://dx.doi.org/10.1016/B978-0-12-394801-4.00014-4

but there is a prevalent viewpoint that turmeric rhizomes collected at night are more potent in their effectiveness. It remains to be scientifically established whether turmeric rhizomes collected at night have higher potency because of their higher alkaloid content and other active ingredients, such as oleoresin and essential oils. There are many aspects of the ancient system of Indian medicine *Ayurveda*, which also has a decided connection with spirituality. It is only of late that many aspects of this thought, hitherto unknown to the scientific world, are coming to light, and one is wonderstruck by the subtle yet deep connections between the two.

Turmeric Varieties Used in *Ayurveda*

Only one species of turmeric is described in Ayurvedic use, which, indisputably, is *C. longa*. However, the cultivated and wild varieties (*Vanaharidra*) of turmeric were distinguished in ancient times. Apart from this, several related and nonrelated species of medicinal plants have been listed as varieties in classical texts. The plant *Berberis aristata* is popularly known as Daru *Haridara* (*Haridara* in Sanskrit referring to turmeric) or woody turmeric, because it also yields a yellow dye similar to that from turmeric, and also because these two plants are used frequently in combination. Turmeric and woody turmeric are known as the Turmeric Duo (*Haridara Dvaya*) in classical Ayurvedic texts. *Coscinium fenestratum* has also come to be known as woody turmeric. Certain other species of *Curcuma* have also been referred to by the epithet *Haridara*. For instance, *C. amada* is known as *Amragandhi Haridara* in the classical Ayurvedic texts. This is on account of the morphological resemblance between the two plants. Another variety known as wild turmeric or *Aranya Haridara* or *Kasturi Manjal* (*C. aromatica*) is used in combination with astringents and aromatics to cure bruises, sprains, hiccoughs, bronchitis, cough, leukoderma, and skin eruptions.

Pharmacological Properties of Turmeric

The authoritative texts on *Ayurveda* have listed the pharmacological properties of turmeric. Turmeric has a pungent and bitter taste, is hot in potency, and dries up secretions. It has a pungent postdigestive taste. It is useful in a wide range of ailments such as diabetes, skin diseases, itching, swelling, poisoning, anemia, wounds, ulcers, sinusitis, loss of appetite, worms, and tumors, as detailed in Table 14.1. It mainly regulates *Kapha* (one of the causes of diseases in *Ayurveda*, which refers to accumulation of fluids in the body), on account of its potency and pungent taste. Due to its bitterness, it also normalizes *Pitta* (another cause of diseases in *Ayurveda*, denoting malfunctioning of the liver). To some extent, it can also balance *Vata* (another cause of diseases in *Ayurveda*, denoting arthritic pain). *Kapha*, *Pitta*, and *Vata* represent the three principal categories of *Ayurveda*, under which all physiological functions can be subsumed. *Kapha* stands for functions that build structure, secretions, immunity, and the like. *Pitta* stands for digestion, metabolism, and other

Table 14.1 Principal Uses of Turmeric (as Single and/or Combination Drug, with or without Adjuncts) in *Ayurveda* (A) and Home Remedies (B)

Main Drug Turmeric Dry (D)/Raw (R)	Combination Drug Combination (C)/Adjunct (Ad)	Uses—External (E)/ Oral (O)	Function	Docum (A) Oral (HR)	References
Powder (D)	Mustard paste (C)	Leech therapy	Weaning leech and to disinfect wound	A	Ibid p.36,404, 433, 434
Paste (R)	Neem (C) leaves	Ringworm, scabies chicken pox	Wormicidal, antifungal skin remedy	HR	–
Paste (D)	Gooseberry juice (C)	Antidiabetic (O)	Microvascular damage/prevention	A	–
Powder (D)	Jaggery (Ad)	Flush the calculus (O)	Dislodge urinary stones	A	Vaidya (2002)
Powder (D)	Cow's urine (Ad)	Antitoxic (O)	Nullifies the plant and animal poisons	A	Ibid. p.36,404, 433, 434
Powder (D)	*Euphorbia* species (C)	Alkalifying effect (E)	Cauterizes the hemorrhoids	A	Trikamji (1980)
Rhizomes (D)	Oil (Ad)	Inhalation (E)	Nasal decongestant	H	–
Powder (D)	Buttermilk (Ad)	Diet (O)	Digestive antidiarrheal, antimicrobial	H	–
Concentrated jelly-like substrate of decoction of the rhizomes	*Berberis aristata* concentrated jelly extract	Application of paste over the wound (E)	Aids wound healing especially over the joints	A	Trikamji (1980)

D = Dry rhizome, R(J)=raw juice, MD = major drug, CD = combination drug, Kal = kalkan, PC = Prakshepa Churnam, O = oral, E = external, Kash = Kashayam (an Ayurvedic concoction) drug, WT = woody turmeric.

biochemical transformations, and *Vata* represents control, movement, and regulation of physiological functions.

The part used in *Ayurvedic* preparations is the rhizome, which is chopped up into small pieces and a paste is made in the fresh state. If the rhizome is dried, it is powdered and used both internally and externally. The rhizome, which is bulky and has a saffron color when sliced, is recommended for medicinal use (Sastri, 2002, p. 409).

Turmeric is unknown to cause any adverse side effects following prolonged use. However, there is a very interesting reference to a method to purify turmeric before use in *Ayurveda*. Turmeric is to be boiled in a broth containing cow's urine (in Indian medicine, cow's urine has a very important role as it contains unique medicinal properties), the juice of *Luffa acutangula* and a combination of five tender leaves each of *Mangifera indica* (mango), *Syzygium cuminii*, *Limonia acidissima*, *Citrus acida*, and *Aegle marmelos*. The broth is then steamed in the vapors of a group of aromatic plants, which include *Abies spectabilis*, *Cinnamomum tamala*, *Vetiveria zizanioides*, *Cyperus rotundus*, *Saussurea lappa*, and *Sida rhombifolia*. This procedure makes it completely nontoxic.

It is to be noted that excessive use of turmeric in food can make the food taste bitter. It can also cause thinning of blood, and hence, must be very cautiously used with anticoagulants, especially when treating the human body with profusely bleeding wounds.

Use of Turmeric in *Ayurveda*

Turmeric is used singly, or in combinations, in Ayurvedic preparations, but instances of its use in combination with other herbs and drugs outnumber those where it is used singly, as seen in Table 14.2.

Turmeric powder is used to wean the leech away after it is made to suck blood from the body of the affected patient for therapeutic purposes (Trikamji, 1980). This practice has a dual purpose. The primary purpose is to release the grip of the leech. Secondarily, turmeric serves as an antiseptic and healing agent to close the wound. A very popular combination of turmeric with neem leaves is used widely by Indians to manage a range of skin eruptions like ringworm, scabies, and chicken pox. The same combination taken internally is effective in managing worm infestations.

Turmeric made into a paste in the juice of gooseberry fruit is recommended for regular use in diabetes (Vaidya, 2002). Clinical observations suggest that turmeric may prevent microvascular damage in diabetes more effectively than in regulating total blood sugar levels. Turmeric with jaggery taken in water is claimed to flush out kidney stones (Joshi, 1939, p. 505). Turmeric is advised to be taken in cow's urine as a general antidote for all kinds of poisons (Sharma, 1987). Turmeric powder mixed in the latex of *Euphorbia* species is advised to be taken as an external application to cure hemorrhoids (Trikamji, 1980).

It is a common practice in India to inhale the smoke generated by burning dry turmeric rhizomes dipped in oil to relieve nasal congestion. In some texts of *Ayurveda*, turmeric is used along with other herbs for the same purpose. Caution is to be

Table 14.2 Important Examples of Reputed *Ayurvedic* Preparations Where Turmeric Is Used as Major/Auxiliary Ingredient

Category of Medicine	Name of Formulation (Medicine)	Form of Turmeric—D, R(J) (with WT)	Turmeric Added as MD/CD/ Kash/Kal/PC	Administered— O/E	Therapeutic Claims	References
Churnam	Haridara Khandam	D	MD	O for 7 days	Antiallergic, useful in urticaria	Vaidya (2002)
	Sudarshana Churnam	D	CD	O	Antipyretic in all kinds of fevers especially due to water-borne infection and causes due to use of incompatible medication	Sastri (2002, p. 784)
Churnam	Rajanyadi Churnam	D	CD	O to be licked	Digestive, appetizer, useful in a range of pediatric disorders especially in duodenal disorders	
Kashayam	Nisha Katakadi Kashayam	D	CD	O	Diabetes	Sharma (1987)
Lehyam	Brahma Rasayanam	D	PC	O	Rasayanam which promotes longevity	Trikamji (1980)
Ghritam	Mahapanc agavya ghritam	D	CD	O	Epilepsy and psychotic disorders, constant use is advocated	Sastri (1980)
Ghritam	Kalyanaka ghritam	D	CD	O	Indicated in a range of psychological disorders and in infertility	Sastri (1980)
Thailam	Lakshadi Thailam	D	CD	E	For pediatric use, prevents cold and upper respiratory disorders, nourishes children	Joshi (1939, p. 505)
Thailam	Jathyadi Thailam	D	CD	E	Cleanses and aids healing. Useful in chronic and fistulous tract	Vaidya (2002)

D = Dry rhizome, R(J)=raw juice, MD = major drug, CD = combination drug, Kal = kalkan, PC = Prakshepa Churnam, O = oral, E = external, Kash = Kashayam (an Ayurvedic concoction) drug, WT = woody turmeric.

exercised when inhaling turmeric smoke, as excessive inhalation can lead to nasal bleeding.

Buttermilk boiled with turmeric powder is a household remedy for diarrhea in India. This recipe acts as a digestive stimulant, antimicrobial, and antidiarrheal. A decoction made out of turmeric and woody turmeric is an excellent agent for healing all kinds of wounds (Joshi, 1939, p. 403).

Excellent Turmeric-Based Ayurvedic Formulations

Turmeric is an important ingredient in many excellent Ayurvedic formulations, which are extensively used in India. *Haridara Khanda* is a formulation that is used in the management of infection from worms and allergies (Sastri, 1980). *Rajanyadi Churnam* is a formulation with turmeric, which is the key ingredient used in the treatment of many pediatric diseases (Vaidya, 2002). *Lakshadi Thailam* is a turmeric-based oil (coconut oil or gingelly oil—*Sesamam*) which is used as an antidote for cold and any upper respiratory problems, especially in children (Joshi, 1939, p. 641). The formulation is gently massaged on the chest. *Jathyadi Thailam* is an important oil for healing chronic wounds (Joshi, 1939, p. 641). *Nisha Katakadi Kashayam* is an excellent formulation in the management of diabetes (Sharma, 2004). Some of the other important turmeric-based formulations are *Brahma Rasayanam*, which enhances memory and body immunity (Trikamji, 1992, p. 378). *Kalyanaka Ghritam* is an antidote to mental disorders and is also an anti-inflammatory agent (Sastri, 1980). *Mahapancagavya Ghritam* is a mental stimulant (Trikamji, 1992, p. 475), *Punarnava Manduram* combats anemia (Trikamji, 1992, p. 475), and *Sudarshana Churnam* combats bodily fever.

The Multifaceted Uses of Turmeric in *Ayurveda*

A review of classical and antique literature on *Ayurveda* reveals that turmeric has multifaceted uses in clinical medicine. It improves skin tone and complexion. It heals wounds and works as an antiseptic and antimicrobial agent. It is useful in the management of chronic wounds and skin lesions from ringworm infection and erysepalis.

Turmeric is both an excellent carminative and a liver tonic. It aids digestion and is an effective antidote against poisoning—both external due to snake and scorpion bite and internal due to toxins in ingested food. It is an ingredient of a formulation that purifies breast milk (Yadavji, 1980, p. 168) and has been known to have immunomodulatory properties (Trikamji, 1992, p. 378).

In addition to improving the blood circulatory process within the body, it also aids in building up body fat.

In Ayurvedic preparations, which are used against both benign and malignant tumors, turmeric is an important ingredient (Trikamji, 1980). Turmeric has also been found to regulate uterine activity and minimize menstrual pain. Along with other herbs, turmeric is used in the management of respiratory disorders. Powder of turmeric, black pepper, raisins, jaggery, galangal root, long pepper, and Kaemferia galangal, mixed with oil, either coconut or sesame, has been found to relieve respiratory problems (Yadavji, 1980, p. 331).

A simple formulation of turmeric with *Terminalia chebula*, *Terminalia belerica*, *Emblica officinalis*, *Azadirachta indica* (the popular neem), *Sida cordifolia*, and Licorice root along with milk and ghee (melted butter) made out of buffalo milk is recommended in the treatment of jaundice (Yadavji, 1980, p. 276).

External application of turmeric, woody turmeric, and red ocher is said to neutralize the poisonous effects inflicted through the nails of animals (Trikamji, 1992, p. 378).

In the event of thirst, due to the derangement of *Kapha* in humans, a drink made of turmeric and sugar is recommended, as it is considered beneficial (Yadavji, 1980, p. 276). An important use of turmeric is found in its ability to detoxify adverse reactions from metals, such as tin and iron and mineral, such as mica present in allopathic medicine (Satpute, 2003). It is also an important ingredient in the formulation of medicated enemas (Trikamji, 1980).

Corroboration with Scientific Evidence

Modern scientific research has corroborated many of the claims made on the unique medicinal value of turmeric, based on the details mentioned in Ayurvedic text. Most of the uses which are now being substantiated by modern research were anticipated in early *Ayurveda*. It was at the initiative of the Director General of the Council of Scientific and Industrial Research in New Delhi, India, that the US Patent given for turmeric products to two Indian scientists working in the University of Mississippi Medical School in the United States, on the "use of turmeric in wound healing" was revoked. This was a globally sensational case of biopiracy because turmeric has been mentioned in the Indian *Ayurveda* for this unique property, as discussed earlier in this chapter. India's successful contesting of the patents given by the US Patent and Trade Mark Office (USPTO) clearly demonstrated how developed countries clandestinely exploit the developing countries for pecuniary benefit. The revocation was the first instance in the history of USPTO and was a watershed in biopiracy.

Scientific research has revealed that curcumin is the most important active ingredient in turmeric. Prior to the advent of green chilies, it was customary in the Indian tradition to combine turmeric with black pepper in the preservation of food articles. Researchers at St. John's Medical College, Bengaluru, India, found that a combination of curcumin and piperine increased the absorption of curcumin 2000-fold, without causing any adverse side effects (Shoba et al., 1998).

Ayurvedic texts indicate the effectiveness of turmeric in containing the onset and spread of skin disease, known in India as *Pandu*, which bleaches the skin in patches or on the entire human body and gives a repelling look to the individual. The disease also includes anemic conditions such as leukemia. Researchers at the Loyola University Medical Center in Chennai, Tamil Nadu, have suggested that regular use of turmeric might prevent the incidence of childhood leukemia. Perhaps this is one very important reason for the lower incidence of leukemia in parts of the world, such as Asia, where turmeric use, internally and externally, is extensive.

Excess of NF-κB, a powerful protein which promotes abnormal inflammatory response in the human body, can lead to malignancy, arthritis, a wide range of other, not easily, curable diseases. Investigations show that curcumin suppresses the activity of NF-κB, suggesting that it is indeed the pathway in which the spice shows its powerful protective properties in sustaining good human health.

It is also seen that turmeric has the unique property in preventing blood vessels from growing in tumors. It has also been shown to inhibit the deposition of fat in the body, which precludes the growth of adipose tissue, in a similar fashion as described above, by inhibiting angiogenesis in fat tissues (Aggarwal et al., 2005).

Indeed, the role of turmeric in human health is vast. Its potential must be further tapped, through systematic and concerted modern research tools, so that its beneficial effects can be brought for the benefit of all of humankind. That is a task cut out for agricultural scientists, as much as for those dealing with modern research in all branches of medicine.

References

Aggarwal, B.B., Shisodia, S., Takada, Y., Banerjee, S., Newman, R.A., Bueso-Ramos, C.E., et al., 2005. Curcumin suppresses the paclitaxel-induced nuclear factor-kappa B pathway in breast cancer cells and inhibits lung metastasis of human breast cancer in nude mice. Clin. Cancer Res. 11 (20), 7490–7498.

Joshi, S.S., 1939. Yogaratnakara. Chaukhambha Sanskrit Series Office, Benares, Uttar Pradesh, India, p. 505.

Sastri, A., 1980. Bhaisajya, Ratnavali, Chaukhambha Sanskrit Sansthan, Varanasi, Uttar Pradesh, India, p. 79.

Sastri, B.R., 2002. Chaukhambha Sanskrit Sansthan, Varanasi, Uttar Pradesh, India, p. 409.

Satpute, A.D., 2003. Rasaratnasamucchaya. Chaukhambha Sanskrit Pratishthan, Delhi, India, pp. 23, 31, 50, 51, 131, 141, 144, 149, 155.

Sharma, P., 1987. Chikitsakalika. Chaukhambha Surabharati Prakashan, New Delhi, India, p. 87.

Sharma, R.N., 2004. Sahasrayogam. Chaukhambha Sanskrit Pratishthan, Delhi, India, p. 275.

Shoba, G., Joy, D., Joseph, T., Majeed, M., Rejendran, R., Srinivas, P.S., 1998. Influence of piperine on the pharmacokinetics of curcumin in animals and human volunteers. Planta Med. 64 (4), 353–356.

Trikamji, Y., 1980. Sushruta Samhita. Chaukhambha Orentalia, Varanasi, Uttar Pradesh, India, p. 57.

Trikamji, Y., 1992. Caraka Samhita. Munshiram Manoharlal Publishers, Delhi, India, p. 378.

Vaidya, H., 2002. Ashtanga Hridayam. Varanasi, Uttar Pradesh, India, p. 943.

Yadavji, T., 1980. Sushruta Samhita. Chaukhambha Oreientalia, Varanasi, Uttar Pradesh, India, p. 168.

15 The Agronomy and Economy of Ginger

Introduction—A Peek into the History of Ginger

Among the spices of the world, ginger assumes considerable importance, along with turmeric, as one of the most important and sought-after medicinal spices. Ginger, botanically known as *Zingiber officinale* Rosc., belongs to the family Zingiberaceae and in natural order Scitamineae (Zingiberales of Cronquist, 1981). Owing to its universal appeal, its spread has been rapid to both tropical and subtropical countries, from the China–India region, where ginger has been cultivated from time immemorial. Ginger was most valued for its medicinal properties, in ancient times, and also played a very important role in primary health care in India and China. As a carminative, it was widely used in European medicines as well.

The Latin term *Zingiber* was derived from the ancient Tamil (one of the regional languages of Southern India—Tamil Nadu) words, *Ingiver*, meaning ginger rhizome. Arab traders, in search of spices, took the term to Greece and Rome, and from there to Western Europe. The present-day name of ginger in most Western European countries is derived from this ancient term. Examples of this are ingefaer (Danish), gember (Dutch), ginger (English), zingibro (Esperanto), barlik ingver (Estonian), inkivaari (Finnish), gingembre (French), and ingver (German). A catalog of these names, along those related to Indian languages, is given in Table 15.1.

Earlier some authors thought that the term *Zingiber* was derived from the ancient Indian Sanskrit, *singavera* (Purseglove et al., 1981; Rosengarten, 1969; Watt, 1872), meaning antler-like or horn-shaped, indicating the shape of the rhizome. It is improbable because Sanskrit was not popular those days in the regions in question. Ginger was exported from the Malabar coast, in Kerala State, India, and the Arab traders might have used only the prevalent local Tamil name for trading the commodity. Mahindru (1982) opined that the original word for ginger was, in all probability, a pre-Dravidian one, and that it is found with minor variations in about 20 languages extending from China and the islands of the Pacific Ocean to England. In some languages, there are separate terms for fresh and dried ginger, which points to the fact that both forms of ginger are put to specific use (Table 15.2).

As early as second century AD, ginger was one among the very few items on which duty was levied at the Alexandria port of entry, during the time of the Roman Empire (Flukiger and Hanbury, 1879). During subsequent periods and in the Middle Ages, ginger was on the list of privileged goods in the European trade, and a duty was levied. In England, it must have been well known even before the Norman Conquest, for it is frequently named in the Anglo-Saxon beech-books of the eleventh century, as well as in the Welsh "Physician of Myddvai" (Parry, 1969). During the

The Agronomy and Economy of Turmeric and Ginger. DOI: http://dx.doi.org/10.1016/B978-0-12-394801-4.00015-6

Table 15.1 Ginger—its Varied Names

Languages	Colloquial Names
Indian Languages	
Assamese (Assam State)	Ada
Bengali (West Bengal)	Ada
Gujarati (Gujarat State)	Adhu (fresh ginger), Sunth, Shuntya (dried ginger)
Hindi (India's national language)	Adi, Adrak (fresh ginger), Sonth (dried giner)
Kannada (Karnataka State)	Alla (fresh ginger), Sunthi (dried ginger)
Kashmiri	Sho-ont
Malayalam (Kerala State)	Inchi (fresh ginger), Chukku (dried ginger)
Marathi (Maharashtra State)	Alha, Aale (fresh ginger), Sunth, Shuntya (dried ginger)
Oriya (Odisha State)	Ada, Adraka
Sanskrit (Ancient Indian script)	Adraka (fresh ginger), Shunthi (dried ginger), Shringaveran, Sringaaran, Nagara
Tamil (Tamil Nadu)	Ingee, Ingiver (fresh ginger), Chukku (dried ginger)
Telugu (Andhra Pradesh)	Allam
Urdu (Indian and Pakistani language)	Adraka
Other Languages	
Arabic	Zanjabil
Brazilian	Mangaratia
Burmese	Gin, Gyin sein, Khyenseing, Ginsi-kyaw
Chinese	Jeung, Sang keong, San geung, Chiang, Jiang, Keong, Shen jiang, Gan jinang, Shengjiang
Czech	Zazvor
Danish	Ingefaer
Dutch	Gember, Djahe
English	Ginger
Esperanto	Zingibro
Estonian	Harilik ingver
Ewe	Nkrawusa, Nkrama, Nkrabo, Agumetakui
Fante	Akakadur, Tsintsimir, Tsintsimin
Farsi	Jamveel, Zanjabil
Finnish	Inkivaari
French	Gingembre
German	Ingver
Hausa	Chitta, Afu
Hebrew	Zangvil
Hungarian	Gyomber
Icelandic	Engifer
Indonesian	Jahe, Aliah, Jae, Lia
Italian	Zenzero, Zenzevero
Japanese	Shouga, Myoga, Kankyo, Shoukyo, Kinkyo
Khmer	Khnehey, Khnhei phlung
Laotian	Khing
Malay	Halia, Atuja, Jahi, Keong phee, Kong Keung
Norwegian	Ingefaer

(Continued)

Table 15.1 (Continued)

Languages	Colloquial Names
Persian	Shangabir, Zangabi
Polish	Imbir
Portuguese	Gengibre
Romanian	Ghimbir
Russian	Imbir
Scandinavian	Ingefaer
Singhalese (originally Ceylonese language)	Inguru
Spanish	Jengibre
Swahili	Tangawizi
Swedish	Ingefara
Tagalog	Luya
Thai	Kinkh, Khing-daen
Tibetanese	Gamug, Sga smug, Sman-sga
Turkish	Zencefil
Twi	Akakaduru, Kakaduru
Vietnamese	Gung, Sinh khuong

Table 15.2 Equivalent Names for Ginger in Some Languages

Language	Plant	Fresh	Dried
Hindi	Adrak	Adrak, Adhruka	South, Saindhi
Bengali	Ada	Adrok	Sont
Assamese	Ada	Adrak	Sonth
Punjabi	Ada, Adrak	Aunjbel	Sanjzabil, South
Marathi	Adu, Aale	Alen, Alem, Adrak	Sonth, Sunta, Sunt
Gujarathi	Adu, Adhu	Adu, Adhu	Sunt
Tamil	Ingee	Ingee, Ingiver	Chukku
Malayalam	Inchi	Inchi	Chukku
Telugu	Allam	Allam	Sonthi
Kannada	Sunthi	Hasisunthi	Vana sunthi
Burmese	Khyenseing	Ginsin	Ginsi-khaiv
Singaporean	Ingru	Ammuingru	Velicha-ingru
Sanskrit	Adraka, Sringavara	Ardrakam	Vishva-bhishakam Nagfara, Sunti Mahaushadha
Arabic	-	Sanjzabile-ratal	Sanjzabile-Yabis
Persian	-	Zanjzabil-tar	Zanjzabil-Khushk

thirteenth and fourteenth centuries, next to pepper, ginger was the most common and most precious of spices, costing nearly seven scrolling per pound, or about the price of a sheep. The merchants of Italy during the thirteenth and fourteenth centuries knew three kinds of ginger, *belledi*, *colombino*, and *micchino*. *Belledi* is an Arabic word meaning "country," and was probably the common ginger. *Colombino* referred probably to Columbum, Kollam, or Quilon (in the State of Kerala), an ancient port on the southern Malabar coast of Kerala State, and *micchino*, the ginger brought from Mecca (which again only comes from the Malabar coast; Mahindru, 1982; Watt, 1872). The literature also indicates that ginger preserved in syrup (called green ginger) was also imported to the Western world during the Middle Ages and was regarded as a delicacy of the choicest kind. In Zanzibar on the east coast of Africa, ginger is regarded as auspicious and is absolutely necessary to the Savaras tribe for their religious and marriage functions.

There is mention of ginger in the Koran (ref: 76: pp. 15–17), which says "Round amongst them (the righteous in paradise) is passed vessels of silver and goblets made of glass...a cup, the admixture of which is ginger." In the Middle Ages, ginger was considered to be so important a spice that the street in Basle where Swiss traders sold spices was named *Imbergasse*, meaning "Ginger Alley" (Rosengarten, 1969). In Henry VIII's time, ginger was recommended against plague. It was during that time that "gingerbread" became popular, and it turned out to be the favorite of Queen Elizabeth I and her court. The legend has it that around 2400 BC, a baker on the Isle of Rhodes near Greece prepared the first gingerbread, which shortly thereafter found its way to Egypt, where the Egyptians savored its excellent flavor and served it on ceremonial occasions. The Romans distributed it to the entire Roman Empire (Farrell, 1985).

In the Middle Ages and until the end of the nineteenth century, English tavern keepers used to have ground ginger in constant supply for the thirsty customers to sprinkle on top of their beer or ale and then stir into the drink with a red-hot poker (Rosengarten, 1969). The Western world herbalists and naturalists knew the great qualities of ginger, as confirmed by the well-known British herbalist John Gerad. He writes in his treatise (Gerad, 1577, cited by Parry, 1969) that "ginger is right good with meat in sauces," and adds further, this spice is "of an eating and digesting quality, and is profitable for the stomach, and effectively opposeth itself against all darkness of the sight, answering the qualities of pepper" (Parry, 1969).

India and Ginger

Ginger was not significant as a spice in ancient India, unlike black pepper or cardamom, but was *mahabbeshaj, mahaoushadhi*, literally meaning the great cure, the great medicine. For the ancient Indian, ginger was the God-given panacea for a number of ailments. That may be the reason why ginger found a place of pride in ancient Ayurvedic texts of Charaka (*Charaka samhita*) and Susruth (*Sushrutha samhita*). In *Ashtangahridayam* of Vagbhatt (a very important ancient Ayurvedic text), ginger is recommended along with other herbs for the cure of elephantiasis, gout, extenuating the juices, and purifying the skin from all spots arising from scorbutic acidities. Ginger is also recommended when exotic faculties were impaired due to indigestion.

Rabbi Benjamin Tudella, who traveled between 1159 and 1173 AD and gave an account of spices grown on the West Coast of India, is credited to be the first to have mentioned ginger cultivation. Tudella gives a vivid description of the place and trade in spices as well as cultivation of spices in and around the ancient port of Quilon (now Kollam) in Kerala State, India (Mahindru, 1982). Marco Polo (AD 1298) in his famous travelogue writes "good ginger also grows here and is known by the name of Quilon ginger. Pepper also grows in abundance throughout the country" (translation by Menon, 1929). Another explorer, Friar Odoric (AD 1322) writes: "Quilon is at the extremity of pepper forests towards the south. Ginger is grown here, better than anywhere else in the world and in huge quantities." In those days, Calicut (now Kozhikode), Cochin (now Kochi), Allepey (now Alappuzha), and Quilon (now Kollam) were the ports through which all the spices were traded with the Western world. Nicolo Corai (AD 1430) describes Calicut as the "Spice Emporium of the East." He described it as a maritime city of eight miles in circumference, a notable emporium for the entire India, abounding in black pepper, aloe, ginger, and a large kind of cinnamon, myrobalans, and zedoary. Linschotten (1596) gives a very interesting account of the spices. He states that ginger grew in many parts of India, but the best and the most exported ones grew on the Malabar coast. He described the method of cultivation and preparation that appear to be similar to the present-day practices. Linschotten also wrote about the ginger trade and mentioned that ginger was mainly brought to Portugal and Spain from the West Indies, indicating the fact that the Portuguese were successful in cultivating ginger extensively in Jamaica and the adjoining West Indies Islands. Flukiger and Hanbury (1879) wrote: "… it (ginger) was shipped for commercial purposes from the Islands of St. Domingo, as early as 1585 and from Barbados in 1654. Reny (1807) mentions that in 1541, 22053 cwt of dry ginger was exported from West Indies to Spain" (Watt, 1872).

The most significant event in the history of spice trade was the landing of Vasco da Gama on the West Coast of India. da Gama started from Lisbon in Portugal, arrived at Mozambique in March 1498, and from there reached Mlinde by the end of April. The King of Mlinde had advised da Gama to sail to Calicut (now Kozhikode in Kerala State, India) and arranged for an Arab pilot to help him. This Arab brought the Portuguese explorer across the Arabian Sea in 20 days, and on March 17, 1498, da Gama anchored in Kappad, a tiny hamlet near Calicut, on the Malabar Coast. Following this, a wave of expeditions arrived on the West Coast of India, known at the time as the Malabar Coast, and the spice trade with Europe flourished. The arrival of the Portuguese also signaled the end of the Arab monopoly on the sea route, and consequently, on the spice trade. da Gama reentered India commanding an armada of 15 ships. Through the technique of intimidation, coercion, and bribe, he entered into an understanding with the then King of the independent Kochi State, to obtain all the rights for a free trade in spices. Subsequently, in 1513 AD a treaty was signed with the King of Calicut as well (known as the Zamorin then and now), ending the decade-long "spice war" between the Arabs and the Portuguese, securing for the Portuguese not only a big stake in spice trade but also in getting a foothold to enter India as part of the future of the colony. Subsequently, the Dutch and the British came into the fray, but the latter succeeded in remaining in India for the longest period. Hence, if one traces Indian history, colonization of India starts with a "spice war." Though the British succeeded in usurping Indian territory for the longest period, when the East India Company handed over the power to the British crown, the Portuguese and French also succeeded in having their colonies within India. Goa, an island city on the western coast near Mumbai (formerly Bombay), was the colony of the Portuguese until about mid-1960s, when the late Prime Minister Jawaharlal Nehru had to use military force to get the Portuguese evacuated. The French managed to keep their stranglehold through clever politics and Mahe, an island town near Calicut, still enjoys special status as a "Union Territory," with its citizens able to opt for French citizenship. Indeed, spice was the focal point in all these historical developments in India. Through the treaty that the Portuguese entered into with the Zamorin Government in Calicut (Kerala State), the former obtained a license to trade spices freely. The inefficient government and tendency to bribery of the Zamorin led to this development. There was then no restriction to procure ginger directly from the growers (Mahendru, 1982). All these developments led to historical consequences.

The enhanced demand for pepper and ginger in Europe made the Portuguese exert pressure on the farmers to cultivate more and more of these crops. This helped the growers in a way, as it freed them from the bondage of big marketers. But the Portuguese could not continue singly for long. The Dutch arrived on the scene and they practically drove out the whole Portuguese from the West Coast. The Dutch controlled spice trade in India only for a short time, because their main focus was on East Asia, namely countries like Indonesia and Suriname. In Kerala State, India, where they made a presence, war broke out between them and the King of Travancore State, and the Dutch were defeated. As time passed, the British arrived on the scene, and they could maneuver to corner the entire spice trade in India.

While these developments took place on the West Coast in India, North India was being ruled by the Emperor Akbar, and he gave a great impetus to spice cultivation both in North and in Western India. This was a policy of the Mogul empire. Ginger obtained a special impetus as it was an ingredient in most vegetarian and nonvegetarian dishes of the Mogul. For instance, in Ain-i-Akbari, written by Akbar's Prime Minister Abdul-Fazl, an account is provided about the various dishes in vogue during the Mogul period. In Ain 27 (f), he records that the market prices of spices and ginger was relatively cheaper than many others. Dried ginger was four dinars per seer (a measure in those times, rightly corresponding to a pound), and fresh ginger was 2.5 dinar per seer. He mentioned that pickled green ginger was available at 2.5 dinar per seer. Ginger was, hence, a common man's spice, unlike black pepper and saffron, the spices of the privileged (Mahindru, 1982). Ginger was widely grown in the West Coast of Kerala State from time immemorial. Subsequently, its cultivation spread to other parts of India, mainly to undivided Bengal and northeastern India. Buchanan (1807), who journeyed through the various heartlands of various kingdoms that existed in southern India in those times, made many references on the cultivation of various spices, including ginger, on the Malabar coast. Ridley (1912) gives a detailed description of agricultural practices prevalent in the nineteenth century, India. About ginger and turmeric he noted: "The planting of ginger and turmeric was preferred under the shade of orchard trees ... The output of ginger was 2500 pounds per acre ... Green ginger was sold at rupees four for 25 pounds. The cost of cultivation worked out to about rupees 250 per acre" in the book *A Hand Book of Agriculture* authored by N. Mukherjee. This translates to the farmer earning Rs 166 per acre as profit (66.4%) from ginger. The quantity of rhizomes required for planting was estimated at 100 pounds per "bigha" (1600 sq. yards). Harvested ginger was processed before being sold in the market. Different methods were followed in the processing of ginger in different regions. In Maharashtra (Khandesh region), the processing was done as follows:

> The rhizomes were dug up, cleaned of dirt and roots and boiled in a wide-mouthed vessel, and then dried. After drying for a few days, the rhizomes were steeped in diluted lime water, sun dried, and again steeped in stronger lime water and buried for fermentation. Later the rhizomes were dried and marketed. The product was known as "Sonth."
>
> *Watt (1872)*

The practice adopted in Bengal was: "Ginger was first brushed with a hand brush to remove dirt and steeped overnight in lime water; subsequently rinsed in clear water and dried slowly on a brick oven." The Bengal province in those days extended to the Himalayan Mountains, and ginger cultivation was prevalent in these parts. Campbell, who wrote the book *Agricultural and Rural Economy of the Valley of Nepal*, states that ginger was carefully grown in Nepal and the produce " … is reckoned by the people of the neighbouring plains of Tirhoot and Sarun of very highest flavor and superior to the produce of their own country" (Watt, 1872). The author also provides details of ginger cultivation prevalent in these regions.

Sir Baden Powell, the legendary founder of the Boy Scout movement, reported the following practice:

> The rhizomes were dried up by placing them in a basket suspended by a rope and shaking for two hours each day for three days. Later, these were sundried for eight days and again shaken in the basket and re-dried for 48 hours in the basket itself. This removed the scales and skins, making the produce suitable for marketing.
>
> *Watt (1882)*

In the nineteenth century in Bombay province, ginger was processed by peeling the rhizome with a piece of metal or tile and later drying it in the sun.

The Cochin ginger (ginger that came from the Cochin principality and exported from Cochin) was processed similarly as the Bombay ginger. Harvested rhizomes were heaped for a few days and then washed thoroughly to remove dust and soil. The outer skin was peeled off using a bamboo splinter, washed again, and dried in the sun. Often, the dried ginger was heaped in lime water for a few hours and redried to improve the appearance.

A *bigha* of ginger crop yielded 10 mounds of produce fit for sale, at the rate of Rs 6 per mound. The prevailing rate for ginger during the end of the nineteenth century was: Bengal Rs 10.6/cwt; Bombay Rs 9/cwt; Sind Rs 11.6/cwt. In Madras Province (including Cochin in Kerala State), ginger was available at 20 paise per kg (Mukherjee, quoted by Ridleey, 1912).

It is also of historical importance to record the first detailed chemical investigations on ginger by J.O. Thresh (*Year Book of Pharmacy*, 1879, 1881, and 1882). He

Table 15.3 Chemical Constituents of Cochin Ginger

Constituent	Percentage
Volatile oil	1.350
Fat (wax) resin	1.205
Neutral resin	0.950
α and β resins	0.865
Gingerol	0.600
Substance precipitated by acids	5.350
Mucilage	1.450
Indifferent substance precipitated by tannins	6.800
Extraction soluble in spirits of wine, not in ether or water	0.280
Alkaloid	Traces
Metarabin	8.120
Starch	15.790
Parabin	14.400
Oxalic acid	0.427
Cellulose	3.750
Albuninoides	5.570
Vasculose	14.463
Moisture	13.530
Ash	4.800

analyzed a sample of Cochin ginger that was found to contain the ingredients listed in Table 15.3.

Global Centers of Ginger Cultivation

There is no wild state in which ginger occurs in nature. The most probable place of its origin is Southeast Asia, but it has been cultivated from time immemorial in India and China. No definite information exists on the primary center of domestication. On account of the ease of transporting the ginger rhizome over long distances, it has spread throughout the tropical and subtropical regions of the Southern Hemisphere. In fact, ginger is the most widely cultivated spice (Lawrence, 1984).

The principal ginger-growing countries are India, China, Jamaica, Taiwan, Sierra Leone, Nigeria, Fiji, Mauritius, Indonesia, Brazil, Costa Rica, Ghana, Japan, Malaysia, Bangladesh, Philippines, Sri Lanka, Solomon Islands, Thailand, Trinidad and Tobago, Uganda, Hawaii, Guatemala, and many other Pacific Ocean islands.

India and Other South Asian Countries

India is the largest producer of ginger, with an annual production of about 263,170t from an area of 77,610ha, contributing approximately 30–40% of world production. Productivity is rather low, at about 3428g/ha. Of the total production, 10–15% is exported to about 50 countries worldwide. The crop occupies the largest area in Kerala State (19%), followed by Odisha State (17%), Meghalaya (12%), West Bengal (12%), and Arunachal Pradesh (6%). Kerala and Meghalaya together account for nearly 40% of India's production. In terms of productivity, Arunachal Pradesh stands first with an average yield of 7164 kg/ha, followed by Meghalaya with an average yield of 5139 kg/ha. Mizoram and Kerala harvest on average 5000 and 3428 kg/ha, respectively.

Nigeria

It was in 1927 that large-scale cultivation of ginger started in Nigeria, in the southern part of Zaria, especially within Jemma's federated districts, as well as in the adjoining parts of the plateau. Nigeria has attempted to widen the genetic base of the crop through introduction of ginger cultivation mainly from India. Currently, Nigeria is one of the largest producers and exporters of split-dried ginger. The annual production is around 90,000 metric tons from a total area of 17,400 ha.

Jamaica

Ginger is grown in the hills of South Central Parish of Manchester and in the Christiana Area Land Authority. There is also some production in the border parishes of Clarendon, Trelawny, and St. Elizabeth, as well as in the hills of St. James, Hanover, and Westmoreland in the North-West. The area involved in ginger cultivation was about 65,000–70,000 acres in the past but now that has dwindled considerably, and the current production is below 1000 t.

Fiji

Early European settlements introduced ginger as an export crop in Fiji in 1890. The Indian migrants started large-scale cultivation subsequently. The major production areas are Suva peninsula, especially in Tamarua, Colo-Suva, and Tacinua districts. Ginger has also spread to Sawani, Waibu Nabukaluka, and Viria districts. The area under cultivation is around 1000 ha.

Ghana

In Ghana, early attempts at growing ginger were not successful, but with the launching of the economic recovery program in 1983, ginger cultivation was promoted by the new government in place. Large-scale production was taken up in the Kadzebi district. The production reached 80,000 t in 1990. However, production declined subsequently. Currently, the country produces over 1000 t.

Australia

It was during the World War II that ginger gained a foothold in Australia as a commercial crop in Queensland. A farmer introduced it in Buderim in 1920, small town north of Brisbane in Queensland, which has been the center of ginger production

ever since. The growers are concentrated in Buderim, Nambour, North Arm, and
Eumundi. Production was over 6200 t in 1974. It has increased since then, and
the entire produce is processed into preserved ginger and other ginger products.
However, ginger production declined since then, and currently the crop occupies
only a very little area, and production is processed mainly by the famous Buderim
Ginger Company into more than a hundred value-added products.

Sierra Leone

Sierra Leone remained a ginger producer for over 100 years. The crop is grown along
the railway lines, laid under British administration, around Freetown, Bola, Kennama,
Pendemba, and Najala, as well as in the Mayamba district and parts of East Kano.
Sierra Leone ginger was traditionally known as African ginger. It is less aromatic but is
more pungent than other varieties grown for commercial purposes (Lawrence, 1984).

Mauritius, Trinidad, and Tobago

In Mauritius, ginger is grown in all districts on the island, although most of the production comes from Pamplemousses and Flacqdisbiets. Guajana has a small-scale
ginger cultivation in the northwestern region. Current production is around 500 metric tons in Mauritius. In Trinidad and Tobago, ginger is a traditional spice which is
grown mixed with other crops.

Southeast Asia

Southeast Asia is a major center of ginger production. It comes mainly from China,
Thailand, Taiwan, Korea, and Vietnam. China produces the most, followed by
Thailand, Korea, and Vietnam. China has an acreage ranging from 5000 to 80,000 ha.
It is cultivated in the provinces of Shandong, Guangdong, Zhejiang, Anhui, Jiagxi,
and Hubai. The largest variability in ginger is seen in China, where many distinctly
different morphotypes have been identified. Available figures indicate a production
of about 2,400,000 t. The country consumes internally the major share of ginger produced, with many ginger products being available in commercial markets.

Taiwan has only 3000–4000 ha under ginger, and the produce is marketed mainly as a
vegetable. It is grown either as an intercrop between tea or as a pure crop on hill slopes.

Thailand and Korea produce ginger only for domestic consumption. The former produces about 3000 t from about 12,000 ha. The Republic of Korea produces
about 8000 t from about 4200 ha, clearly indicating that Thailand is way behind the
Republic of Korea in ginger production.

Indonesia

With more than 10,000 ha under ginger cultivation and a production level of around
77,000 metric tons, which is a high-yield level, Indonesia is another important producer of ginger. The main ginger belt in Indonesia is the Java–Sumatra island region.

Sri Lanka

Sri Lanka grows ginger mixed with turmeric, cocoa, coffee, jackfruit, arecanut, coconut, or green vegetables, mostly in a haphazard manner. It is cultivated mostly in the central eastern provinces of the island in Yatinurwara, Harispatta, Siambolagoda, and Girijama. Ginger production is mostly consumed locally and goes into the production of ginger beer and ginger ale.

Philippines

In Philippines, ginger is produced in Las Banos, Laguna, Tanavan, Bantagas, Silag, and Carite. The current area under ginger cultivation extends to about 5000 ha with a total production of about 29,000 metric tons.

Many other countries, such as Nepal, Bangladesh, Bhutan, Cameroon, Costa Rica, Kenya, Reunion Islands, and the United States, produce ginger in small quantities for domestic consumption.

According to the Food and Agriculture Organization (FAO) in Rome, ginger production is looking up because of increase in production area and higher productivity, which are definitely bound to increase in the coming years. The crop is coming under a growing demand in the State of Kerala, India, with a huge domestic market for both fresh and dried ginger.

Uses of Ginger

A unique plant, ginger, is used universally. Ancient Indians considered ginger as the "mahashoudha" (the great medicine). It is an ingredient in several drinks and sweet-meat products. The plant thus possesses a combination of many attributes and properties. As ginger contains a variety of important constituents (Table 15.3), its use can be wide ranging. The characteristic organoleptic properties are contributed by the volatile oil and nonvolatile solvent-extractable pungent compounds. Among the many components, α-zingiberene is the predominant component of the oil. Gingerol and shogaolare are the pungency-contributing constituents. The refreshing aroma and the pungent taste make ginger an essential ingredient of most world cuisine and of the food-processing industry. The solvent-extracted oleoresin is available in convenient consumer packets. Ginger powder is also an ingredient in many *masala* (Indian curry powder) mixes. In Western countries, ginger finds its most widespread use in the making of gingerbread, biscuits, cakes, puddings, soups, and pickles. Ginger ale, ginger beer, and ginger wine are widely used soft drinks. Ginger is one of the most widely used medicinal plants in the traditional Indian systems of medicine, namely, *Ayurveda*, and also in Chinese and Japanese systems of medicine. According to *Ayurveda*, ginger has both carminative and digestive properties. It is believed to be useful in anorexia, dyspepsia, and suppression of inflammation. Dry ginger is useful in dropsy, otalgia, cephalgia, asthma, cough, colic, diarrhea, flatulence, nausea, and vomiting. Pharmacological investigations have indicated its usefulness in preventing nausea and vomiting associated with chemotherapy, pregnancy, travel, and seasickness. Ginger also has antiplatelet activity, hypolipidemic activity, and an anxiolytic effect. In addition to *Ayurveda*, ginger is in wide use in Indian folklore medicine as a cure for indigestion, fever, colic, and any ailment associated with the digestive system.

One of the most important value-added products to emerge is "Ginger Tea," brought out by the Tata group (biggest Indian corporate house).

Ginger is an important constituent of both Chinese and Japanese medicine. In the Chinese *Materia Medica*, ginger is indicated, for example, in the treatment of vomiting, diarrhea, light-headedness, blurred vision, dyspepsia, tremors, decrease in body temperature, and high blood pressure. In both Chinese and Japanese systems of medicine, fresh and dry ginger are used for different purposes.

Botany of Ginger

Genus *Zingiber* belongs to the family Zingiberaceae, which is distributed in tropical and subtropical Asia and Far East Asia and consists of about 150 species. Zingiberaceae is of considerable importance as a member of the "Spice Family." The Spice Family includes, most importantly of all, black pepper, the "King" of spices, cardamom, the "Queen" of spices, turmeric, nutmeg, allspice, cinnamon, and so on. These crops have great economic and medicinal values. Zingiberaceae was earlier divided into subfamilies Costoideae and Zingiberoideae, which were subsequently given independent family status as Costraceae and Zingiberaceae. Three tribes were recognized in the subfamily Zingiberoideae by investigators such as Peterson (1889) and Schumann (1904). And the genus *Zingiber* was included in the tribe Zingibereae along with *Alpinia*, *Amomum*, and some others. This tribe is characterized by the absence of lateral straminodes or staminodes that are united to the labellum, in comparison with the tribe Hedychieae, in which the lateral staminodes are well developed. Subsequently Holttum (1950) removed *Zingiber* from Zingibereae and renamed it as Alpimeae. His argument was that *Zingiber* is closer to the genera under Hedychieae as their lateral staminodes appear as lobes at the base of the labellum, whereas in *Alpinia*, these staminodes are well developed. Many subsequent investigators accepted the opinion of Holttum. Burtt and Smith (1983), however, felt that the contention of Holttum is nomenclaturally incorrect, and hence, proposed that *Zingiber* should be in an independent tribe.

The first documentation of ginger was by Van Rheede (1692) in his *Hortus Indicus Malabaricus* (Vol 11), the first-ever written account of the plant species of India. The author described the cultivated *Zingiber* species, *Zingiber officinale*, under the local name *inschi* (*inchi*). The Indian species was first botanically described by Roxburg (1810), who reported eleven species, and placed them in two sections based on the nature of the spike: Section 1, Spikes Radical and Section 2, Spikes Terminal.

Baker (1882) carried out an exhaustive survey of the Zingiberaceae of the Indian Peninsula for *The Flora of British India* (J.D. Hooker). In this he recognized the following four sections:

1. **Crytanthermum** Horan—Spikes are produced directly from the rhizome and are very short and dense; peduncle very short comprising of 11 species.
2. **Lampuzium** Horan—Spikes are produced from the rhizome on more or less elongated peduncles with sheathing scariose bracts comprising of 10 species.
3. **Pleuranthesis** Benth—Spike peduncle arising from the side of the leafy stem comprising of just one species.
4. **Dymczewiczia** (Horan) Benth—Spikes terminal on the leafy stem comprising of two species.

Zingiber Boehmer

Boehmer and Ludwig, def Gen. PI 89, 1760, nom. cons. Benth & Hook. f. Gen. PI.

The above classification was accepted by subsequent investigators including Schumann (1904). Holttum (1950) provided the following description of the genus: Rhizomes at or near the surface of the ground, bearing leaf shoots close together. Leaf shoots short to moderately tall, often with many leaves. Leaves thin in texture, never very large (rarely more than 50 cm long), sessile or with quite short petioles, the ligule short to long, deeply bilobed or entire. Inflorescence on a separate shoot without normal leaves (rarely at the apex of the shoot); scape usually erect, short or long, clothed with two-ranked sheaths that are often colored red; spike short or long, slender or thick, cylindrical, ovoid, or tapering to a narrow apex, elongating gradually. Bracts fairly large, usually bright colored, red or yellow, usually thin fleshy, closely imbricating or with apices free, margins plane or inflexed. One flower in the axil of each bracht; flowers fragile or short lived. Bracteoles one to each flower, facing the bracht, thin and narrower than bracht, usually persisting and enclosing the fruit, split to the base, never tubular.

Calyx thin, tubular spathaceous usually shorter than the bracteole but sometimes longer. Corolla tube slender, usually about as long as the bracht; dorsal lobe usually broader than the others, erect, narrowed to the tip, and hardly hooded; edges inflexed, lateral lobes usually below the tip and on either side of it, sometimes joined partly together by their adjacent sides and to the tip; color usually white or cream. Labellum deeply three-lobed (the side lobes representing staminodes), or rarely the side lobes hardly free from the mid-lobe, side lobes erect on either side of the stamen, mid-lobe shorter than or not greatly longer than the lateral corolla lobes, its apex usually retuse or cleft; color cream to white or more or less deeply suffused with crimson or purple. Filament of stamen short and broad, anther rather long, narrow; connective prolonged into a slender curved beak-like appendage as long as the pollen sac, with inflexed edges, containing the upper part of the style. Stigma protruding just below the apex of the appendage, not thickened, with a circular apical aperture surrounded by stiff hairs. Stylodes usually slender and free, not surrounding the base of the style. Ovary glabrous or hairy, trilocular with several ovules in each loculus. Fruit with a fleshy wall when fresh, more or less leathery when dry, smooth, hairy, enclosed by a persistent bracht or bracteole, dehiscent loculicidally within the persistent brachts. Seed ellipsoid, black or dark brown, covered by a thin saccate white aril with irregularly lacerate edges.

The main distinguishing features of the genus are: (A) long, curved, anther-appendage embracing the style; (B) the three-lobed lip (the side lobes are staminodes, which are relatively broad and fused more or less to the mid-lobe or lip proper); and (C) the relatively large brachts, each with a single flower and nontubular bracteole, more or less imbricating on a lengthening inflorescence (*Z. clarkei* from Sikkim State, India, is an exception that has 2–4 flowers to each bract). The bracts are often, but not always, colored; in some species, they change color as they grow older. The color of the lip is an important distinguishing character.

The genus contains about 150 species, of which 34 have been traced to China (Shu, 2003) and 24 to India (Baker, 1882). The main centers of diversity are South China, Malaysia, Northeast India, Myanmar region, and the Java–Sumatra region of Indonesia. Shu (2003) has recently revised the Chinese species. The only species extensively used as a flavoring agent for food is the true ginger, *Z. officinale*. Some species, such as *Z. zerumbet* and *Z. cassumunnar* are well known for their uses in native medicine. *Z. mioga* is used as a spice and its flower buds are in great demand in Japan as a vegetable.

Zingiber officinale Rosc.

Roscoe, New arrangements of the plants of the monandrian class usually called "Scitaminea," Trans. Linn. Soc. 8:348, 1807; Valeton, Bull. Buitenz, 2nd Ser; xxvii, 128, 1818; Flukiger and Hanbury, Pharmacographia, 574, 1874; Engler, Pflanzenw. Ost.-Afrikes and Nachbargebiete, B. Natzpflanzen; 264, 1895; Schumann, Zingiberaceae, in Das Pflanzenrich, 4, 46, 179, 1904. *Inschi*, Rheede, *Hort. Malabaricus*, 11, 23–25, 1692.

Rhizome entirely pale yellow within or with a red external layer. Leafy stems to about 50 cm tall, 5 mm in diameter, glabrous except for short hairs near base of each leaf blade; leaf blades commonly about 17 cm×1.8 cm; rather dark green, narrow evenly to slender tip; ligule broad, thin, glabrous, 5 mm tall, slightly bilobed. Scape slender, 12 cm tall, the upper sheaths with or without short leafy tips; inflorescence approximately 4.5 cm long and 15 mm in diameter; brachts approximately 2.5 cm × 1.8 cm; green with pale submarginal band and narrow translucent margin; margins incurved, lower brachts with slender white tip. Bracteoles as long as bracht; calyx with ovary 12 mm long; corolla tube 2.5 cm long, lobes yellowish, dorsal lobe 18 mm × 18 mm (flattened), curving over the anther and narrowed to the tip, laterals narrower. Lip (mid-lobe) nearly circular, approximately 12 mm long, and wide, dull purple with cream blotches and base, side lobes about 6 mm × 4 mm; free almost to the base, colored at mid-lobe; anther cream, 9 mm long, appendage dark purple, curved, 7 mm long (Holttum, 1950). The species is sterile and does not set seeds.

Taxonomical Notes

Roscoe (1807) described *Z. officinale* from a plant in the Botanic Garden at Liverpool as "*Bracteis ovato-lanceolatis, laciniis corolla revolutis, nectario trilobata*" and referred to it as *Amomum zingiber* Willd. Sp.Pl. 1:p6. Willdenow (1797) extended

Linnaeus's description "*Amomum scapo nude, spica ovata*" with "*squamis ovatis, foliis, lanceolatisbad apicem margine ciliates.*" Linnaeus's (1753) *Amomum zingiber* is the basionym for the species. The genus *Amomum* of Linnaeus is a nomenclatural synonym of the conserved generic name, *Zingiber* Boehm (Burt and Smith, 1968). The specific epithet *zingiber* could not be used in the genus *Zingiber*. Thus, *Z. officinale* was adopted as the correct name for ginger. The specimens available in most of the herbaria are without flowers, and it is assumed that Linnaeus based his description on the account and figure given by Rheede in *Hortus Malabaricus*. The figure given by Rheede is the designated lectotype of the species *Z. officinale* Rosc. (Jansen, 1781). The species epithet *officinale* is derived from Latin, meaning "work shop," which in early Latin was used to mean pharmacy, thereby implying that it had medicinal value.

The Morphology and Anatomy of Ginger

The ginger plant is a herbaceous perennial grown as an annual crop. The plant is erect, with many fibrous roots, aerial shoots (pseudostem) with leaves, and underground stem (rhizome). The roots of ginger are of two types, fibrous and fleshy. After planting, many roots having indefinite growth grow out of the base of the sprouts. These are the fibrous roots, and the number of such roots keeps on increasing with the growth of tillers. The fibrous roots are thin, with root hairs, and their function is mainly absorption of plant nutrients and water from soil. When the plant grows further, several fleshy roots of indefinite growth are produced from the lower nodes of the mother ginger rhizome and primary fingers. These roots are thicker, milky white in color, with few root hairs, and with no lateral roots. Such roots carry out functions of anchorage as well as conducting vessels for water and nutrient absorption. During the initial growth, the apical bud of the rhizome piece planted grows out and becomes the main tiller or mother tiller. As this tiller grows, its base enlarges into a rhizome. This is the first formed rhizome knob and is frequently referred to as the "mother rhizome." From either side of the mother rhizome, branches arise and they grow out and become the primary tillers. The bases of these tillers become enlarged and develop into the primary fingers. The buds on these primaries develop in turn into secondary tillers and their bases into secondary fingers. The buds on the secondary fingers in turn can develop into tertiary tillers and tertiary fingers.

The aerial shoots have many narrow leaves borne on very short petioles and with sheaths that are long and narrow, and the overlapping sheaths produce the aerial shoot. A pair of ligules is formed at the junction of leaves and sheath. The leaves are arranged in a distichous manner.

Ginger is a subterranean stem (rhizome) modified for the vegetative propagation and storage of food materials. The stem has nodes with scale leaves and internodes. Except for the first few nodes, all others are axillary buds. When the rhizome bit is used to plant ("seed rhizome" or "set"), there may be one or more apical buds on it; however, normally only one bud becomes active. When large pieces are used, more than one bud may develop simultaneously. If more than one branch from the parent rhizome is responsible for the ultimate growth and development of the adult rhizome, the branches of the mature rhizome lie in the same plane (Shah and Raju, 1975a).

The pattern of rhizome branching shows that the main axis developing from the apical bud, which is the first developing branch, has 7–15 nodes, which later becomes an aerial shoot. Once this axis becomes aerial, the subsequent growth of the rhizome is due to the development of the axillary buds situated above the first 2–3 nodes of the underground main axis. These axillary branches are plagiotropic and quickly show orthotropic growth at their distal region and subsequently become aerial shoots. A similar pattern of growth continues for successive branches to form a sympodial growth pattern. A few axillary buds at the distal end of the branch remain dormant. The number of primary branches may be two, three, or four. These primary branches arise on either side of the main axis. Subsequent development of the secondary, tertiary, and quaternary branches is on the abaxial side of the respective branches. Irrespective of the number of primary branches, the subsequent branches lie in the same place, although alteration of this scheme is seen sometimes. A mature rhizome may consist of 6–26 axillary branches with foliage leaves or only with sheath leaves, and they show negative geotropic response (Shah and Raju, 1975a).

The number of nodes in each rhizome branch varies. The main axis (mother rhizome) and the subsequent branches (primaries) have 6–15 nodes. The internal length of the rhizome branches ranges from 0.1 to 0.15 cm, and varies even in a single branch. The internodal length is more in the secondary, tertiary, and quaternary branches, and in the aerial stem it ranges from 3 to 7 cm. In the underground stem, the nodes have scale leaves which ensheath and protect the axillary buds. These scale leaves fall off, or may be lost, so that in mature rhizomes, only the scars remain. Young scale leaves have pointed tips which help in the penetration into the soil.

The distal few nodes of the rhizome have sheath leaves. At the early stage of development, they lack an apparent slit due to overlapping of their margins. Subsequently, a longitudinal slit is formed through which the shoot tip projects. After the development of 6–12 scale leaves and 3–5 sheath leaves, the foliage leaves are produced. A foliage leaf consists of a leaf sheath, a ligule, and an elliptical-lanceolate blade. The leaf sheath is about 15–18 cm long and the lamina is about 12–15 cm long. Above its region of insertion, the sheath encircles the internode; and from the side opposite to its origin, up to the ligule, the sheath is open longitudinally. A distinct midrib is present only in the lamina. The phyllotaxy of the scale leaves on the rhizome and foliage leaves on the aerial stem is distichous, with an angle of divergence of about 180°. Within the bud, leaves have imbricate aestivation (Shah and Raju, 1975a).

Rhizome Anatomy

The early investigations on the anatomy of ginger were carried out mainly by the pharmacognosists, and they concentrated on the officinal part, the rhizome, either dry or fresh (Futterer, 1896). A comprehensive survey on the anatomy of the plants belonging to Zingiberaceae was that of Solereder and Meyer (1930), in the classical work *Systematische Anatomie der Monocotyledonen* (Systematic Anatomy of the Monocotyledons). They provided anatomical notes on 18 genera and some 70 species (Tomlinson, 1956). Subsequently, Tomlinson (1956) supplemented the information and filled in the gaps. However, no information was available on the developmental anatomy. Some investigations were carried out by Aiyer and Kolammal (1966), Pillai et al. (1961), and Shah and Raju (1975b). Recently, investigations on developmental anatomy of the rhizomes, oil cells, and associated aspects were carried out (Ramashree et al., 1997, 1998, 1999; Ravindran, 1998). The following discussion is based on the studies of the above-mentioned researchers.

The transection (TS) of a fresh, unpeeled rhizome is almost circular or oval, about 2 cm in diameter, with the outline almost regular. The TS shows a light brown-colored outer border and a central zone 1.2 cm in diameter marked off by a yellowish ring from an intermediate cortical zone. A distinct continuous layer of epidermis is generally present, consisting of a single row of rectangular cells; in some cases, it may be ruptured. Within this is the cork, varying in thickness, from 480 to 640 μm, and differentiated into an outer region of 300–400 μm in thickness, composed of irregularly packed, tangentially elongated, slightly brown-colored cells; then there is an inner zone of 6–12 regular rows of thin-walled, rectangular to slightly tangential elongated cells arranged in radial rows. They measure 30 μm × 30 μm to 114 μm × 48 μm. (Note: Cork tissue develops after the harvest and during storing. Hence, when a rhizome is cut soon after harvest, one may not encounter much cork tissue.) A cork cambium is not evident. Inner to the cork is the cortex that is about 4 mm in thickness, composed of thin-walled large hexagonal to polygonal parenchymal cells. The cortical cells are heavily loaded with starch grains. These grains are large, simple, and ovoid, in length varying from 15 to 65 μm. Scattered within the cortex are numerous oil cells which contain large globules of yellowish green color. The outermost 3–5 rows of cortical cells are not rich in oil content. Many scattered, collateral, closed vascular bundles are present, of which the greater number is seen in the inner cortical zone. The large bundles are partially or entirely enclosed in a sheath of separate fibers, whereas the smaller bundles are devoid of any fiber. Each vascular bundle consists of phloem, composed of small thin-walled polygonal cells with well-marked sieve tubes and xylem composed of 1–9 vessels with annular, spiral, or reticulate thickenings. These vessels have a diameter varying from 21 to 66 μm. In the enclosing sheath of fibers, the number of cells varies a lot. There are 4–48 fibers or occasionally more. These fibers are very long, but less than 1 mm, have a diameter from 10 to 40 μ, and are not straight but undulative in character. The inner limit of the cortex is marked by a single-layered endodermis composed of thin-walled, rectangular cells, much smaller than the cortical cells, with their radial walls slightly thickened and free from starch grains. The endodermis is lined by a pericycle composed of a single row of thin-walled, slightly tangentially elongated cells devoid of any starch grains.

The stele that forms the bulk of the rhizome consists of parenchymal cells similar to those of the cortex, with starch grains and oil globules and a large number of irregularly scattered vascular bundles. Just within the pericycle, a number of very small vascular bundles are arranged in a ring. These bundles have only 1–3 vessels and a small phloem. No fibers are present enclosing these small bundles. Generally, the vascular bundles present within the stele are slightly larger than those present in the cortex. The stele contains more oil cells and starch grains than the cortex (Aiyer and Kolammal, 1966).

Rhizome Enlargement

Rhizome enlargement in ginger is by the activity of three meristematic zones. Very early in the development of the rhizome, a zone of meristematic cells is formed at the base of a young scale leaf primordium of the developing rhizome. These meristematic cells develop into the primary thickening meristem (PTM) and procambial stands. The meristematic activity of the PTM is responsible for the initial increase in the width of the cortex. The second type is the actively dividing ground parenchyma. The third type is the secondary thickening meristem (STM), in which fusiform and ray initials are clearly visible. The STM develops just below the endodermoidal layer.

At a lower level, in the rhizome from the shoot bud apex, the PTM can still be identified. The scattered vascular bundles develop from the PTM or procambial cells. Such groups of cells can be identified by the plane of cell division. The differentiation of procambial cells into vascular tissue takes place at different stages of rhizome growth. Unlike in many monocots, in the ginger rhizome there is a special meristematic layer along with the endodermoidal layer, and this layer consists of cambium-like cells. The cells are thin-walled and arranged in a biseriate manner. In certain loci, where the vascular bundles develop, these cells are elongated with tapered ends and appear similar to the fusiform initials with an average of $62.34\,\mu m$ length and $8.12\,\mu m$ width in mature stages. Between these fusiform initials, some cells show transverse divisions to form ray initials. The presence of the cambium-like layer is an important feature in rhizome development. From this layer, inverted and irregularly distributed groups of xylem and phloem are formed along the intermediate layer. The cells outer and inner to the cambial layer become filled with starch gains.

Development of Oil Cells and Oil Ducts

Oil cells are present in the epidermis or just below the epidermis of the leaf, petiole, rhizome, and root. In the rhizome, oil cell initials are present in the meristematic region. They are spherical and densely stainable. The initiation of oil cells and formation of ducts occur in the apical parts of shoots and roots and start much before the initiation of vascular elements. Secretory ducts are formed both schizogenously and lysigenously (Ravindran et al., 1998; Remashree et al., 1999).

The Schizogenous Type

The schizogenous type of secretory duct originates in the intercalary meristem of the developing regions. The ducts are initiated by the separation of a group of densely stained meristematic cells through dissolution of the middle lamella. Concurrent separation of the cells leads to the formation of an intercellular space bordered by parenchymal cells. These ducts are anastomose and appear branched in longitudinal section. Further separation of the bordering cells along the radial wall leads to widening of the duct lumen.

The Lysigenous Type

The lysigenous type of duct formation is more frequent in the meristematic region, but occurs in mature parts as well. There are four stages involved in its development, which are initiation, differentiation, secretion, and quiescence. These steps are a gradual process that occurs acropetally.

Initiation and Differentiation

In shoot apex, the meristematic cells are arranged in tiers. In between these cells, certain cells in the cortical zone are distinguishable from the rest by their large size, dense cytoplasm, and prominent nucleus. Such cells act as the oil cell mother cell. Anticlinal and periclinal divisions of these cells result in a group of oil cell initials. Cytoplasmic vacuolation initiates in the oil cells at a distance of about 420 μm from the shoot apex. Subsequently, the surrounding cells also enlarge in size, showing cytoplasmic and nuclear disconfigurations. Further development leads to the disintegration of nuclear content of the central cell, which stretches toward the intercellular space. Later, the central cell disintegrates and the contents of the cell spill into the cavity thus formed. This process that takes place in adjacent cells leads to the formation of a duct. The duct can be either articulated or nonarticulated and becomes gradually filled up with the cell contents of the lysed cells. Once the lysogeny of the central cell is completed, the adjacent cells also lyse gradually in a basipetal manner, resulting in the widening of the duct lumen. These stages occur between 1500 and 3000 μm from the apex.

Secretion

The differential oil cells start a holocrine type of secretion and expel their contents into the duct. Then the next cell (in acropetal order) becomes differentiated into an oil cell and starts elimination of its contents, followed by lysis. Simultaneously, the primary tissues continue to become differentiated into new oil cells and reach the secretory stage. The secretion fills the duct in young stages, but the quantity becomes reduced gradually, and finally the ducts appear empty. This could happen because of the diffusion of oil basipetally and radially; such oil particles are deposited in the cells and can be seen as black masses inside cells as well as in the intercellular space. Such stages are noticed about 3250 μm from the root tip (Ravindran et al., 1998; Remashree et al., 1999).

Quiescence

In the mature rhizome, the ground parenchyma does not undergo further division and differentiation into the duct. In this stage, the cells adjacent to the duct become storage cells, containing numerous starch grains and large vacuoles. An empty cell or cells with distorted cytoplasm appear along the duct lumen. Quiescence and secretory stages are visible from the third month onward after planting. In primary tissues, the oil duct development is schizogenous, whereas further development proceeds both schizogenously and lysigenously.

Root Apical Organization

The root apical organization in ginger together with many other zingiberaceous taxa was first reported by Pillai et al. (1961). They found that the structural organization of ginger root apex differs from that of other taxa (such as *Curcuma, Elettaria, and Hedychium*). In ginger, all zones in the root apex originate from a common group of initials. From the rim of this common group, calyptrogen, dermatogen, periblem, and plerome become differentiated. Raju and Shah (1977) also reported a similar observation in ginger and turmeric. The following discussion is adapted from Pillai et al. (1961).

The root cap is not differentiated into columella and a peripheral zone, and hence, there are no separate initials for these regions. The cells in this region are arranged in vertical superimposed files. The cells arise by the activity of a meristem, which can be easily differentiated from the rest of the region. Pillai et al. (1961) named this meristematic region columellogen. In TS, the cells of the columella form a compact mass of polygonal cells in the center with the cells of the peripheral region arranged in radiating rows around it.

In the root body, two histogens could be distinguished: (i) the plerome concerned with the formation of stele and (ii) the protoderm–periblem complex concerned with the formation of the outer shell to the stele including periblem and dermatogen. The protoderm–periblem complex is located outside the plerome and is composed of a single tier of cells. The cells of this zone located on the flanks exhibit T-divisions, which help the tissue to widen out. Periblem consists of the initials of the cortex extending from the hypodermis to the endodermis. The hypodermis arises from the inner layer of the protoderm–periblem initials. The cells composing this tissue vacuolate earlier than the outer cells of the cortex.

Endodermis differentiates from the innermost periblem cells. Outside the plerome dome, all cells of the periblem exhibit T-divisions initially, but, subsequently in development show anticlinal divisions, and the endodermis is differentiated at that time.

Plerome has at its tip a group of more or less isodiametric cells. On the sides of the plerome dome is the uniseriate pericycle. Near the dome, cells take less stain because of their quiescent nature. The metaxylem vessel elements with wider lumens can be seen near the plerome dome. The isodiametric cells at the very center of the plerome divide like a rib meristem to give rise to the pith. In TS, passing near the tip of the plerome dome, the initials can be distinguished as a compact mass of isodiametric cells, surrounded by radiating rows of periblematic cells.

Cytophysiological Organization of the Root Tip

The root tip can be distinguished into two zones on cytological ground:

1. The quiescent center: This zone is found at the tip of the root body, characterized by its cells having (a) cytoplasm highly strained with pyronin-methyl green and hematoxylin, (b) smaller nuclei and nucleoli, (c) cell divisions less frequent, and (d) vacuolation noticeable in most.

The median longisection of this group of cells is in the shape of a cup with the rim forward. The above characteristics show their state of rest and are called the quiescent center. This zone includes cells belonging to all the structural histogens of the root body (i.e., not structurally delimitable). It gradually merges with the zone outside, the meristematic zone. Raju and Shah (1977) studied the root apices of ginger, mango ginger, and turmeric with azure B staining to localize DNA and RNA contents in order to identify the quiescent center. A quiescent center was present in all three cases as indicated by the light stainability of its cells. In longisection, the quiescent center resembles the inverted cup.

2. The meristematic zone: This zone is shaped like an arch surrounding the quiescent center on the sides of the root body. The cells of this zone have the following features: (a) cytoplasm deeply stained with pyronin-methyl green and hematoxylin, (b) divides more frequently, (c) have larger nuclei and nucleoli, and (d) vacuolation is absent or not prominent.

The meristematic zone includes the cells of all the structural histogens of the root body. The percentage of cell division is much lower in the quiescent center compared to the meristematic zone. This character combined with the response of these cells to stains such as pyronin-methyl green indicates that these cells are in a state of comparative repose and hence are not synthesizing nucleic acids (Pillai et al., 1961). The distance between the tip of the root body and the nearest mature phloem element, which carries the metabolic products required in the active cells, was reported to be 480–490 µm. This led to the suggestion that the cells at the tip of the root body go into quiescence because of the dearth of sufficient metabolites (Pillai et al., 1961).

Ontogeny of Buds, Roots, and Phloem

The ontogeny of ginger was investigated by Shah and Raju (1975b), Remashree et al. (1998), and Ravindran et al. (1998). In a longisection, the shoot apex is dome shaped with a single tunica layer, below which the central mother cell zone is present. The flank meristem is situated on either side of the central mother zone. The leaf is initiated from the outer tunica layer and from the flank meristem. The shoot apical organization and acropetal differentiation of procambial strands are closely related to the phyllotaxy. At an even lower level basipetally in the rhizome axis, additional inner cortical cells are produced by a lateral PTM or procambium in which the resulting cells are radial rows.

The Nature of the Shoot Apex

Shah and Raju (1975a) studied the nature of the shoot apex in ginger. In all the stages of growth in the shoot apex, a single layer of tunica occurs, showing only anticlinal divisions. Cytohistological zonation based on staining affinity is not observed at any stage. The distal axial order (cr) includes the central group of corpus cells dividing periclinally and anticlinally and the overlying cells of the tunica. The peripheral zone is concerned with the initiation of the next leaf primordium and formation of the leaf sheath on the opposite side. It is delimited by the shell zone on the rhizome apices, which appears as an arc of narrow cells in median longitudinal section. The peripheral zone (pr_2) is associated with the initiation of the next leaf primordium. In the rhizome apices, it is also associated with the initiation of the axillary buds. As the phyllotaxy is distichous, this zone is opposite to pr_1 in median longisections. Pith cells differentiate in the inner axial zone (rr).

Seven developmental stages of the apical bud have been recognized (Shah and Raju, 1975b). In the first stage (dormant apex), the shoot apex lies in a shallow depression, the apex measures 116–214 μm × 45–70 μm. A few cells toward the flank showed increased concentrations of DNA as evidenced by dense staining. Some cells of pr_1 and pr_2 showed dense stainabilty for C-RNA (cytoplasmic RNA). The outer corpus cells show peripheral divisions. In the second stage, the apex is dome shaped and its width and height are 94–165 μm and 35–75 μm, respectively. Zones pr_1 and pr_2 show denser histological staining than cr and rr zones. A biochemical zonation is present at pr_2 that shows deep staining for DNA. The apex at stage three measures 74–140 μm in width, 53–86 μm in height and is dome shaped. The cells of the inner axial zone are vacuolated. The shoot apex dome at stage four is 140–160 μm high and 90–116 μm wide. Outer corpus cells are vertically elongated. At stage five, the apex is a low dome having 214–248 μm height and 53–75 μm width. Cells of the pr_2 zone show dense staining. The apex of stage six is prominently dome shaped having a width of 169–200 μm and height of 87–96 μm. During stage seven, the underground branch reaches the soil level. The shoot apex is 91–112 μm in width and 134–167 μm in height.

All of the underground branches of ginger show a negative geotropic response. Two kinds of apices are found in ginger: the apices which are low dome and surrounded by either scale leaves or leaf bases; and they are dome shaped and raised on an elongated axis. In the base of the rhizome apices, cells derived from the inner axial zone elongate tangentially and contribute to the widening of the axis. In certain cases, these cells extend up to the base of the axillary buds. In a dormant apex, they are thick walled and contain starch grains. These cells are distinct in the dormant or early active rhizome apex and constitute latitudinal growth meristem. During the vascular differentiation a few cells of this meristem develop into procambium. During subsequent development of the rhizome apex, the cells derived from the inner axial elongate and contribute to the pith.

Procambial Differentiation

The peripheral or flank meristem divides periclinally and produces parenchymal cells. Some of the cells are distinguishable from the rest by deeper stainability, smaller size, less or no vacuolation, and darkly stained nuclei. These are the procambial initials and each such group contains 15–20 cells. Subsequently, these cells elongate, vacuolation increases, and they develop gradually into sieve tubes. Protophloem differentiation precedes that of protoxylem. The collateral differentiation of phloem and xylem with parenchymal bundle sheaths becomes distinct after an intermediate stage of random differentiation of the bundles. Ultimately, the vascular bundles are found scattered in parenchymal ground tissue. In TS, an endodermoidal layer is also visible during the development (Ravindran et al., 1998; Remashree et al., 1998).

Axillary Bud

The development of leaves and scale leaves that encircle the shoot apex in ginger rhizomes is in a clockwise direction. The axillary bud meristem is first discernible in the axillary position on the adaxial sides of the third leaf primordium from the apical meristem as a distinct zone by the stainability of the constituent cells and multiplane division of the cells in the concerned peripheral meristem sectors. The axillary buds thus originate as a cellular patch in the adaxial side of a leaf or scale leaf of the node. In a fully developed axillary bud, the cytohistological zones akin to the main shoot apex can be distinctly observed. The development of a new rhizome is by the enhancement of a dormant axillary bud, which acts just like the main shoot apex. The procambial cells and the ground meristem cells divide, and parenchyma as well as vascular tissues add thickness to the newly enhanced axillary bud. Likewise, many buds become active during favorable conditions, each of which produces secondary or tertiary rhizomes. The axillary buds show vascularization by the activity of the procambial strands of the mother rhizome and procambial cells originated from the differentiation of the parenchymal cells.

Development of the Root

The adventitious root primordial becomes different endogenously from the endodermoidal layer of the rhizome. Roots always develop just below the nodal region. The TS of the rhizome reveals that the endodermoidal layer and the pericycle become meristematic and undergo periclinal and anticlinal divisions resulting in a group of root initials. This is in direct connection with the vascular ring situated beneath the endodermoidal layer. The root primordial is of the open type having common initials for the cortical meristem, root cap, and protoderm. The actively dividing and deeply staining central cylinder shows vascular connections with the rhizome vasculature. As the enlarging root primordial emerges through the cortex, the cortical cells are crushed and torn apart. Normally, these roots originate from the lateral or opposite side of the axillary bud and scale leaf.

Phloem

As a rule, there is no secondary growth in monocots. However, the rhizome structure of ginger provides evidence of both primary and secondary growth having a well-developed endodermoidal layer and cambium. The vascular bundles are collateral, closed, and scattered in the ground parenchyma. The phloem element consists of the sieve tube, companion cells, parenchyma, and fiber.

Sieve Tube

Phloem cells originate from a group of actively dividing procambial cells of PTM. These cells can be distinguished from the surrounding cells by their meristematic activity, stainability, and size of the nucleus. During development, a procambial cell elongates and becomes thick walled with cytoplasm and a prominent nucleus; this is the sieve tube mother cell. It undergoes a longitudinal unequal division, and the resulting smaller cell gives rise to the companion cell. This cell continues to divide, forming 4–8 cells. The large cell is the sieve cell. It has cytoplasm and nucleus in early stages, which degenerate during the development into the sieve tube. During further development, the vacuolation increases and the cytoplasm shrinks and appears like a thread along the wall. At the same time, the nucleus disintegrates and the cell assumes the features of the enucleated sieve tube element. The transverse wall of the sieve tube changes to simple sieve plates with many pores and with very little callose deposition. The first sieve tube element can be distinguished at a distance of 720–920 µm from the shoot apex.

In the ginger rhizome, 4–8 companion cells per sieve tube element are arranged in the vertical lines with transverse end walls. They may vary from 18 to 32 µm in length, and 7–19 µm in width. The sieve tube elements are arranged end to end to form columns of sieve tubes. The length of a sieve tube element varies from 57.5 to 103.8 µm, the average being 76.8 µm. The width varies from 5.29 to 10.35 µm, the average being 8.76 µm (Remashree et al., 1998). At the early stage of development, the slime body is present in the sieve tube, which appears to be amorphous but homogenous. Subsequently, the slime body disintegrates.

Development of a sieve tube in ginger is pycnotic, similar to the second type of nuclear degeneration reported by Esau (1969). The sieve element passes through a "fragmented multinucleated stage," a unique feature in the ontogeny of the multinucleated sieve tubes as reported by Esau (1953).

Phloem Parenchyma

The phloem parenchyma cells are comparatively larger than the companion cells and smaller than the normal cortical parenchymal cells. The increase in size of the phloem element is proportional to the growth of the rhizome. Some older phloem parenchymal cells become lignified into thick phloem fibers.

Anatomical Features of Ginger in Comparison to Related Taxa

Ginger has many species-specific anatomical variations. These variations were shown in a comparative investigation of ginger and three other related species (Ravindran et al., 1998) and the salient features are given in Table 15.4.

The salient features in Table 15.4 present important anatomical similarities and differences among the four species investigated. Ginger has distinct anatomical features compared to other species, such as the absence of periderm, short-lived functional cambium, the presence of xylem vessels with scalariform thickening, helical and scalariform type xylem tracheids, scalariform perforation plate, outer bundles with a collenchymatous bundle sheath, and high frequency of oil cells. The oil cell frequency was found to be $17.8/mm^2$ in ginger, whereas the corresponding frequency in the other species was 9.5, 5.3, $2.8/mm^2$ in *Z. zerumbet*, *Z. marostychum*, and *Z. roseum*, respectively. Species differences were also noticed in fiber length, fiber width, and fiber wall thickness. Histochemical studies indicated *Z. zerumbet* has greater amount of fibers than the others.

In general, xylem elements in *Zingiber* consist mainly of tracheids and rarely of vessels. The secondary wall thickening in the tracheids of ginger is of two types, scalariform and helical. The rings, or helices, are arranged in either a loose or a dense manner. The helical bands are found joined in certain areas giving ladder-like thickening. The width of helical tracheids is less than that of scalariform tracheids. Similar tracheids are present in *Z. macrostachyum*, whereas in *Z. zerumbet* and *Z. roseum* only scalariform thickening occurs (Ravindran et al., 1998). Xylem vessels occur in ginger and not in other species. Snowden and Jackson while studying the microscopic characters of ginger powder, recorded that the vessels are fairly large, reticulately thickened, less commonly spiral, and annularly thickened.

Leaf Anatomical Features

The leaves of ginger are isobilateral. The upper epidermal cells of leaf are polygonal and predominantly elongated at right angles to the long axis of the leaf. In the lower epidermis, the cells are polygonal and irregular, except that at the vein region, where they are vertically elongated and thick walled. The epidermal cells in the scale and sheath leaves (the first 2–5 leaves above the ground are without leaf blades) are elongated and parallel to the long axis of the leaf. Oil cells in the upper and lower epidermis are rectangular, thick walled, and suberized. Unicellular hairs are present in the lower epidermis of the foliage leaves. Occasionally, a hair is present at the polar side of the stomata. Ginger leaves are amphistomatic. A distinct substomatal chamber is present. Stomata are either diperigenous or terraperigenous. Occasionally, anisocytic stomata are also observed. The subsidiary cells are completely aligned longitudinally with the guard cell. The lateral subsidiary cells may divide to form anisocytic stomata. Occasionally, 3–5 lateral subsidiary cells are formed by further division (Raju and Shah, 1975).

The guard cells on the foliage leaves are 40.6 µm long, whereas those on the sheath and scale leaves are only 28.9 µm long. The stomata on the scale leaves are rarely on the sheath leaves, which show pear-shaped guard cells with a large central pore. The nuclei of the guard cells are smaller than those in the subsidiary cells. Raju and Shah (1975) also reported the uncommon wall thickening at the polar ends of the guard cells. This wall thickening may be restricted to the outer wall at the polar regions or may also be extended to the common inner cell wall.

Tomlinson (1956) provided a brief note on the petiolar anatomy of ginger. The shorter petiole shows a swollen pulvinus-like appearance. The TS just above the pulvinus shows typical structure with two bundle arcs, air canals, and collenchymas. The TS through the pulvinus shows a different structure. Here, air canals and assimilating tissue are absent, there is extensive hypertrophy of ground tissue parenchymal cells and abundant deposition of tanniferous substances. The most striking feature according to the author is the collenchymatous thickening of the cells of the bundle sheath. Below the pulvinus the structure is again normal as that of the pulvinus region. Table 15.5 gives the comparative leaf anatomical features of the four species of *Zingiber*.

Table 15.4 Comparative Anatomy of Four Species of *Zingiber*

Tissue	Z. officinale	Z. roseum	Z. zerumbet	Z. macrostachyum
Epidermis	Single-layered	Single-layered	Single-layered	Single-layered
Periderm	Absent	Periderm with lenticels	Periderm present	Absent
Cortex (outer cylinder)	Not wide	Not wide	Not wide	Wide
Endodermis	Present	Present	Present	Present
Casparian strips	Present	Present	Present	Present
Cambium	Present	Not found	Not found	Not found
Central cylinder	Wider than the outer zone	Wider than the outer zone	Comparatively less wider than the outer zone	Not wider than the outer zone
Number of vascular bundles	Less in the outer cylinder than in inner zone	Less in the outer zone than in the inner zone	Less in the outer zone than in the inner zone	More in the outer cylinder than the other three species, but, lesser than in the inner zone
Nature of vascular bundles	Collateral closed	Collateral closed	Collateral closed	Collateral closed
Vascular bundles distribution	More toward inner cortex and scattered in the central zone	More toward inner cortex and scattered in the central zone	More bundles in the middle cortex and number of bundles is very less compared to other three species	Bundles are arranged in two rows in the middle cortex and only a few bundles in the inner cortex and the bundles are uniformly distributed in the central zone

Pith	Present	Present	Present	Present
Xylem elements	Vessels, tracheids, and fibers	Tracheids, fibers	Tracheids, fibers	Tracheids, fibers
Vessels	Vessels few with scalariform/reticulate thickening	Not found	Not found	Not found
Xylem tracheids thickening	Helical and scalariform type	Scalariform	Scalariform	Helical and scalariform
Perforation plate	Scalariform type	None	None	None
Phloem	Sieve tube, companion cells, phloem parenchyma, and phloem fiber	Sieve tube, companion cells phloem parenchyma, and phloem fiber	Sieve tube, companion cells, phloem fiber, and phloem parenchyma	Sieve tube, companion cells, phloem fiber, and phloem parenchyma
Metaxylem width	Outer zone: 57 μm Inner zone 84 micron	20–53 μm	25–53 μm	32–76μm
Bundle sheath	Outer vascular bundles possess collenchymatous bundle sheath	Absent	Absent	Collenchymatous sheath is present only in outer bundles
Oil cell frequency	Very high	Least	High	Less
Curcumin cell	None	Present	Present	None

Table 15.5 Leaf Anatomical Characteristics in Four Species of Ginger

Tissue	Z. officinale	Z. machrostachyum	Z. zerumbet	Z. roseum
Epidermis	Upper larger than lower	Both epidermis equal	Both epidermis equal	Both epidermis equal
Hypodermis	Two layers on upper side, one layer on lower side	Two-layered on both sides	Two-layered on both sides	Upper cells are larger, lower, smaller
Mesophyll palisade	Single-layered on upper side	No palisade tissue	No palisade tissue	Single-layered on upper side
Spongy tissue	Three to four layers, closely packed	Four to five layers, loosely packed	Four to six layers, loosely packed	Four to five layers, closely packed
Air cavities	Absent in lamina, present in the midrib region	Present in the mesophyll tissue and more in the midrib region	Few cavities in lamina, more in the midrib region	Absent in lamina, present in the midrib region
Vascular bundle sheath	Present on both sides and extend to both epidermis	Present on both sides and extend to upper epidermis only	Present on both sides and extend to upper epidermis only	Present on both sides and extend to both epidermis
Stromatal Index	5.8–8.9, 7.45, 1.4	7.8–10.3, 8.15,1.08	8.9–13.2, 10.23, 1.4	8.01–12.03,9.11
Range, Mean Standard Deviation	1.4	1.2		

The stomata are tetracyclic in all the species. The first two subsidiary cells are parallel to the guard cells and the other two lie at right angles. In *Z. officinale*, *Z. roseum*, and *Z. macrostachyum*, there is a special thickening of the upper and lower sides of the guard cell, but *Z. zerumbet* showed some extra thickening on the corners of subsidiary cells. The stomatal index was higher in *Z. zerumbet*. Guard cells were the largest in *Z. zerumbet*, followed by *Z. officinale* and *Z. macrostachyum*. In *Z. roseum*, the guard cells were shorter and broader. The leaf anatomical characteristics in four species of ginger are given in Table 15.5.

Stomatal Ontogeny

Raju and Shah (1975) described the structure and ontogeny of stomata of ginger. Here, the differentiation of a guard cell, mother cell, or a meristemoid occurs by an asymmetrical division of protodermal cells. The meristemoid is distinguished from the adjacent protodermal cells by its small size, dense sustainability of cytoplasm, and less vacuolation. The anticlinal wall of the meristemoid appears lightly stained with periodic acid-Schiff (PAS) reaction than the lateral walls of the epidermal cell and the meristemoid. The epidermal cell on either side of the meristemoid divides to form a small subsidiary cell. This epidermal cell shows dense sustainability for nuclear DNA. The young lateral subsidiary cells are smaller than the epidermal cells. Subsequently, the meristemoid divides to form a pair of guard cells. The epidermal cells that lie at the polar region of the guard cell may divide and occasionally completely about the stomatal complex and appear as subsidiary cells (Raju and Shah, 1975).

Anatomical Features of Dry Ginger

In commercial ginger rhizome (peeled dried rhizome), the outer tissue consisting of cork, epidermis, and the hypodermis is scraped off. Hence, the TS of processed rhizome consist of cortex, endodermis, pericycle, and central cylinder or the vascular zone. The epidermis (of dry unpeeled ginger) is frequently disorganized, consisting of longitudinally oblong-rectangular cells; the cork consists of several layers of oblong-rectangular, thin-walled suberized cells. The cortex is made of (i) thin-walled parenchymal cells containing plenty of starch grains, (ii) brown-colored oleoresin and oil cells scattered throughout the cortex, and (iii) fibrovascular bundles. There is an unbroken endodermis made of tangentially elongated cells with thickened suberized radial walls. Below the endodermis, there is a pericycle that consists of an unbroken ring of tangentially elongated cells.

The central cylinder consists of an outer and an inner zones. In the outer zone, adjoining the pericycle, there is a vascular bundle zone with fibers. Fibrovascular bundles and oleoresin cells occur in the central zone of the central cylinder. The ground tissue of the central cylinder consists of thin-walled parenchymal cells containing starch.

The fibrovascular bundles are large. In longisections, the fibers are long with moderately thick walls and a wide lumen. The vessels are large and scalariform, except in the vascular bundle zone, adjoining the pericycle, where large reticulate vessels, scalariform vessels, and some special vessels occur.

Starch grains are present in abundance. The granules are ovate and many are characterized by a protuberance at one end. They vary in size to about 45 μm in length and 24 μm in width. Under polarized light, the granules exhibit a distinct cross through the hilum at the tapering end (Parry, 1962).

Microscopic Features of Ginger Powder

Ginger rhizome powder is pale yellow or cream in color with a pleasant, aromatic odor, and a characteristic and pungent taste. The diagnostic characteristics of ginger powder are as follows:

1. The abundant starch granules are mostly simple, fairly large, flattened, oblong to subrectangular to oval in outline with a small pointed hilum situated at the narrower end; infrequent granules show very faint transverse striations. Compound granules with two components occur very rarely.
2. The fibers usually occur in groups and may be associated with the vessels; they are fairly large and one wall is frequently dentate; the walls are thin and marked with numerous pits,

which vary from circular to slit-shaped in outline; very thin transverse septa occur at intervals. The fibers give only a faint reaction for lignin.

3. The vessels are fairly large and usually occur in small groups associated with the fibers; they are reticulately thickened, frequently showing distinct, regularly arranged rectangular pits, and are often accompanied by narrow, thin-walled cells, containing dark brown pigment; a few smaller, spirally thickened vessels also occur. All the vessels give only a faint reaction to lignin.

4. The oleoresin cells in uncleared preparations are seen as bright yellow ovoid to spherical cells occurring singly or in small groups in the parenchyma.

5. The abundant parenchyma is composed of thin-walled cells, rounded to oval in outline with small intercellular spaces; many of the walls are characteristically wrinkled; the cells are filled with starch granules or oleoresin. Occasionally, groups of parenchyma are associated with thin-walled tissue composed of several rows of collapsed cells.

Floral Anatomy

Rao and Gupta (1961), Rao and Pai (1959, 1960), and Rao et al. (1954) investigated the floral anatomy of the members of Scitamineae, in which a few species of *Zingiber* were also included. The floral anatomy of *Z. ottensi*, *Z. macrostachyum*, *Z. cernuum*, and other *Zingiber* species was reported by these investigators. On account of the basic similarities in floral characters, it is presumed that the floral anatomical features will also be identical. The following discussion is based on the reports of the above-mentioned investigators. The floral anatomical features of *Z. cernuum* (which is different from *Z. officinale* only by the absence of staminodes) show that the peduncle contains two rings of vascular bundles with a few strands in the central pith. The inner ring gives off three dorsal bundles of the carpels outward and the latter then divide into three large strands alternating in position with the dorsal bundles of the carpels. The central strands unite first into one bundle for a short length and fuse with the vascular tissue immediately to the outside. The three large bundles divide first into smaller inner placental bundles and a large outer parietal bundle. The parietal bundle travels into the septa and sends a few outward branches into the ovary wall. The placental bundles in the axile area bear the ovular traces. The posterior parietal bundle is larger and divides even at a lower level than the other two into two or three. A transverse section through the basal part of the ovary at this level shows: (i) a comparatively thick ovary wall in which there are numerous vascular bundles almost irregularly scattered, (ii) in each of the three septa there is a prominent bundle that may divide into two, and (iii) in the placental zone there are 6–10 strands that bear traces for the ovules. Most of the potential bundles are exhausted in supplying the ovules, while one or two may fuse with the nearest parietal bundle. The loculi extend for a considerable distance above the ovuliferous zone, and in this terminal part of the ovary, the number of bundles in the ovary wall is reduced by fusions among themselves, and all of them form almost a single ring near the level where the loculi end. Just on top of the ovary, the three parietal strands, which have already divided into two or three bundles, extend laterally and form a broad network-like cylinder of vascular tissue. This network establishes vascular connections (anastamoses) with the peripheral bundles. The three loculi continue upward into a

Y-shaped stylar canal. After the anastamosis, the vascular tissue directly forms (i) an outermost ring of about 15 small bundles for the calyx, (ii) a next inner ring of about 25 larger strands for the corolla and androecial members, and (iii) toward the center a number of small scattered strands arrange somewhat in the form of an arc. The stylar traces are given off from the two margins of this arc-like group and they stand close to the two arms of the Y-shaped stylar canal. The numerous small bundles, arranged at first as an arc, break up into two groups, which supply the two epigynous glands present in anteriolateral positions. The tubular basal parts of the calyx containing the sepal traces referred to earlier are at first separated, and, at the same level, the two epigynous glands also separate. A very short distance above the style also separates.

The basal part of the floral tube contains a ring of vascular bundles, an additional bundle in the median posterior position, and a pair of closely placed bundles on either side. The median posterior strand and the double strands on either side constitute the supply to the functional stamen. One of the component bundles of each double strand divides into two in such a way as to result in a third bundle that lies toward the inner side with its xylem pointing to the outside. On the anterior side of the floral tube, the vascular bundles divide and form two rings, whereas on the posterior face, external to the stamen traces, there is only one ring of bundles. The latter are for the labellum, whose margins are fused to a short distance with those of the filament. The outer ring of bundles is for the corolla.

The flat filament receives: (i) a small median bundle; (ii) a triple strand on either side of it, the constituent bundles of which more or less fuse together; and (iii) two or four minute strands toward either margin. The lateral triple strands are opposite the line of attachment of the anther lobes to the filament. The minute marginal traces disappear quickly, leaving only a small median bundle and the two lateral large composite ones. These run in parallel manner upward, and the composite strands of each lateral group fuse together more or less completely, so that the anther connective contains a small median and two large lateral bundles. Above the level of the anther, the connective is continued upward as a narrow flat plate with margins incurved and enclosing the style. Each of the two composite lateral strands becomes smaller and divides into two. Thus, in the terminal part of the filament, five bundles are present, of which one is the median one. The median bundles fade out first, leaving a pair of bundles on either side. The bundles of each pair then fuse together giving only two bundles, which run right up to the tip and disappear.

The style receives only two traces and these run throughout its length without any branching. The styled canal is narrow, Y-shaped, the arms of the Y pointing to the posterior side. Toward the tip, the arms of the stylar canal spread out so that the canal appears as a curved slit in transverse sections. It then widens out into a large canal, which opens freely to the outside. The two vascular bundles of the style become more prominent in this terminal part and then disappear (Rao and Pai, 1959).

Floral Biology

Ginger flowers are produced in the peduncled spikes arising directly from the rhizomes. The oval or conical spike consists of overlapping brachts, from the axils of which flowers arise, each bracht producing a single flower. The flowers are fragile, short lived, and surrounded by a scariose, glabrous bracteole. Each flower has a thin tubular corolla that widens up at the top into three lobes. The colorful part of the flower is the labellum, the petalloid stamen. The labellum is tubular at the base, three-lobed above, pale yellow outside, dark purple inside the top and margins, and mixed with yellow spots. The single fertile anther is ellipsoid, two celled, cream colored, and dehisces by longitudinal slits. The inferior ovary is globose, the style is long and filiform, and the stigma is hairy. Flowering is not common and is probably influenced by climatic factors and photoperiod. On the West Coast of India (Kerala State), most of the cultivars of ginger flower if sufficiently large rhizome pieces are used for planting. When rhizomes are left unharvested in pots, profuse flowering occurs in the following growing season. Flowering is also reported from the East Coast of India (Bhubaneshwar in Odisha State). However, ginger does not usually flower or flowers very rarely in the growing areas of such locations as Himachal Pradesh, Uttar Pradesh, West Bengal, and Northeast India. Holttum (1950) reported that ginger seldom, if at all, flowers in Malaysia. Flowering is reported from South China, but not from North China, and also from Nigeria. In general, ginger, does not flower under subtropical or subtemperate climatic conditions. Japanese investigators reported that flowering leads to yield reduction. Ginger is shown to be a quantitative short-day plant (Adaniya et al., 1989).

Jayachandran et al. (1979) reported that the flower bud development took 20–25 days from the bud initiation to full bloom and 23–28 days to complete flower opening in an inflorescence. Flower opening takes place in an acropetal succession. Anthesis is between 1:30 PM and 3:30 PM under the climatic conditions prevailing in the West Coast (Malabar region) of Kerala State, India. Anther dehiscence almost coincides with flower opening. The flowers fade and fall the following day in the morning. There is no fruit setting.

Das et al. (1999) reported floral biology in four cultivars of ginger (Bhaisey, Ernad, Chernad, Gurubathan, and Turia local). They observed that anthesis under greenhouse and field conditions took place at around 1 PM–2 PM under coastal conditions of Odisha State, India. Flowers were hermaphroditic with pin-and-thrum-type incompatibility, and dehisced pollen grains did not reach the stigma. Selfing and cross-pollination did not produce any seed set.

Self-Incompatibility

Dhamayanthi et al. (2003) investigated the self-incompatibility system in ginger. They reported that heterostyly with a gametophytically controlled self-incompatibility system exists in ginger. Flowers are distylous, there are long ("pin") and short ("thrum") styles. The "pin" type has a slender style that protrudes out of the floral parts, which are short, covering not even half the length of the style. The stigma is receptive before the anthesis, whereas the anthers dehisce after 15–20h. The anthers are situated far below and hence the pollen grains cannot reach the stigma. In case of the "thrum" style, the stigma is very short and the staminodes are long and face inward. However, the occurrence of thrum styles is very rare among cultivated ginger. According to the above-mentioned investigators, this heterostyly situation may be a contributing factor to the sterility in ginger. However, this may not be very important as almost all cultivars are the "pin" type and pollination is entomorphilous, mostly by honeybees. Dhamayanthi et al. (2003) have also reported inhibition of pollen tube growth in the style, and this was interpreted to be due to incompatibility. Adaniya (2001) reported the pollen germination in a tetraploid clone of ginger, 4×Sanshu. Pollen germination was highest at around 20°C and pollen tube growth in the style was greatly enhanced at 17°C. At this temperature, the pollen tubes penetrated into the entire length of the style in 66.7% of the styles analyzed. Pollen stored for 3 h at 40–80% RH completely lost its viability, whereas pollen incubated at 100% RH retained relatively high germinability. When the RH was low, the pollen tube in the style stopped growing. Hence, for pollen to germinate and grow in the stylar tissue, relatively low temperature (approximately 20°C) and 100% RH are essential.

Embryology

The embryology of ginger has not been investigated critically until now, and it is rather amazing that such an economically important species has been ignored by embryologists from their investigations. Possibly, absence of flowering and seed set in most growing regions could have been an important contributory factor for this lack of interest and investigation in this specific field of investigation. However, some scanty information is available in a related species, namely *Z. macrostachyum*. The embryological features of the genera in Zingiberaceae are similar, and hence, the information on *Z. macrostachyum* may as well be applicable to ginger.

The embryo sac development follows the *Polygonum* type (Panchakshrappa, 1966). The ovules are anatropous, bitegmic, and crassinucellate and are borne on an axil placentation. The inner integument forms the micropyle. In the ovular primordium, the hypodermal archesporial cell cuts off a primary parietal cell and a primary sporogenous cell. The former undergoes anticlinal division. The sporogenus cell enlarges into a megaspore mother cell, which undergoes meiosis forming megaspores. The chalazal spore enlarges and produces the embryo sac. Its nucleus undergoes three successive divisions resulting in an eight-nucleate embryo sac. Prior to fertilization in *Z. macrostachyum*, the synergids and antipodals degenerate. The fate of the nuclei in the embryo sac of ginger (which is a sterile species) is not known. However, some studies have indicated that a postmeiotic degeneration of the embryo sac can happen (Pillai, personal communication).

Cytology, Cytogenetics, and Palynology

Mitotic Studies

The chromosome number of ginger was reported as $2n = 22$ by Moringa et al. (1929) and Sugiura (1936). Darlington and Janaki Ammal (1945) cited a report from Takahashi (1930) who claimed $2n = 24$ for *Z. officinale*. A more detailed study was carried out by Raghavan and Venkatasubban (1943) on the cytology of three species,

Table 15.6 The Chromosome Reports on *Zingiber*

Species	*n*	*2n*	Reference
Z. officinale		22	Sugiura (1936)
		22	Moringa et al. (1929)
		22	Raghavan and Venkatasubban (1943)
		22	Chakravorti (1948)
		22	Sharma and Bhattacharya (1959)
		22 + 2B	Darlington and Janaki Ammal (1945)
		24	Takahashi (1930)
	11	22	Ramachandran (1969)
	11	22	Ratnambal (1979)
Z. roseum	11	22	Ramachandran (1969)
Z. wightianum	11	22	Ramachandran (1969)
Z. spectabile		22	Mohanty (1970)
Z. cylindricum		22	Mohanty (1970)
Z. cassumunnar		22	Raghavan and Venkatasubban (1943)
Z. clarkei		22	Holttum (1950)
Z. ottensi		22	Holttum (1950)
Z. mioga		55	Moringa et al. (1929), Sato (1948)
Z. zerumbet	11	22	Ratnambal (1979)

namely *Z. officinale*, *Z. cassumunnar*, and *Z. zerumbet*, and all the three had the somatic chromosome number $2n = 22$. Based on the differences in ideogram morphology, the above-mentioned researchers concluded that the chromosome morphology of *Z. officinale* was different from that of the other two species. Darlington and Janaki Ammal (1945) reported two "B" chromosomes in certain types of ginger in addition to the normal complement of $2n = 22$. Chakravorti (1948) also found $2n = 22$ in ginger. He concluded that in view of the normal pairing of 11 bivalents in species, such as *Z. cassumunnar* and *Z. zerumbet*, *Z. mioga* having a somatic chromosome number $2n = 55$ is to be considered as a pentaploid. Table 15.6 gives a summary on the question of chromosome number.

Ratnambal (1979) investigated the karyotype of 32 ginger cultivars (*Z. officinale*) and found that all of them possess a somatic chromosome number of $2n = 22$. The karyotype was categorized based on Stebbin's classification (Stebbins, 1958), which recognizes three degrees of differences between the longest and the shortest chromosome of the complement and four degrees of differences with respect to the proportion of the chromosome that are acro-, meta-, and telocentric. An asymmetrical karyotype of "1B" was found in all the cultivars except cultivars of Bangkok and Jorhat, which have a karyotype asymmetry of 1A (Ratnambal, 1979). Table 15.7 details karyotypes of various cultivars, which only exhibited minor differences.

Meiosis

Ratnambal (1979) and Ratnambal and Nair (1981) investigated the process of meiosis in 25 cultivars of ginger. These cultivars exhibited much intercultivar variability

Table 15.7 Karyotype Variation in Ginger Cultivars

S. No.	Karyotype Character	Range	Cultivars with Lowest and Highest Values
1.	Total chromatin length (μm)	22.4–37.4	cv. Jorhat, cv. China
2.	Length of longest chromosome (μm)	2.8–4.8	cv. Jorhat, cv. China
3.	Length of shortest chromosome (μm)	1.2–2.2	cv. Rio de Janeiro, cvs. China and Poona
4.	Number of median chromosomes	1–9	cv. Jugijan, cvs. Mananthody and Arippa
5.	Number of submedian chromosomes	2–10	cv. Mananthody, cv. Jugijan
6.	Number of subterminal chromosomes	0–2	cvs. Kuruppumpadi, Poona and Himachal Pradesh
7.	Number of satellite chromosomes	1	In all cultivars
8.	Type of symmetry	1A	cvs. Jorhat, Bangkok, Z. macrostachyum, Z. zerumbet, Z. cassumunnar
		1B	In all other cultivars

in meiotic behavior. The presence of multivalent and chromatin bridges was found to be a common feature in most cultivars investigated by Ratnambal (1979). The presence of multivalent in a diploid species indicates structural hybridity involving segmental interchanges, and 4–6 chromosomes are involved in the translocations as evidenced by quadrivalents and hexavalents. This structural hybridity might be contributing to the sterility in ginger.

Structural chromosomal aberrations have occurred at all stages of microsporogenesis in ginger. The predominant aberrations were laggards, bridges, and fragments at anaphase as follows: (i) laggards, bridges, and fragments, irregular chromosome separation, and irregular cytokinensis at anaphase and (ii) micronuclei and supernumerary spores at the quartet stage. Ratnambal (1979) had shown a positive linear regression between pollen sterility and chromosomal aberrations at anaphase II and aberrant quarters. Structural chromosomal aberrations have been attributed to as the cause of sterility in ginger. But how such a diploid species such as ginger came to acquire a complicated meiotic system which led to chromosomal sterility is not well understood. A hybrid origin followed by continuous vegetative propagation can be one reason for the abnormal chromosomal behavior (Ratnambal, 1979). Beltram and Kam (1984) studied meiotic features of 33 species in Zingiberaceae, including the nine species of *Zingiber*. They observed various abnormalities such as aneuploidy, polyploidy, and B chromosomes. They also confirmed the diploid nature of the Malaysian ginger ($x = 11$) and the pentaploid nature of the Japanese ginger *Z. mioga*.

Das et al. (1998) studied meiosis and sterility in four ginger cultivars (Bhaisey, Ernad Chernad, Gorubathany, and Thuria local) and reported a 30.35–40.5% meiotic

index in them. Pollen mother cells showed incomplete homologous pairing at metaphase I and spindle abnormalities (e.g., late separation, laggards, sticky bridges) at
anaphase, leading to high pollen sterility. Das et al. (1999) opined that the sterility
might be due to nonhomology of bivalents, with irregular separation of genomic
complements leading to sterile gamete formation. The absence of germination pores
on the pollen grains has also been indicated as an impediment to seed set.

Pollen Morphology

The earlier investigators (Dahlgen et al., 1985; Stone et al., 1979; Zavada, 1983)
were of the opinion that the pollen grains of the family are extineless, possessing
a structurally complex intine (Hesse and Waha, 1982). However, subsequent studies indicated that in the majority of the Zingiberaceae, an extinous layer does exist,
although it is poorly developed in many taxa (Chen, 1989; Kress and Stone, 1982;
Skavaria and Rowley, 1988). Recent palynological investigations have demonstrated
differences in pollen structure between sections of *Zingiber*: The Sect. *Zingiber*
has spherical pollen grains with cerebroid sculpturing, whereas Sect. *Cryptanthium*
has ellipsoid pollen grains with spirostriate sculpturing (Chen, 1989; Liang, 1988).
The pollen of Zingiberaceae is usually classified as inaperturate, but *Zingiber* is an
exception. Some investigators have described the pollen of ginger plant as monosul-
cate (Dahlgen et al., 1985; Zavada, 1983), while others have described it as inaperturate (Chen, 1989). The pollen is spherical or ellipsoidal. The spherical pollen grains
have a cerebroid or reticulate sculpturing. The grains have a length of 110–135 μm
and width of 60–75 μm. The pollen grains have a 2–3 μm thick coherent extine. The
intine consists of two layers, a 5 μm thick outer layer and 2–3 μm thick inner layer
adjacent to the protoplast. The outer layer is radially striated. The inner layer has a
distinct, minute fine structure. No apertures are present. It has been indicated that the
entire wall functions as a potential germination site (Hesse and Waha, 1982). Nayar
(1995) studied the germinating pollen grains of 22 taxa in Zingiberales including *Z. roseum* and *Z. zerumbet* and reported that the pollen grains possess an extine
containing sporopellenin. Inside this layer there is a well-defined lamellated cellulosic layer (described as the outer layer of intine by earlier researchers), which is
the medine. The intine is membraneous and consists of cellulose and protein and is,
in fact, the protoplasmic membrane. At germination, a solitary pollen tube develops
that has the protoplasmic membrane (intine) as its wall and pierces the outer layers
smoothly even in the absence of a germpore or aperture. The stainability percentage
ranges from 14.7 in cultivar Thingpuri to 28.5 in cultivar Pottangi and cv. China.
Usha (1984) observed 12.5% and 16.4% stainability in cultivars Rio de Janeiro and
Moran, respectively. Pollen germination ranged from 8% (in cultivar Sabarimala)
to 24% in cultivar Moran (Dhamayanthi et al., 2003). Pillai et al. (1978) reported
17% pollen germination in cultivar Rio de Janeiro. The pollen tube growth *in vitro*
was maximum in cultivar China (488 μm) and minimum in cultivar Nadia (328 μm).
The number of pollen tubes ranged from 6.5 (in cultivar Nadia) to 16.7 (in cultivar
Varada) (Dhamayanthi et al., 2003).

Physiology of Ginger

Effect of Day Length on Flowering and Rhizome Enlargement

Ginger is grown under varying climatic conditions and in many countries in both the northern and the southern hemispheres. It is generally regarded as being insensitive to day length. Adaniya et al. (1989) carried out a study to determine the influence of day length on three Japanese ginger cultivars (Kinoki, Sanshu, and Oshoga) by subjecting the plants to varying light periods in comparison with natural daylight. In three cultivars, as the light periods decreased from 16 to 10 h, there was inhibition of vegetative growth of shoots and the underground stem. The rhizome knobs became more rounded and smaller. As the day length increased to 16 h, the plants grew vigorously and the rhizome knobs became slender and larger and active as new sprouts continued to appear. When the light period was extended to 19 h, there was reduction in all of the growth parameters, and they were on par when the light period was just 13 h. It appears that the vegetative growth was promoted by exposure of the plant to a longer light period, up to a certain limit, whereas rhizome enlargement was accelerated under a relatively lower light period. The results also suggested that a relatively short-day length accelerated the progression of the reproductive growth, whereas relatively long day length decelerated it. Ginger plant is therefore described as a quantitative short-day plant for flowering and rhizome enlargement (Adaniya et al., 1989). These researchers have also observed intraspecific variations in photoperiodic response; cultivar Sanshu responded more sensitively, while Kinoki was more sensitive than Oshoga. They concluded that such an intraspecific response to the photoperiod could be related to their traditional geographical distribution. Kinoki and Sanshu are early-duration cultivars adapted to the northern part of Japan in Kanto district and Oshoga is a late-duration cultivar adapted to the southern Japan (Okinawa to Shikoku districts).

Sterling et al. (2002) investigated the effects of photoperiod on flower bud initiation and development in *Zingiber mioga* (myoga or Japanese ginger). Plants grown under long day conditions (16 h) and short-day conditions (8 h) with a night break produced flower buds, whereas those under short-day conditions (8 h) did not. This failure of flower bud production under short-day conditions was due to abortion of developing floral bud primordia rather than a failure to initiate inflorescence production. It was concluded that although for flower development in myoga, a quantitative long day requirement must be met, flower initiation was day neutral. Short-day conditions also resulted in premature senescence of foliage and reduced foliage dry weight.

Chlorophyll Content and Photosynthetic Rate in Relation to Leaf Maturity

Xizhen et al. (1998b) investigated the chlorophyll content, photosynthetic rate (Pn), malondialdehyde (MDA) content, and the activities of the protective enzymes during leaf development. Both chlorophyll content and Pn increased with leaf expansion and reached a peak on day 15 and then declined gradually. In the first 40 days of leaf growth, the MDA content of leaves remained constant and superoxide dismutase (SOD) activity showed a little decrease. After 40 days, the MDA content increased markedly and SOD activity dropped substantially. Peroxidase (POD) and catalase activities exhibited a steady increase during 60 days. Xizhen et al. (1998c) concluded that senescence in ginger leaf sets in when leaf age reaches about 40 days.

Xizhen et al. (1998) also investigated the photosynthetic characteristic of different leaf positions, and reported that the Pn of mid-position leaves was the highest followed by the lower leaves and Pn was lowest in upper leaves. The light compensation point of different leaf positions was from 18.46 to 30.82 μmol/1 m². It was highest in mid-position leaves and lowest in lower leaves. The light saturation point ranged from 624.8 to 827.6 μmol/1 m². The values were 624.8, 827.6, and 799.5 μmol/1 m², respectively, in the upper, middle, and lower leaves. CO_2 compensation points in upper, middle, and lower leaves were 1543.3, 1499.0, and 1582.0 μl/l. The diurnal variation of Pn in different leaf positions gave a double-peak curve, the first peak was at about 9 AM and the second appeared from 1 PM to 2 PM.

Stomatal Behavior and Chlorophyll Fluorescence

Dongyun et al. (1998) studied the chlorophyll fluorescence and stomatal behavior of ginger leaves. Ginger leaves were enclosed individually in cuvettes and studied to find out the relationship between photosynthesis and changes in microclimate. Stomatal conductance (gsc) increased and was saturated at relatively low values of high intensity (400 μmol/l). At different leaf temperatures, gsc peaked at 29°C, but transpiration (tr) increased with increased irradiance and temperature. Increasing external concentrations of CO_2 caused gsc to increase but were relatively insensitive to increasing soil moisture availability until a threshold was attained (0.5–2.0 g/g). At a soil moisture content of 2–3.5 g/g, gsc increased approximately linearly with increased transpiration. Fluorescence (Fv/Fm, electron transfer in PS II) decreased with increasing photon flux density (PFD). In leaves exposed to high PFD, and varying temperatures, Fv/Fm was the lowest at 15°C and the highest at more than 25°C. In leaves exposed to low PFD, Fv/Fm remained at a similar value over all temperatures tested.

Photosynthesis and Photorespiration

Zhenxian et al. (2000) measured using a portable photosynthetic system and a plant efficiency analyzer, the photosystem inhibition of photosynthesis and the diurnal variation of photosynthetic efficiency under shade and field conditions. There were marked photoinhibition phenomena under high light stress at midday. The apparent quantum yield (AQY) and photochemical efficient of PS II (Fv/Fm) decreased at midday, and there was a marked diurnal variation. The extent of photoinhibition due to higher light intensity was severe in the seedling stage. After shading, AQY and Fv/Fm increased and the degree of photoinhibition declined markedly. However, under heavier shade, the photosynthetic rate declined because of the decline in carboxylation efficiency after shading. Shi-jie et al. (1999) investigated the seasonal and diurnal changes in photorespiration (Pr) and the xanthophylls cycle (L) in ginger leaves under field conditions in order to understand the role of L and Pr in protecting leaves against photoinhibitory damage. The seasonal and diurnal changes of Pr and L of ginger leaves were marked, and Pr showed diurnal changes in response to PFD, and its peak was around 10 AM to 12 PM. Pr declined with increasing shade intensity. The L cycle showed a diurnal variation in response to PFD and xanthophylls cycle pool. Both increased during the midday period and peaked around noon. The results, in general, indicated that Pr and the xanthophyll cycle had positive roles in dissipating excessive light energy and in protecting the photosynthetic apparatus of ginger leaves from midday high light stress.

Xizhen et al. (2000) have also investigated the role of SOD in protecting ginger leaves from photoinhibition damage under high light intensity. They observed that on a sunny day, the photochemical efficiency of PS II (Fv/Fm) and AQY of ginger leaves declined gradually in the morning, but rose progressively after noon. The MDA content in ginger leaves increased but the Pn declined under midday high light stress. SOD activity in ginger leaves increased gradually before 2 PM and then decreased. At 60% shading in the seedling stage, Fv/Fm and AQY of ginger leaves increased, but the MDA content, SOD activity, and Pn decreased. Pn, AQY, and Fv/Fm of ginger leaves treated with diethyldithio carbamic acid (DDTC) decreased whether shaded or not, but the effect of DDTC on shaded plants was less than that on unshaded plants. These researchers concluded that midday high light intensity imposed a stress on ginger plants and caused photoinhibition and lipid peroxidation. SOD and shading played important roles in protecting the photosynthetic apparatus of ginger leaves against high light stress.

Xizhen et al. (1998a) have investigated the effect of temperature on photosynthesis of the ginger leaf. They showed that the highest photosynthetic rate and apparent quantum efficiency was under 25°C. The light compensation point of photosynthesis was in the range of 25–69 μmol/m^2; it increased with increasing temperature. The light saturation point was also temperature dependent. The low light saturation point was noted at temperatures below 25°C. The CO_2 compensation point and the saturation point were 25–72 and 1343–1566 μl/l, respectively, and both increased with the increase in leaf temperature.

Xianchang et al. (1996) investigated the relationship between canopy, canopy photosynthesis, and yield formation in ginger. They found that canopy photosynthesis was closely related to yield. In a field experiment using a plant population of 5000–10,000 plants/666.7 m^2 area, they obtained a yield increase from 1.733 to 2.626 kg, which is almost a twofold increase. The Pn increased from 8.16 μmol CO_2/m^2 1 (ground) s^{-1} to 14.66; the LAI from 3.21 m^2/m^2 to 7.02 m^2/m^2. The unit area of branches (tillers) and LAI were over 150/m^2 and 6 m^2/m^2, respectively, in the canopy of the higher yield class. The canopies over 7000 plants per 666.7 m^2 satisfied these two criteria and among them there were no significant differences in height, number of tillers, LAI, canopy photosynthesis, and yield. Diurnal changes in the canopy Pn showed a typical single-peak curve, which was different from the double-peak curve obtained from the single leaf Pn.

Effect of Growth Regulators

Investigations have been carried out to find out the various growth regulators on ginger growth, flowering, and rhizome development. The principal objective of such investigations is to break the rhizome dormancy, to induce flowering and seed set, and to enlarge the rhizome, followed by enhanced yield. Islam et al. studied the influence of 2 chloroethyl phosphonic aid (Ethrel or Ethephon) and elevated temperature treatments. Exposure of ginger rhizome pieces to 35°C for 24 h or to 250 ppm Ethrel for 15 min caused a substantial increase in shoot growth during the first 23 days of growth. Ethrel was more effective in increasing the number of roots per rhizome piece by a factor 4.0 and the number of shoots having roots by a factor of 3.7 (both at day 16). Relatively low concentrations of Ethrel (less than 250 ppm) were sufficient to produce maximum responses in terms of shoot length parameters, although significant increases in the number of shoots per seed piece, the number of rooted shoots, and the total number and length of roots per seed piece, occurred even up to the highest concentrations of 1000 ppm studied by Islam et al. Treatment of Ethrel was found to be effective in reducing the variability in root growth, but shoot growth variability had increased particularly at concentrations below 500 ppm.

Furutani and Nagao (1986) investigated the effect of daminozide, GA, and Ethephon on flowering, shoot growth, and yield of ginger. Field-grown ginger plants were treated with three weekly foliar sprays of GA (0, 1.44, and 2.88 mM); Ethephon (0, 3.46, and 6.92 mM), or daminozide (0.3, 1.3, and 6.26 mM). GA inhibited flowering and shoot emergence, whereas Ethephon and daminozide had no effect on flowering but promoted shoot emergence. Rhizome yields were increased with daminozide and decreased with GA and Ethephon.

Ravindran et al. (1998) tested three growth regulators—triacontanol, paclobutrazole, and GA—on ginger to find out their effect on rhizome growth and developmental anatomy. Paclobutrazole- and triacontanol-treated rhizomes resulted in thicker-walled cortical cells compared to GA and control plants. The procambial activity was higher in plants treated with triacontanol and paclobutrazole. In the cambium layers, the fusiform cells were much larger in paclobutrazole-treated plants. Growth regulator treatments did not affect the general anatomy, although dimensional variations existed. The numbers of vascular bundles were more in plants treated with paclobutrazole and triacontanol. Paclobutrazole-treated plants exhibited greater deposition of starch grains than other treatments. The fiber content in the rhizome was less in GA-treated rhizome. A higher oil cell index and higher frequency oil cells were observed in paclobutrazole-treated rhizomes. GA treatment also led to considerable increase in the number of fibrous roots.

Growth-Related Compositional Changes

Baranowski (1986) studied the cultivar Hawaii for 34 weeks and recorded the growth-related changes of the rhizome. The solid content of the rhizome increased throughout the crop season, but there was a decline in the acetone extractable oleoresin content of dried ginger. However, the oleoresin content on a fresh weight basis was roughly constant.

The (6)-gingerol content of ginger generally increased with the age of the rhizome on a fresh weight basis. These results indicate the basis for the gradual increase in pungency with maturity. On a dry weight basis, gingerol generally exhibited a linear increase with maturity up to 24 weeks, followed by a steady decline through the rest of the period. The results in general indicate that it may be advantageous to harvest ginger early (i.e., by 24 weeks) for converting to various products.

Genetic Resources

The history of domestication of ginger is not definitely known. However, the crop is known to have been under cultivation and use in India and China for the last 2000 years or even more. China is probably the region where domestication started, but little is known about the center of origin of the plant, although the largest variability in the crop exists in China. Southwestern India, known as the "Malabar Coast" (in Kerala State) in ancient times traded ginger with the Western world, which definitely indicates its cultivation in this region. This long period of domestication might have played a major role in the evolution of this crop which is sterile and propagated only vegetatively. Ginger has rich cultivar diversity, and most major growing tracts have cultivars which are specific to the area. These cultivars are mostly known by the region's name. Cultivar diversity is richest in China. In India, the diversity is more in the Kerala State and in northeastern India. Being clonally propagated, the population structure of this species is determined mainly by the presence of isolation mechanisms and the divergence that might have resulted through the accumulation of random mutants. Presently, there are more than 50 ginger cultivars possessing varying quality attributes and yield potential being cultivated in India, although the spread of a few improved and high-yielding ones cause the disappearance of the traditional land races. The cultivars popularly grown (cultivar diversity) in the various ginger-growing states in India are given in Table 15.8. Some of these cultivars were introduced in India, and the cultivar Rio de Janeiro, an introduction from Brazil, has become very popular in Kerala State. Introductions such as China, Jamaica, Sierra Leone, and Taffin Giwa are also grown occasionally.

Among the ginger-growing countries, China has the richest cultivar diversity. Among the cultivars, Zaoyang of Hubei province is a very important one. Zunji big white ginger of Giuzhou province is another important cultivar. Also, Chenggu Yellow of Shaxi province, Yulin round fleshy ginger of Guangxi province, Bamboo root ginger, and Mian yang ginger of Sichuan are other important ones. Xuanchang ginger of Ahuii, Yuxi yellow ginger of Yunan, and Taiwan fleshy ginger may also be counted as important cultivars of China. Many of these cultivars have unique morphological markers for facilitating identification.

In general, the cultivar variability is much less in other ginger-growing countries. Tindall (1968) reported that there were two main types of ginger grown in West Africa. These differ in color of the rhizome, one with a purplish red or blue tissue below the outer scaly skin, whereas the other has a yellowish white flesh. Graham (1936) reported that there were five kinds of ginger recognized in Jamaica known as

Table 15.8 Major Ginger-Growing States of India and Their Popular Cultivars Indicating Diversity in the Ginger Plant

State	Cultivar Name and Yield/ha	Specific Trait/ Character	References
Kerala	Rio de Janeiro (32.55 t/ha—fresh)	High yielder	Thomas (1966)
	Burdwan, Jamaica	High yielder	Muralidharan and Kamalam (1973)
	Nadia (28.55 t/ha—fresh and 6.54 t/ha—dry) Maran, Bajpai and Narasapatam, Rio de Janeiro	High yielder	AICSCIP (1978); Khan (1959)
	Kuruppumpady	Ratoon crop	Sreekumar et al. (1982)
	SG-666	Fresh rhizome	Rattan (1989)
	Rio de Janeiro (21.80 t/ha—fresh, 3.27 t/ha—dry), Assam (17.23 t/ha—fresh), Maran (3.27 t/ha—dry), and Thingpuri (2.79 t/ha—dry)	High yielder	
	Thingpuri, Rio de Janeiro, China, IISR-Varada (22.6 t/ha—fresh)	High yielder	Sreekumar et al. (1982)
	IISR Mahima (23.2 t/ha—fresh), IISR-Rejatha (22.4 t/ha—fresh)	Wider adaptability, high yielder	
	V2E$_{3-2}$ (33.83 t/ha), Rio de Janeiro (27.38 t/ha—fresh), Ernadan (25.11 t/ha—fresh), and Mananthavady (22.94 t/ha—fresh)	High yielder	Pradeep Kumar et al. (2000)
Himachal Pradesh	Himachal Selection, Rio de Janeiro	High yielder	Jogi et al. (1978)
	SG-646 and SG-666, Kerala local (3.76 t/ha—dry), B-1 (3.83 t/ha—dry), Himachal selection (local) (10.9 t/ha—fresh) and Kerala local (9.6 t/ha—fresh), SG-534 (10.35 t/ha—fresh), V$_1$ E$_8$-2 (8.92 t/ha—fresh)	High yielder	Rattan (1989)
	Acc. No.64 (8.9 t/ha—fresh)	High altitude areas	AICRPS (2000)
Assam	Nadia (6.7 t/ha—dry), Chekerella (5.7 t/ha—fresh)	High yielder, fresh rhizome yield	Aiyadurai (1966), Saikia and Shadeque (1992)
Nagaland	Thinladium, Nadia, and Khasi local (>30 t/ha—fresh)	High yielder, fresh rhizome yield	Singh et al. (1999)
Odisha	SG-666	High fresh rhizome yield	Rattan (1989)
	Rio de Janeiro and China 239 g/plant), Vingra selection, Ernad Manjeri, UP, Thingpuri Kuruppampadi, Wynad, Kunnamangalam, Thingauri (2.20 t/ha)	High yield	Mohanty et al. (1981), Panigrahi and Patro (1985)
	V$_1$E$_{G-2}$ (25.13 t/ha), V$_3$S$_1$-8 (22.12 t/ha)	High altitude area	AICRPS (2000)

(Continued)

Table 15.8 (Continued)

State	Cultivar Name and Yield/ha	Specific Trait/ Character	References
Andhra Pradesh	IISR-Varada	High altitude area	Naidu et al. (2000)
Karnataka	Himachal Pradesh (19.97 t/ha), Jorhat (18.88 t/ha), Waynad local (18.68 t/ha—fresh yield)	High fresh rhizome yield	Gowda and Melanta (2000)
Meghalaya	Tura (26.69 t/ha—fresh yield), Poona (25.04 t/ha—fresh yield), Basar local (24.88 t/ha—fresh yield)	Midhills area	Chandra and Govind (1999)
West Bengal	Gurubathan (27.9 t/ha—fresh yield), Acc. No.64 (18.93 t/ha—fresh yield)	High yielder	AICRPS (2001)
Madhya Pradesh	V_3S_1-8 (17.4 t/ha)	High yielder	AICRPS (1999)

St. Mary, Red eye, Blue Tumeric, Bull blue, and China blue. But Lawrence (1984) reported that only one cultivar is grown widely in Jamaica. According to Ridley, three forms of ginger were known in Malaysia in earlier times. These were *Halyia betel* (true ginger), *Halyia bara* or *padi*, a smaller-leaved ginger with a yellowish rhizome used only in medicine, and *Halyia udanf*, red ginger having red color at the base of the aerial shoot. A red variety of ginger, *Z. officinale* var. rubra (also called pink ginger) has been found in Malaysia and described. In this cultivar, the rhizome skin has a reddish color. A variety called *Withered Skin* also has been reported. In the Philippines, two cultivars are known, the native and the Hawaiian (Rosales, 1938). In Nigeria, the cultivar Taffin Giwa (bold yellow-colored ginger) is the most common one encountered, the other being Yasun Bari, the black ginger.

In many cases, the major production centers are far from the areas of origin of the crop concerned (Simmonds, 1979). This is true of ginger as well. The Indo-Malayan region is very rich in Zingiberaceous flora (Holttum, 1950). Considering the present distribution of genetic variability, it is only logical to assume that the Indo-Malayan region is probably the major center of genetic diversity for ginger. It may be inferred that geographical spread accompanied by genetic differentiation into locally adapted populations caused by mutations could be the main factor responsible for variations encountered in cultivated ginger (Ravindran et al., 1994). In India, the early movement of settlers across the length and breadth of the Kerala State and adjoining regions, where the maximum ginger cultivation is found, and the story of shifting cultivation in northeastern India (the second major ginger-growing sector in India), are well-documented sociological events. The farmers invariably carried small samples of the common crops that they grew in their original place, along with them and domesticated the same in their new habitat, in most cases virgin forestlands. Conscious selection for different needs such as high fresh ginger yield, good dry recovery, and less fiber content over the years has augmented the spread of differentiation in this crop. This would have ultimately resulted in the land races of ginger of today (Ravindran et al., 1994).

Conservation of Ginger Germplasm

In India, major collections of ginger germplasms are maintained at the Indian Institute of Spices Research (IISR) in Calicut, Kerala State, India, under the administrative control of the Indian Council of Agricultural Research in New Delhi, India. The other global collection and conservation center is the Research Institute for Spices and Medicinal Plants in Bogor, Indonesia. Concerted efforts in collection and conservation of ginger germplasms is under way in India. At present, the ginger germplasm conservatory at IISR has 645 accessions which include exotic cultivars, indigenous collections, improved cultivars, mutants, tetraploids, and related species (IISR, 2002). In addition, 443 accessions are maintained at different centers of the All India Coordinated Research Project on Spices (AICRPS) and the National Bureau of Plant Genetic Resources (NBPGR), Regional Station, Thrissur AICRPS (in Kerala State, India), details of many collections centers are given in Table 15.9. The principal constraint involved in the conservation of germplasms of ginger is the widespread prevalence of soil-borne diseases, of which rhizome rot caused by *Pythium* spp. (such as *P. aphanidermatum, P. myriotylum,* and *P. vexans*) and the bacterial wilt caused by *Ralstonia solanacearum (Pseudomonas solanacearum)* are the most devastating. Additionally, infection by the leaf fleck virus is also causing major worry for conservation. In field conditions, these diseases are extremely difficult to control, once the onslaught occurs, and subsequently they spread. Hence, in the National Conservatory for Ginger at IISR, ginger germplasm is conserved in specially made cement tubs under 50% shade, as excessive shade assists the spread of the disease. This is a nucleus gene bank to safeguard the material from the deadly diseases and to maintain the purity of germplasm from adulteration, which is very common during field planting. Every year, part of the germplasm is planted in the field for their evaluation of performance and the recording of important growth traits (Ravindran et al., 1994). The collections are harvested every year and replanted in the following crop season in fresh potting mixture. On harvest of the rhizomes, each accession is cleansed and dipped in fungicidal and insecticidal solutions to control disease spread, and stored in individual brick-walled cubicles lined with saw dust or sand in a well-protected building to preclude any damage by rodents or other intruding animals.

In Vitro Conservation

In vitro conservation of ginger germplasm is a safe and complementary strategy to protect the genetic resources from epidemic diseases and other natural disasters. This is also an excellent method to supplement the conventional conservation strategies. Conservation of ginger germplasm under *in vitro* conditions by slow growth was standardized at IISR, Calicut, Kerala State, India (Geetha, 2002; Geetha et al., 1995; Nirmal Babu et al., 1996). By this method, ginger could be stored up to 1 year without subculture in half-strength MS medium 10 gel/l each of sucrose, and mannitol in sealed culture tubes. The survival percentage of such stored material is 85%. At IISR, over 100 unique accessions of ginger are being conserved under *in vitro* gene bank as medium-term storage of germplasm (Geetha, 2002; Ravindran et al.,

Table 15.9 Germplasm Collections of Ginger in India

S. No.	Institution/University	Number of Accessions	References
1.	Indian Institute of Spices Research, Calicut, Kerala State, India	645	IISR (2002)
2.	Orissa University of Agriculture and Technology, Pottangi, Odisha State, India	172	AICRPS (2003)
3.	Dr. Y.S. Parmar, University of Horticulture and Forestry, Solan, Himachal Pradesh, India	271	AICRPS (2003)
4.	Rajendra Agricultural University, Dholi, Bihar State, India	103	AICRPS (2003)
5.	Uttara Bangala Krishi Viswa Vidyalaya Pundibari, West Bengal State, India	31	AICRPS (2003)
6.	Narendra Dev University of Agriculture and Technology, Kumarganj, Faizabad, Uttar Pradesh State, India	29	AICRPS (2003)
7.	Indira Gandhi Krishi Viswa Vidyalaya, Regional Station, Raigargh, Uttar Pradesh State, India	35	AICRPS (2003)
8.	National Bureau of Plant Genetic Resources, Regional Station, Thrissur, Kerala State, India	173	Ravindran et al. (2004)
9.	Department of Horticulture, Sikkim, India	58	Kumar (1999)
10.	Central Agricultural Research Institute, Port Blair, Andamans, India	33	Shiva et al. (2004)

1994). The possibility of storage at relatively high ambient temperatures (24–29°C) by subjecting the ginger and related taxa to stress factors was explored by Dekkers et al. (1991). The increase in the subculture period was better with an overlay of liquid paraffin. After 1 year, 70–100% survival was observed.

Ravindran and associates (Anon, 2004) standardized the use of synthetic seeds in the conservation process. Synthetic seeds, developed with somatic embryos encapsulated in 5% sodium alginate gel, could be stored in MS medium supplemented with 1 mg/l g/l can be substituted for gl^{-1} benzyladenine (BA) at $22 \pm 2°C$ for 9 months with 75% survival. The encapsulated beads on transfer to MS medium supplemented with 1.0 mg/l Benzylaminopurine (BAP) and 0.5 mg/l naphthelene acetic acid (NAA), germinated and developed into normal plantlets. The conservation of germplasm through microrhizome production was also investigated, and it was found that microrhizomes can be induced *in vitro* when cultured in MS medium supplemented with higher levels of sucrose (9–12%). Such microrhizomes can be easily stored for more than 1 year in culture. Without any acclimatization, 6-month-old microrhizomes can be directly planted in the field. The microrhizomes can thus be used as a disease-free seed material and for propagation, conservation, and exchange germplasm (Geetha, 2002). This microrhizome technology is amenable for automation and scaling up.

Cryopreservation is a strategy for long-term conservation of germplasm (Ravindran et al., 1994). Efforts are continuing at IISR and NBPGR for developing such strategies. Cryopreservation of ginger shoot buds through an encapsulation–dehydration method was attempted by Geetha (2002). The shoot buds were encapsulated in 3% sodium alginate beads and pretreated with 0.75 M sucrose solution for 4 days and dehydrated in an air current from laminar airflow and then immersed in liquid nitrogen. Beads conserved like this on thawing and recovery exhibited 40–50% viability. The cryopreserved shoot beads were regenerated into plantlets. The studies carried out at IISR showed that vitrification and encapsulation–vitrification methods are more suitable for the cryopreservation of ginger shoot buds (Nirmal Babu, personal communication/unpublished data).

Characterization and Evaluation of Germplasm

A clear knowledge of the extent of genetic variability is essential for formulating a meaningful breeding strategy. Under a low variability situation, selection programs will not yield worthwhile benefits. In any vegetatively propagated species, the extent of genetic variability will be limited unless samples are drawn from distinctly different agroecological situations. Studies on genetic variability for yield and associated characters in ginger indicated the existence of only moderate variability in the germplasm. Little variability exists among the genotypes that are grown in the same area; however, good variability has been reported among cultivars that came from widely divergent areas.

One hundred accessions of ginger germplasm were characterized based on morphological, yield, and quality parameters (Ravindran et al., 1994). Moderate variability was observed in many yield and quality traits (Table 15.10). The number of tillers per plant had the highest variability, followed by rhizome yield/plant. Among the quality traits, the shogaol content recorded the highest variability, followed by crude fiber and oleoresin. None of the accessions possessed resistance to the causal organism of leaf spot disease, caused by *Phyllosticta zingiberi*. Quality parameters such as dry recovery and oleoresin and fiber contents are known to vary with soil type, cultural conditions, and climatic pattern (Ravindran et al., 1994).

Mohanty and Sarma (1979) reported that expected genetic advance and heritability estimates were high for the number of secondary rhizome and total root weight. Genetic coefficient of variation was high for weight of root tubers. Rhizome yield was positively and significantly correlated with number of pseudostems (tillers), leaves, secondary rhizome fingers, tertiary rhizome fingers, total rhizome, plant height, leaf breadth, girth of secondary rhizome fingers, and number and weight of adventitious roots. Studies indicated that straight selection was useful to improve almost all the characters except the number of tertiary fingers and straw yield. Rattan et al. (1998) reported that plant height was positively and significantly correlated with the number of leaves, leaf length, rhizome length and breadth, and yield per plot. The number of leaves per plant was positively and significantly correlated with rhizome length, breadth, and yield. The rhizome length was also related to rhizome breadth and yield. Positive correlation of rhizome weight with plant height, tiller number, and leaf number was reported by Sreekumar et al. (1982). Mohanty et al. (1981) observed

Table 15.10 Mean, Range, and Coefficient of Variation (CV%) for Yield Attributes and Quality Traits in Ginger Germplasm

Trait	Mean	Range	CV (%)
Plant height (cm)	59.2	23.1–88.6	19.00
Leaf number/plant	37.1	17.0–52.0	18.22
Tiller number/plant	16.8	2.80–35.5	45.90
Leaf length (cm)	23.8	17.0–36.5	10.90
Leaf width (cm)	2.6	1.90–3.70	10.80
Days to maturity	226.0	214–236	13.50
Dry recovery (%)	21.7	14.0–28.5	14.30
Rhizome yield/plant (g)	363.1	55.0–770.0	39.30
Crude fiber (%)	4.31	2.1–7.0	23.30
Oleoresin (%)	6.10	3.2–9.5	21.70
Gingerol (%) in oleoresin	19.9	14.0–27.0	15.20
Shogaol (%) in oleoresin	4.1	2.7–7.5	24.30

Source: Ravindran et al. (1994).

significant varietal differences for all the characters except for the number of tiller per plant and number of leaves per plant. Pandey and Dobhal (1993) observed a wide range of variability for most of the characters investigated by them. Rhizome yield per plant was positively associated with plant height, number of fingers per plant, weight of fingers, and primary rhizome weight/yield.

At IISR, Sasikumar et al. (1992) studied the 100 accessions of ginger germplasm for variability, correlation, and path analysis. They found that rhizome yield was positively correlated with plant height, tiller and leaf number, and leaf length and width (Table 15.11). Plant height also had a significant and positive association with leaf and tiller number as well as length and width of leaf. The association of leaf number with tiller number, leaf length, and width was also positive and statistically significant. Tiller number had a significant negative association with dry recovery. Leaf width had a positive significant association with dry recovery.

Yadav (1999) reported a high genotypic coefficient of variation for length and weight of secondary rhizomes, weight of primary rhizomes, number of secondary and primary rhizomes, and rhizome yield per plant. High heritability, coupled with high genetic advance as a percentage of mean was observed for plant height, leaf length, suckers per plant, number of mother and secondary rhizomes, weight of primary rhizome, and rhizome yield per plant, indicating that desirable improvement in these traits can be brought about through straight selection. Plant height followed by number of tillers per plant and leaf length had a maximum direct effect on rhizome yield (Singh, 2001).

Nybe and Nair (1979) suggested that morphological characters are not reliable to classify the types, although some of the types can be distinguished to a certain extent from rhizome characters. All the morphological characters were found to vary among types except for leaf breadth, LAI, and number of primary fingers. Mohandas et al. (2000) found that all the cultivars differed significantly in tiller number and

Table 15.11 Path Analysis in Ginger

Character	Leaf Number	Tiller Number	Leaf Length	Leaf Weight	Days to Maturity	Dry Recovery	Rhizome/ Plant
Plant height	0.69[*]	0.32[**]	0.59[**]	0.51[**]	0.12	0.18	0.47[**]
Leaf number		0.26[**]	0.56[**]	0.36[**]	0.01	0.07	0.38[**]
Tiller number			0.30[**]	0.13	0.03	0.29[**]	0.26[**]
Leaf length				0.42[**]	0.04	0.04	0.49[**]
Leaf width					0.01	0.42[**]	0.23[**]
Days to maturity						0.06	0.03
Dry recovery							0.10

Source: Sasikumar et al. (1992).
*Significant at 5% level.
**Significant at 1% level.

leaf number. Yield stability analysis revealed that cultivars Ernad and Kuruppumpadi were superior, indeed, as they showed high rhizome yield, nonsignificant genotype–environment interaction, and stability in yield.

Biochemical Variability

Oleoresin of ginger is the total extract of ginger containing all the flavoring principles as well as the pungent constituents. The oleoresin contains two important compounds—gingerol and shogoal—the two constituents which contribute to the pungency of ginger. On long-term storage, gingerol gets converted to shogaol. The quality of ginger thus depends on the relative content of gingerol and shogoal. Zachariah et al. (1993) classified 86 ginger accessions into high-, medium-, and low-quality types based on the relative contents of the quality components. There are many ginger cultivars with high oleoresin content. A few of them, such as Rio de Janeiro, Ernad Chernad, Wynad, Kunnamangalam, and Meppayyur, have high gingerol content as well. The intercharacter association showed a positive correlation with oleoresin, gingerol, and shogaol.

Shamina et al. (1997) investigated the variability in total free amino acids, proteins, total phenols, and isozymes using 25 cultivars. Moderate variations were recorded in the case of total free amino acids, proteins, and total phenols. Isozyme variability in the case of polyphenol oxidase, peroxidase, and SOD was reported to be low, indicating only a low level of polymorphism. The quality parameters of ginger cultivars are detailed in Table 15.12.

Table 15.13 contains information on germplasm evaluation. The screening details of ginger varieties/cultivars/accessions are given in Table 15.14.

Path Analysis

The partitioning of phenotype correlation between yield and morphological characteristics into direct and indirect effects by the method of path coefficient analysis

Table 15.12 Range, Mean, and Coefficient of Variation in Quality Components in Ginger Cultivars

Quality Constituent	Range	Mean	Coefficient of Variation (%)
Oleoresin (%)	3.2–9.5	6.1	21.5
Gingerol (%) in oleoresin	14–25	19.9	15.0
Shogaol (%) in oleoresin	2.8–7.0	4.1	23.7

Source: Zachariah et al. (1993).

Table 15.13 Evaluation of Ginger Germplasm for Rhizome Yield and Yield Attributes

S. No.	Character/Trait	Variety/Cultivar/Accession	References
1.	High yield (fresh and dry)	UP, Rio de Janeiro, Thingapuri, Suprabha, Anamika, Jugijan, Karakkal	Mohanty and Panda (1994)
		SG-646 (Kerala), (159 g/plant) and SG-666 (HP) (151 g/plant)	Rattan (1989)
		Rio de Janeiro, Suprabha, Suruchi, Suravi, Jugijan, Thingpuri, Waynad local, Himachal, Karakkal, Varada, Maran, Acc. Nos. 64, 117, and 35	Sasikumar et al. (1994)
		Rio de Janeiro (average yield 21 t/ha—fresh), Maran, (average yield 20 t/ha—fresh, 4.4 t/ha—dry), Nadia, (average yield 19 t/ha—fresh, 3.8 t/ha—dry), Naraspattnam (average yield 3.8 t/ha—dry)	Poulose (1973)
		Rio de Janeiro (32.55 t/ha), China (16.76 t/ha), Ernad Chernad (15.84 t/ha)	Thomas (1966)
		Waynad (9 kg/3 m^2), SG-700, 705, and BDJR-1226 (7.5–7.7 kg/3 m^2), V$_2$E$_4$-5, and PGS 43 (7.8 kg/3 m^2), SG-876, SG-882 (9.2 kg/3 m^2)	AICRPS (1999, 2000, 2001)
2.	Bold rhizome	China, Taffingiva, SG-35, Varada, Gurubathan, Bhaise, China, Acc. Nos. 117, 35, 15, 27, and 142	Sasikumar et al. (1994, 1999)
3.	Slender rhizome	Suruchi, Kunduli local	Mohanty and Panda (1994)
4.	Short duration	Sierra Leone	Mohanty and Panda (1994)
5.	High dry recovery (%)	Tura local-2 (29.7%)	Mohanty (1984)
		Tura (28%)	Sasikumar et al. (1982)

(Continued)

Table 15.13 (Continued)

S. No.	Character/Trait	Variety/Cultivar/Accession	References
		Thodupuzha (22.6%), Kuruppampadi (23.0%), and Nadia (22.6%)	Nybe et al. (1982)
		Tura and Maran	CPCRI (1973), Muralidharan (1972), Nair (1969)
		Tura (22.07%), Thinladium (21.03%), and Jorhat (20.60%)	Muralidharan (1972)
		Vengara (25%), Ernad (24.3%), Himachal Pradesh, and Sierra Leone (23.12% each)	Thomas (1966)
		Zhahirabad, Jorhat Local, Kuruppampadi, Ernad, Suruchi, Maran, Assam, China, Mowshom Thingpuri, Varada, Acc. Nos. 27, 117, 204, and 294	Sasikumar et al. (1994, 1999)
		SG-685	AICRPS (2000)
6.	High oleroresin (%)	Assam (9.3%) and Mananthody (9.2%)	Krishnamurthy et al. (1970), Natarajan et al. (1972)
		Kuruppampadi (7.1%)	Muralidharan (1972)
		Rio de Janeiro (10.5%), Maran (10%)	Nybe et al. (1982)
		Waynad local (9.1%), Rio de Janeiro (10.8%)	Sasikumar et al. (1980)
		Waynad, Kunnamangalam, Ambalavayalan, Ernad, Santhing Pui, Rio de Janeiro, Kuruppampadi, Himachal, Varada, and China	Sasikumar et al. (1994)
		Rio de Janeiro, Kunnamangalam Waynad, Meppayyur, Santhing Pui (Manipur-1), Ernad, Erattupetta, Tamarassery local, PGS-33, and PGS-11 (>7.4%)	Zachariah et al. (1993)
		Acc. Nos. 14 (9.0%) and 118 (6.0%), Nadan, Pulpally, Acc. No. 57	Sasikumar et al. (1999)
		Acc. Nos. 110, 582, 236, 388, 414, 6, and 3 (6.2—8.9%)	Zachariah et al. (1999)
		V_1S_1-8, BDJR-1226, and Chanog-11 (8.3–8.7%)	AICRPS (1999)
7.	High essential oil (%)	Mananthody (2.2%)	Krishnamurthy et al. (1970), Lewis et al. (1972)
		Karakkal (12.4%), Rio de Janeiro (2.3%), Vengara (2.3%), and Valluvana (2.2%) Elakallan and Sabarimala	Nybe et al. (1982)

Table 15.13 (Continued)

S. No.	Character/Trait	Variety/Cultivar/Accession	References
		Acc. Nos. 118 (2.6%), 14 (2.5%), Pulpally, Sabarimala, Nadan, Pulpally, and Thodupuzha	Sasikumar et al. (1999)
		Acc. Nos. 418, 399, 389, 205, 110, 236, 104, and 296 (2.9–3.2%)	Zachariah et al. (1999)
		BLP-6, SG-723, BDJR-1054, SG-55, and Maran (2.0–2.8%)	AICRPS (1999)
		Shilli, Bangi, Himgiri, Acc. No. 64, V_1E_4-4, PGS-23, and SG-706	AICRPS (2000)
8.	Low crude fiber (%)	China (3.4%), UP (3.7%), Himachal Pradesh (3.8%), Nadia (3.9%), Himachal Pradesh	Nybe et al. (1982)
		Tura (3.5%)	Sreekumar et al. (1980)
		China (3.43%), Ernad (4.43%)	Thomas (1966)
		Zahirabad, Kuruppampadi, Mizo, PGS-16	Sasikumar et al. (1994)
		China, UP, Nadia, Poona, and Jamaica, Acc. Nos. 287 (3.0%), 288, 22, and 18 (3.2%), Varada, Acc. Nos. 15 and 27	Sasikumar et al. (1999)
		Poona (4.62%), Nadia (4.84%), and Thinladium (5.01%)	Jogi et al. (1978)
		Acc. Nos. 419, 386, 415, 200, 110, and 336 (2.2–3.3%)	Zachariah et al. (1999)
9.	High yield of dry ginger (t/ha)	Rio de Janeiro, Maran (3.27 t/ha)	Muralidharan (1972)
		Thingpuri (2.79 t/ha), Maran (4.40 t/ha), Nadia (3.80 t/ha), Narasapatam (average 3.80 t/ha)	Poulose (1973)
10.	High gingerol and shogoal	Waynad, Kunnamangalam, Ambalavayalan, Ernad, Thingpuri, and Rio de Janeiro	Sasikumar et al. (1994)
		Mizo, Nadia, Maran, Ernad, Kada, and Narianpara (high gingerol—22% of oleoresin), Rio de Janeiro, Santhing Pin (Manipur-1), PGS-37, S-641, Maran, Erattupetta, Nadan, Pulpally, Jorhat local, PGS-16, Mizo, and Nadia (high gingerol- 5% oleoresin)	Zachariah et al. (1993)
		Baharica and Amaravathy	Sasikumar et al. (1999)
11.	High gingiberene and (6)-gingerol	Baharica and Amaravathy	Sasikumar et al. (1999)

Table 15.14 Screening of Ginger Germplasm Against Pests and Diseases Incidence

S. No.	Character/ Trait	Variety/Cultivar/ Accession	Reaction	References
A.	**Reaction to Pests**			
1.	Shoot borer	Rio deJaneiro	Tolerant	Nybe et al. (1982)
2.	Rhizome scale	Wild-2	Least infestation	Mohanty (1984)
		Anamika	Least incidence	Sasikumar et al. (1994)
3.	Storage pest	Varada, Acc. Nos. 212 and 215	Resistant	
4.	Root knot nematode	Valluvanad, Tura, and HP	Least infestation	Charles and Kuriyan (1982)
		Acc. Nos. 36, 59, and 221	Resistant	
B.	**Reaction to Diseases**			
1.	Rhizome rot	Jorhat and Sierra Leonne	Least incidence (11.25%)	AICSCIP (1975)
		Maran	Least infection	Nybe and Nair (1979)
		Narasapattam	Least susceptible (1–20%)	Mohanty (1984)
		Burdwan-1, Anamika, Poona and Himachal	Less susceptible (1–20%)	Sasikumar et al. (1994)
		BDJR-1226, Jamaica, BLP-6	Less susceptible	AICRPS (1999)
2.	Bacterial wilt	V_2E_5-2, Rio de Janeiro	Least incidence	Pradeep Kumar et al. (2000)
3.	Leaf spot	Taffingiva, Maran, Bajpai, and Nadia	Most tolerant	Nybe and Nair (1979)
		Maran, Kunduli local	Less susceptible	Sasikumar et al. (1994)

revealed that plant height exhibited a high direct effect as well as high indirect effect in the establishment of correlation between yield and other morphological characters (Nair et al., 1982; Ratnambal, 1979). Rattan et al. (1998) indicated that the number of leaves per plant had maximum direct contribution to yield per plant, followed by rhizome breadth.

Das et al. (1999) reported very high positive direct effects of stomatal number, leaf area, leaf number, and plant height on rhizome yield. Leaf temperature, RH of leaves, stomatal resistance, and rate of transpiration showed negligible effects. The direct effect of leaf number on rhizome yield was very high (0.631), and this trait is recommended for use as a selection criterion for improving rhizome yield. The study of Pandey and Dobhal (1993) revealed that the strongest forces influencing yield are weight of fingers, width of fingers, and leaf width. Singh et al. (1999) grouped 18 cultivars into 3 clusters under Nagaland conditions based on D^2 analysis. The major forces influencing divergence of cultivars were rhizome yield per plant, oleoresin, and fiber contents.

Sasikumar et al. (1992) carried out path analysis using 100 accessions of ginger. They reported that plant height followed by leaf length exhibited the highest direct effect on the rhizome yield. Dry recovery had a negative direct effect on yield. All other direct effects were negligible. The highest indirect effect was for leaf number through plant height, followed by leaf length, again through plant height. In turn, plant height exerted a moderately good indirect effect on rhizome yield. Moderate indirect effects were also noticed in the case of leaf width (through plant height), leaf length, and leaf number (through leaf length). However, the researchers noticed a residual effect of 0.8217, thereby indicating that the variability accounted for in the study was only 18%. They concluded that plant height should be given prime importance in a selection program, as this character had positive and significant correlation as well as a good direct effect on rhizome yield.

Multiple regression analysis using morphological characters indicated that the final yield could be predicted fairly accurately taking into consideration the plant height, leaf number, and breadth of last fully opened leaf on the 90th and 120th days after planting (Ratnambal et al., 1982). Rattan et al. (1998) found that to improve the yield per plant, emphasis should be laid on the number of leaves per plant and rhizome length by employing partial regression analysis. Rai et al. (1999) reported that higher rhizome yields were strongly associated with chlorophyll-a, carbohydrate, and lower polyphenol levels in the leaf. Leaf protein contents showed significant correlation with carbohydrates and the chlorophyll a:b ratio. The chlorophyll a:b ratio also showed a highly positive correlation with the leaf carbohydrate content. However, polyphenols showed a significant positive correlation with chlorophyll-b and carotenoids with chlorophyll-a and chlorophyll-b.

Crop Improvement

Crop improvement research efforts on ginger are constrained due to the absence of seed set. As a result, clonal selection, mutation breeding, and induction of polyploidy were the crop improvement methods employed. More recently, somaclonal variations arising through the callus regeneration is also being made use of in crop improvement work. Most of the research in this area was carried out in India. The major breeding objectives are: high yield, wide adaptability, resistance to pests and diseases (such as rhizome rot and bacterial wilt and *Fusarium* yellows), improvement in quality parameters (oil, oleoresin), and low fiber. Work in this area is carried out mainly at the IISR, Calicut, the AICRIPS center at High Altitude Research Station, Pottangi, under the administrative control of the Orissa University of Agriculture and Technology, Bhubaneshwar, Odisha State, and AICRPS (All India Coordinated Research Project on Spices) Center at the Y.S. Parmar University of Horticulture and Forestry at Solan, Himachal Pradesh, India.

Crop improvement research carried out until now is confined mainly to germplasm collection, evaluation, and selection. A large number of collections have been assembled at IISR, and these collections have been evaluated for yield and quality characters. In addition, a few introductions from other countries have also been made use of for breeding research. Some of the indigenous cultivars have been known to be high yielders and of good quality. In general, variability was found to be limited in cultivars grown in the same region, but wider variability is met within cultivars growing in geographically distant locations.

Khan (1959) reported the high-yielding capacity of Rio de Janeiro. In a trial with 18 cultivars, the yield of Rio de Janeiro was found to be double of China in Kerala (Thomas, 1966). Kannan and Nair (1965), Muralidharan and Kamalam (1973), and Thomas and Kannan (1969) also found that cultivar Rio de Janeiro was found to be superior compared to other cultivars. However, the percentage of ginger recovery was lesser than that from cultivar Maran. Randhawa and Nandpuri (1970) evaluated 15 cultivars during 4 years and reported that none could outyield the local cultivar, Himachal local under colder conditions of Himachal Pradesh. Jogi et al. (1978) also reported that the local cultivar Himachal produced the highest yield, followed by cultivar Rio de Janeiro.

Trials carried at Kasaragod in Kerala State under the All India Coordinated Research Project on Spices indicated that the yield potential of cultivars Rio de Janeiro, Burdwan, and Jamaica (AICSCIP, 1978) was quite high. In Assam, cultivar Nadia outyielded other cultivars (Aiyadurai, 1966).

Nybe et al. (1982) evaluated 28 cultivars for their yield potential and noted that the fresh and dry ginger yields among them varied significantly. Fresh rhizome yield was highest in the cultivar Nadia, followed by cultivars Maran, Bajpai, and Narasapattam. Cultivar Nadia also gave the highest yield of dry ginger. Sreekumar et al. (1982) found that cultivars Rio de Janeiro and Kuruppampadi were the very best yielders.

Muralidharan (1972) investigated the varietal performance of ginger in Wynad, Kerala State and concluded that the cultivar Rio de Janeiro gave the highest yield, whereas the yield of dry ginger was lower compared to that from other cultivars. Dry ginger yield was highest in the cultivar Tura. Cultivars Maran, Nadia, and Thingpuri are the other high yielders and were, more or less, on par with cultivar Rio de Janeiro in their performance. This author recommended cultivar Rio de Janeiro for fresh ginger production, while cultivars Maran, Nadia, and Thingpuri when the aim was to obtain high quantities of dry ginger.

Evaluation and Selection for Quality

Jogi et al. (1978) evaluated 14 cultivars and reported that the fiber content ranged from 4.62% (Poona) to 6.98% (Narasapattam). Cultivar Karakkal was lowest in dry ginger recovery followed by cultivars Wynad local and Rio de Janeiro. Cultivar Rio de Janeiro had the highest oleoresin content, whereas cultivar Karakkal had the highest oil content. Crude fiber was least in cultivars Nadia and China.

Nybe et al. (1982) evaluated 28 cultivars and reported that cultivars Rio de Janeiro and Maran had higher oleoresin content, 10.53% and 10.05%, respectively. Essential oil was highest in cultivar Karakkal (2.4%) and crude fiber was highest in Kuruppumpadi (6.47%). Sreekumar et al. (1982) found that the dry ginger recovery ranged from 17.7% in cultivar China to 28.0% in cultivar Tura. Cultivars having more than 22% dry recovery (Maran, Jugijan, Ernad Manjeri, Nadia, Poona, Himachal Pradesh, Tura, and Arippa) are suitable for dry ginger production.

Breeding Strategies

Conventional Method: The Clonal Selection Pathway

The clonal selection pathway has been the most successful breeding method in the absence of seed set. The steps involved are: the collection of cultivars from diverse sources and their assemblage in one or more locations, evaluation of cultivars for superiority in yield, quality or stress resistance, selection of promising lines, replicated yield trials in multilocations, selection of the best performers, their multiplication and testing in large evaluation plots, and finally their release for commercial cultivation. For a cultivar to be released, it should give a yield increase of 20% or more over the ruling standard cultivar. This strategy has been used successfully to evolve the present-day cultivars, which have been developed mainly for higher yield adaptability and quality, as depicted in Table 15.15.

The general breeding objectives in most breeding programs have been high yield, high quality, resistance to fungal and bacterial pathogens, bold rhizomes, high dry recovery, and low fiber content. Resistance to *Pythium* (the causal organism of rhizome rot), and *Ralstonia solanacearum* (fungus causing bacterial wilt) has so far not been encountered. In one such selection program carried out at IISR, 15 cultivars short listed from the germplasm evaluation program were tested in replicated trials for 4 years in 5 locations. Results are shown in Table 15.16. This effort led to the selection of Varada, one of the most important cultivated cultivars now in south and central India. The data presented in Table 15.16 also demonstrate the influence of genotype–environment interaction. The quality characters of these accessions are detailed in Table 15.17.

In another trial for increasing the rhizome size, 15 bold rhizome accessions short listed from the germplasm were tested in multilocation plots. Results are given in Table 15.17. Based on the overall superior performance, accessions 35 and 107 were selected, multiplied, and released for cultivation under the names IISR Rejitha, and IISR Mahima, respectively (Sasikumar et al., 2003). Clonal selection programs for crop improvement were carried out at the High Altitude Research Station in Pottangi, Odisha State, and the Department of Vegetable Crops at the Y.S. Parmar University of Horticulture and Forestry, in Solan, Himachal Pradesh. The former came out with the selections Suprabha and Suravi, and the latter with the selection *Himgiri*. Details on yield and recovery of bold rhizome selections are given in Table 15.18.

Mutation Breeding

Ginger is not amenable to any conventional recombination breeding programs due to its inherent sterility. Induction of variability through mutations, chemical mutagens, ionizing radiation, and tissue culture (somaclonal variations) has been attempted by a few researchers (Gonzalez et al., 1969; Raju et al., 1980). In a general scheme

Table 15.15 Elite Cultivars Developed and Released for Commercial Cultivation

Cultivar Name	Pedigree	Mean Yield (t/ha)	Dry Recovery (%)	Oil (%)	Oleoresin (%)	Crude Fiber (%)	Mean Days to Maturity
Suprabha	Selection from Kunduli local	16.6	20.5	1.9	8.9	4.4	230
Suruchi	Induced mutant of Rudrapur local	11.6	23.5	2.0	10.0	3.8	220
Suravi	Selection from germplasm	17.5	23.0	2.1	10.2	4.0	225
IISR-Varada	Selection from germplasm	22.6	19.5	1.7	6.7	3.3	200
IISR Mahima	Selection from germplasm	23.2	23.0	1.7	4.5	3.3	200
IISR Rajitha	Selection from germplasm	22.4	19.0	2.4	6.2	4.0	200
Himgiri	Clonal selection from Himachal local	14.0	20.6	1.6	4.3	6.0	230
Buderim gold[a]	Induced tetraploid of cultivar Queensland local	NA	NA	NA	NA	NA	NA
4× Sanshu[b]	Induced tetraploid of Sanshu	NA	NA	NA	NA	NA	NA

[a]Developed by the Buderim Ginger Company of Queensland, Australia. Reported to be a high yielder.
[b]Developed in Japan. Reported to be a high yielder.

Table 15.16 Yield and Dry Recovery of Ginger at Different Ginger-Growing Locations in India

Accession No.	Mean Fresh Yield (kg/3 m²) bed					Dry Recovery (%)			
	Peruvanna-muzhi	Muvattu-puzha	Ambalavayal	Peechi	Niravilpuzha	Peruvanna-muzhi	Ambalavayal	Peechi	Niravilpuzha
51	9.5	9.43	6.28	7.17	11.08	19.5	24.0	24.0 16.0	
64	11.17	11.5	7.38	9.83	11.00	21.0	23.0	24.0	20.0
141	9.83	9.83	6.78	8.00	10.00	20.5	18.0	20.0	19.0
251	12.33	8.17	6.09	8.83	9.47	20.0	23.0	19.0	19.5
222	10.17	8.00	5.17	7.83	6.92	20.5	22.0	22.0	22.0
63	10.83	9.00	6.87	7.67	10.83	18.5	21.0	1.0	14.0
151	11.0	9.27	6.30	8.17	8.10	20.0	19.0	24.0	20.0
53	11.0	10.33	6.41	9.83	9.60	20.0	15.0	20.5	19.4
11	10.6	9.00	6.47	7.17	9.67	17.5	14.0	20.5	19.5
249	10.1	10.00	6.00	8.33	9.16	17.5	17.0	20.5	20.5
65	9.83	10.5	5.33	7.33	11.00	20.0	20.0	21.5	20.0
250[a]	10.17	10.5	6.10	8.16	10.23	21.0	22.5	25.0	21.4
293[b]	11.17	9.67	7.25	7.83	11.28	18.5	15.0	19.0	15.1
295[c]	10.17	8.83	7.23 9.0	7.36	21.5	19.0	20.0	22.2	
252[d]	11.0	8.83	6.70	8.16	7.83	20.0	16.0	19.0	19.27
CD	0.62	0.51	0.54	0.75	0.90				
CV	47.48	17.19	11.27	12.1	12.6				

Source: IISR Annual Report (1994–1995).
CD = Critical difference, CV = coefficient of variation.
[a]Himachal
[b]Suprabha
[c]Maran
[d]Muvattupuzha

Table 15.17 Yield, Dry Recovery, and Quality of Promising Ginger Accessions at IISR

Accession No.	Quality		
	Essential Oil (%)	Oleoresin (%)	Crude Fiber (%)
51	2.1	6.8	5.7
64	1.9	6.0	5.4
141	1.9	6.5	4.0
251	2.4	9.0	6.6
222	2.0	7.0	3.9
63	2.3	7.0	4.9
151	2.0	7.0	6.0
53	2.3	9.9	5.1
11	2.0	7.0	4.0
249	2.4	9.0	3.5
65	2.7	8.0	5.3
HP Local	1.2	5.8	8.5
Suprabha	1.9	6.3	4.4
Maran	2.0	7.5	6.1
Muvattupuzha Local	1.9	6.3	NA

Source: IISR Annual Report 1903–1904.
NA: Value not available.

Table 15.18 Yield and Recovery of Bold Rhizome Selections

Accession No.	1996		1997	
	Yield (Fresh, t/ha)	Dry Recovery (%)	Mean Yield (Fresh, t/ha)	Dry Recovery (%)
117	13.3	22.0	9.9	25.5
35	14.7	17.5	11.9	21.2
49	12.3	22.0	10.3	21.3
27	13.0	22.0	11.3	26.3
3573	5.0	23.0	9.6	25.5
142	7.8	23.0	6.9	26.8
15	9.6	19.5	13.1	24.8
415	12.3	22.0	11.4	24.3
116	7.3	15.0	10.6	22.0
294	12.2	22.0	10.5	27.0
204	11.4	23.0	10.9	25.5
64	13.2	19.5	11.4	24.5
179	13.0	23.0	10.9	26.5
71	8.5	21.5	7.4	23.8
244	13.1	17.5	9.9	22.0
CD	1.13	–	1.86	–

Source: IISR Annual Report 1997–1998.
CD = Critical difference.

for a mutation, rhizome bits were treated with chemical mutagens or irradiated with gamma rays. Ginger buds are sensitive to irradiation and the LD_{50} was reported to be below 2 Krd. The LD_{50} (50% of the lethal dose) for germination was reported to be between 1.5 and 2.0 Krd (Giridharan, 1984). Jayachandran (1989) treated the ginger cultivar Rio de Janeiro with gamma rays ranging from 0.5 to 1.5 Krd and Ethylmethane sulfonate (EMS) at 2.0–10.0 mM and studied VM1 and VM2 generations to isolate useful mutants. This investigation revealed that the percentage of sprouting, survival, and the height of plants decreased as the mutagen dose was increased. The LD_{50} in the study for sprouting and survival was found to be between 0.5 and 1.0 Krd of gamma rays and below 8 mM of EMS.

Mutagen treatment affected tiller production; in the 1.5 Krd treatment, there was 45% reduction, whereas in 10 mM EMS there was 61% reduction in tiller production. The mutagen treatment did not affect pollen fertility or improve seed set. Rhizome yield was affected in a dose-dependent manner.

Jayachandran (1989) analyzed the VM2 generation and found a significant reduction in plant height as the dose increased. The mean tiller number indicated transgression to either side of the control treatment. Similarly, the mean rhizome yield in the VM2 generation indicated shifts in both the directions, with the lower doses of the mutagens giving positive shifts and the higher doses giving negative shifts. The variation in rhizome yield ranged from 1 to 1320 g/plant. This same author found that lower doses of gamma rays (0.5–0.75 Krd) and EMS (2–4 mM) are more effective in inducing wider variations. Screening against the soft rot pathogen, *Pythium aphanidermatum*, and bacterial wilt caused by *Ralstonia solanacearum* did not reveal any change in pathogenic susceptibility. Jayachandran (1989) observed that the effects of mutagen treatment in the subsequent generations vanished, indicating the operation of strong diplontic selection.

Nwachukwu et al. (1994) irradiated rhizomes of two Nigerian cultivars (Yatsun Biri and the yellow ginger Tafin Giwa) at a dose of 2.5–10.0 Gy gamma rays (Gy—Gray—is the unit of absorbed dose; 1 Gy = 100 Krd). In these cultivars, the GR 50 (50% growth reduction) was found to be at 5 and 6 Gy in Tafin Giwa and Yatsun Biri, respectively. The LD_{50} was found to be 8.75 Gy for both cultivars.

Mohanty and Panda (1991) reported the isolation of a high-yielding mutant from the VM3 generation. They employed EMS, sodium azide, colchicine, and gamma rays as mutagenic agents, and five cultivars, namely UP, Rio de Janeiro, Thingpui, PGS-10, and PGS-19, were treated and investigated in V1, V2, and V3 generations. Twenty promising individual clumps ("mutants") were selected for evaluation. One of them V1K1-3 gave the highest rhizome yield of 22.08 t/ha, which was found to be significantly higher than the top yielder Suprabha. Six top yielders were further tested in comparative yield trials and multilocational field trials. The results indicated that V1K1-3 was superior and has been subsequently released for commercial cultivation in India under the name "Suravi."

The genotype differences were consistent over the locations tested and V1K1-3 was found to outyield all the others tested at all the locations. This line has a dry recovery of 23%. The rhizomes are plump with cylindrical fingers having dark glazed skin and dark yellow flesh with bulging oval tips and finger nodes, which are covered with deep brown scales. This genotype has an oil content, oleoresin content, and crude fiber content of 2.1%, 10.2%, and 4.0%, respectively.

Tashiro et al. (1995) investigated induced isozyme mutations to find out the possible use of isozyme analysis as markers for detecting mutants at an early stage or under an *in vitro* culture system. They employed cultivars Otafuku, Kintoki, and Shirome Wase and excised shoot tips were treated with 5 mM methyl nitrosourea (*N*-methyl-*N*-nitrosourea—MNU) for 5–20 min and cultured on MS medium supplemented with 0.05 mg NAA and 0.5 mg BA/l. Regenerated plants were analyzed to locate mutations in the following isozymes: glutamate dehydrogenase (GDH), glutamate-oxaloacetate transaminase (GOT), malate dehydrogenase (MDH), 6-phosphogluconate dehydrogenase (6-PGDH), phosphoglucomutase (PGM), and shikimate dehydrogenase (SKDH). Analysis of the untreated control gave uniform isozyme profiles, in the case of all the three cultivars. Five of the 21 MNU-treated plants had isozyme profiles which differed from the basic pattern of GOT, 6-PGDH, PGM, and SKDH. All of these isozyme mutants expressed morphological variations, such as multiple shoot formation, dwarfing, and abnormal leaves. The results indicated that treating shoot tips with MNU and then culturing them in appropriate media can recover mutants and that isozyme analysis is a good technique in detecting the rate of mutation, and hence, is useful in mutation breeding programs.

Polyploidy Breeding

Induced polyploidy has been tried in ginger to introduce variability, improve pollen and ovule fertility, growth, and yield. Ratnambal (1979) reported induction of polyploidy in the cultivar Rio de Janeiro through colchicines treatment. The tetraploids showed stunted growth and had reduced length and breadth of leaves. However, in this case a stable polyploidy line could not be established and all the plants reverted to diploidy in the succeeding generations.

Ramachandran and Nair (1982) reported successful production of stable tetraploid lines in cultivars Maran and Mananthody. The polyploids were more vigorous than the diploids and flowered during the second year of induction. The stable tetraploid lines ($2n = 44$) had larger, plump rhizomes and gave high yield (198.7 g/plant). However, the essential oil content was lower (23%) than in the original diploid cultivar. There was considerable increase in pollen fertility in the tetraploids. These tetraploids are maintained in the germplasm collection, at IISR, Calicut, India.

Adaniya and Shirai (2001) induced tetraploids under *in vitro* conditions by culturing shoot tips in MS solid medium containing BA, NAA, and 0.2% w/v colchicines for 4, 8, 12, and 14 days and transferred the shoot tips to medium without colchicines for further growth. More tetraploids were recovered from buds cultured for 8 days. Induced tetraploid line of the cultivars (4×Kintoki, 4×Sanshu, and 4×Philippinecebu 1) were later transferred to the field where they flowered. These tetraploids produced pollen with much higher fertility and germinability than the diploid plants (0.0–1.0% in the diploid plants as against 27.4–74.2% in the tetraploids).

Buderim Ginger Co., the commercial ginger company in Queensland, Australia, has developed and released for commercial cultivation a tetraploid line from the local cultivar. This line, named Buderim Gold, is much higher yielding and has plump rhizomes that are ideally suitable for processing (Buderim Ginger Co., 2002). Nirmal Babu et al. (1996) developed a promising line of cultivar Maran from soma-clonal variants. This line is high yielding with bolder rhizomes and taller plants. In addition to somaclonal variation, other biotechnological approaches have been initi-ated to evolve disease-resistant genotypes.

The breeding strategies currently in use will not be useful to solve many of the serious problems besetting the ginger crop. Despite extensive search, no genes resist-ant to the most devastating disease of ginger, the rhizome rot, caused by *Pythium* or *Fusarium* wilt or bacterial wilt could be located in the germplasm collection. The absence of sexual reproduction and seed set imposes a severe restriction in the efforts of plant breeders to develop disease-resistant cultivars. Recourse to bio-technological approaches might provide solutions. However, efforts in this line are scanty. Resorting to recombinant DNA technology, using resistant genes to the tar-get pathogen from other related plants, might be a viable path in evolving resistant ginger plants. This is, indeed, a long drawn-out effort, and it is hazardous to guess when a reliable solution would evolve. Until such times, cultural methods, such as good phytosanitation, crop rotation, recourse to biocontrol agents, would be the only alternative. Even ginger nutrition is based on classical textbook knowledge. Recent advances in soil fertility and plant nutrition has shown that there are alternatives to classical methods of fertilizing crop plants. A well-nurtured ginger plant will be less susceptible to pest infestation. The case in point is with reference to the devastat-ing "Quick Wilt" in *Piper nigrum* (black pepper), caused by the *Phytophthora* fun-gus, where Zn was found to be intimately involved (Nair, 2002). A revolutionary soil testing program, now globally known as "The Nutrient Buffer Power Concept," developed by Nair (1996), as opposed to classical textbook knowledge, has shown important alternatives (Nair, 1969). It is for the ginger plant breeders, physiologists, agronomists, soil scientists, and especially biotechnologists, to evolve a strategy, by pooling and sharing knowledge, as opposed to the watertight approaches currently being employed in India, and also perhaps, elsewhere in the world, which will open up a new chapter in stable ginger production, currently ravaged by diseases and pests, and lead to production instability. This will open up a new chapter in ginger production across the world, where the crop is an important economic mainstay for millions of poor, marginal, and often rich farmers.

References

Adaniya, S., 2001. Optimal pollination environment of tetraploid ginger (*Zingiber officinale* Rosc.) evaluated by *in vitro* pollen germination and pollen growth in styles. Sci. Hortic. 90, 219–226.

Adaniya, S., Shirai, D., 2001. *In vitro* induction of tetraploid ginger (*Zingiber officinale* Rosc.) and pollen fertility and germinability. Sci. Hortic. 83, 277–287.

Adaniya, S., Ashoda, M., Fujieda, K., 1989. Effect of day length on flowering and rhizome swelling in ginger (*Zingiber officinale* Rosc.). J. Jpn. Soc. Hortic. Sci. 58, 649–656.

AICRPS, 1999. All India Coordinated Research Project on Spices, Annual Report for 1999–2000. IISR, Calicut, Kerala State.

AICRPS, 2000. All India Coordinated Research Project on Spices, Annual Report for 1999–2000. IISR, Calicut, Kerala State.

AICRPS, 2001. All India Coordinated Research Project on Spices, Annual Report for 2000–2001. IISR, Calicut, Kerala State.

AICRPS, 2003. All India Coordinated Research Project on Spices, Annual Report for 2002–2003. IISR, Calicut, Kerala State.

AICSCIP, 1975. All India Cashew and Spices Crop Improvement Project, Annual Report for 1974–1975. Central Plantation Crops Research Institute, Kasaragod Kerala.

AICSCIP, 1978. All India Cashew and Spices Crops Improvement Project, Annual Report for 1977–1978. Central Plantation Crops Research Institute, Kasaragod, Kerala State.

Aiyadurai, S.G., 1966. A Review of Research on Spices and Cashewnut in India. ICAR, New Delhi.

Aiyer, K.N., Kolammal, M., 1966. Pharmacognosy of Ayurvedic Drugs Kerala. Department of Pharmacognosy, University of Kerala, Trivandrum, Kerala State, Series I, No.9.

Anon, 2004. Conservation of Spices Genetic Resources in *In Vitro* Gene Bank. Indian Institute of Spices Research, Calicut, Kerala State, Project Report submitted to the Department of Biotechnology, Government of India, New Delhi, India.

Baker, J.G., 1882. Scitamineae In: Hooker, H.D. (Ed.), The Flora of British India, vol. VI Bishen Singh Mahendrapal Singh, Dehra Dun, India, pp. 198–264. Rep. 1978.

Baranowski, J.D., 1986. Changes in solids, oleoresin, and (6)-gingerol content of ginger during growth in Hawaii. Hortic. Sci. 21, 145–146.

Beltram, I.C., Kam, Y.K., 1984. Cytotaxonomic studies in the Zingiberaceae. Edinb. J. Bot. 41, 541–557.

Buchanan, E., 1807. A Journey from Madras Through the Countries of Mysore. Canara and Malabar Directors of East India Co., London (Reprint).

Bunderim Ginger Co., 2002. Bunderim Gold (Corporate author). Plant Var. J. 15, 85.

Burtt, B.L., Smith, R.M., 1983. Zingiberaceae In: Dasanayake, M.D. (Ed.), A Revised Handbook to the Flora of Ceylon, vol. 1V Amerind Pub., New Delhi India, pp. 488–532.

Chakravorti, A.K., 1948. Multiplication of chromosome numbers in relation to speciation in Zingiberaceae. Sci. Cult. 14, 137–140.

Chandra, R., Govind, S., 1999. Genetic variability and performance of ginger genotypes under mid-hills of Meghalaya. Indian J. Hortic. 56, 274–278.

Charles, J.S., Kuriyan, K.J., 1982. Relative susceptibility of ginger cultivars to the root knot nematode, *Meloidogye incognita*. In: Nair, M.K., Premkumar, T., Ravindran, P.N., Sarma, Y.R. (Eds.), Proceedings of the National Seminar CPCRI, Kasaragod, Kerala, pp. 133–134.

Chen, Z.Y., 1989. Evolutionary patterns in cytology and pollen structure of Asian Zingiberaceae. In: Holm-Nielson., L.B., Nielsen., I.C., Balslev, H. (Eds.), Tropical Forests. Botanical Dynamics, Speciation and Diversity Academic Press, Kasaragod, Kerala, pp. 185–191.

CPCRI, 1973. Central Plantation Crops Research Institute, Annual Report for 1972–73, Kasaragod, Kerala.

Cronquist, A., 1981. An Integrated System of Classification of Flowering Plants. Columbia University Press, New York, NY.

Dahlgen, R.M.T., Clifford, H.T., Yeo, P.F., 1985. The Families of the Monocotyledons. Springer, Berlin, pp. 350–352.

Darlington, C.D., Janaki Ammal, E.K., 1945. Chromsome Atlas of Cultivated Plants. George Allen & Urwin, London, p. 397.

Das, P., Rai, S., Das, A.B., 1999. Cytomorphology and barriers in seed set of cultivated ginger (*Zingiber officinale* Rosc.) II. Cytologia 69, 133–139.

Das, A.B., Rai, S., Das, P., 1998. Estimation of 4C DNA and karyotype analysis in ginger (*Zingiber officinale* Rosc.) II. Cytologia 63, 133–139.

Dekkers, A.J., Rao, A., Goh, C.J., 1991. *In vitro* storage of multiple shoot cultures of gingers at ambient temperature of 24°C to 29°C. Sci. Hortic. 47, 157–167.

Dhamayanthi, K.P.M., Sasikumar, B., Ramashree, A.B., 2003. Reproductive biology and incompatibility studies in ginger (*Zingiber officinale* Rosc.). Phytomorphology 53, 123–131.

Dongyun, H., Ki Young, K., Inlok, C., Soo Dong, K., Moonsoo, P., 1998. Stomatal behavior and chlorophyll fluorescence to environmental conditions in ginger (*Zingiber officinale* Rosc.). J. Korean Soc. Hortic. Sci. 39, 115–118.

Esau, K., 1953. Plant Anatomy. McGraw Hill, New York.

Esau, K., 1969. The phloem. In: Linsbaur, K. (Ed.), Handbauch der Pflanzenanatomie Gebruder Borntraeger, Berlin.

Farrell, K.T., 1985. Spices, Condiments and Seasoning. The AVI Pub.Co., Westport, CN.

Flukiger, F.A., Hanbury, D., 1879. Pharmacographia: A History of the Principal Drugs of Vegetable Origin Met Within Great Britain and British India. Macmillan & Co, London.

Furutani, S.C., Nagao, M.A., 1986. Influence of daminozide, gibberllic acid and etherpon on flowering, shoot growth and yield of ginger. Hortic. Sci. 21, 428–429.

Futterer, 1896. Cited from Tomlinson, P.B. 1956. Studies in the systematic anatomy of Zingiberaceae. J.Linn. Soc. (Bot.) 55, 547–592.

Geetha, S.P., 2002. *In Vitro* Technology for Genetic Conservation of Some Genera of Zingiberaceae. Unpublished Ph.D. thesis, University of Calicut, Kerala State.

Geetha, S.P., Manjula, C., Sajina, A., 1995. *In vitro* conservation of genetic resources of spices In: Proceedings of the 7th Kerala Science Congress. State Committee on Science, Technology and Environment, Kerala, pp. 12–16.

Giridharan, M.P., 1984. Effect of Gamma Irradiation in Ginger (*Zingiber officinale* Rosc.). M.Sc. (Hort.) thesis, Kerala Agricultural University, Vellanikkara, Kerala State.

Gonzalez, O.N., Dimaunahan, L.B., Pilac, L.M., Alabastro, V.Q., 1969. Effect of gamma irradiation on peanuts, onions and ginger. Phlippine J. Sci. 98, 279–292.

Gowda, K.K., Melanta, K.R., 2000. Varietal performance of ginger in Karnataka. In: Muraleedharan, N., Rajkumar, R. (Eds.), Recent Advances in Plantation Crops Research Allied Pub, New Delhi. pp. 92–93.

Graham, J.A., 1936. Methods of ginger cultivation in Jamaica. J. Jamaica. Agric. Sci. 40, 231–232.

Hesse, M., Waha, M., 1982. The fine structure of the pollen wall in *Strelitzia reginae* (Musaceae). Plant Syst. Evol. 141, 285–298.

Holttum, R.E., 1950. The Zingiberaceae of the Malay peninsula. Gard. Bull. (Singapore) 13, 1–50.

IISR, 2002. Indian Institute of Spices Research, Annual Report for 2001–2002. IISR, Calicut, Kerala.

IISR Annual Report 1994–1995. Calicut, Kerala.

Janson, P.C., 1981. Spices, Condiments and Medicinal Plants in Ethiopia. Centre for Agricultural Publishing and Documentation, Wageningen, the Netherlands.

Jayachandran, B.K., 1989. Induced Mutation in Ginger. Unpublished thesis, Kerala Agricultural University, Vellanikkara, Kerala State.

Jayachandran, B.K., Vijayagopal, P., Sethumadhavan, P., 1979. Floral biology of ginger, *Zingiber officinale* Rosc. Agric. Res. J. Kerala 17, 93–94.

Jogi, B.S., Singh, I.P., Dua, N.S., Sukhiya, P.S., 1978. Changes in crude fibre, fat and protein content in ginger (*Zingiber officinale* Rosc.) at different stages of ripening. Indian J. Agric. Sci. 42, 1011–1015.

Kannan, K., Nair, K.P.V., 1965. Ginger (*Zingiber officinale* Rosc.) in Kerala. Madras Agric. J. 52, 168–176.

Khan, K.I., 1959. Ensure twofold ginger yields. Indian Farming 8 (2), 10–14.

Kress, W.J., Stone, D.E., 1982. Nature of the sporoderm in monocotyledons with special reference to pollen grains of *Canna* and *Heliconia*. Grana 21, 129–148.

Krishnamurthy, N., Nambudiri, E.S., Mathew, A.G., Lewis, Y.S., 1970. Essential oil of ginger. Indian Perf. 14, 1–3.

Kumar, S., 1999. A note on conservation of economically important Zingiberaceae of Sikkim, Himalaya. In: Biodiversity Conservation and Utilization of Spices, Medicinal and Aromatic Plants. Indian Society of Spices (IISR), Calicut, Kerala State, pp. 201–207.

Lawrence, B.M., 1984. Major tropical spices: ginger (*Zingiber officinale* Rosc.). Perf. Flav. 9, 1–40.

Lewis, Y.S., Mathew, A.G., Nambudiri, E.S., Krishnamurthy, N., 1972. Oleoresin ginger. Flav. Ind. 3 (2), 78–81.

Liang, Y.H., 1988. Pollen morphology of the family Zingiberaceae in China—pollen types and their significance in the taxonomy. Acta Phytotax. Sin. 26, 265–286.

Mahindru, S., 1982. Spices in Indian Life. Sultan Chand & Sons, New Delhi.

Menon, K.P.P., 1929. History of Kerala. Asian Educational Services, New Delhi, (Reprint).

Mohandas, T.P., Pradeep Kumar, T., Mayadevi, P., Aipe, K.C., Kumaran, K., 2000. Stability analysis in ginger (*Zingiber officinale* Rosc.) genotypes. J. Spices Aromat. Crops 9, 165–167.

Mohanty, D.C., 1984. Germplasm Evaluation and Genetic Improvement. I Ginger. Unpublished thesis, Orissa University of Agriculture Technology, Bhubaneshwar, Orissa State.

Mohanty, D.C., Panda, B.S., 1991. High yielding mutant V1K1-3 ginger Idian Cocoa, Arecanut and Spices, J. 15, 5–7.

Mohanty, D.C., Panda, B.S., 1994. Genetic resources in ginger. In: Chadha, K.L., Rethinam, P. (Eds.), Advances in Horticulture, Vol. 9: Plantation Crops and Spices Part 2 Malhotra Pub, New Delhi, pp. 151–168.

Mohanty, D.C., Sarma, Y.N., 1979. Genetic variability and correlation for yield and other variables in ginger germplasm. Indian J. Agric. Sci. 49, 250–253.

Mohanty, D.C., Das, R.C., Sarma, Y.N., 1981. Variability of agronomic characters of ginger. Orissa J. Hortic. 9, 15–17.

Mohanty, H.K., 1970. A cytological study of the Zingiberales with special reference to their taxonomy. Cytologia 35, 13–49.

Moringa, T., Fukushina, E., Kanui, T., Tamasaki, Y., 1929. Chromosome numbers of cultivated plants. Bot. Mag. (Tokyo) 43, 589–594.

Muralidharan, A., 1972. Varietal performance of ginger in Waynad, Kerala. J. Plantation Crops 1973 (Suppl.), 19–20.

Muralidharan, A., Kamalam, N., 1973. Improved ginger means foreign exchange. Indian Farming 22, 37–39.

Naidu, M.M., Padma, M., Yuvraj, K.M., Murty, P.S.S., 2000. Evaluation of ginger varieties for high altitude and tribal area of Andhra Pradesh Spices Aromat. Plants. ISSC (ISSR), Calicut, Kerala, pp. 50–51.

Nair, K.P.P., 1996. The buffering power of plant nutrients and effects on availability. Adv. Agron. 57, 237–287.

Nair, K.P.P., 2002. Sustaining crop production in the developing world through the nutrient buffer power concept. In: Proceedings of the 17th World Soil Science Congress, vol. 2, 14–21 August 2002, Bangkok, Thailand, p. 652.

Nair, M.K., Nambiar, M.C., Ratnambal, M.J., 1982. Cytogenetics and crop improvement of ginger and turmeric. In: Nair, M.K., Premkumar, T., Ravindran, P.N., Sarma, Y.R. (Eds.), Ginger and Turmeric CPCRI, Kasaragod, Kerala, pp. 15–23.

Nair, P.C.S., 1969. Ginger cultivation in Kerala. Arecanut Spices Bull. 1 (1), 22–24.

Natarajan, C.P., Kuppuswamy, S., Shankaracharya, N.B., Padma Bai, R., Raghavan, B., Krishnamurthy, M.N., et al., 1972. Chemical composition of ginger varieties and dehydration studies of ginger. J. Food Sci. Technol. 9 (3), 120–124.

Nayar, J., 1995. On the nature of pollen of Zingiberiflorae based on pollen germination. In: Second Symposium on the Family Zingiberaceae, 9–12 May, Ghuangzhan, China, p. 21 (Abst).

Nirmal Babu, K., Samsudeen, K., Ravindran, P.N., 1996. Biotechnological approaches for crop improvement in ginger, Zingiber officinale Rosc. In: Ravishankar, G.A., Venkataraman, L.V. (Eds.), Recent Advances in Biotechnological Applications on Plant Tissue and Cell Culture Oxford University Press, New Delhi, pp. 321–332.

Nwachukwu, E.C., Ene, L.S.O., Mbanaso, E.N.A., 1994. Radiation sensitivity of two ginger varieties to gamma irradiation. Tropenlandwirt 95, 99–103.

Nybe, E.V., Nair, P.C.S., 1979. Studies on the morphology of the ginger types. Indian Cocoa Arecanut Spices J., 7–13.

Nybe, E.V., Nair, P.C.S., Mohanakumaran, N., 1982. Assessment of yield and quality components in ginger. In: Premkumar, M.K., Ravindran, P.N., Sarma, Y.R. (Eds.), Ginger and Turmeric CPCRI, Kasaragod, Kerala, pp. 24–29.

Panchakshrappa, M.G., 1966. Embryological studies in some members of Zingiberaceae. II. Elettaria cardamomum, Hitchenia caulina and Zingiber micro stachyum. Phytomorphology 16, 412–417.

Pandey, G., Dobhal, V.K., 1993. Genetic variability, character association and path analysis for yield components in ginger (Zingiber officinale Rosc). J. Spices Aromat. Crops 2, 16–20.

Panigrahi, U.C., Patro, G.K., 1985. Ginger cultivation in China. Indian Farming 33 (5), 3–4. 17.

Parry, J.W., 1962. Spices: The Morphology, Histology and Chemistry, vol. 2. Chemical Pub., New York, NY.

Parry, J.W., 1969. Handbook of Spices, vol. 1. Chemical Pub, New York, NY.

Peterson, O.G., 1889. Zngiberaceae. In: A. Engler, K. Prant's (Eds.), Die Naturilichen Pflanzenfamilien, 2 (6) pp. 10–30 (cited from Tomlinson, 1956).

Pillai, P.K.T., Vijayakumar, G., Nambiar, M.C., 1978. Flowering behavior, cytology and pollen germination in ginger (Zingiber officinale Rosc.). J. Plantation Crops 6, 12–13.

Pillai, S.K., Pillai, A., Sachadeva, S., 1961. Root apical organisation in monocotyledons—Zingiberaceae. Proc. Indian Acad. Sci. 53B, 240–256.

Poulose, T.T., 1973. Ginger cultivation in India Proceedings of Conference on Spices. TPI, London, pp. 117–121.

Pradeep Kumar, T., Mohandas, T.P., Jayarajan, M., Aipe, K.C., 2000. Evaluation of ginger varieties in Waynad. Spice India 13 (1), 13.

Purseglove, J.W., Brown, E.G., Green, C.I., Robbins, S.R.J., 1981. Spices, vol. 2. Longman, London.

Raghavan, T.S., Venkatasubban, K.R., 1943. Cytological studies in the family Zingiberaceae with special reference to chromosome number and cytotaxonomy. Proc. Indian Acad. Sci 17B, 118–132.

Rai, S., Das, A.B., Das, P.B., 1999. Variations in chlorophyll, carotenoids, protein and secondary metabolites amongst ginger (Zingiber officinale, Rosc.) cultivars and their association with rhizome yield. New Zealand J. Crop Hortic. Sci. 27, 79–82.

Raju, E.C., Shah, J.J., 1975. Studies in stomata of ginger, turmeric and mango ginger. Flora 164, 19–25.

Raju, E.C., Shah, J.J., 1977. Root apical organization in some rhizomatous spices: ginger, turmeric and mango ginger. Flora 166, 105–110.

Raju, E.C., Patel, J.D., Shah, J.J., 1980. Effect of gamma radiation in morphology of leaf and shoot apex of ginger, turmeric and mango ginger. Proc. Indian Acad. Sci. 89, 173–178.

Ramachandran, K., 1969. Chromosome numbers in Zingiberaceae. Cytologia 34, 213–221.

Ramachandran, K., Nair, P.C.S., 1982. Induced tetraploidy of ginger (*Zingiber officinale* Rosc.). J. Spices Aromat. Crops 1, 39–42.

Ramashree, A.B., Sherlija, K.K., Unnikrishnan, K., Ravindran, P.N., 1997. Histological studies on ginger rhizome (*Zingiber officinale* Rosc.). Phytomorphology 47, 67–75.

Ramashree, A.B., Unnikrishnan, K., Ravindran, P.N., 1998. Developmental anatomy of ginger rhizome II. Ontogeny of buds, roots and phloem. Phytomorphology 48, 155–156.

Ramashree, A.B., Unnikrishnan, K., Ravidran, P.N., 1999. Development of oil cells, and ducts in ginger (*Zingiber officinale* Rosc.). J. Spices Aromat. Crops 8, 163–170.

Randhawa, K.S., Nandpuri, K.S., 1970. Ginger in India: review. Punjab Hortic. J. 10, 11–112.

Rao, V.S., Gupta, K., 1961. The floral anatomy of some Scitamineae. IV. J. Univ. Bombay 29, 134–150.

Rao, V.S., Pai, R.M., 1959. The floral anatomy of some Scitamineae. II. J. Univ. Bombay 28, 82–84.

Rao, V.S., Pai, R.M., 1960. The floral anatomy of some Scitamineae. III. J. Univ. Bombay 29, 1–19.

Rao, V.S., Karnick, H., Gupta, K., 1954. The floral anatomy of some Scitamineae. Part I. J. Indian Bot. Soc. 33, 118–147.

Ratnambal, M.J., 1979. Cytological studies in ginger (*Zingiber officinale* Rosc.). Unpublished Ph.D. thesis, University of Bombay, India.

Ratnambal, M.J., Nair, M.K., 1981. Microsporogenesis in ginger (*Zingiber officinale* Rosc.). In: Proc. Placrosym. VI. CPCRI, Kasaragod, Kerala State, pp. 44–57.

Ratnambal, M.J., Balakrishnan, R., Nair, M.K., 1982. Multiple regression analysis in cultivars of *Zingiber officinale* Rosc. In: Nair, M.K., Premkumar, T., Ravindran, P.N., Sarma, Y.R. (Eds.), Ginger and Turmeric Central Plantation Crops Research Institute, Kasaragod, Kerala, pp. 30–33.

Rattan, R.S., 1989. Improvement of ginger. Proceedings of Ginger Symposium, February 1998, Nahan, Himachal Pradesh, India.

Rattan, R.S., Korla, B.N., Dohroo, N.P., 1998. Performance of ginger varieties in Solan area of Himachal Pradesh. In: Satyanarayana., G., Reddy, M.S., Rao, M.R., Azam, K.M., Naidu, R. (Eds.), Proceedings on National Seminar on Chillies, Ginger and Turmeric Spices Board, Cochin, pp. 71–73.

Ravindran, P.N., 1998. Genetic resources of spices and their conservation. In: Sasikumar, B., Krishnamurthy, B., Rema, J., Ravindran, P.N., Peter, K.V. (Eds.), Biodiversity, Conservation and Utilization of Spices, Medicinal and Aromatic Plants Indian Institute of Spices Research, Calicut, Kerala State, pp. 16–44.

Ravindran, P.N., Sasikumar, B., George, J.K., Ratnambal, M.J., Nirmal Babu, K., Zachariah, T.J., 1994. Genetic resources of ginger (*Zingiber officinale* Roc.) and its conservation in India. Plant Genet. Resour. Newsl. 98, 1–4.

Ravindran, P.N., Nirmal Babu, K., Peter, K.V., Abraham, C.Z., Tyagi, R.K., 2004. Genetic resources of spices—the Indian scenario. In: Dhillon, B. (Ed.), Crop Genetic Resources: An Indian Perspective Indian Society for Plant Genetic Resources, New Delhi. 2005.

Ravindran, P.N., Remashree, A.B., Sherlija, K.K., 1998. Developmental Morphology of Rhizomes of Ginger and Turmeric Final Report of the ICAR (Indian Council of Agricultural Research, New Delhi, India) ad Hoc Scheme, IISR. Indian Institute of Spices Research, Calicut, Kerala State, India.

Ridley, H.N., 1912. Spices. MacMillan & Co. Ltd, London.

Rosales, P.B., 1938. An agronomic study of the native and Hawaiian gingers. Philipp. Agric. Sci. 26, 807–822.

Rosengarten, F.J., 1969. The Book of Spices. Livingston Pub. Co, PA.

Saikia, I., Shadeque, A., 1992. Yield and quality of ginger (*Zingiber officinale* Rosc.) varieties grown in Assam. J. Spices Aromat. Crops 1, 131–135.

Sasikumar, B., Nirmal Babu, K., Abraham, J., Ravindran, P.N., 1992. Variability, correlation and path analysis in ginger germplasm. Indian J. Genet. 52, 428–431.

Sasikumar, B., Ravndran, P.N., George, K.J., 1994. Breeding ginger and turmeric. Indian Cocoa Arecanut Spices J. 18, 10–12.

Sasikumar, B., Saji, K.V., Ravindran, P.N., Peter, K.V., 1999. In: Rema, J., Ravindran, P.N., Peter, K.V. (Eds.), Biodiversity Conservation and Utilization of Spices, Medicinal and Aromatic Plants IISR, Calicut, Kerala State, pp. 96–100.

Sasikumar, B., Saji, K.V., Antony, A., George, J.K., Zachariah, T.J., Eapen, S.J., 2003. IISR Mahima and IISR Rejatha—two high yielding and high quality ginger (*Zingiber officinale*) varieties. J. Spices Aromat. Crops 12, 34–37.

Sato, D., 1948. The karyotype and phylogeny of Zingiberaceae. Jpn. J. Genet. 23, 44. (cited from Sharma 1972).

Schumann, K.N., 1904. Zingiberaceae. In: A. Engler's (ed.), Pflanzenreich, vol. 4, pp. 1–428.

Shah, J.J., Raju, E.C., 1975a. General morphology, growth and branching behavior of the rhizomes of ginger, turmeric and mango ginger. New Bot. 11, 59–69.

Shah, J.J., Raju, E.C., 1975b. Ontogeny of the shoot apex of *Zingiber officinale*. Norw. J. Bot. 22, 227–236.

Shamina, A., Zachariah, T.J., Sasikumar, B., George, J.K., 1997. Biochemical variability in selected ginger (*Zingiber officinale* Rosc.) germplasm accessions. J. Spices Aromat. Crops 6, 119–127.

Sharma, A.K., Bhattacharya, N.K., 1959. Cytology of several members of Zingiberaceae and a study of the inconsistency of their chromosome complement. La Cellule 59, 279–349.

Shi-jie, Z., Xizhen, A., Shaohui, W., Zhenxian, Z., Qi, Z., 1999. Role of xanthophyll cycle and photorespiration in protecting the photosynthetic apparatus of ginger leaves from photoinhibitory damage. Acta Agric. Boreali-Occidentalis Sin. 8 (3), 81–85.

Shiva, K.N., Suryanarayana, M.A., Medhi, R.P., 2004. Genetic resources of spices and their conservation in Bay Islands. Indian J. Plant Genet. Resour. 2005.

Shu, E.J., 2003. Zingiber Miller. In: E.J. Ke, D. Delin, K. Larson, (Eds.), Zingiberaceae. http://www.servicedirect.com/service?Ob = articleURL-udi = B6VSC.4876DKY-9:-8/1/ 2003. accessed from the Web.

Simmonds, N.W., 1979. Principles of Crop Management. Longman Group Ltd, New York.

Singh, A.K., 2001. Correlation and path analysis for certain metric traits in ginger. Ann. Agric. Res. 22, 285–286.

Singh, P.P., Singh, V.B., Singh, A., Singh, H.B., 1999. Evaluation of different ginger cultivars for growth, yield and quality character under Nagaland condition. J. Med. Aromat. Plant Sci. 21, 716–718.

Skavaria, J.J., Rowley, J.R., 1988. Adaptability of scanning electron microscopy to studies of pollen morphology. Aliso 12, 119–175.

Solereder, H., Meyar, F.J., 1930. Cited from Tomlinson, P.B. 1956.

Sreekumar, V., Indrasenan, G., Mammen, M.K., 1982. Studies on the quantitative and qualitative attributes of ginger cultivars. In: Nair., M.K., Premkumar., T., Ravindran, P.N., Sarma, Y.R. (Eds.), Proceedings of National Seminar on Ginger and Turmeric CPCRI, Kasaragod, Kerala State, pp. 47–49.

Stebbins, G.L., 1958. Longevity, habitat, and release of genetic variability in higher plants. Cold Spring Harb. Symp. Quant. Biol. 23, 365–378.

Sterling, K.J., Clark, R.J., Brown, P.H., Wilson, S.J., 2002. Effect of photoperiod on flower bud initiation and development in myoga (*Zingiber mioga* Rosc.). Sci. Hortic. 95, 261–268.

Stone, D.E., Sellers, S.C., Kress, W.J., 1979. Ontogeny of exineless pollen in *Heliconia*, a banana relative. Ann. Mol. Bot. Gard. 66, 701–730.

Sugiura, T., 1936. Studies on the chromosome numbers in higher plants, with special reference to cytokinesis. Cytologia 7, 437–595.

Takahashi, R. 1930. Cited from Darlington and Janaki Ammal, 1945.

Thomas, K.M., 1966. Rio de Janeiro will double your ginger yield. Indian Farming 15 (10), 15–18.

Thomas, K.M., Kannan, K., 1969. Comparative yield performance of different types of ginger. Agric. Res. J. Kerala 1 (1), 58–59.

Tashiro, V., Onimaru, H., Shigyo, M., Isshiki, S., Miyazaki, S., 1995. Isozyme mutations induced by treatment of cultured shoot tips with alkalytating agents in ginger cultivars (*Zingber officinale* Rosc.). Bull. Fac. Agri. Saga. Uni. No. 79, 29–35.

Tindall, H.D., 1968. Commercially grown vegetables. In: Commercial Vegetable Growing. Oxford University Press, London.

Tomlinson, P.B., 1956. Studies in the systematic anatomy of the Zingiberaceae. J. Linn. Soc. (Bot.) 55, 547–592.

Usha, K., 1984. Effect of Growth Regulators on Flowering, Pollination and Seed Set in Ginger (*Zingiber officinale* Rosc.). Unpublished M.Sc. (Ag) thesis, Kerala Agricultural University, Vellanikkara, Trichur.

Watt, G., 1872. The Commercial Products of India. Toay & Tomorrow's Pub., New Delhi, (Reprint 1966).

Watt, G., 1882. Dictionary of the Economic Products of India Today and Tomorrows Pub. Delhi, India (Reprint).

Xianchang, Y., Kun, X., Xizeng, A., Liping, C., Zhenxan, Z., 1996. Study on the relationship between canopy, canopy photosynthesis and yield formation in ginger. J. Shandong Agric. Univ. 27 (1), 83–86.

Xizhen, A., Zhenxian, Z., Shaouhui, W., 1998a. Effect of temperature on photosynthetic characters of ginger leaf. China Vegetables 3, 1–3.

Xizhen, A., Zhenxian, Z., Zhifeng, C., Liping, C., 1998b. Changes in photosynthesis rate, MDA content and the activities of protective enzymes during development of ginger leaves. Acta Hortic. Sin. 25, 294–296.

Xizhen, A., Zhenxian, Z., Zhifeng, C., Liping, C., 1998c. Changes in photosynthetic rate, MDA content and activities of protective enzymes during development of ginger leaves. Acta Hort. Sinica. 25, 294–296.

Xizhen, A., Zhenxian, Z., Shaouhui, W., Zhifeng, C., 2000. The role of SOD in protecting ginger leaves from photoinhibition damage under high light stress. Acta Hortic. Sin. 27 (3), 198–201.

Yadav, R.K., 1999. Genetic variability in ginger (*Zingiber officinale* Rosc.). J. Spices Aromat. Crops 8, 81–83.

Zachariah, T.J., Sasikumar, B., Ravindran, P.N., 1993. Variation in ginger and shogaol contents in ginger accessions. Indian Perf. 37, 87–90.

Zachariah, T.J., Sasikumar, B., Nirmal Babu, K., 1999. Variations for quality components in ginger and turmeric and their interaction with environment. In: Sasikumar, B., Krishnamurthy, B., Rema, J., Ravindran, P.N., Peter, K.V. (Eds.), Biodiversity Conservation and Utilization of Spices: Medicinal and Aromatic Plants Indian Society of Spices, IISR, Calicut, Kerala State, pp. 116–120.

Zavada, M.S., 1983. Comparative morphology of monocot pollen and evolutionary trends of apertures and wall structure. Bot. Rev. 49, 331–379.

Zhenxian, Z., Xizhen, A., Qi, Z., Shi-jie, Z., 2000. Studies on the diurnal changes of photosynthetic efficiency of ginger. Acta Hortic. Sin. 27 (2), 107–111.

16 The Chemistry of Ginger

The Composition of Ginger Rhizome

Natarajan et al. (1972) reported the following composition for ginger rhizome. The sample was from Kerala State, India. It contained essential oil (2.7%), acetone extract (3.9–9.3%), crude fiber (4.8–9.8%), and starch (40.4–59%). Percentages of volatile oil and nonvolatile extract in ginger from various countries are summarized in Table 16.1.

Some years later, Haq et al. (1986) studied the composition of ginger from Bangladesh. They found that the rhizome contains:

1. Essential oil—4% on the basis of rhizomes dried at 60°C for 8 h, and 0.8% on raw rhizome basis
2. Mixture of mainly sesquiterpene hydrocarbons (10–16%) based on dry ginger
3. Ash (6.5%)
4. Proteins (12.3%) and water-soluble proteins (2.3%)
5. Starch (45.25%)
6. Fat (4.5%) including free fatty acids (acid number: 10.38, as oleic acid: 5.2). Achinewhu et al. (1995) reported free fatty acid (as % of (dry matter) oleic acid) content and peroxide number (peroxide value) of ginger from Nigeria as 0.48±0.04 and 3.2, respectively
7. Phospholipids (traces) determined from petroleum extract
8. Sterols (0.53%)
9. Crude fiber (10.3%)
10. Cold alcoholic extract (7.3%) as oleoresin
11. Vitamins (as given in Table 16.2)
12. Reducing sugars (glucose, fructose, arabinose), traces
13. Water solubles (10.5%)
14. Minerals (in g/100 g): Ca (0.025), Na (0.122), K (0.035), Fe (0.007), P (0.075), Mg (0.048), Cl (1.5 ppm), F (5.0 ppm).

Afzal et al. (2001) have summarized the data on the mineral composition of ginger in Table 16.3.

In summary, ginger rhizomes contain the two following kinds of constituents:

1. Volatile compounds constituting the essential oil.
2. Nonvolatile constituents (heavy) products, including oleoresin (gingerols, shogaols, and related products which are the pungent principles of ginger) and the other usual organic and inorganic compounds found in foods. The high contents of vitamin C, manganese (Mn), sodium (Na), chlorine (Cl), and iron (Fe) is noteworthy.

Characteristics of two kinds of fresh Brazilian ginger rhizomes have been reported and are summarized in Table 16.4.

The Agronomy and Economy of Turmeric and Ginger. DOI: http://dx.doi.org/10.1016/B978-0-12-394801-4.00016-8

Table 16.1 Percentages of Volatile and Nonvolatile Extracts of Ginger from Various Countries

Origin	Percentage (w/w)	
	Volatile	**Oil Nonvolatile Extract**
India (Kochi, Kerala State)	2.2	4.25
Sierra Leone	1.6	7.20
Jamaica	1.0	4.40
Nigeria	2.5	6.50

Source: Akhila and Tewari (1984).

Table 16.2 Vitamins in Ginger Rhizome Powder from Bangladesh

Vitamin	Percentage in Rhizome Powder
Thiamine	0.035
Riboflavin	0.015
Niacin	0.045
Pyridoxin	0.056
Vitamin C	44.0
Vitamin A	Traces
Vitamin E	Traces
Total	41.15

Source: Haq et al. (1986).

Table 16.3 Inorganic Elements in Trace Amounts in Ginger

Constituent Element	Quantity (ppm on Dry Weight Basis)
Cr	0.89
Mn	358
Fe	145
Co	18 ng/g
Zn	28.2
Na	443
K	12.9
As	12 ng/g
Se	0.31
Hg	6 ng/g
Sb	39
Cl	579
Br	2.1
F	0.07
Rb	2.7
Cs	24 ng/g
Sc	42 ng/g
Eu	44 ng/g

Source: Zaidi et al. (1992).

Table 16.4 Characteristics of the Two Kinds of Fresh Brazilian Rhizomes[a]

Constituent Compound (g/100 g of Dry Material)	Compound	
	Gigante	Calpira
Proteins[b]	5.55–13.84	7.23–7.70
Ether extracts	3.24–8.35	3.60–7.29
Carbohydrates	76.82–84–86	78.54–81.62
Crude fiber	5.50–11.72	9.56–13.17
Ash	4.30–7.99	6.94–7.08
Alcohol extract	1.66–6.01	5.42–8.01
Volatile extract (E.O)	0.40–2.68	1.49–2.64
Nonvolatile extract	2.02–5.99	2.11–4.65

[a]The moisture content ranged from 80% to 90%.
[b]Taveira et al. (1997).

Okwu (2001) reported the chemical evaluation, nutritional, and flavoring potential of ginger stem. Results are in Table 16.5.

Ginger possesses a high nutritional value. However, α-acids, reducing sugars, and vitamin C can give rise to the Maillard reaction upon heating (similarly as in other foods) with the formation of off-flavors (mainly heterocyclic compounds) and the formation of melanoidins (Rogacheva et al., 1998; Vernin et al., 1992). Ginger does not contain aflatoxin (Martins et al., 2001).

Extraction, Separation, and Identification Methods

Extraction Methods

Van Beek (1991) has reviewed all of the extraction methods. In addition to the usual hydrodistillation, steam distillation, leaching, and pressing, extraction with supercritical CO_2 also has been widely used during the last almost three decades to obtain essential oils. For instance, solvent extraction with acetone gives the ginger oleoresin, which contains the essential oils as well as the pungent principles and other nonvolatile compounds present in ginger. When compared with other methods, this one gives the best results (He et al., 1999; Roy et al., 1996).

Hydrodistillation and Steam Distillation

Hydrodistillation is principally used for laboratory purposes in a glass, copper, or steel reactor connected to a cooling and decanting flask. Krishnamurthy et al. (1970) investigated the water-distilled oil from green and dry ginger. Green ginger oil has a spicier odor and is considered superior to the oil from dry ginger. This is probably due to the higher amount of α-zingiberene in green oil. Because green ginger

Table 16.5 Overall Composition of the Ginger Stem and
Ginger Oil

Mineral[a]	Amount Found
P (g/100 g dry matter)	0.11 ± 0.01
Mg (%)	0.80 ± 0.10
Ca (%)	0.50 ± 0.10
K (%)	1.00 ± 0.50
Na (%)	0.04 ± 0.20
N (%)	2.50 ± 0.01

Others	Percentage
Moisture	16.10 ± 0.10
Ash	9.52 ± 0.02
Fat/Oil	17.20 ± 0.10
Carbohydrates	57.00 ± 0.77
Protein N×6.25%	15.69 ± 0.06
Crude fiber	4.53 ± 0.05
Food energy (FE)[b]	400.56 g/cal

Various Indices for the Ginger Oil	
Iodine number (mg/100 g)	87.69 ± 0.10
Peroxide number (mg/g oil)	9.80 ± 0.10
Saponification number (mg/KOH/g oil)[c]	4.80 ± 0.20
Acid number (mg/COOH/g oil)	190.74 ± 0.10

Source: Okwu (2001).
[a]The minerals were determined according to the standard AOAC methods
(1984).
[b]Was estimated using the equation FE=(% Crude Protein×4)+
(% Lipids×9)+(% Carbohydrates×4).
[c]The saponification number was obtained by William's method. The iodine
number was determined by Strong and Koch's method. The acid and peroxide
numbers were calculated using Pearson's method.

is perishable, the distillation must be done locally. More α-zingiberene and smaller amounts of other sesquiterpene alcohols are present in green oil than in oils from Cochin ginger and peeled ginger.

Steam distillation, an old and well-known method, is commonly used for commercial isolation of ginger oils. Yields ranged from 0.2% to 3.0% according to the origin and the state of the rhizome (fresh or dried) (Anzaldo et al., 1986; Connell, 1971; Krishnamurthy et al., 1970). It leads to high levels of monoterpenes and low amounts of nonvolatile compounds, in part by thermal degradation of gingerols giving rise to straight-chain aldehydes and 2-alkanones (Badalyan et al., 1998).

Solvent Extraction

This method is used to obtain oleoresin extracts. Several solvents have been recommended. Oleoresin from Australian ginger rhizome was prepared by acetone or ethanol extraction of dried ground ginger. Gingerols constituted about 33% of the freshly prepared oleoresin. It decomposes to afford shogaols and zingerone. Ethyl ether, acetone, and hexane were used by Jo (2000), Mathew et al. (1973), and Nishimura (2001), respectively, as well as pentafluoropropane (Hill et al., 1999) and heptafluopropane (Dowdle et al., 2002). Antioxidant compounds in ginger rhizome from Korea were extracted using ethyl acetate from a crude methanol extract and separated by TLC. Ethyl acetate was also used by Harvey (1981). Dry root ginger from Jamaica (1.5 g) was crushed and left to stand with acetyl acetate for 30 min. The solution was filtered and evaporated to dryness to yield 200 mg of oil. This was dissolved in 20 ml of ethyl acetate to give the stock solution. A kinetic study of extraction of gingerols using acetone as solvent was carried out. The drawbacks and advantages of the method were reviewed by Koedam (1987) and Bicchi and Sandra (1987).

Solid-Phase Microextraction Method

Solid-phase microextraction (SPME) from the sampling in aromatic analysis was carried out by Faulhaber and Shirey (1998). They described an extraction and desorption process and method of SPME for fruit, juice drinks, peppermint oil in chocolate, spearmint oil, gum ginger oil, and citrus oil. The latter method was presented as a quick and solvent-free alternative to conventional extraction methods.

Extraction by Supercritical Carbon Dioxide

For almost two decades, supercritical carbon dioxide (CO_2) has been employed by extraction of natural products and particularly for ginger powder and other spices (Chen et al., 1986, 1987; Meyer-Warnod, 1984). It is used mainly because supercritical carbon dioxide is a safe, noncombustible, inexpensive, odorless, colorless, tasteless, nontoxic, and readily available solvent. Its low viscosity enables it to penetrate the matrix to reach the material extracted, and its low latent heat of vaporation and high volatility mean that it can easily be removed without leaving a solvent residue. Several reviews have been devoted to the CO_2 extraction of essential oils (Meireles and Nikolov, 1994; Moyler et al., 1994). According to Moyler et al. (1994), a distinction has to be made between subcritical liquid CO_2 ($SLCO_2$) extraction and supercritical fluid CO_2 ($SFCO_2$) extraction. In the first process, temperature and pressure ranged between 0°C and 10°C and 50–80 bar pressure, respectively. It is mainly used selectively to extract essential oils from ground plants. Supercritical CO_2 extraction is not currently used commercially to extract flavor oleoresins because of cost constraints. However, some supercritical CO_2 is available for commercial use. Brogle (1982) showed that a fractionated extract can be obtained by reducing the pressure of a CO_2 solution of a supercritical extract, while still in the condenser. Insoluble components such as waxes, resins, and alcohols can be separated to give an essential oil similar to that of the subcritical CO_2 extract. A third method consists of using CO_2 with entraining solvents such as ethanol in order to obtain a specific flavor profile. The apparatus is generally used for 500–600 kg of powdered ginger under a blanket CO_2 gas to prevent surface oxidation. Liquid CO_2 at low temperature is pumped around the

Table 16.6 GC Comparison of Some Sesquiterpene Hydrocarbons and Zingerone Obtained from Indian Ginger Extracts by Three Extraction Methods

Volatile Compound	Percentage		
	Steam Distillation	**Supercritical CO_2**	**Hexane Extraction**
ar-Curcumene	10.0	3.7	2.3
α-Zingiberene	44.0	19.6	12.1
β-Zingiberene	8.0	3.4	2.0
β-Bistabolene	8.3	3.7	2.4
β-Sesquiphellandrene	17.8	7.9	4.9
Total	88.1	38.3	23.7
Zingerone	0.8	0.7	0.3

Source: Pellerin (1991).

circuit and the extract is collected in a condenser evaporator. The pressure is released, extract trapped, and the CO_2 recycled. By varying the temperature and pressure during extraction, the flavor and odor components can be selectively extracted (Roy et al., 1996). Chen et al. (1986) extracted freeze-dried ginger powder with liquid CO_2 (600–700 psi) for 48 h. The oil was fractionated into hydrocarbon fraction and oxygenated fractions using silica gel chromatography. Each fraction was then analyzed by GC on a Carbowax 20 M capillary column (60 m×0.32 m i.d.). Previously, Chen et al. (1986) also analyzed the oil by TLC on silica gel and HPLC on a reverse phase column. The ginger oil had both the pungent and the aromatic properties of ginger. Two cultivars of ginger from Korea were treated by simultaneous stream distillation and CO_2 extraction (Kim et al., 1991). The oil from the latter process (6.96%) was fractionated into one hydrocarbon fraction and another oxygenated hydrocarbon fraction by using silica gel column chromatography. Each fraction was analyzed by GC and GC/MS.

Pellerin (1991) compared the extracts of Indian ginger obtained by the conventional processes (steam distillation and hexane extraction) and that obtained by supercritical CO_2 extraction. The results are given in Table 16.6. These results show a great difference in the sesquiterpene hydrocarbons percentages between steam distillation and supercritical CO_2 extraction, particularly for α-zingiberene.

Several cooked dishes (soup, fish, poultry, etc.) were seasoned with ginger prepared by the same three methods. Tasters commented favorably on the flavor balance and fresh characteristics of food seasoned with supercritical CO_2 extract of ginger. In a study of a Fijian ginger oil extracted with CO_2, the content of gingerols increased with rising CO_2 pressure (Zhu et al., 1995).

The separation conditions were as follows:

1. In the first separator, temperature and pressure were 60°C and 25 MPa (1.1 g of ginger wax was obtained).
2. In the second separator, temperature and pressure were 70°C and 20 MPa, respectively (2.1 g of hot gingerin was obtained).
3. In the third separator, 9.2 g of ginger essential oil was isolated at 60°C and 14 MPa, and in the final separator, 33.5 g water was separated at 15°C and 3.5 MP$_a$.

The effects of extraction ($25\,MP_a$), temperature, time (2 h), CO_2, flow rate (0.09 l/h), and material size (40–60 μm) on the extraction rate of ginger oleoresin from China were explored under industrial conditions. Other methods have been published by Wen et al. (2001).

In conclusion, the supercritical CO_2 extraction method was found to have many advantages over normal extraction methods for the following reasons:

1. Shorter extraction time
2. Lower energy consumption
3. Better quality of sensory properties.

The selective solubility of components in CO_2 enables it to extract all the useful aromatics from a flavoring source. However, according to Van Beek (1991), its use should be limited in the industry because of the high pressures needed and the high cost of the appropriate apparatus.

Analytical and Isolation Methods

Prior to 1970, fractional vacuum distillation, analytical and preparative chromatographic procedures such as column chromatography (CC), thin-layer chromatography (TLC), gas chromatography (GC), and chemical methods were used. These procedures have been extended to high-performance liquid chromatography (HPLC), high-performance gas chromatography (HPGC), associated with specific detectors, direct vaporization in the GC apparatus, and dynamic headspace techniques.

Liquid Column Chromatography

Liquid column chromatography (LCG) is the oldest method used in organic chemistry since Tiswett's discovery in 1906. Essential oils from ginger can be fractionated on a silica gel column into a hydrocarbon fraction and an oxygenated fraction. Elution solvents are nonpolar hydrocarbons such as pentane, and, for the latter, a mixture of pentane with ether (2:1 v/v) or acetone (4.1 v/v), respectively (Van Beek, 1991). Thus, essential oils with removed terpenes can be obtained on a large scale. However, terpenes containing a furan ring, such as perillene and rosefuran, found in ginger oil occur in both fractions. More polar compounds such as aliphatic acids can be eluted with a more polar solvent (pentane/ethanol: 9.1 v/v). Chromatographic procedures on alumina were used by Herout et al. (1953) for the isolation of some sesquiterpene hydrocarbons, such as (+)-ar-curcumene, bisabolene, farnesene, and α-zingiberene. But this class of compounds was better separated by CC on silver nitrate-treated supports. Balladin and Headley (1999) isolated essential oil and pungent principles of West Indian ginger by liquid chromatography using silica gel (70–230 mesh) and a mixture of petroleum ether (60–80°C) and diethyl ether (3:7 v/v) as mobile phase. They isolated seven fractions: the first 15 ml contained the very volatile and less polar compounds present in the extracted oleoresin from sun-dried ginger rhizome; that is, the essential oil accounts for 25.6% (w/w) of the total oleoresin charge to the column. The next 5 ml aliquot was without any compound. The following 25 ml contained the shogaol fraction and represented 47.3% of the sample. The next 5 ml aliquot was without any compound. The following 35 ml contained the gingerol fraction and represented 27.1% of the sample. Each fraction was subjected to GC/MS and HPLC/MS analyses. Two techniques of CC have been used to separate pungent principles of ginger: vacuum and flash chromatography, with toluene/methanol (16:10 v/v) as the mobile phase. Whereas chromatography is not very successful for separation of these compounds, vacuum chromatography is more rapid and effective.

Purification and characterization of cysteine proteinase from fresh ginger rhizome has been carried out by CC using diethylaminoethyl (DEAE)-cellulose and Sephadex G 75 (Kitamura and Naguno, 2000). Two CPI fractions of 11,000–11,800 and 15,000–16,000 molecular weights were recorded. Both fractions showed potent papain inhibitory activities and were stable at <40–60°C, but the activities decreased and disappeared when exposed to higher temperatures.

Thin-Layer Chromatography

Since 1962 when Stahl published his work on TLC, a rapid, easy, and inexpensive method in organic chemistry, the same has been in wide use (Vernin, 1970). Unfortunately, the method is unsuitable for complex mixture analysis, such as for essential oils. However, it is adequate for the preparative separation of some compounds or a set of compounds having the same retention time (R_t). The extract after solvent extraction can be submitted to GC and GC/MS analyses. Quantitative determination by ultraviolet (UV) densitometry can also be used in a simple case. Analyses for the ginger oleoresin have been reported by Connell (1970) using TLC on silica gel plates with hexane-diethyl ether mixtures as eluent. Quantitative determination by densitometry allowed him to separate three main groups: gingerols, shogaols, essential oils, and more polar and heavy compounds as a trailing. Some sesquiterpene hydrocarbons were separated by TLC on silica gel plates treated with silver nitrate (Connell, 1970). Analysis of gingerol compounds of raw ginger and its paste was carried out by TLC (Jo, 2000). Antioxidant compounds in ginger rhizomes from Korea, extracted with ethyl acetate from crude methanol extract, were separated through TLC. Ten phenolic antioxidant bands were visualized through color reactions using ferric chloride, potassium ferrocyanide, and 1,1-diphenyl-2-picrylhydrazyl (DPPH) and were purified through preparative TLC and HPLC. Among them five antioxidants were identified as (4)-, (6)-, and (10)-gingerols and (6)-shogaol on the basis of their molecular weight determination through LC/MS. As shown in experiments using DPPH free radicals, (6)-gingerol and PT_4-HPS were revealed to be more efficient than BHT. Total gingerol content (determined through reversed phase HPLC in rhizomes of different ginger varieties varied significantly. Two varieties collected in Korea (HG 55) and in Brazil (HG 52) showed the highest content of gingerol.

High-Performance Liquid Chromatography

HPLC has supplanted the TLC for the preparative separation of essential oil compounds and for the quantitative determination of important and heavy compounds.
Two examples have been given for ginger oil by Van Beek (1991). The first example concerns the separation of the more important sesquiterpene hydrocarbons
of ginger oil from India, accounting for 70% of the oil. The following conditions
were used: a 25 cm × 1 cm column fitted with 5 μm C-18 silica gel reversed phase
eluted with MeCN/H$_2$O (88:12) with a flow rate of 4 ml/min, and UV detection at
215 and 245 μm. After the four preparative runs, ar-curcumene was obtained in
>99% purity and (E,E)-α-farnesene in 84% purity. Other sesquiterpene hydrocar-
bons (β-sesquiphellandrene, α-zingiberene, β-bisabolene) have been separated under
different analytical conditions. Two fractions were collected. The first consisted of
53% α-zingiberene, 19% β-bisabolene, and 9% β-sesquiphellandrene. They were further purified by means of preparative capillary GC. Using a reversed phase system:
HPLC column 15 cm × 0.46 cm fitted with Microsorb 5 μm C-18 silica gel, solvent:
MeCN/H$_2$O (6/1, 1 ml min to MeCN/H$_2$O (0.5/5 in 30 min) and a detection UV at
236 μm, the geranial and neral content of any ginger oil can be measured in minutes (the
minimum detectable quantity was 1 ng) (Van Beek, 1991). Other compounds detected
were: myrcene, β-phellandrene, (E,E)-α-farnesene, and β-sesquiphellandrene + α-zingi
berene. The comparison by HPLC of the extracts of Indian ginger root obtained by the
conventional methods and that obtained by supercritical CO$_2$ shows that steam distillation is not suitable for the extraction of the pungent principles of ginger oleoresin.
The supercritical CO$_2$ method gives better results than the hexane extracts (Table
16.7; Pellerin, 1991). Paradol has not been taken into account.

A quantitative method by HPLC of pungent principles of ginger was developed.
The content of (6)-, (8)-, and (10)-gingerols, (6)- and (8)-shogaols, 6-dihydroging-
erdione, and galanolactone in 20 kinds of rhizomes originating from China, Taiwan,
Vietnam, and Japan, and fresh ginger root cultivated in Shizuoka Prefecture of Japan

Table 16.7 HPLC Comparison of Gingerols and Shogaols obtained from Indian Ginger
Extracts by Three Different Extraction Methods

Pungent Principle[a]	Percentage		
	Steam Distillation	**Supercritical CO$_2$ Extraction**	**Hexane Extraction**
(6)-Gingerol	0.2	16.4	0.9
(8)-Gingerol	0.3	3.1	0.7
(10)-Gingerol	–	3.8	0.8
(6)-Shogaol	0.3	2.8	6.3
(8)-Shogaol	–	–	1.6
Total	0.8	26.1	10.3

Source: Pellerin (1991).
[a]The supercritical CO$_2$ method gives better results than the hexane extraction.

were examined. It was found that Japanese fresh ginger root contained gingerols, shogaols, 6-dehydrogingerdione, and galanolactone as the major constituents.

A HPLC method was developed by Sane et al. (1998) to study the geographical variation in the ginger samples obtained from different states of India with respect to their gingerol and ginger oil contents. Analyses of gingerol compounds of raw ginger and its paste were carried out by a combination of TLC and HPLC with Licrosorb RP-18 column by Jo (2000). (6)-, (8)-, and (10)-gingerols were identified by HPLC/MS and nuclear magnetic resonance (NMR). The content of (6)-, (8)-, and (10)-gingerols were 635.3, 206.3, and 145.7 mg%, respectively, in raw ginger from Korea, and they were 418.2, 142.6, and 103.3 mg% in ginger paste, respectively. Another HPLC method was applied by Chen et al. (2001) for the determination of pungent constituents in ginger and to evaluate ginger extracts obtained by supercritical CO_2 or anhydrous alcohol extraction. The effective content of pungent constituents was 13.84% and 1.46%, respectively. Ginger oleoresins extracted by CO_2 were of especially high quality and contained higher amounts of natural gingerols and related compounds.

Gas Chromatography

Both analytical and preparative GC have undergone a considerable development since the discovery by James and Martin (1952). For about two decades, stainless steel-packed columns with various polar and nonpolar stationary phases (3–15%) on silica gel GC (80–100 mesh) were used, which were subsequently replaced by capillary columns (high-resolution GC (HRGC)). Ginger oils were first analyzed on packed columns (Jain et al., 1962; Nigam et al., 1964), for example, Reoplex 400, silicone nitrile, Apiezon M and L, Carbowax 400 and 20 M, and SE 30. Monoterpenes of ginger oil have been analyzed using Reoplex 400 at 70°C and Carbowax 400 at 144°C as stationary phases (Connell, 1970). Apiezon M has been also used at isothermal or programmed temperature. The same authors reported the separation of five sesquiterpenes: ar-curcumene, α, α-zingiberene, (−)-β-sesquiphellandrene, and *trans*-β-franesene on a packed column (6 ft × 1/8 in. i.d) containing 15% of Apiezon M on silica gel GC (80–100 mesh). The pungent constituents of ginger were analyzed as trimethylsilyl (TMS) derivatives by Harvey (1981) using a 3% silicone SE 30-packed column (100–120 mesh), GC Q, programmed from 100°C to 300°C at 4°C/min, with nitrogen at 30 ml/min as the carrier gas. Injector (FID) and detector temperature were kept at 300°C. TMS derivatives were used in a combination of chemical reactions (lithium aluminum hydride reduction, deuterium exchange, deuterium exchange reduction, oxidation). Capillary GC chromatogram on Carbowax 20 M of Japanese ginger oil was reported by Masada (1976).

Gas chromatograms of an Indian ginger oil with a high citral content were obtained with DB-1 and DB-Wax capillary columns (Van Beek, 1991). On the nonpolar column (DB-1), a good separation of monoterpene hydrocarbons, oxygenated monoterpenes, sesquiterpene hydrocarbons, and oxygenated sesquiterpens were found. However, limonene, β-phellandrene, and 1,-8-cineole overlap. On the polar column (DB-Wax), the three latter compounds were well separated, but oxygenated monoterpene and sesquiterpene hydrocarbons were poorly separated. Application of GC in the analysis of essential oils was clarified by a committee formed in 1993. A rapid and simple isolation of α-zingiberene from ginger oil was developed. A sesquiterpene-enriched fraction was treated with the dienophile-4-phenyl-2,5-oxazolinedione A, which selectively formed a Diels–Alder adduct with zingiberene. The adduct was purified by flash chromatography and then hydrolyzed to afford zingiberene in good yield (99% purity). Quality assessment of flavors and fragrances by HPLC has been widely used (Mosandhl, 1992).

Enantiomers also can be separated by capillary columns coated with β-cyclodextrin derivatives. Thus, Takeoka et al. separated sesquiterpene hydrocarbons using a per-methylated β-cyclodextrin (P-β-CD) as a GC stationary phase. Sesquiterpenes include ar-curcumene, α- and β-cyclodextrin and β-elements, α-copaene, cadinene, cis- and trans-calamenes, and bicyclogermacrene. Four years later, Koenig et al. (1994) separated the two enantiomers and ar-curcumene and β-bisabolene in ginger oil, using a fused capillary column coated with heptakis (2,3-O-methyl-6-O-t-butyldimethylsily-β-cyclodextrin) in polysiloxane OV-1701 (50% w/w). The column temperature was 115°C and carrier gas:hydrogen (0.5 bar). The chromatogram shows that ar-curcumene exists in the (+) and (−) in equal quantity. Enantiomeric separation of the characteristic aromatic compounds in fresh rhizomes of Japanese ginger was carried out using the off-line multidimensional GC (MDGC) system and confirmation of the odor character of each enantiomer by GC/olfactometry (Nishimura, 2001). GC has been widely used for preparative purposes as well. Packed columns giving poor separation were replaced by megabore columns that are wide-bore capillary columns (WBCC) of 0.5–0.8 mm i.d. coated with thick films of 3–5 μm of stationary phases. Van Beek (1991) reported the quantitative separation of some sesquiterpene hydrocarbons (SQHC) using 30 m/3 μm DB-1 megabore column. From SQHC fraction containing α-zingiberene (69%), β-bisabolene (19%), β-sesquiphellandrene (9%), and minor compounds (3%), and obtained by preparative HPLC and then injected several times in the WBCC, β-bisabolene could be obtained sufficiently pure for ^{1}H- and ^{13}C-NMR analyses. β-sesquiphellandrene was purified as well. This type of column is also useful in the sniffing method (at the end of the column).

Other GC Methods: Dynamic Headspace

The thermal desorption cold trap injector (TCT) method was used as part of the off-line MDGC system. It was shown that the TCT can be used not only for headspace analysis but also as a part of an MDGC system. Direct vaporization of a pulverized sample of ginger can be subjected to heating at 250°C for 1 min in a vaporizer directly connected to the GC on a GC/MS apparatus. Out of 54 constituents, 25 were identified (Chen et al., 1987). They found little decomposition regardless of the heating time and temperatures and a similar composition with the essential oil obtained from the same sample. GC of headspace vapor from dry ice-cooled trap of low-boiling compounds from steam-distilled ginger was reported by Kami et al. (1972). The identification of peaks was carried out by comparing the retention time with authentic samples. The identification was supported by a chemical reaction including 2,4-dinitrophenylhydrazones (for carbonyl compounds), 3,5-dintrobenzoates (for aliphatic alcohols), a mercuric complex (for sulfide derivatives), and hydroxamic acid (for monoterpenes). They were analyzed directly or after regeneration by TLC, GC, and combined GC/MS.

The GC dynamic headspace is a very suitable technique and useful to analyze liquid or solid aromatic materials and has been widely used in flavors and fragrances. It uses an inert gas (helium or nitrogen) to flush a small flask containing the aromatic product for 10–15 min at room temperature. Volatile compounds are trapped on Tenax GC and then thermally desorbed at 250°C for a few seconds in the capillary column, under the usual conditions. De Pooter et al. (1985) used this technique for ginger powder, and the GC pattern was found to be quantitatively similar to that of the corresponding essential oil prepared by hydrodistillation.

It is an excellent and powerful method for the comparison of different samples of ginger powder from different origins. The method can be indirectly used with a separated flask connected to a Tenax GC trap (in glass or stainless steel). Volatiles are extracted by solvent extraction (ethyl ester). The extract after evaporation of the solvents is then injected in the GC column and GC/MS apparatus.

GC Artifacts

Under gas chromatographic conditions, gingerols are decomposed into zingerone, aldehydes, and shogaols. Xanal and minor amounts of the other aldehydes and zingerone were also formed on treatment with hot alkali of an extract of gingerols obtained by dry CC. Connell (1970) used two packed columns (3 ft×1/8 in.): 1/6% SE 30 at 188°C and Apiezon M at 200°C on Embacel, respectively.

Chen et al. (1987) used HPGC on Carbowax 20M and OV-1 columns to study the thermal degradation products of gingerols in steam-distilled oil from ginger. Significant higher concentrations of aliphatic aldehydes (C_6–C_{12}) and 2-alkanones in the steam-distilled sample confirmed that thermal degradation of nonvolatile gingerols occurred during steam distillation. On the other hand, during steam distillation or hydrodistillation, artifacts can be formed either by hydrolysis of labile esters or by isomerization (or rearrangement) of some hydrocarbons. In the headspace technique, these artifacts will be greatly diminished (Van Beek, 1991). Examples of artifact formation by chromatographic techniques (GC, LC, GC/MS) have been reviewed by Garnero and Tabacchi (1987).

To conclude this section, it should be pointed out that packed columns should be strongly discouraged owing to the limited number of separated compounds (about 40). This problem can be partially alleviated by using better performance capillary columns of two different polarities and specific detectors for compounds containing nitrogen and sulfur atoms. Even in this case, some peaks overlap. Thus, it is necessary to proceed to a prefractionation of the ginger oil on a short column of silica gel in a hydrocarbon and an oxygenated fraction using pentane (or hexane) and pentane/ether (2:1 v/v) as solvents, respectively. Each fraction is again analyzed by GC and GC/MS. The method does not require a large amount of the product, just 20–50 µl will do.

Gas Chromatography/Mass Spectrometry Coupling

Since 1967, HRGC and GC/MS coupling have been successfully used in the analysis of essential oils in general, and in particular, in the analysis of ginger oils. A number of researchers have investigated this aspect and Table 16.8 summarizes these.

Table 16.8 Principal Research Publications on Ginger (*Zingiber officinalis* Roscoe) Essential Oils as per Their Origin

Origin	Extraction/Identification Methods[a]	References
India	Steam distillation/GC/MS	Nigam et al. (1964)
India (Kerala State)	Steam distillation	Natarajan et al. (1972)
	Dynamic headspace/GC/MS, Steam distillation	Narayanan and Mathew (1985)
India/Australia	Steam distillation/GC/MS	Erler et al. (1988)
India/China	Steam distillation/GC/MS	Vernin and Parkanyi (1994)
West India	Steam distillation/GC/MS	Balladin and Headley (1999)
China (Szechaun)	Steam distillation/GC, TLC	Chen and Guo (1980)
China	Dynamic headspace/CO_2 supercritical/GC/MS, RI, thermal treatment	Chen et al. (1986, 1987)
	Steam distillation/GC/MS, different methods/GC/MS, steam distillation (dry and fresh rhizomes)/GC/MS UV/TLC	Li et al. (2001)
Taiwan	Steam distillation/CO_2, supercritical/CC GC, GC/MS	Chen and Ho (1988)
Sri Lanka	Steam distillation/GC/MS, RI	Macleod and Pieris (1984)
Vietnam	Steam distillation/GC/, GC/MS	Van Beek et al. (1987)
Japan	Steam distillation/GC, GC/MS, SIM	
Philippines	Hydrodistillation (yield 0.2–1%), TLC, GC:IR	Anzaldo et al. (1986)
Malaysia	Steam distillation/GC	Ibrahim and Zakaria (1987)
Korea	CO_2 supercritical/GC/MS	Kim et al. (1992)
Fiji	Steam distillation (yield 0.3% fresh rhizomes) GC, GC/MS	Duve (1980)
Tahiti	Steam distillation/GC/MS	Vahira-Lechat et al. (1996)
Australia	Steam distillation/GC	Connell (1970), Connell and Jordan (1971)
Australia/Africa	Steam distillation/GC	Connell (1970)
Argentina	Review	Rosella et al. (1996)
Brazil (Rio de Janeiro)	Steam distillation/GC	Taveira et al. (1997a,b)
Gigante and Calpira	Steam distillation (yield 2.2%)/ GC, GC/MS	Taveira and Magalhaes (1997a)
Jamaica	Steam distillation/GC	Gopalam and Ratnambal (1989)
Mauritius Island	Hydrodistillation/GC, GC/MS	Gurib-Fakim et al. (2002)
Poland	Solvent extract/GC, GC/MS	Kostrzewa and Karwowska (1976)
Nigeria	Steam distillation (dried rhizomes)/GC/MS	Ekundayo et al. (1988)
	Steam distillation/GC, GC/MS	Dambatta et al. (1998)
	Hydrodistillation (yield 2.4%)/GC, GC/MS	Onyenekwe and Hashimoto (1999)

[a]CC: Column chromatography, TLC: thin-layer chromatography, GC: gas chromatography, GC/MS: gas chromatography/ mass spectrometry coupling, EO: essential oil, RI: retention index (Kovats Indices).

Pungent principles of ginger can also be analyzed as such or as TMS derivatives. In the first case, after extraction using 0.85 g supercritical CO_2, the extract was directly submitted to electron-spray mass spectrometric identification. Gingerols and shogaols have been identified and their concentrations determined. A low concentration of shogaols is characteristic to this extraction procedure (Bartley, 1995). In the second case, Harvey (1981) reported GC/MS data for TMS derivatives of gingerols, methyl gingerols, shogaols, methylshogaols, paradol, gingerdiols, hydroxylcurcumins, and demthylated hexahydrocurcumin. GC/MS identification of ginger components was made by comparison of the unknown mass spectrum with a great number of mass spectra of known molecules stored on computer disks equipped with a data acquisition system that is an integral part of the instrument. Several papers, reviews, books, and data banks are available (Vernin et al., 1992).

Retention Indices as Filters (or Relative Retention × α_g)

Retention ties R_t and mass spectra of unknown compounds have been compared with those of authentic standards in the earlier studies of ginger oils (Connell, 1970). However, these retention values are no reproducible and vary greatly with temperature on a given chromatographic column. Kovats indices (KI) do not suffer from this disadvantage (Kovats, 1958, 1965). At isothermic temperatures, Kovats uses the following logarithmic equation with linear alkanes as reference compounds:

$$KI = \frac{100n + 100 \text{x} (\log t \, R_x) - \log t \, R_n}{\log ti \, R_{(n+1)} - \log t \, R_{(n)}} \quad (16.1)$$

where $t^1 R_{(x)}$, $t^1 R_{(n)}$, $t^1 R_{(n+1)}$, are reduced retention times $(t_g) - t_m$ of the unknown compound and the linear alkanes with n and $n+1$ carbon atoms, respectively, which are eluted just before and after compound x. $t_{(m)}$ is the retention time of the air or that of the more volatile solvent such as pentane. Their reproducibility on polar (Carbowax 20M, DB-Wax, FFAP, BP 20, HP 20) and nonpolar columns (SE 30, SF 96, OV 1, OV 101, OV 117, DB-1, DB 5) is good under similar GC conditions on a given column. In linear programmed temperature, which is the usual case, Van den Dool and Kratz (1963) used the following formula:

$$KI = \frac{100n + 100I \, x(x - M)_{(n)-(n)}}{(M_{(n+1)}) - M_{(n)}} \quad (16.2)$$

where x, $M_{(n)}$, and $M_{(n+1)}$ are either the retention temperatures or the adjusted retention times of the unknown and straight-chain aliphatic esters.

The calculation of KI values starts with the injection of either a standard alkane mixture (C_6–C_{22}) for Carbowax 20M or a DB-wax and (C_6–C_{30}) for OV 101 or DB 5 (or linear ethyl esters) under the same linear programming temperature (2–4°C/min).

The software of the integrator can detect each peak and calculates each KI value automatically and prints it out with other GC data. All GC and GC/MS data are stored on a disk and visualized on a screen with the possibility to make a zoom on a wanted part of the chromatogram. Kovats indices (or retention indices) have been compiled in various reviews, books, and many other publications (Jennings and Shibamoto, 1980; Sadtler Research Laboratories, 1985). Another method for the calculation of Kovats indices has been suggested by Boniface et al. (1987) using scans (S) instead of retention times. A simple program called MBASIC.SCAN 1 using liner relationship is described as which follows, has been used to calculate all KI of a listing:

$$KI = a \times S + b \tag{16.3}$$

However, a prerequisite for this is to find at least 10 compounds for which KI are known and which are uniformly distributed in the reconstructed chromatogram. The interest in KI is evident as on polar and nonpolar columns different compounds can give the same mass spectra (Vernin and Petitjean, 1982). KI sesquiterpene hydrocarbons have also been reported by Anderson and Falcone (1969). KI as a preselection routine in mass spectra library searches of flavor and fragrance volatiles have been used by many investigators (Alencar et al., 1984). Furthermore, these indices possess a number of properties that can be suitable as a route of identification. Differences between KI (DKI) on polar and nonpolar columns are characteristic of a particular family of compounds. They have been summarized by Vernin et al. (1992). Modern GC/MS techniques can also be used to obtain more information about either a series of compounds or molecular weights of unknown compounds.

Selected Ion Monitoring Technique

A complement to the use of KI is the selection of a certain number of characteristic fragments of a particular compound or a homologous series. This technique is called selected ion monitoring (SIM) and can be used with both electron impact (EI) and chemical ionization (CI). Some of these ions for different groups of products have been reported by Vernin et al. (1992). For instance, sesquiterpene hydrocarbons is an essential oil that can be visualized qualitatively and quantitatively by selection of ions at m/z = 93, 121, 161, 204, and monoterpene hydrocarbons (with the exception of limonene, which gives EI a base peak at m/z = 68 by selection of ions at m/z = 93, 136, and so on). Thus, it is of particular importance to compare these two fractions obtained from different extraction methods and countries. This method was applied to the study of a "kintoki" Japanese ginger extract. Selected ions or base peaks (BP) of some compounds are reported (Table 16.9). Each compound has been quantified using naphthalene as the internal standard.

Chemical Ionization Technique

Because some molecular masses at EI at 70 eV are not always visible, various investigators used either recording mass spectra at 20 eV or gentler ionization methods. The theory and application of these latter techniques were developed in Harrison (1983) and in several other research papers (Bruins, 1987).

Table 16.9 Example of Separated Compounds Obtained by SIM Technique

Base Peak (m/z)	Separated Compound[a]
69	Neral, geranial, geranyl acetate, β-bisabolene, β-sesquiphellandrene
81	1,8-Cineole
93	α-Pinene, camphene, myrcene, β-phellandrene
95	Borneol and isoborneol
119	ar-Curcumene, α-zingiberene

[a]Each compound has been quantified using naphthalene as an internal standard.
Source: Tanabe et al. (1991).

In a chemical ionization chamber, the pressure is higher (007–1.5 torr) than in a source under EI. In many cases, it is possible to obtain quasimolecular ions both in positive chemical ionization (PCI), using isobutene or ammonia as gaseous reagents $(M+H)^+$, and in negative mode (NCI) using hydroxyl OH^- or NH^{2-} ion as a reagent giving $(M-H)^-$. Numerous other reagents are employed for this purpose, and, their applications in analysis of essential oils and aromas have been summarized by Vernin et al. (1992). PCI (NH_3) gives better results with carbonyl compounds (terpenes, aliphatics, alicyclics, and aromatics) than in the NCI mode. Another important fragment at $m/z=(M+NH_4)^+$ is observed. Upon PCI/i-C_4H_{10} sesquiterpene hydrocarbons give an intense quasimolecular ion $(M+1)^+$. Under the same conditions, mass spectra of sesquiterpene alcohols show intense fragment ions $(M+H-H_2O)^+$ at $m/z = 205$. Upon PCI (NH_3)$^+$, a typical set of ions is formed: $(M+H-H_2O)^+$, $(M+NH_4-H_2O)^+$, $(M+NH_4)^+$, and $(M+NH_3+NH_4)^+$, respectively. NCI (OH^-) is particularly suitable for the identification of aliphatic and terpenic esters of essential oils that give rise to the formation of $RCOO^-$ ions. In the case of alcohols, the RO^- ion is the base peak. In the NCI technique, correct analysis of unstaturated monoterpenes is complicated because of the secondary reactions between $(M-H)$ ions and nitrous oxide. On the other hand, molecular ions have low intensity because protons are difficult to abstract. Saturated terpenes cannot be ionized by this technique.

Miscellaneous Methods

More and more sophisticated spectroscopic methods such as UV, high-resolution infrared (HRIR), ^{1}H-NMR, and ^{13}C-NMR are usually employed for the identification of unknown compounds previously extracted from a complex mixture. But these methods are very time consuming and quite expensive to run.

IR spectra as a film on a KBr disk of some naturally occurring sesquiterpene hydrocarbons (β-sesquiphellandrene and zingiberene) have been reported by Wenninger et al. (1967). IR spectra of zingiberene have also been published by Herout et al. (1953) and Pliva et al. (1960). But the assigned structure was that of β-zingiberene instead of the α-isomer. Prior to 1970, a summary of these investigations was reported by Connell (1970). Van Beek (1991) claimed that IR spectroscopy can distinguish between various types of ginger oils by comparing the intensity of peaks at $3470\,cm^{-1}$ (alcohols), $1743\,cm^{-1}$ (esters), and $1680\,cm^{-1}$ (conjugated alde-hydes). Addition of large quantities of extraneous materials can also be detected. Coupling GC/high-performance infrared (HPIR) constitutes another interesting method of identification, but it has only limited use in the case of complex mixtures.

UV spectroscopy at $375\,\mu m$ has been used as a method for determination of (6)-gingerol in ginger oil consignments from China (Li, 1995).

^{13}C-NMR of ginger oils is very complex and the spectra are difficult to interpret because of the great number of peaks. On the other hand, compounds occurring below 0.2% are not measurable. The method can be used in addition to the usual technique based on retention indices and mass spectra. ^{1}NMR and ^{13}C-NMR are extremely valuable techniques for elucidation of the structure of an unknown iso-lated compound.

Atomic emission spectroscopy (AES) methods have also been shown to be effec-tive and accurate in determining the components of food spices and essential oils from Zingiberaceae and other families. A new fluorimetric assay at $481-486\,\mu m$ for the determination of pungency of fresh ginger from China (gingerols) was found to be better than the standard TLC/densitometric method (Variyar et al., 2000).

Oleoresins: Gingerols, Shogaols, and Related Compounds

Oleoresins contain the nonvolatile pungent principles of ginger in addition to some essential oils and other nonvolatile compounds such as carbohydrates and fatty acids (Connell, 1970). Several reviews have been devoted to the chemistry and proper-ties of gingerols and shogaols as pungent compounds of ginger rhizomes that are considered responsible for its medicinal properties (Nakatani, 1995). Oleoresins are obtained by solvent extraction (acetone, ethanol, dichloromethane, dichloroethane, and trichloroethane). Yields range from 3% to 11% and sometimes can reach 20%. However, they are greatly dependent on the solvent extraction conditions, the state of rhizomes (fresh or dried), the country of origin, the various areas within the same country, and harvest season.

The major pungent component of ginger oleoresin is (6)-gingerol, 1b (1-(4-hydroxy-3-methoxyphenyl)-5 hydroxydecane-3-ol), first identified by Lapworth (1917) and by Connell and Sutherland (1969). These investigators established the S-configuration for the hydroxyl group. The name (6)-gingerol was derived from the fact that alkaline hydrolysis of gingerol afforded n-hexanal, a six-carbon aldehyde. Besides the presence of the well-known (6)-, (8)-, and (10)-gingerols, Masada et al. (1973) pointed out the presence of lower homologs, that is, (3)-, (4)-, (5)-gingerols, and possibly also (12)-gingerol. Gingerdiol acetates and methyl ginger diacetates also occur in Japanese extracts.

The oleoresin and gingerol contents in nine popular cultivars of ginger from India, Brazil, and Jamaica were evaluated by Gopalam and Ratnambal (1989). The gingerol content was determined according to the Indian Standards Institute (ISI) specifications and methods. The minimum gingerol content in the oleoresin is 18% by weight as per ISI (1975) requirements.

Cultivars can be divided into three groups: (i) those that contain a minimum amount of gingerol (i.e., 17.7–19.25% corresponding to cultivars Waynad local, Narasapattam, and Maran); (ii) those that contain medium gingerol levels (i.e., 20.09–21.32% including Nadia, Karakkal, and Acc. No. 646 cultivars); and (iii) those containing the highest gingerol content (i.e., 24.66–26.67%, corresponding to Ernad, Chernad, Rio de Janeiro, and Jamaica).

The mean of oleoresin content is 6.74%, the minimum and maximum being 5.30% for Waynad local (Kerala State) and 8.59% for Acc. No. 646 (Himachal Pradesh), respectively. The amount of (6)-gingerols in fresh rhizomes of ginger cultivated in China and Japan was approximately 0.3–0.5%. The two other homologs, (8)- and (10)-gingerols (1c and 1d), were 5–20%, relative to (6)-gingerol, respectively. Dehydration of (6)-gingerol leads to (6)-shogaol, and its oxido-reduction to (6)-gingerdione 1e and (6)-gingerdiol 1f, which upon acetylation affords (6)-gingerdiacetate 1g. (6)-Methyl gingerols 2b with a methoxy group instead of a hydroxyl group at C-4 and (6)-dimethoxy gingerol 2a lacking a methoxyl group at C-3 of the benzene ring were also reported in the ginger rhizomes (Harvey, 1981). Their structures have been established by the usual spectroscopic methods.

During prolonged storage and according to the extraction process, large amounts of shogaols, and to a lesser extent of zingiberone, were formed (Connell, 1971). They affect the quality of the pungent flavor of ginger causing volatile off-flavors.

Besides zingerone (1a), pungent compounds of extracts are gingerols (1, 2), shogaols (3 and 4), which arise from dehydration of the corresponding gingerols, and related compounds, such as paradol (1i), gingerdiones (1e), gingerdiols (1f), and gingerdiol acetates (1g).

Zingerone, (6)-gingerol, and (6)-shogaol displayed moderate antioxidant activity (Nakatani and Kikuzaki, 2002). Activity diminished with the increasing chain length. Diarylheptanoids (or curcuminoids), another related category of products, have also been reported.

They are characterized by a heptane chain with two terminal 1,7-diphenyl rings substituted by a hydroxyl and methoxyl substituents in the *para* and *meta* positions, respectively. The 3,5-positions of the alkyl chain contain either a β-ketohydroxy (5), a dihydroxy (6), and the corresponding diacetate (7) groups or the dehydroxylated compounds from hexahydrocurcumin (5a) giving rise to 4-one-3 oil called gingerenone (8a) and its derivatives (8b) and (8c). The level of hydroxydrocurcumin is less than 0.005%. These compounds were isolated from ginger rhizome between 1969 and 1996 (Kikuzaki, 2000). Gingerenone A (8a) exhibits a moderate anticoccidial activity and a strong antifungal effect against *Pyricularia orysae*, a plant pathogen (Endo et al., 1990).

Related diarylhaptanoids isolated from turmeric (*C. domestici* L., synonym *C. longa* L.) which also belongs to the Zingiberaceae family, exhibit antioxidant activity as well. Connell and Sutherland (1969), using an acetone extract of ginger rhizome from Australia, found the following ratio for (6)-, (8)-, and (10)-gingerols: 56:13:31, respectively. This accounts for 1:3 of the extract. Analysis of the natural pungent compounds of ginger from oleoresin was also carried out by Lewis et al. (1972), Connell and McLachlan (1972), and Haq et al. (1986). Gingerol and shogaol levels

Table 16.10 Percentages of Oleoresins and Brazilian Essential Oils
Extracted with Different Solvents

Solvents	Oleoresin (%)	Essential Oils (%)
Ethanol	6.91–10.9	3.92–12.64
Acetone	2.53–5.62	8.05–18.89
Methylene chloride	3.35–3.91	13.67–27.72

Source: Taveira Magalhaes.

in a Polish extract were 31.1% and 20% against 14.8% and 11.2%, respectively, when compared with foreign extracts (Kostrzewa and Karwowska, 1976).

Pungent constituents of ginger, separated as TMS derivatives, were reported by Harvey (1981). Five gingerols ($n=2$, 4, 6, 8, and 10), two methylshogaols ($n=4$, 8), paradol (Li), and three gingerdiones were identified. Also found in the low-temperature region of the chromatogram were ar-curcumene, several sesquiterpne hydrocarbons and their hydroxylated derivatives, zingerone and fatty acids (palmitic, oleic, and stearic). Odoragram and aromagram of gingerol have been reported by Ney (1990a) using the sniffing method. As pepper, chilies, or mustard, they are characterized by a pungent taste. The content of oleoresins extracted with three different solvents has been studied from various areas of Brazil. Results for the oleoresins are given in Table 16.10. The highest percentage of oleoresins was observed with ethanol as the solvent, whereas essential oils are better extracted with methylene chloride.

Geographical variation of the content of gingerols was studied for different areas of India: Gujarat, Madhya Pradesh, Maharashtra, Kerala, and West Bengal (Sane et al., 1998). Analyses were carried out by HPLC using a Camag Limonat IV apparatus, benzene/methanol (10:1 v/v) as the eluent, 310 µm scanning wavelength, and analysis of variance (ANOVA) calculations in order to determine whether the variation was significant or not. From these results, it can be concluded that there is no significant variation in the gingerol content in ginger grown in different states.

Comparison between the pungent compounds in ginger powder, ginger skin, and baked ginger was carried out by Huang et al. (1999a). The content of gingerol in the three samples was 1.02, 0.28, and 0.25 g/100 g, respectively. Baked ginger contained 0.45 g/kg shogaols, but there were only traces in the other two samples. The gingerols and shogaols were stable, but there were only traces in the other two samples. The stability of gingerols and shogaols in aqueous solution was studied by Bhattarai et al. (2001). Because gingerols are biologically active compounds, they may make a significant contribution toward medicinal applications of ginger and other products derived from ginger. They are thermally labile due to the presence of a β-hydroxyl-keto group in the structure and undergo dehydration readily to form the corresponding shogaols are previously shown. Their stability was studied in the temperature range of 33–100°C in aqueous solutions at pH 1.4–7.0.

Modern techniques of extraction and qualitative and quantitative analyses (HPLC, GC/MS as TMS derivatives) provide a powerful identification tool. Their stability according to the medium is an important problem, not only for flavor quality but also for medicinal uses.

Synthesis and Biosynthesis of Pungent Compounds of Ginger Rhizomes

Several syntheses of gingerone, shogaols, and gingerols have been identified and described in the literature (Connell, 1970). An example is the synthesis of gingerone and (6)-shogaol from 3-(4 hydroxy-3-methoxy)-2-proenoic acid. The name shogaol is derived from the word "shoga" in Japanese, meaning ginger in that language.

Zingiberone is obtained as colorless needles (m.p. = 40–41°C) with a salicy-laldehyde odor (Molyneux, 1971). The first report on the synthesis of (6)-shogaol found in Japanese ginger was by Nomura and Tsurami (1926) by condensation of gingerone and hexanal. The structure of (6)-shogaol was determined by UV, IR, NMR, and MS data (Connell, 1970). The same investigator reported the structure of (6)-gingeryl methyl ether using the same spectroscopic methods. The S-*trans* conformation in the liquid state at room temperature was established.

Biosynthesis of gingerols (18) from dihydroferulic acid (13) has been reported by Harvey (1981). It involves the condensation of dehydroferulic acid (14), first with melavonic acid and then with a short-chain carboxylic acid such as hexanoic acid to give the intermediate gingerdiones (19). All the compounds have been identified in ginger oleoresin by Connell (1971), Connell and Sutherland (1969), and others.

1-Dehydrogingerdione (21) was isolated and identified as well (Charles et al., 2000). The structures of five-membered ring diarylheptanoids (22, 23) isolated from purified dichloromethane extracts of ginger were elucidated by spectroscopic (^{1}H-NMR, ^{13}C-NMR, IR, and MS) and chemical methods (Kikuzaki and Nakatami, 1996).

Essential Oils of Ginger

Essential oils in ginger are obtained by hydrodistillation or steam distillation, by extraction with supercritical CO_2, or by solvent extraction of dried rhizomes. Yields and chemical composition vary greatly according to the region in which the crop is cultivated, both within a country and between countries, the cultivar/variety cultivated, the agronomic practices under which the crop grew, and the experimental and analytical conditions. Ginger oil is a light yellow to yellow liquid with a characteristic lemon color. India is the major producer of ginger oil followed by China, Japan, Jamaica, Australia, and Africa.

Physicochemical Properties

The physicochemical properties of ginger are given in Tables 16.11 and 16.12. Badalyan et al. (1998) investigated the effect of a solvent-free ratio (SF) on the

Table 16.11 Physicochemical Properties of Ginger Essential Oils from India, ISI[a] and EOA[b]

Characteristics	India (Kerala)	Cochin[c]	ISI[a]	EOA[b]
Specific gravity[d]	0.8690	0.8718	0.868–0.880	0.87–0.88
X=(°C)	27°	28°	30°	25°
Optical rotation	−54°	−40.1°	−28 to −45°	−28 to −45°
X=(0°C)	30	28	20	20
Refractive index	1.4891	1.4872	1.4840–1.4894	1.4880–1.4940
X=(°C)	28°	18°	30°	20°
Saponification value	7.4	–	20 max.	20 max.

[a]ISI (1975).
[b]Essential Oil Association (USA); also reported by Masada (1976).
[c]Reported by Akhila and Tewari (1984).
[d]Mathew et al. (1973) gave the following values at 30°C: 0.8718, 1.4803, and −0.4°C.

Table 16.12 Characteristic Constants of Two Ginger Varieties from Brazil

Characteristics	Gigante	Calpira
D^{20}_{20}	0.8815–0.8873	0.8826–0.8860
$\alpha^{25} D$	−15.3° to −6.8°	−22.0° to −32.0°
$N^{20} D$	1.4874–1.4897	1.4895–1.4908
Yields [a]	1.2–2.5	2.5–2.8

[a]From dried rhizomes (Taveira et al., 1997a).

refractive index (n_D) of an extract of Australian-grown air-dried ginger at 10 MO, from 9°C to 35°C. The n_D value increases with SF ratio, with most of the essential oil components being recovered when the SF ratio was less than 2. The SF ratio was found to be the major factor affecting the composition of the ginger extract. The temperature had little effect on the composition within the range of 9–35°C. Extraction of the sample with supercritical CO_2 revealed that solubility was the dominant limiting factor in the extraction procedure. Introducing small amounts of ethanol as a cosolvent increased the overall yield from liquid CO_2 and the recovery of ginger oleoresin.

Chemical Composition

Used in perfumery and cosmetics, the essential oils of ginger have been widely investigated in producer countries. These investigations have been reviewed by several investigators (Afzal et al., 2001; Connell, 1970; Guenther, 1952). Before 1970, (Connell, 1970) and the use of packed columns, the number of identified compounds remained low. The main components were the sesquiterpene hydrocarbons, (−)-α-zingiberene, (+)-ar-curcumene, β-bisabolene, β-sesquiphellandrene, farnesene, γ-selinene, β-elemene, and β-zingiberene. Other compounds identified at this time were monoterpene, hydrocarbons: α-pinene, β-pinene, myrcene, β-phellandrene, limonene, p-cymene, cumene, and oxygenated compounds: 1,8-cineole, d-borneol, linalool, neral, and geranial, bornyl acetate, in addition to some aliphatic aldehydes (nonanal, decanal), ketones (methylheptenone), alcohols (2-heptanol, 2-nonanol), esters of acetic and caprylic acid, and chavicol. After 1967, the development of capillary GC and its coupling with MS and data banks has provided a suitable technique for a good separation and identification of a greater number of compounds.

Essential Oils from India

In a ginger oil from India, Nigam and Levi (1963) found the following: α-pinene, myrcene, p-cymene, 2-heptanol, 2-nonanol, bornyl acetate, and other compounds. According to these investigators, these data are important to establish the authenticity of the essential oil. A year later, Nigam et al. (1964) identified several sesquiterpene hydrocarbons (68.5%), (−)-α-zingiberene (38.6%), β-zingiberene, (+)-ar-curcumene (17.7%), α-farnesene (9.8%), γ-selinene (1.4%), β-elemene (1.0%), β-bisabolene in mixture with α-zingiberene, β-sesquiphellandrene, and sesquiterpene alcohols (e.g., zingiberenol, elemol), accounting for 16.7% of the oil.

Natarajan et al. (1972) studied 26 varieties of essential oil from Kerala State, India. Mitra (1975) has described the physical and chemical properties of essential oils. The major components of the ginger oil are α-zingiberene and β-bisabolene, and those of the oily resin are zingirone, gingerols, and shogaols. Terhune et al. reported the presence of epizonarene, zonarene, and sesquithujene. Lawrence (1983) reviewed the investigations devoted to ginger oils and reported his own results on essential oil from India (Table 16.13). Of the 115 constituents isolated and identified, 43 were

Table 16.13 GC Analyses of Ginger Oils from India, Brazil, and Jamaica

Compounds	Percentage		
	India [a]	Brazil (var. Rio de Janeiro)	Jamaica
α-Pinene	2.9	2.7	2.4
Camphene	2.0	1.5	1.4
β-Pinene	1.1	0.07	1.3
Aldehyde C9	1.9	4.4	0.1
1,8-Cineole	3.9	5.1	1.0
p-Cymene	4.0	6.9	10.8
Linalool	3.4	3.1	4.8
1-Nonanol	5.2	2.1	2.8
α-Zingiberene	31.4	21.8	27.7
ar-Curcumene	8.1	0.26	10.9
β-Sesquiphellandrene	4.6	2.8	2.2
Nerolidol	4.7	Traces	6.8
Zerumbone	5.1	8.8	3.4

Source: Adapted from Gopalam and Ratnambal (1989).
[a]Mean of values (rounded to the first decimal) of seven regions of India. Six compounds remain unidentified.

previously identified, whereas 72 had been identified for the first time as components of the ginger oil.

In a review on the chemistry of ginger, Akhila and Tewari (1984) reported about 30 volatile constituents. Among sesquiterpene hydrocarbons, the most important are: α-zingiberene, ar-curcumene, and β-farnesene, previously reported by Nigam et al. (1964). The percentages of neral, geranial, linalool, and borneol were 0.8%, 1.4%, 1.3%, and 2.2%, respectively. Narayanan and Mathew (1985) investigated 15 cultivated varieties of ginger, which contain 1.4–2.6% oil. Major constituents were: α-zingiberene (16.6–28.7%), neral (6.6–15.1%), and geraniol (5.8–11.5%). Zerumbone (traces—5.6%) and limonene (0.3–1.7%) were also found in most varieties but were absent in three others. Alcohols and aldehydes account for 7.5–12.9% and 12.9–26.6%, respectively.

Erler et al. (1988) compared the composition of ginger oils from India and Australia. They found noticeable differences in their terpenoid components. The main constituents were the usual sesquiterpene hydrocarbons. However, the essential oils from the Australian ginger consisted mainly of monoterpene hydrocarbons (camphene, β-phellandrene) and the oxygenated derivatives, such as neral, geranial, and 1,8-cineole.

There is a great difference in the percentages of the 13 identified compounds in the samples grown in the 7 regions of India. The amounts of α- and β-pinenes were higher in cultivars Karakkal and Ernad (4.2%) and Chernad (4.5%). Typically, the lower the contents of essential oils, the better the quality. The 1,8-cineole content was higher in cultivar Nadia (13.3%) followed by cultivars Waynad local (8.6%). α-Zingiberene and β-sesquiphellandrene are the most important compounds of freshly distilled oil. They can be converted into ar-curcumene during long storage. Its percentage varied from as low as 0.1% (Waynad local) to 32.9% in Narasapattam. The ratio of β-sesquiphellandrene and α-zingiberene is indicative of the age of the oil. The percentage of zerumbone, a well-known compound in wild ginger, is the highest in Waynad local (19.8%), followed by Brazil (Rio de Janeiro, 8.8%), Ernad and Chernad (8%), and Acc. No. 656 (Himachal Pradesh, India, 7.4%). Only traces were observed in cultivar Narasapattam. Indian ginger oils obtained by hydrodistillation of coarsely ground ginger rhizomes are valued for their flavor and perfume.

Essential Oils from China

A review of the raw and dry ginger produced in Szechaun shows that the essential oil contains α-zingiberene, ar-curcumene, α-farnesene, β-farnesene, linalool, gingerol, β-sesquiphellandrene, zingerone, dehydrogingerol, hexahydrocurcumin, and other compounds (Chen and Guo, 1980). Dry ginger contained 1–2% essential oil as compared to 0.2–0.4% in raw ginger found in Peikin. TLC indicated that most of the constituents were identical. The citral content for Chinese oil was very low (0.22%) (Lin and Hua, 1987) compared to that found in Japanese oils or in a Fijian oil (64%). Chen and Ho (1988) compared the chemical composition of ginger oils from Taiwan obtained by steam distillation and supercritical CO_2 extraction. These results are given in Table 16.14.

Table 16.14 Comparison of the Hydrocarbon Fraction of Ginger Oils from Taiwan Obtained by Steam Distillation and Supercritical CO_2 Extraction

Compound	Steam Distillation	Liquid CO_2 Extraction
Monoterpene Hydrocarbons		
α-Pinene	0.16	0.98
Camphene	0.40	3.08
β-Pinene	0.02	0.16
Myrcene	0.08	0.90
β-Phellandrene	0.09	0.89
Limonene	0.31	2.75
r-Terpinolene	0.11	0.11
Total	1.17	8.87
Sesquiterpne Hydrocarbons		
β-Elemene	0.29	0.31
α-Zingiberene	14.19	24.15
β-Bisabolene	6.37	5.60
r-Bisabolene	3.47	8.40
ar-Curcumene	16.30	7.96
Total	51.24	56.56

Source: Chen and Ho (1988).

Essential Oils from Other Countries in Asia (Vietnam, Korea, Japan, Indonesia, Fiji, Philippines, and Malaysia)

Vietnamese ginger oil obtained by steam distillation of dried rhizomes was studied using GC, GC/MS, and ^{13}C-NMR (Van Beek et al., 1987). The oil included 28% monoterpene hydrocarbons, 37% oxygenaeted monoterpenes, 25% sesquiterpene hydrocarbons, 8% oxygenated sesquiterpenes, and 2% nonterpenoid compounds. Geranial (16%), neral (8%) gave a lemon-like character to the oil. Among other identified compounds were: furfural, 2,5-dimethylhept-5-enal, dihydroperillene, p-cymene-8-ol, allo-aromadendrene, y-muurolene, lauric acid, methyl isoeugenol, y-eudesmol, farnesol, and xanthorrhizol. Decontamination of the dried ginger by gamma irradiation did not affect the yield and composition of the oil, which was similar to that of fresh Sri Lankan ginger.

Ginger from Korea was extracted with liquid CO_2 by Kim et al. (1991) and the extract was fractionated into two fractions using a silica gel column. The major compounds in ginger oil were: α-zingiberene, citronellol+, β-sesquiphellandrene, geraniol, y-bisabolene, and ar-curcumene + geranyl acetate. Ginger oil contained 68.1% sesquiterpene hydrocarbons and 31.9% of oxygenated hydrocarbons (1,8-cineole, neral, geranial, gernyl acetate, citronellol, geraniol, α-terpineol+, borneol). The volatile oil content was 0.33%.

Seven varieties of small ginger (Type Kintoki and Yanaka), medium ginger (Type Sansyu) and large ginger (Type Otafuku, Tosaichi, Cambo, and Jumbo) were studied. The comparison of three extraction methods (solvent extraction with acetone, steam distillation, and freeze-dried under vacuum) showed the first method to be the best for Kintoki sample. Prolonged exposure to heat may cause decomposition of some components. The different types contained α-pinene, camphene, myrcene, β-phellandrene, 1,8-cineole, borneol, neral, geranial, gernyl acetate, ar-curcumene, α-zingiberene, β-bisabolene, and β-sesquiphellandrene.

Van Beek (1991) reported gas chromatograms of an Indonesian ginger oil with a high citral content on DB-1 and DB-Wax capillary columns. The major compounds were: the usual sesquiterpene hydrocarbons, geranial, camphene, neral, α-pinene, myrcene, limonene, β-phellandrene, geraniol, and 1,8-cineole.

Essential oil of ginger from Fiji was studied and compared with essential oil of ginger from other countries like, India, China, Japan, and Australia. The yield of the oil from Fiji obtained by steam distillation ranged from 0.1% to 0.2%. Analyses were carried out on packed columns. Among the 25 identified compounds, a high content of neral ($15 \pm 5\%$), geranial ($27 \pm 9\%$), and 1,8-cineole ($8 \pm 2\%$) was found when compared with the contents of ginger oil from other countries. The oil from Fiji was found to be similar to those of the Japanese samples from Taneshoga and Oyashoga, which had a low geraniol–gernyl acetate content, and 17–20% neral, 23–35% geranial, respectively. Nevertheless, other Japanese oils differed significantly depending on the location. A low citral (neral + geranial) content has been found in oils from India (<3%) and Africa. It is higher in oil from Australia and Japan and highest in that from Fiji.

By hydrodistillation of the fresh rhizomes of ginger from Philippines, Anzaldo et al. (1986) obtained 0.2–1.0% oil yield. By using TLC, GC, and IR spectroscopic data, 10 components were identified, with citral being the major component. Geraniol and linalool were also present. Physicochemical constants of the oil were also reported.

Some important metabolites from Malaysian ginger have been reported by Hasnah and Ahmad (1993). Among several categories of identified compounds were: sesquiterpenes (germacrene, humulene, zerumbone, zerumbodienone, and humulene epoxide), diterpenes including coronarin C, coronarin E, isocoronarin D, the isomer of isocoronarin D and labd-8(17), 12-dien-15, 16-dial, phenolics, and piperene. Some of these compounds seemed to be potent inasmuch as certain biological activities are concerned.

Essential Oils in Ginger Grown in Africa (Nigeria)

Nigerian ginger essential oils were analyzed by Ekundayo et al. (1988). Samples of fresh rhizomes were purchased from a local market. Fresh and dried rhizomes after homogenization, were hydrodistilled (yields were 1.02% and 1.84%, respectively in fresh and dried ginger). Results showed a usual percentage of α-zingiberene in essential oil obtained from dried rhizomes (28%), and a high content of neral and geranial as in Australian and Japanese oils. These percentages are higher in oil obtained from fresh rhizomes. The rhizomes of ginger were also examined for their oil and moisture contents by Dambatta et al. (1988). Some characteristics of the oil, such as acid value, the saponification index, and the iodine value, showed slight variations from the values reported for ginger oils obtained in other countries. GC/MS analysis of the oil revealed the presence of many components including camphene, myrcene, α-phellandrene, α-copaene, α-farnesene, β-caryophyllene, α-zingiberene, and germacrene, all being well-known components in ginger oils. The following year, Onyenekwe and Hashimoto (1999) studied the composition of the Nigerian essential oil obtained from dried rhizomes. The hydrodistilled oil (2.4%) consisted of 64.4% carbonyl compounds, 5.6% alcohols, 2.4% monoterpene hydrocarbons, and 1.6% esters. The main components were α-zingberene (29.5%) and sesquiphellandrene (18.4%). Besides the usual compounds, they also identified the following: 2,6-dimethylhepten-1-ol, α-gurjunene, linalool oxide, isovaleraldehyde, 2-pentanone, α-cadinol, α- and γ-calacorenes, eremophyllene, γ-muurolol, α-himachalene, α-cubebene, acetic acid, pinanol, α-santalene, gernyl propionate, geranic acid, (E,E)-α-farnesene, and N-methylpyrrole. The identification of this latter heterocyclic compound is of interest.

Essential Oils in Ginger Grown in Other Countries (Australia, Brazil, Poland, Mauritius Island, and Tahiti)

GC analysis of ginger oil from Australia obtained by hydrodistillation or steam distillation of dried rhizomes indicated that it was similar to that from other regions (Connell, 1970). The major constituents are usual sesquiterpene hydrocarbons with a bisabolene skeleton—α-zingiberene (20–30%), ar-curcumene (6–19%), and β-sesquiphellandrene (7–12%). These compounds are accompanied by usual monoterpene hydrocarbons and oxygenated derivatives. Of particular interest is the presence of a relatively high proportion of neral and geranial responsible for the distinctive citrus-like aroma of the oil. GC chromatograms on a column coated with Apiezon M at 130°C and 147°C were reported as well as the chromatogram of an African oil. Zingiberone and zingiberenol were also identified.

Taveira et al. (1997a) studied the essential oils and oleoresins of ginger from Brazil. Two kinds of ginger are characterized by the size of their rhizomes; *Gigante* commercial crop for export and *Calpira* for household consumption and of a Japanese origin. Steam distillation of dried rhizomes gave essential oils that were characterized by their physicochemical constants and GC analysis. Comparative analysis showed differences between *Calpira* and *Gigante* gingers depending on the distillation method and the type of rhizomes (fresh or dried). The strong lemony scent due to its neral and geranial contents is more pronounced in the case of oil from *Gigante* than in the case of oil from *Calpira*. Agronomical and market aspects of ginger cultivation of the two varieties were also reported. The same investigators reported the chemical composition of the essential oils of these two kinds of ginger. Most of the important components are given in Table 16.15. Yields of essential oils obtained from dried rhizomes were 1.2–2.5% for *Gigante* and 2.5–2.8% in *Calpira*.

Essential Oils from Wild Ginger (Zingiber zerumbet Smith)

Wild ginger is a perennial herb that grows in subtropical climate such as that in India, Southeast countries, South Pacific islands, and Okinawa. It has been used

Table 16.15 Gas Chromatography Analyses of *Gigante* and *Calpira* Ginger Oils from Brazil

Compound	Percentage of Fresh Ginger	
	Gigante	Calpira
α-Pinene (+ camphene)	5.3–9.3	7.4–9.2
β-Phellandrene (+ 1,8-cineole)	6.4–11.2	6.2–9.3
Neral + Geranial	14.3–20.7	6.2–7.0
(+) ar-Curcumene	4.0–6.0	7.0–8.4
(−) α-Zingiberene	18.4–27.8	25.3–32.6
β-Bisabolene (+ *E,E*-α-farnesene)	8.4–13.6	8.4–12.4
β-Sesquiphellandrene	7.0–9.7	9.2–12.2
Total SQHC	40.4–57.1	51.3–64.2

Source: Taveira et al. (1997a).
GC analyses were carried out on packed columns (3 m×1 m/8 in. i.d.) coated with 5% SE on Chromosorb W—AW DMCS 80–100 mesh.

locally for gardening and also in the production of folk (native) medicines. Since the publication of the first report on the essential oil of Indian wild ginger by Nigam and Levi (1963), essential oils have been widely studied for their medicinal properties due to their high content of zerumbone. The oil originating from Fiji, said to have medicinal properties, differs from the cultivated variety in India in its zerumbone content, which is much higher (57.7%) as compared to 37.5% which is the normal content (Duve, 1980). Steam distillation of the local variety yielded 0.3% of the clear oil, which compared favorably with that of the cultivated variety overseas.

Chemical investigation of the aerial parts (stems, leaves, and flowers) of wild ginger (*Z. zerumbet*) of Vietnamese origin was subjected to GC and GC/MS analyses (Dung et al., 1995). About 40 compounds have been identified in the stem and leaf oils, accounting for more than 83±1% of the oils. The major compounds appeared to be (Z)-nerolidol (16.8–22.3%), β-caryophyllene (10.4% and 11.2%), zerumbone (21.3% and 2.4%), and *trans*-phytol (7.0% and 12.6%), respectively. Predominant minor constituents included β-pinene (5.4% and 5.2%), α-humulene (2.5% and 2.9%), caryophyllene oxide (1.1% and 5.5%), and linalool (1.1% and 2.4%). The volatile flower oil was found to contain more than 60 compounds with 45 compounds making up to 85% of the oil. (Z)-nerolidol (36.3%) and β-caryophyllene (13.2%) were the major constituents.

GC/MS analysis of the essential oil obtained from dried rhizomes of wild ginger from India resulted in the identification of 36 compounds. Curzerenone (14.4%), zerumbone (12.6%), camphor (12.8%), isoborneol (8.9%), and 1,8-cineole (7.1%) were found as the major components. GC, GC/MS, IR, and ^{1}H-NMR analyses showed that the main constituents found in the hydrodistilled oil from *Zingibere spectabile*. Valert were terpin-4-ol (23.7%), labda-8 (17), 12-diene-15, 16-dial (24.3%), α-terpineol (13.1%), and β-pinene (10.3%).

Zerumbone isolated from wild ginger has been claimed to be a potent inhibitor of 12-D-O-tetradecanoylphorbol-13 acetate, which induces Epstein–Barr virus activation (Murakami et al., 1999).

Diterpenoids

Besides the usual mono- and sesquiterpene derivatives, several diterpenoids (24–26), have been identified in ginger rhizomes (Lee et al., 1982). The galanolactone (27), a furan diterpene, has been isolated from the Japanese ginger "Kintoki" by Kano et al. (1990) but was not found in ginger from China and India.

Characteristic Flavor and Odor in Ginger

Odor quality of the oils for several imported and Australian commercial varieties of ginger have been compared by Mathew et al. (1973). Hot air drying was the best method for raw ginger products. The identification and evaluation of the flavor-significant components of ginger essential oil was made by Bednarczyk (1974). They employed step-wise multiple regression analysis, taking into account individual peak intensities and taste panel scores for ginger flavor intensity as a dependent variable. Taste panel evaluation of the isolated components indicated that β-sesquiphellandrene and ar-curcumene are the prime contributors to the characteristic ginger attribute. α-Terpineol, neral, and geranial contribute to the lemony aroma of ginger oil and may therefore be desirable additives to whole ginger oil to intensify its lemony or citrus character. Nerolidol contributes to the woody or soapy flavor and does not appear to be a good potential additive to ginger oil. A trained sensory panel judged a mixture of α-terpineol, neral, geranial, β-sesquiphellandrene, ar-curcumene, and nerolidol to be characteristic of ginger oil.

Kostrzewa and Karwowska (1976) investigated the characteristic flavor and odor of a Polish extract of ginger by comparing the yield of the essential oil and it levels of α-zingiberene and zingberol as well as total gingerols and shogaols with two foreign extracts.

Purseglove et al. (1981) and Heath and Reineccius (1988) reported the impact compounds of ginger. Organoleptic properties of ginger oils from various origins have been described by Akhila and Tewari (1984).

Aromagrams and key compounds of various spices including ginger have been outlined and relations between their chemical structure and flavor have been described by Ney (1990a). Chemically speaking, ginger contains 2-methoxyphenols (or 4-substituted guaiacols). From the study of Nishimura (2001), it can be concluded that flavor dilution factors (FD), according to Grosh (1994), compounds identified in fresh ginger are the highest for 1,8-cineole (15), linalool (20), citronellyl acetate (16), borneol (16), geranial (17), and geraniol (20), which are the most important contributors to the odor of ginger. Enantiomer separation of the characteristic odor compounds in Japanese fresh rhizomes of ginger was described in an excellent investigation of Nishimura (2001). Using MDGC and chiral analytical HRGC/olfactometry (or sniffing), the odor character of each enantiomer was confirmed.

The fresh rhizomes of ginger were fractionated into a hydrocarbon fraction (hexane extract) and an oxygenated hydrocarbon fraction (methylene chloride extract).

Table 16.16 Composition and the Odor Description of Enantiomers of Four Potent Odor Components of Japanese Fresh Ginger

Compound	Configuration	Percentage Content	Odor
Linalool	R-($-$)	66	Floral
	S-($+$)	34	Black tea-like
			Weaker floral note
4-Terpineol	R-($+$)	71	Musty
	S-($-$)	29	Musty, dusty
Isoborneol	($1R$, $2R$, $4R$)-($-$)	100	Camphoraceous, India ink-like
	($1S$, $2S$, $4S$)-($+$)		
Borneol	($1S$, $2R$, $4S$)-($-$)	8	Camphoraceous, India ink-like
	($1R$, $2S$, $4R$)-($+$)	92	Camphoraceous, India ink-like, fatty putrid

Source: Nishimura (2001).

This latter fraction was subjected to chiral GC/olfactometry (GC/O) and GC/MS analyses. It was considered that monoterpinoids such as linalool, 4-terpineol, isoborneol, and borneol as well as geranial and neral might contribute to the characteristic odor of the fresh ginger. Each enantiomer was easily separated using the off-line MDGC system. The odor character of each enantiomer was confirmed by GC/O. The results are reported in Table 16.16.

Chromometrics

Fingerprint analysis of ginger oils by chromometrics methods using individual peak quantities as independent variables and their flavor intensities as dependent variables was performed by Bednarczyk and Kramer (1975). α-Terpineol, neral, and geranial, β-sesquiphellandrene, ar-curcumene, nerolidol, and *cis*-β-sesquiphelandrol accounted for 85% of the panel's flavor response. Identification of each isolated peak was carried out by IR, ¹H-NMR, and MS data and comparison with authentic samples. Unfortunately, α-zingiberene was not included in this study. Organoleptic evaluation included:

1. Tasting the mixture of compounds in each peak, individually
2. Tasting combinations of the mixtures
3. Tasting the compounds in each selected peak in combination with ginger oil
4. Tasting combination of compounds in selected peaks with ginger heat chemicals added.

The results can be summarized as follows:

1. Neral, geranial, and α-terpineol are responsible for the characteristic lemony flavor and odor of ginger. β-Sesquiphellandrene and ar-curcumene partly contribute to the characteristic ginger flavor intrinsic to the spice. Owing to their high threshold values, they have more of a tendency to dilute the flavor intensity of the oil rather than to increase it when they are added to the oil.

2. Nerolidol possesses woody, soapy, or green notes, not very reminiscent of ginger. *Cis-* and *trans*-β-sesquiphellandrol presumably contribute to the significant flavor of ginger as well as other contributors selected by the regression analysis. Bednarczyk and Kramer (1975) concluded that this step-wise regression analysis makes it possible to preselect flavorfully significant compounds in a flavor essence.

More recently, Chau et al. (2001) described a fingerprint analysis by GC/MS of dried and fresh ginger rhizomes from China using chemometric techniques. Unfortunately, essential oil classification by chemometrics according to its origin and flavoring properties seems to be missing. For such studies, the percentages of the following compounds have to be taken into account: limonene, camphene, 1,8-cineole, neral, geranial, nerol, geraniol, α-terpineol, gernyl acetate, α-zingiberene, β-sesquiphellandrene, β-bisabolene, (*E,E*)-α-farnesene, ar-curcumene, nerolidol, zigiberenoids, and sesquiphellandrols.

Synthesis of Some Authentic Samples

Synthesis of terpenoids and sesquiterpenoids in flavors and fragrances has been widely investigated since the end of the nineteenth century. Synthetic procedures have been reviewed in several books, particularly in Teisseire's excellent work. In the following section, only zingiberene isomers and its derivatives, as well as other important constituents of ginger oils will be briefly described. Spectral data (MS, ^{1}H-NMR, ^{13}C-NMR, and KI) on a nonpolar column sesquiterpne hydrocarbons have been reported by Joulain and Konig (1998). α-Zingiberene was first isolated in a concentrated form by Von Sodem and Rojahn by a distillation procedure.

Physical properties of α-zingiberene were reported by Pliva et al. (1960) are as follows:

b.p.   =   134–135°C/15 mm Hg
d^{20} =   0.8713
n^{20} =   1.4937
$(\alpha)^{20}$ =   −119.6°C.

It was identified in other essential oils such as Santal Amgris and in *Zizyphus spinachristi* from Egypt by El-Hamouly and Mohamad (2001). (+)α-Zingiberene, an enantiomer of the above, has been synthesized for the first time from (−)menthol in nine steps via the crucial intermediate 7(R,S)-isopropyl-10(R)-methyl-1(S)-4-oxobicyclo (4.4.0) dec-5-ene (Bhonsle et al., 1994). 7-epi-Zingiberene, a diastereoisomer of the natural product, was isolated from wild tomato leaves (Breeden and Coates, 1995). An easy synthesis of ar-(or α)-curcumene was carried out by Birch (1965). By reacting the 6-methyl-5-hepten-2-one with 4-methylphenyl-magnesium bromide, the intermediate alcohol was obtained. Upon treatment with sodium–ammonia in ethanol, it gave the expected compound.

Synthesis of Zingiberenol and β-Sesquiphellandrols

The structure of zingiberenol extracted from higher boiling fraction of ginger oil and isolated by chromatography was tentatively established by Varma et al. (1962). NMR studies show a close similarity with β-eudesmol. In reality, the extracted compound was a mixture of β-eudesmol with 70% *trans*-ring and 30% *cis*-ring juncture. Some years later, the structure of zingiberenol was determined by its chemical analysis and spectroscopic data (IR, NMR, MS).

Zingiberenol and oxygenated bicyclo (3.3.1) nonanes were synthesized by Paquette and Kinney (1982) using the regiospecific y-alkylation of the sulfonylcyclohexanone ketal. The latter was prepared in 85% yield by Diels–Alder reaction of sulfonylphenylethylene with TMS derivative of the 1-methoxy-1,3-butanediene-3-ol followed by ketalization with ethylene glycol. This ketal was alkylated by RBr ($RH_2CCHCH_2^-$, $PhCH_2$) in the presence of sodium hydride in dimethyl formamide to give the corresponding alkyl cyclohexanone. Ketals were obtained in 77% and 81%, respectively.

Desulfonation and hydrolysis of the y-alkylated derivatives ($R = PhCH_2$) gave a mixture of α, β- and β,γ-cyclohexenones in 83% of a 68/32 mixture of 4-benzyl-α, β- and β, γ-cyclohexenones, respectively.

α-Zingiberenol and β-sesquiphellandrene were synthesized from 4-(6-methyl-5-hepten-2-yl)-2-cyclohexen-1-one by Flisak and Hall (1986).

The unsaturated ketone was prepared using a one-pot tandem arylation multistep reaction-hydrolysis sequence. *Cis*- and *trans*-β-Sesquiphellandrols have been isolated from ginger oil and their structures determined by IR, NMR, UV, and MS data by Bednarczyk and Kramer (1975). They are stereoisomers of 5-(1,5-dimethyl-4-hexenyl)-2-methylene-3-cyclo-hexanol.

Precursors of Aroma and Flavoring Compounds

Biosynthesis of terpenoids and sesquiterpenoids in essential oils has been widely studied and references are cited therein. Biosynthesis of the main sesquiterpene hydrocarbons with a bisabolene skeleton has been reported by Rani (1999) in an Indian ginger oil. Cyclization of farnesyl pyrophosphate isomers affords ar-curcumene, α-zingiberene, and β-bisabolyl cation.

The atomic ratios ($^{14}C/^3H$) in ar-curcumene, α-zingiberene, and β-bisabolene, which were fed with (2-^{14}C, 2-$^3H_2$), (2-^{14}C, 4R–$^3H_1$), and (2-^{14}C, 5-$^3H_2$) mevalonic acid and (1-$^3H_2$) farnesyl pyrophosphate (FPP) revealed:

1. (2*E*, 6*E*)-isomer of FPP is isomerized to (2*E*, 6*E*)-isomer without loss of epimeric hydrogen, that is, without a redox process
2. (2*Z*, 6*E*) FPP is cyclized to a bisabolyl cation, that is, the penultimate precursor of α-zingiberene, ar-curcumene, and β-bisabolene
3. Two, 1,2-hydrogen shifts take place during the formation of α-zingiberene, whereas one 1,2-hydrogen shift was observed during the formation of ar-curcumene.

Another interesting investigation is that of the isolation of geranyl disaccharides from ginger and their relation to aroma formation.

The precursors and enzyme activities involved in the formation of geraniol and related compounds in ginger were investigated. Repeated chromatography afforded the isolation of a glucoside and three kinds of disaccharides of geraniol from fresh rhizomes of ginger. Their structures were determined by spectroscopic analyses.

After incubating, each glycoside with a crude enzyme solution prepared from ginger powder released significant quantities of geraniol. These data suggest that the glycosides exist as precursors or intermediates of geraniol-related compounds in ginger aroma. The nomenclature of these glycosides and amounts of geraniol liberated are given below:

1. Geranyl 6-O-α-L-arabino pyranosyl-β-D-glucopyranoside: 76.6% geraniol
2. Geranyl 6-O-β-D-xylopyranosyl-β-D-glucopyranoside: 19.3% geraniol
3. Geranyl 6-O-β-D-apiopyranosyl-β-D-glucopyranoside: 76.2% geraniol (relative to β-D-pyranoside = 100).

Properties of Ginger

The compounds responsible for the biological activity of ginger extracts are gingerols, shogaols, (6)-paradol, (6)-gingerdiol, gingerone A, zingerone, diarylheptanoids, hexahydrocurcumin, and its derivatives (curcuminoids), and other compounds. It is interesting to note that the action of these compounds is quite different according to the state of the rhizome. The fresh rhizome called "Shoukyo" in China is considered to be antiemetic, cough and cold remedy, an antitoxic, and a digestive stimulant. The dried rhizome "Kankyo" in China is considered a good remedy for stomach ache. Roasted ginger led to a marked decrease in blood coagulation in mice. Ginger essential oils have interesting properties not only in perfumery but also in cosmetology as an antiaging agent. Tables 16.17 and 16.18 summarize this important information.

Following the great strides in biochemistry and pharmacology during the last almost five decades, the noteworthy bioactivity of ginger has been well authenticated scientifically.

Processing of Ginger

Deterpenation

Ginger deterpenation by liquid chromatography using silica gel column was achieved by Shankaracharya and Shankarnarayana. They claimed that the terpeneless or deterpenated essential oils are valued for their stability and enhanced flavor strength. Therefore, distilled and pepper oils were deterpenated by CC. They were slurried with petroleum ether (60–80°C) and eluted with acetone and ethyl acetate. The yield of terpeneless ginger was 16.6%.

Preservation and Encapsulation

Preservation of grated ginger is done in the presence of 0.1, 1.0 wt% ginger spicy oil at a pH of 3.4–4.5; ethanol and organic acids are added. The preserved grated ginger has its freshness, flavor, and taste well kept. Grated ginger (80 g), ginger oil (0.6 g), NaCl

Table 16.17 Biological and Physiological Activities of Ginger [a]

Property	Responsible Compound	Reference
Antioxidant Effects		
Inhibition of lipid peroxidation	Ginger extracts (zingirone)	Krishnakantha and Lokesh (1993)
Inhibition of fulvic acid-induced hydrogen peroxide production in chardocyte	Ginger oil	Guo et al. (1997)
Antioxidant effect in ground pork patties	Ginger extracts	
Protective effect against oxygen-free radical damage	A new cyclic diarylheptanoid[b]	He et al. (2001)
Antioxidant properties	Gingerols, shogaols	Wang (2001)
Antitumor Activities		
Antitumor activity	Ginger extracts	Masada et al. (1973)
	Malaysian ginger metabolites, vanillyl ketones (6)-gingerol, (6)-shogaol	Hasnah and Ahmad (1993)
Antitumor activity (antiulcer)	(6)-Ginger sulfonic acid	
	Ginger extracts *Amitra Bindu*	Ahn et al. (1993)
Chemoprevention of cancer (colonic mucosa)	Zerumbone (from wild ginger)	Duve (1980)
Chemoprevention of cancer	Curcumin and curcuminoids	Masuda et al. (1998)
Chemoprevention of skin cancer	Ginger components+ antioxidants	Ahmad et al. (2001)
Antiinflammatory Activities		
Antiinflammatory activity	Proteolytic enzymes in ginger Gingerols, shogaols, gingerenone A, (6)-gingerdiol, hexahydrocurcumin, zingerone	Thomson et al. (1974)
Antiulcer effects	Gingerols, shogaols	
Against arthritis and related disorders	Ginger extracts	
Antiinflammatory and antirheumatic activity	Ginger oil (+eugenol)	Sharma et al. (1997)
Inhibition of arachidonic acid, prostaglandins and leukotrienes	(6)-Shogaol, (6)-gingerol, (6)- and (8)-gingerdiols	Kiuchi et al. (1982)
Inhibition effect on the leukotriene biosynthesis	Ginger oil	Ma et al. (1990)
Effects on Blood and Heart		
Decreasing blood pressure	(6)-Shogaol and (6)-gingerol in large concentration	
Cardiovascular effect	Gingerols, shogaols	
Ca++ spike suppression in portal veins of mice	(±)(6)-Gingerols and (±) yakuchinone	

(Continued)

Table 16.17 (Continued)

Property	Responsible Compound	Reference
Ca ++ pumping activity in rabbit Skeletal and Dog Cardiac Muscles	Gingerols	Kobayashi et al. (1987)
Inhibition of platelet aggregation	Gingerols or powdered ginger in large dose	Verma et al. (1993), Kawashi et al. (1994)
Cholesterolemic and Hepatic Effects		
Cholesterolemic activity	Pungent principles of ginger	Gujral et al. (1978)
Reduction of hypercholesterolemia in mice	(E)-8 β, 17-epoxylabd-12-ene 15, 16-dial	
Antispasmodic effect (easier gastric mobility) and neuromuscular effect in mice	(6)-Shogaol and (6)-gingerol	
Antihepatotoxic activity Miscellaneous	Gingerols and diarylheptanoids	Hikino et al. (1985)
Cough suppressant	(6)-Shogaol (identical effect to that of dihydrocodein phosphate)	Kiuchi et al. (1982)
Serotonin-antagonistic effect (hypothermia and diarrhea inhibition)	Galanolactone	Huang et al. (1999b)
Antileukemic	Ginger extracts	Hasnah and Ahmad (1993)
Antiallergic	Gingerols and shogaols	Yamahara et al. (1995)
Bactericidal against Salmonella and Staphylococcus	Ginger extracts	Duke (1994)
Reduction of rhinoviral activity	Sesquiterpenes of ginger	Denyer et al. (1994)
Increasing production of ceramide in the epidermis of mice	Ginger extracts	Ohkubo et al. (2000)
Antimicrobial activity	Ginger oleoresins	Chen et al. (2001)
Strong antibacterial effect	Ginger essential oil	Singh et al. (2001)
Prevention and treatment of induced disease β-myloid protein	Gingerols and shogaols	Kim (2001)
Reducing symptoms of osteoarthritis (knee pain)	Highly purified ginger extracts	Altman and Marcussen (2001)
Increasing activity of peroxidase and β-1,3-glucanase enzymes in the callus cultures	Callus cultures of ginger (+ salicylic acid)	Prachi et al. (2002)
Metabolism of (6)-gingerol in rats		Nakazawa and Ohsawa (2002)

[a]Besides these properties, ginger also possesses antimigraine effects, antinausea effects, antimotion sickness, and anti-postoperative nausea, as well as antiemetic, antipyretic, and analgesic effects (Afzal et al., 2001).
[b]New cyclic diarylheptanoid.

Table 16.18 Agricultural Properties of Ginger

Property	Responsible Compound	Reference
Antifertility	(−)-α-Zingiberene superior to (−)-zingiberol and β-sesquiphellandrene	Ni et al. (1988)
Ground regulators (effect of NAA and BAP)	Ginger extracts	Arimura et al. (2000)
Insecticidal activities against the shoot borer *D. punctiferalis* in ginger (Melathion)		Koya et al. (1988)
Insecticidal synergistic effect	Ginger oil + garlic oils	Hus et al. (1999)
Insecticidal effect against cockroach (*Blattela germanica*)	Ginger oil + insecticidal products	Kawara (1998)
Antifungal	–	Hasnah and Ahmad (1993)
Insecticidal, antifeedant, and antifungal activities against *Rhizoctonia solani*	(6)-Dehydroshogaol (insecticidal) dehydrogingerone (antifungal)	Agarwal et al. (2001)
Reduction of *Dirofilaria immitis* by microfilarial in dogs	–	Data and Sukul (1987)
Destruction of *Anisaki*s larvae	(6)-Gingerol and (6)-shogaol	Goto et al. (1990)

(2.8 g), and sugar (16 g) were mixed, adjusted to pH 3.5 with citric acid, and preserved for 12 months without losing the original taste or flavor.

Four ginger preparations were compared by Ding and Ding (1988):

1. Fresh ginger
2. Dried ginger under sunlight
3. Dried by heating at 220°C in a sand bath
4. Dried by heating at 300°C in a sand bath.

The essential oil decreased by approximately 57% after 300°C treatment. Heating also decreased the levels of gingerols and shogaols in ginger. The effect of the drying process on the composition of the essential oil from Australian ginger showed that the major effects are a reduction in gingerol content, an increase in terpenehydrocarbons, and the conversion of some monoterpene alcohols to their corresponding acetates, for both fresh and dried ginger samples (Bartley and Jacobs, 2000).

Molecular encapsulation of fresh ginger flavor was investigated by Sankarikutty and Narayanan. Encapsulation of ginger flavor by inclusion of β-cyclodextrin was better by addition of steam-distilled ginger oil to a saturated aqueous solution of the cyclodextrin, followed by agitating, holding at 5°C for 2–3 h, and drying at room temperature.

Irradiation Effects

Effects of ethylene dioxide and gamma irradiation on the chemical sensory and microbial quality of ground spices and their essential oils (ginger, cinnamon, fennel, and fenugreek) were investigated by Toofanian and Stegeman (1988). Irradiation of ground ginger with a dose of 5 KGy resulted in a slight decrease of 14%, whereas fumigated ginger showed no significant loss in volatile oil content. No major differences in sensory properties were found while comparing the untreated irradiated or fumigated species.

Andrews et al. (1995) studied the chemical and microbial activity of irradiated fresh ground ginger using 10 KGy ionizing radiation dose from cobalt 60 source. Gamma irradiation decreases the amounts of most of the extractable flavor components. Those key traits known to contribute to the typical ginger flavor, namely α-zingiberene, β-sesquiphellandrene, β-bisabolene, farnesene isomers, ar-curcumene, and α-cubebene, decreased from 25% up to 59% following irradiation.

Formulations and Uses

Several reviews have been devoted to various ginger formulations as a spice for food in ready-cooked dishes and fish, not only in India and other Asian countries but also throughout the world (Ho et al., 1989). Goku (1983) describes an excellent and very palatable ginger-based formulation.

In a study of the chemical and nutritional quality of fermented fish silage containing potato extracts, formalin, and ginger extracts, it was found that ginger extracts proved to be an effective antioxidant in fermented *tilapia* silage (*Oreochromis niloticus*) (Fagbento and Jauncey, 1994).

A preparation process for seasoning oil consists of:

1. Mixture of cysteine (0.1–5%), lysine (0.5–10%), serine (1–15%), arginine (2–20%), glucose (0.05–5%), fructose (0.1–5%), xylose (0.01–2%) at 10–50°C to obtain seasoning liquid mixture (I)
2. Soaking seeds of sunflower or powders of peanuts or soybean in 1 to 5 volumes of (I) for 5–180 min
3. Drying to 5–20% water content
4. Roasting at 170–220°C for 10–40 min
5. Pressing to obtain oil
6. Adding vegetable extract: The preparation method for the vegetable extract consists of stirring one of the dried vegetables (garlic, red pepper, ginger, cabbage, carrot, mustard, parsley, or pepper) in 20–30 volumes of the oil of sunflower or soy at 40°C to −60°C for 6–120 min. The roasting step (4 above) can give rise to Maillard reaction between α-amino acids and reducing sugars (Vernin, 1982). This reaction has not been studied in the case of ginger.
   Okwu (2001) reported the chemical evaluation and nutritional flavoring potentials of ginger and five other indigenous spices. Preserved ginger is prepared by harvesting and drying the mature rhizome. Dried ginger is used directly as a spice and also for the preparation of ginger oil and ginger oleoresin. The ginger oleoresin possesses the full organoleptic properties of the spice, that is, aroma, flavor, and pungency and finds similar applications as the ground spice in flavoring of processed foods (Ebwele and Jimoh, 1988)).

Ginger extracts in mixture with other compounds can be used in products such as antiaging cosmetics and skin protectants. A lotion containing ginger extract, ethanol, guanidine compounds, trimethylglycine, methyl-*p*-hydroxy benzoate, triethoxy-ethoxyethyl phosphate, hydrolyzed *Prunus amygdalus* extract, glycine, and purified water to 100 wt% was suggested by Sane et al. (1998).

Anecdote

In a tribe of the Mollucas Islands, also called the "Spices Islands," it is usual for girls of 12–13 years of age to have their faces and arms smeared with a yellow-colored cream containing dried ginger prior to wedding festivities.

References

Achinewhu, S.C., Ogbonna, C.C., Hart, A.D., 1995. Chemical composition of indigenous wild herbs, spices, fruits, nuts and leafy vegetables used as foods. Plant Foods Hum. Nutr., Dordrecht, Netherlands 48 (4), 341–348.

Afzal, M., Al-Hadidi Menon, M., Pesek, J., Dhami, M.S., 2001. Zinger: an ethno-medical, chemical and pharmacological review. Drug Metab. Drug Interact. 18 (3–4), 159–190.

Agarwal, M., Walia, S., Dhingra, S., Khambay, B.P.S., 2001. Insect growth, inhibition, antifeedant and antifungal activity of compounds isolated/derived from *Zingiber officinale* Roscoe (ginger) rhizomes. Pest Manag. Sci. 57 (3), 289–300.

Ahmad, N., Katiyar, S.K., Mukhtar, H., 2001. Antioxidants in chemoprevention of the skin cancer. Curr. Probl. Dermatol. 29, 128–139.

Ahn, B., Lee, D.H., Yeo, S.G., Kany, J.H., Do, J.R., Kim, S.B., et al., 1993. Inhibitory action of natural food components on the formation of carcinogenic nitrosomine. Bull. Korean Fish Soc. 26, 289–295.

Akhila, A., Tewari, P., 1984. Chemistry of ginger: a review. Curr. Res. Med. Aromat. Plants 6 (3), 143–156.

Alencar, J.W., Craveiro, A.A., Matos, F.J.A., 1984. Kovats indices as a preselection routine in mass spectra library searches of volatiles. J. Nat. Prod. 47 (3), 890–892.

Altman, R.D., Marcussen, K.C., 2001. Effects of a ginger extract on knee pain in patients with osteoarthritis. Arthritis Rhem. 44 (11), 2531–2538.

Anderson, N.H., Falcone, M.S., 1969. The identification of sesquiterpene hydrocarbons from GC retention data. J. Chromatogr. 44, 52–59.

Andrews, I.S., Cadwallader, K.R., Grodner, R.M., Chung, H.V., 1995. Chemical and microbial quality of irradiated ground ginger. J. Food Sci. 60 (4), 829–832.

Anzaldo, F.E., Coronel Violera, Q., Manalo, J.B., Nuevo, C.R., 1986. Chemical components of local (Philippines) ginger oil. Natl. Inst. Sci. Technol. 11 (3), 11–19.

Arimura, C.T., Finger, F.L., Casali, Vincente, W.D., 2000. Effect of NAA and BAP on ginger (*Zingiber officinale* Roscoe) sprouting in solid and liquid medium. Rev. Bras. Plant. Med. 2 (2), 23–26.

Association of Official Analytical Chemists (AOAC), 1984. Methods of analysis. Fourteenth Edition.

Badalyan, A.G., Wilkinson, G.T., Chun, B.S., 1998. Extraction of Australian ginger root with carbon dioxide and ethanol entrainer. J. Supercrit. Fluids 13 (1–3), 319–324.

Balladin, D.A., Headley, O., 1999. Liquid chromatographic analysis of the main pungent principles of solar dried West Indian ginger (*Zingiber officinale* Roscoe.). Renewable Energy 18 (2), 257–261.

Bartley, J.P., 1995. A new method for the determination of pungent compounds in ginger (*Zingiber officinale* Roscoe). J. Sci. Food Agric. 68, 215–222.

Bartley, J.P., Jacobs, A.L., 2000. Effects of drying on flavor compounds in Australian-grown ginger (*Zingiber officinale*). J. Sci. Food Agric. 80 (2), 209–215.

Bednarczyk, A.A., 1974. Identification and evaluation of the flavor-significant components of ginger essential oil. Diss. Abstr. Int. B 35 (1), 306.

Bednarczyk, A.A., Kramer, A., 1975. Identification and evaluation of the flavor-significant components of ginger essential oil. Chem. Senses 1 (4), 377–386.

Bhattarai, S., Tran Van, H., Duke, C.C., 2001. The stability of gingerol and shogaol in aqueous solutions. J. Pharm. Sci. Australia 80 (10), 1658–1664.

Bhonsle, J.B., Deshpande, V.H., Ravindranathan, T., 1994. Synthesis of (+)-zingiberene. Indian J. Chem. B 33B (4), 313–316.

Bicchi, C., Sandra, P., 1987. In: Bicchi, C., Sandra, P. (Eds.), Capillary gas chromatography in essential oil analysis Huethig Verlag, Heidelberg, pp. 85–121.

Birch, E.J., 1965. Reduction/hydrolysis of p-substituted phenols (cited from Lewis, K.G., Williams, G.J.1965) Tetrahedron Lett. 4573 and Teisseire 1991.

Boniface, C., Vernin, G., Metzger, J., 1987. Identification informatisee de composes par analyse combinee: spctres de masse-indices de Kovats. Analusis 15, 564–568.

Breeden, D.C., Coates, R.M., 1995. 7-Epizingiberene, a novel bisabolene sesquiterpene from wild tomato leaves (Erratum to document cited in CA 121: 276729). Tetrahedron 51 (6), 1533.

Brogle, H., 1982. Supercritical fluid CO$_2$ extract. Chem. Ind. (London) 12, 385–390.

Bruins, A.P., 1987. In: Sandra, P., Bicchi, C. (Eds.), Capillary Gas Chromatography in Essential Oil Analysis Huethig Verlag, Heidelberg, pp. 329–357.

Charles, R., Garg, S.N., Kumar, S., 2000. New gingerdione from the rhizomes of *Zingiber officinale*. Fitoterapia 71 (6), 716–718.

Chau, F.T., Mok, D.K.W., Gong, F., Tsui, S.K., Wong, S.K., Huang, L.Q., et al., 2001. Fingerprinting analysis of raw herb: application of chromometrics techniques for finding out chemical fingerprint of Chinese herb. Analyt. Sci. 17 (Suppl), 419–422.

Chen, C.C., Ho, C.T., 1988. GC Analysis of volatile components of ginger oil (*Zingiber officinale* Roscoe) extracted with liquid carbon dioxide. J. Agric. Food Chem. 36 (2), 322–328.

Chen, Y.H., Guo, H.Z., 1980. A survey of the raw and dry ginger produced in Szechaun (China). Yao Hsueb Tung Pao 15 (10), 12–13.

Chen, C.C., Kuo, M.C., WU, C.M., Ho, C.T., 1986. Ginger oil extracted by liquid carbon dioxide. Shib Pin ko Hsueb 13 (3–4), 188–197.

Chen, Y., Li, Z., Xue, D., Qi, L., 1987. Determination of volatile constituents of Chinese medicinal herbs by direct vaporization capillary gas chromatography–mass spectrometry. Anal. Chem. 59 (5), 744–749.

Chen, Y., Cai, T., Fu, L., Shan, J., 2001. Improved high performance liquid chromatography (HPLC) determination of pungent constituents of ginger. Shipin Kexue (Beijing) 22 (4), 60–63.

Connell, D.W., 1970. Chemistry of the essential oil and oleoresin of ginger (*Zingiber officinale*). Flavour Ind. 10, 677–693.

Connell, D.W., 1971. Chemical composition of certain products from ginger (*Zingiber officinale*). Aust. Chem. Process. Eng. 24 (11), 27.

Connell, D.W., Jordan, R.A., 1971. Composition and distinctive volatile flavor characteristics of the essential oils from Australian-grown ginger (*Zingiber officinale*). J. Sci. Food Agric. 22 (2), 93–95.

Connell, D.W., McLachlan, B., 1972. Natural pungent compounds IV Examination of the gingerols, shogaols, paradol and related compounds by TLC and GC. J. Chromatogr. 67 (1), 29–35.

Connell, D.W., Sutherland, M.D., 1969. A re-examination of gingerol, shogaol and zingerone, the pungent principles of ginger (*Zingiber officinale* Roscoe). Aust. J. Chem. 22, 1033–1043.

Dambatta, B.B., Kazaure, M.A., Tapley, K.N., 1998. Extraction and characterization of essential oils from Nigerian ginger. Adv. Colour Sci. Technol. 1 (3), 80–82.

Data, A., Sukul, N., 1987. Antifilarial effect of *Z. officinalis* on *Dirofilaria immitis*. J. Helminthol. 61, 268–270.

De Pooter, H.L., Coolsack, B.A., Dirinck, P.J., Schamp, N.M., 1985. GLC of the headspace after concentration on Tenax GC of the essential oils of apples, fresh celery, fresh lovage, honeysuckle and ginger extracts. In: Berkeim, A., Sweden, J., Scheffer, J.C. (Eds.), Essential Oils and Aromatic Plants Martinus Nijhoff/Dr W.Junk, Dordrecht, the Netherlands.

Denyer, C., Jackson, P., Loakes, D.M., Ellis, M.R., Young, D.A., 1994. Isolation of anti-rhinoviral sesquiterpenes from ginger (*Zingiber officinale* Rosc.). J. Nat. Prod. 57, 658–662.

Ding, A., Ding, Q., 1988. Comparison of the contents of main chemical constituents in different processed preparations of ginger. Zhongyao Tongbao 13 (11), 657–659.

Dowdle, P.A., Corr, S., Harris, H., 2002. Solvent Extraction Process. Patent WO 2002 036232 A1, Date 20020510, Appl. WO 2001 GB4904.

Duke, J.A., 1994. Biologically active compounds in important spices In: Charalambous, G. (Ed.), Spices, Herbs and Edible Fungi, 34 Elsevier Science, Amsterdam, the Netherlands, pp. 225–250.

Dung, N.X., Chin, T.D., Leclercq, P.A., 1995. Chemical investigation of the aerial parts of *Zingiber zerumbet* (L.) Sm. from Vietnam. J. Essential Oil Res. 7 (2), 153–157.

Duve, R.N., 1980. Highlights on the chemistry and pharmacology of wild ginger (*Zingiber zerumbet* Smith). Fiji Agric. J. 42 (1), 41–43.

Ebwele, R.O., Jimoh, A.A., 1988. Local processing of ginger: prospects and problems Proceedings of the First National Ginger Workshop. National Root Crops Research Institute, Umudike, Nigeria, pp. 22–33.

Ekundayo, O., Laasko, I., Hiltunen, R., 1988. Composition of ginger (*Zingiber officinale* Roscoe) volatile oils from Nigeria. Flavour Fragr. J. 3 (2), 85–90.

El-Hamouly, M.M.A., Mohamad, 2001. Phytochemical and biological evaluation of volatile constituents of *Zizyphus phlembristi* (L) wild leaves and flowering tops, cultivated in Egypt Al-Azhar. J. Pharm. Sci. 28, 370–379.

Endo, K., Kanno, E., Oshima, Y., 1990. Structures of antifungal diarylheptones, gingerenones A, B, C and isogingerenone B, isolated from the rhizomes of *Zingiber officinale*. Phytochemistry 29, 797–799.

Erler, J., Vostrowsky Strobel, H., Knobloch, K., 1988. Essential oils from ginger (*Zingiber officinale* Roscoe). Z. Lebensm. Unters. Forsch. 186 (3), 231–234.

Fagbento, O., Jauncey, K., 1994. Chemical and nutritional quality of fermented fish silage containing potato extracts, formalin or ginger. Food Chem. 50 (4), 383–388.

Faulhaber, S., Shirey, R., 1998. Solid-phase microextraction for the sampling in aromatic analysis. Lab. Praxis 22 (5), 52. 55–58.

Flisak, J.R., Hall, S.S., 1986. Alkylation-reduction of carbonyl systems. 15. Efficient syntheses of beta-sesquiphellandrene and zingiberenol employing a tandem arylation-multistep reduction-hydrolysis sequence. Synth. Commun. 16 (10), 1217–1228.

Garnero, J., Tabacchi, R., 1987. Examples of artifact formation by chromatographic techniques. In: Sandra, P., Bicchi, C. (Eds.), Capillary Gas Chromatography in Essential Oil Analysis Huethig Verlag, Heidelberg, the Netherlands, pp. 359–366.

Goku Kazuo, 1983. Tablets as Breath Refreshners. Jpn. Kokai Tokyo Kobo Patent Jp 58,088, 308 A2, Date 1983 05 26 Appl. JP 1981–185, 655.

Gopalam, A., Ratnambal, M.J., 1989. Essential oils of ginger. Indian Perfum. 33 (1), 63–69.

Goto, C., Kasaya, S., Koga, K., Ohmoto, H., Kagei, N., 1990. Lethal efficacy of extract from Z. officinale or (6)-shogaol and (6)-gingerol in Anisakis larvae in vitro. Parasitology, 10, 653–656.

Grosh, W., 1994. Determination of potent odorants in foods by aroma extract dilution analysis (AEDA) and calculation of odor quality values (COAVS). Flavor Fragr. J. 9, 147–158.

Guenther, E., 1952. The essential oils, second ed. Individual Essential Oils of the Plant Families, vol. 5 van Nostrand, New York, NY.

Gujral, S., Bhumra, H., Swaroop, M., 1978. Cholesterolemic activity of pungent principles of ginger. Nutr. Rep. Int. 17, 183–189.

Guo, P., Xu, J., Xu, S., Wang, K., 1997. Inhibition of fulvic acid-induced hydrogen peroxide production in chondrocyte by ginger volatile oil. Zbongguo Zbongyao Zasshi 22 (9), 559–561.

Gurib-Fakim, A., Mandarbaccus, N., Leach, D., Doimo, L., Wohlmuth, H., 2002. Essential oil composition of Zingiberaceae species from Mauritius. J. Essential Oil Res. 14 (4), 271–273.

Haq, F., Faruque, S.M., Islam, S., Ali, E., 1986. Studies on Zingiber officinale Roscoe. Part 1. Chemical investigation of the rhizome. Bangladesh J. Sci. Ind. Res. 21 (1–4), 61–69.

Harrison, A.G., 1983. Chemical Ionization Mass Spectrometry. CRC Press, Boca Raton, FL.

Harvey, D.J., 1981. Gas chromatographic studies of ginger constituents. J. Chromatogr. 212, 75–84.

Hasnah, M.S., Ahmad, A.R., 1993. Some important metabolites from Malaysian ginger. In: Sharman, N.A. (Ed.), Applications of Plants In Vitro Technology. Proceedings of the International Symposium, Serdand, Malaysia, 16–18 November 1993 Department of Biochem and Microbiology, Universiti Pertanian Malaysia, Serdang, Malaysia, pp. 191–196.

He, W., Li, L., Guo, S., Li, Y., 1999. Extraction of ginger oil and its anti-oxidative activity for edible oils and fats. Zbongguo Youzhi 24 (1), 42–44.

He, W., Li, L., Li, Y., Guo, S., Guo, B., 2001. Anti-oxidative activity of a new compound from ginger. Zbongguo Bingli Shenli Zashi 17 (5), 461–463.

Heath, H.P., Reineccius, G., 1988. Flavour Chemistry and Technology. Avi Publishing, Westport, CT.

Herout, V., Benesova, V., Pliva, J., 1953. The sesquiterpenes of ginger oil. Cull. Czeck Chem. Comm. 18, 248–256.

Hikino, H., Kiso, Y., Kato, N., Hamada, Y., Shiori, T., Aiyama, R., et al., 1985. Antiepatotoxic activity of gingerols and diarylheptanoids. J. Ethnopharmacol. 14, 31–39.

Hill, C.E., Dowdle, P.A., Corr, S., 1999. Solvent extraction process PCT. Int. Appl. WO 0064,555 (Cl B 01 D1/00), 2 Nov 2000. G.B.Appl. 1999/9 , 136, 22 April 1999.

Ho, C.T., Zhang Shi, H., Tang, J., 1989. Flavor chemistry of Chinese foods. Food Rev. Int. 5 (3), 53–87.

Huang, X., Wang, J., Zhang, X., 1999a. Essential oil of Zingiber officinalis. Huaxue Shijie 30 (9), 420–433.

Huang, X., Wang, J., Zhang, X., 1999b. Determination of the pungent principles in ginger powder, ginger skin and baked ginger. Zbongcaoyae 30 (6), 423–425.

Hus, H.J., Chang, H.L., Jian, N., 1999. Synergistic Natural Pesticides Containing Garlic. Eur. Pat. Appl. 945,066, A1, 1999 09 29, Appl. 1999-302.286 (EP) 17P. US 6231865, BI, 2001 05 15, Appl. 1999-273636.

Ibrahim, H., Zakaria, M.B., 1987. Essential oils from three Malaysian Zingiberaceae species. Malaysian J. Sci. 9, 73–76.

ISI, 1975. Specification for Ginger Oleoresin. Indian Standards Institution, New Delhi.

Jain, T.C., Varma, K.R., Bhattacharya, C.S., 1962. Terpenoids XXVII GLC analysis of monoterpenes and its application to essential oils. Perf. Essent. Oil Res. 53, 678–684.

James, A.T., Martin, A.J.P., 1952. Gas liquid chromatography. Analysis 77, 198.

Jennings, W., Shibamoto, T., 1980. Qualitative Analysis of Flavour and Fragrances Volatiles by Glass Capillary Gas Chromatography. Academic Press, New York.

Jo, K.S., 2000. Analysis of gingerol compounds of raw ginger (*Zingiber officinale* Roscoe.) and its paste to high performance liquid chromatography mass spectrometry (LC-MS). Hanguk Sikpum Yongyang Kwabak Hoechi 29 (5), 747–751.

Joulain, D., Konig, W.A., 1998. The Atlas of Spectral Data Sesquiterpene Hydrocarbons. E.B. Verlag, Hamburg, Germany.

Kami, T., Nakayama, N., Hayashi, S., 1972. Volatile constituents of *Zingiber officinale*. Phytochemistry 11, 3377–3381.

Kano, Y., Tanabe, M., Yasuda, M., 1990. On the evaluation of the preparation of Chinese medicinal prescriptions (V): diterpenes from Japanese ginger Kinto. Shoyakugaku Zasshi 44 (1), 55–57.

Kawara, H., 1998. Insecticidal Baits Containing Ginger Oil Against Cockroach. Japan Kokat Tokyo, 4 p. No 10, 017,405, A2, 20 Jan 1998, 1996-172, 020.

Kawashi, S., Morimitsu, Y., Osawa, T., 1994. Chemistry of ginger components and inhibitory factors of the arachidonic acid cascade. In: Ho, C.T., Osawa, T., Huang, M.T., Rosen, R.T. (Ed.), Food Phytochemicals for Cancer Prevention II ACS Symposium Series 547.

Kikuzaki, H., 2000. Ginger for drug and spice purposes. In: Maza, G.O., Dave, B. (Eds.), Herbs, Botanicals and Teas Technomic Publishing, Lancaster, UK, pp. 75–105.

Kikuzaki, H., Nakatami, N., 1996. Cyclic diarylheptanoids from rhizomes of *Zingiber officinale*. Phytochemistry 43 (1), 273–277.

Kim, D.S.H.I., 2001. Pharmaceutical composition with natural products or synthetic analogs which are useful in the prevention and treatment of beta-amyloid protein-induced disease PCT. Int. Appl.Pat WO 0,130, 335 A2 date 2001 05 03, Appl. WO 2000-US41, 436

Kim, J.S., Koh, M.S., Kim, Y.H., Kim, M.K., Hong, J.S., 1991. Volatile flavor components of Korean ginger (*Zingiber officinale* Roscoe.). Hanguk Sikpum Kwabakhoechi 23 (2), 141–149.

Kim, M.K., Na, M.S., Hong, J., Jung, S.T., 1992. Volatile flavor components of Korean ginger (*Zingiber officinale* Roscoe) extracted with liquid carbon. Hanguk Nongbua Hakborchi 35 (1), 59–63.

Kitamura, Y., Naguno, Y., 2000. Purification and characterization of cysteine proteinase inhibitor from fresh ginger rhizome Kenkyu Kiyo-Tokyo Kasei Daigaku 2. Shizen Kagaku 40, 53–56.

Kiuchi, F., Shibuya, M., Sankawa, U., 1982. Inhibitors of the biosynthesis of prostaglandins. Chem. Pharm. Bull. Tokyo 30, 754–757.

Kobayashi, M., Shoji, N., Ohizumi, Y., 1987. Gingerol, a novel cardiotonic agent, activates the Ca^{2+} pumping ATPase in skeletal cardiac sacroplasmic reticulum. Biochem. Biophys. Acta 903, 96–102.

Koedam, A., 1987. Some aspects of essential oil preparation In: Sandra, P. Bicchi, C. (Eds.), Capillary Gas Chromatography in Essential Oil Analysis, 903 Biochem Biophys Acta, Huethig Verlag, Heidelberg, pp. 13–28.

Koenig, W.A., Rieck, A., Hardt, I., Gehrcke, B., Kubeczka, K.H., Muhle, H., 1994. Enantiomeric composition of the chiral constituents of essential oils. Part 2: sesquiterpene hydrocarbon. J. High Resolut. Chromatogr. 17 (5), 315–320.

Kostrzewa, E., Karwowska, K., 1976. Characteristics of a flavor and odor extract of ginger. Proc. Inst. Lab. Badav. Przem. Sozyw. 26 (1), 63–73.

Kovats, E., 1958. Gas chromatographic characterization of organic compounds I. Retention indexes of aliphatic halides, alcohols, aldehydes, and ketones. Helv. Chim. Acta 41, 1915–1932.

Kovats, E., 1965. In: Giddings, J.C., Keller, R.A. (Eds.), Advances in Chromatography Marcel Dekker, New York, NY, pp. 119–127.

Koya, K.M.A., Premkumar, T., Gautam, S.S.S., 1988. Chemical control of shoot borer Dichocrocis punctiferalis Guen. on ginger Zingiber officinale Roscoe. J. Plantation Crops 16 (1), 58–59.

Krishnakantha, T., Lokesh, B., 1993. Scavenging for superoxide anions by spice principles. Indian J. Biochem. Biophys. 30, 133–134.

Krishnamurthy, N., Nambudiri, E.S., Mathew, A.G., Lewis, Y.S., 1970. Essential oils of ginger. Indian Perfum. 14 (1), 1–3.

Lapworth, 1917. Cited from Connell and Sutherland, 1969.

Lawrence, B.M., 1983. Recent studies on the oil of Zingiber officinale Roscoe. Perfum. Flavours 9, 2–40.

Lee, C.Y., Chiou, J.W., Chang, W.H., 1982. Labdane-type diterpene: Galanolactone. J. Chinese Agri. Chem. 20, 61–67.

Lewis, Y.S., Mathew, A.G., Nambudiri, E.S., Krishnamurthy, N., 1972. Oleoresin ginger. Flavour Ind. 3 (2), 78–81.

Li, A., 1995. Spectrometric determination of gingerol in ginger oil condiments. Zbongguo Tiaoweipin 11, 30–32.

Li, J., Wang, Y., Ma, H., Hao, J., Yang, H., 2001. Comparison of chemical components between dry and fresh Zingiber officinale. Zbongguo Zbongyao Zassbi 26 (11), 748–751.

Lin, Z.K., Hua, Y.F., 1987. Chemical constituents of the essential oil from Zingiber officinale Roscoe. Sichuan Youji Huaxue 6, 444–448.

Ma, X., Gu, Y., Fu, J., 1990. Biosynthesis of LTB4 and selection of its inhibitors. Baiquien Yike Daxue Xuebao 16 (3), 222–225.

Macleod, A.J., Pieris, N.M., 1984. Volatile aroma constituents of Sri Lankan ginger. Phytochemistry, 9, 353–359.

Martins, A.P., Salqueiro, L., Goncalves, M.J., de Cunha, A.P., Vila, R., Canigueral, S., et al., 2001. Essential oil composition and antimicrobial activity of three Zingiberaceae from S Tome principe. Planta Med. 67 (6), 580–584.

Masada, Y., 1976. Analysis of Essential Oils by Gas Chromatography and Mass Spectrometry. John Wiley & Sons, New York, NY, pp. 251–255.

Masada, Y., Inoue, T., Hashimoto, K., Fujika, M., Shiraki, K., 1973. Studies on the pungent principles of ginger (Zingiber officinale Roscoe) by GC–MS. Yakugaku Zasshi 93 (3), 318–321.

Masuda, T., Matsumura, H., Oyama, Y., Takeda, Y., 1998. Synthesis of (+) cassumunins A and B, new cucuminoids antioxidants having protective activity on the living cell against oxidative damage. J. Nat. Prod. 61, 609–613.

Mathew, A.G., Krishnamurthy, N., Nambudiri, E.S., Lewis, Y.S., 1973. Oil Ginger Flavour Ind. 4 (5), 226–229.

Meireles, M.A.A., Nikolov, Z.I., 1994. Extraction and fractionation of essential oils with liquid carbon dioxide (LCO2) In: Charlambous, G. (Ed.), Spices, Herbs and Edible Fungi, 34 Elsevier Science, Amsterdam, the Netherlands, pp. 171–199.

Meyer-Warnod, B., 1984. Natural essential oils. Perfum. Flavor 9, 93.

Mitra, C.R., 1975. Important Indian spices III. Ginger (Zingiberaceae). Rechst. Aromen. Koerperpflegem 25 (6), 170.

Molyneux, F., 1971. Ginger—a natural flavor essence. Aust. Chem. Process. Eng. 24 (3), 29. 31, 33–34.

Mosandhl, A., 1992. Capillary gas chromatography in quality assessment of flavours and fragrances. J. Chromatogr. 624, 267–292.

Moyler, D.A., Browning, R.M., Stephens, M.A., 1994. Carbon dioxide extraction of essential oils In: Charlambous, G. (Ed.), Spices, Herbs and Edible Fungi, 34 Elsevier Science, Amsterdam, the Netherlands, pp. 145–170.

Murakami, A., Nakamura, Y., Ohto, Y., Tanaka, T., Makita Koshimizu, K., Ohigashi, H., 1999. Cancer preventive phytochemicals from tropical Zingiberaceae. In: Whitaker, H.R. (Ed.), Food for Health in Pacific Rim. International Conference on Food, Science and Technology Food & Nutrition Press, Turnbull, CT, pp. 125–133. 1997.

Nakatani, N., 1995. Chemistry and properties of pungent compounds. Koryo 185, 59–64.

Nakatani, N., Kikuzaki, H., 2002. Antioxidants in ginger family. ACS Symposium Series Quality Management of Nutraeceuticals 803 230–240.

Nakazawa, T., Ohsawa, K., 2002. Metabolism of (6)-gingerol in rats. Life Sci. 70 (18), 2165–2175.

Narayanan, C.S., Mathew, A.G., 1985. Chemical investigation on spice oils. Indian Perfum. 29 (1–2), 15–22.

Natarajan, C.P., Bai, R.P., Krishnamurthy, M.N., Raghavan, B., Shankaracharya, N.B., Kuppuswamy, S., et al., 1972. Chemical composition of ginger varieties and dehydration studies on ginger. J. Food. Sci. Technol. 9 (3), 120–124.

Ney, K.H., 1990a. Aromagrams of spices. Alimenta 29 (5), 91–93. 95–100.

Ni, M., Chen, Z., Yan, B., 1988. Synthesis of optically active sesquiterpenes and exploration of their anti-fertility effect. Huadong Huagong Xueyuan 14, 675–679.

Nigam, I.C., Levi, L., 1963. Column- and gas-chromatographic analysis of the oil of wild ginger. Identification and estimation of some new constituents. Can. J. Chem. 41 (7), 1726–1730.

Nigam, M.C., Nigam, I.C., Levi, L., Handa, K.L., 1964. Essential oils and their constituents XXII. Detection of new trace components in oil of ginger. Can. J. Chem. 42 (11), 2610–2615.

Nishimura, O., 2001. Enantiomer separation of the characteristic odorants in Japanese fresh rhizomes of Zingiber officinale Roscoe (ginger) using multi-dimensional GC system and confirmation of the odor character of each enantiomer by GC-olfactometry. Flavour Frag. J. 16 (1), 13–18.

Nomura, H., Tsurami, S., 1926. Structure of shogaol. Proc. Imp. Acad. Tokyo 2, 229.

Ohkubo, K., Tagaki, Y., Takatoku, H., Hori, K., Kumoku, H., Shibuya, Y., 2000. Ceramide Production-Accelerating Agent. Eur. Pat. Appl. EP 993,822 (CL. AK61K7/48) 19 Apr 2000 JP Appl. 1999/122,402 28 Apr 1999.

Okwu, D.E., 2001. Evaluation of the chemical composition of indigenous spices and flavoring agents. Global J. Pure Appl. Sci. 7 (3), 455–459.

Onyenekwe, P.C., Hashimoto, S., 1999. The composition of the essential oil of dried Nigerian ginger (Zingiber officinale). Z. Lebensm. Unters Forsch (A Food Res. Technol.) 209 (6), 407–410.

Paquette, L.A., Kinney, W.A., 1982. A new synthon for the regiospecific y-alkylation of 2-cyclohexenones. Application to the synthesis of zingiberenol and oxygenated bicyclo (3.3.1) nonanes. Tetrahedron Lett. 23 (2), 131–134.

Pellerin, P., 1991. Supercritical fluid extraction of natural raw materials for the flavor and perfume industry. Perfum. Flavor 16 (4), 37–39.

Pliva, J., Horak, M., Herout, V., Sorm, F., 1960. T.I. Sesquiterpene, S10–S11 Sammlung der Spectrum und Physikalischen Konstanten. Akademie Verlag, Berlin.

Prachi, S., Tilak, R., Singh, B.M., 2002. Salicylic acid induced insensitivity to culture filtrate of *Fusarium oxysporum* f. sp. *Zingiberi* in the calli *of Zingiber officinale* Roscoe. Eur. J. Plant Pathol. 108 (1), 31–39.

Purseglove, J.W., Brown, G.G., Green, C.L., Robbins, S.R.J., 1981. Spices, vols I and II. Longman, New York, NY.

Rani, K., 1999. Cyclization of farnesyl pyrophosphate into sesquiterpenoids in ginger rhizomes (*Zingiber officinale*). Fitoterapia 70 (6), 568–574.

Rogacheva, S., Kuntcheva, M., Obretenov, T., Vernin, G., 1998. Formation and structure of melanoidins in foods and model systems. In: O'Brien, J., Nurstsen, E.E., Crabbe, M.J.C., Ames, J.M. (Eds.), The Maillard Reaction in Foods and Medicine Royal Society of Chemistry, Cambridge, UK, pp. 89–93.

Rosella, M.A., de Pfirter, G.B., Mandrile, E.L., 1996. Ginger (*Zingiber officinale* Roscoe Zingiberaceae): ethnopharmacognosy, cultivation, chemical composition and pharmacology. Acta Pharm. Bonaerense 15 (1), 35–42.

Roy, B.C., Goro, M., Hirose, T., 1996. Extraction of ginger oil with supercritical carbon dioxide experiments and modeling. Ind. Eng. Chem. Res. 35, 607–612.

Sadtler Research Laboratories, 1985. The Sadtler Standard Gas Chromatography Retention Index Library. Sadtler Research Laboratories Division of Bio-Rad laboratories, Philadelphia, PA.

Sane, R.T., Phadke, M., Hijli, P.S., Shah, M., Patel, P.H., 1998. Geographical variation study on gingerol (a pungent principle from *Zingiber officinale*) and ginger oil, using HPTLC technique and accelerated stability study on gingerol from *Zingiber officinale* using HPTLC method. Indian Drugs 35 (1), 37–44.

Sharma, R.K., Misra, B.P., Sarma, T.C., Bordoloi, A.K., Pathak, M.G., Leclercq, P.A., 1997. Essential oils of *Curcuma longa* L. from Bhutan. J. Essent. Oil Res. 9, 589–592.

Tanabe, T., Kami, T., Hayashi. 1992. Volatile separated compounds obtained by SIM Technique. Phytochemistry. 12, 3388–3390.

Taveira et al., 1997a. Chemistry of the essential oil and oleoresin of ginger (*Zingiber officinale*). Flavor Ind. 10, 677–693.

Thomson, E.H., Wolf, I.D., Allen, C.E., 1974. Ginger rhizome: a new source of proteolytic enzyme. J. Food Sci. 38, 652–655.

Toofanian, F., Stegeman, H., 1988. Comparative effect of ethylene oxide and gamma irradiation of the chemical, sensory and microbial quality of ginger, cinnamon, fennel and fenugreek. Acta Aliment. 17 (4), 271–281.

Vahira-Lechat, I., Menut, C., Lamaty, G., Bessiere, J.M., 1996. Huiles essentielles de Polynesie Francaise Rivista Ital. EPPOS, (Special Edition) 627–638.

Van Beek, T.A., 1991. Special methods for the essential oil of ginger In: Linkens, H.F. Jackson, J.F. (Eds.), Modern Methods of Plant Analysis: Essential Oils and Wax, 12 Springer, Berlin, Heidelberg, pp. 79–97.

Van Beek, T.A., Posthumus Lelyveld, G.P., Hoang, V.P., Yen, B.T., 1987. Investigation of the essential oil of Vietnamese ginger. Phytochemistry 26 (11), 3005–3010.

Van den Dool, H., Kratz, P.D., 1963. A generalization of the retention index system including linear temperature programmed gas–liquid partition chromatography. J. Chromatogr. 11, 463–471.

Variyar, P.S., Gholap, A.S., Thomas, P., 2000. Estimation of pungency in fresh ginger: a new fluorimetric assay. J. Food Chompost. Anal. 13 (3), 219–225.

Varma, K.R., Jain, T.C., Bhattacharya, S.C., 1962. Terpenoids. XXXIV. Structure and stereo chemistry of zingiberol and juniper camphor. Tetrahedron 18, 974–984.

Verma, S.K., Singh, J., Khamesra, R., Bordia, A., 1993. Effect of ginger on platelet aggregation in man. Indian J. Med. Res. 98, 240–242.

Vernin, G., 1970. La Chromatographie en Couche Mince. Techniques et Applications en Chimie Organique. Dunod, Paris, Hungarian translation, originally published in 1970 as Vekonyreteg-Kromatographia: A Serves Kemiaban, Muskaki Konyvkiado, Budapest.

Vernin, G. (Ed.), 1982. The Chemistry of Heterocyclic Flavouring and Aroma Compounds Ellis Horwood, Chichester.

Vernin, G., Parkanyi, C., 1994. Ginger oil (*Zingiber officinale* Roscoe) In: Charalambous, G. (Ed.), Spices, Herbs and Edible Fungi, 34 Elsevier Science, Amsterdam, the Netherlands, pp. 579–594.

Vernin, G., Petitjean, M., 1982. Mass spectrometry of heterocyclic compounds used for flavourings The Chemistry of Heterocyclic Flavouring and Aroma Compounds. Ellis Horwood, Chichester, pp. 305–342.

Vernin, G., Debrauwer, L., Vernin, G.M.F., Zamkostian, R.M., Metzger, J., Larice, J.L., et al., 1992. Heterocycles by thermal degradation of some Amadori intermediates. In: Charalambous, G. (Ed.), Off-Flavours in Foods and Beverages Elsevier Science, Amsterdam, the Netherlands, pp. 567–624.

Wang, W., 2001. Antioxidant properties of four vegetables with sharp flavor. Shipin Yu Fajiao Gomgye 27, 28–31.

Wen, Z., Yu, D., Lu, Q., 2001. Study on antioxidation of ginger oil in concentrated fish oil. Zbongguo Youzhi 26 (4), 58–60.

Wenninger, J.A., Yates, R.L., Dolinsky, M., 1967. High resolution infrared spectra of some naturally occurring sesquiterpene hydrocarbons. J. Assoc. Anal. Chem. 50 (6), 1313–1335.

Yamahara, J., Matsuda, H., Yamaguchi, S., Shimoda, H., Murakami, N., Yoshikawa, M., 1995. Pharmacological study on ginger processing I. Antiallergic activity and cardiotonic action of gingerols and shogaols. Nat. Med. Tokyo 49 (1), 76–83.

Zaidi, V.H., Variyar, P.S., Gholap, A.S., 1992. Estimation of inorganic elements in trace amounts in ginger. J. Food Chompost. Anal. 9, 220–229.

Zhu, L.F., Li, Y.H., Li, B.L., Ju, B.Y., Zhang, W.L., 1995. Aromatic Plants and Essential Constituents (Supplement I) South China Institute of Botany Chinese Academy of Sciences Hai Feng Publ. Co. Distributed by Peace Book Co. Ltd., Hong Kong.

17 Ginger Physiology

Ginger is the rhizome of the plant *Zingiber officinale*. It lends its name to its genus and family (Zingiberaceae). It is a tuber and is consumed in whole as a delicacy, medicine, or spice. India, with over 30% of its share in ginger production, is now the leading producer of ginger globally. The characteristic odor and flavor of ginger are due to the volatile oils, zingerone, shogaols, and gingerols, which constitute 3% of the weight of fresh ginger. The information on physiological factors contributing to rhizome growth, productivity, and quality are sparse and the available information on these aspects is reviewed in this chapter.

Growth

There are three distinct phases, which can be identified in ginger. The first phase lasts for about 35–45 days (i.e., from the time of planting to the emergence of the shoot). It is characterized by root development and growth, as well as shoot development. The second phase lasts approximately for 150 days, and it comprises of development from the emergence of shoot to flowering. During this stage, shoot and leaf growth takes place rapidly, while rhizome growth is relatively slow. The third phase comprises the period from flowering to harvest of ginger late in the months of June/July (in India) and lasts for about 90 days. During this phase, both shoot and leaf growth ceases and the rhizomes grow rapidly (Anderson, 1991). The crop has no dormancy under 14 and 12 h photoperiods, but enters dormancy under 8 and 10 h natural photoperiods. Highest fresh rhizome yield was produced under 12 h photoperiod, but rhizome yield for propagation purposes was highest under 10 h photoperiod and natural photoperiod (Pandey et al., 1996).

Starch and soluble sugar contents of fresh rhizomes increased with growth. The amylase activities of seed ginger increased sharply at seedling stage, but decreased rapidly during vigorous growth stage. The invertase activity of fresh rhizomes increased at early stage and was then maintained at the higher level. Nitrate reductase activity was highest during the vigorous growth stage. Total nitrogen content of seed ginger and rhizomes were gradually decreased with growth (Xu et al., 2008).

Kandiannan et al. (2009) have proposed a model for LA estimation in ginger, where LA is given by LA = 0.0146 + 0.6621 × L × W. The model predicts values very proximal to the actual values with an r^2 of 0.997. This model can be used reliably to estimate LA of ginger nondestructively. The same equation can be extrapolated to all the varieties and land races of ginger as it is a vegetatively propagated plant with a narrow genetic base.

The Agronomy and Economy of Turmeric and Ginger. DOI: http://dx.doi.org/10.1016/B978-0-12-394801-4.00017-X

Carbon Assimilation and Photosynthesis

Radioactive labeling studies showed that at seedling stage, assimilates were translocated mainly to the shoots and leaves. More of the label was translocated to the rhizomes than the aboveground parts proportionally as the plants aged. The main shoots contributed more to the rhizome formation. Secondary shoots were not so efficient and the youngest shoots contributed little to assimilation or export of ^{14}C (Zhao et al., 1987). During the vigorous growth stage of rhizomes, carbon assimilation was mainly transported from the leaves to the rhizomes, thus the rhizomes become the growth centers. The absorption and utilization of nitrogen were the same as those of carbon assimilates. Over 48% of the nitrogen absorbed from the fertilizer applied at the seedling stage was distributed to the shoots and leaves. Meanwhile, 65.43% of the nitrogen derived from the fertilizer applied at vigorous growth stage of rhizomes was distributed to the rhizomes, and only 32.04% was distributed to the shoots and leaves. The utilization rate was highest (45.24%) when fertilizer was applied during the middle period of vigorous growth stage, while it was only 28.1% when fertilizer was applied at the seedling stage. Stored nutrients in the ginger seed were partly transferred to the new plants during growth. A certain proportion of the nutrients remained in the ginger seed itself. At the same time, a part of the carbon and nitrogen assimilated by the leaves and roots was transported back to the ginger seed (Xu et al., 2004a).

The chlorophyll content and Pn of ginger leaves increased with leaf expansion and reached a peak in the 15-day-old leaves but was subsequently declined (Xi Zhen et al., 1998). The Pn was highest in the middle leaves. Basal leaves had the higher rate than the apical leaves. The Pn of leaves at the eleventh to thirteenth nodes was higher than that at the third to fourth nodes, and it was lowest at the top leaves. The heavy midday depression of photosynthesis was observed in single leaf, at both leaf vigorous growth stage and rhizome vigorous growth stage (Xu et al., 2004). Photosynthesis was saturated at a light intensity of 642–867 μmol/m^2/s, which depended on leaf temperature. The low-light saturation point was found at temperatures above or below 25°C. CO_2 compensation and saturation points were in the range of 25–72 and 1343–1566 μl/l, respectively and increased with increasing leaf temperature. The light compensation point was in the range of 25–69 μmol/m^2/s[1], and it increased with increasing leaf temperature (Xi Zhen et al., 1998).

The diurnal pattern of photosynthesis rate showed a peak at about 9 h and a second smaller peak at about 15 h. The rate increased from mid-June to mid-August, then decreased during mid-October. The rate decreased as the temperature increased from 24°C to 40°C and was very low at a light intensity of 500 lx. The diurnal pattern of photosynthetic rate increased with increasing light intensity up to 30,000 lx and thereafter slightly decreased up to 60,000 lx. Wilting markedly decreased the rate of photosynthesis (Zhao et al., 1991). With increased light intensities from 15.61 to 104.81 klx, rhizome yield decreased from 13,487 to 2666 kg/ha (Shankar and Swamy, 1988).

The daily course of CPn showed a single peak occurring at midday. CPn changed greatly with changes in LAI. As LAI increased, the Pn per unit of ground area increased, but the Pn per unit of LA decreased (De Wan et al., 1995). CPn was shown to have a close relationship with yield in ginger. The diurnal pattern of CPn rate showed a typical single peak, which differed from the double peak shown by a single leaf (Xian Chang et al., 1996). No midday depression of CPn was noted at the rhizome's vigorous growth stage, maximum Pn occurred around 13 h, but midday depression of CPn occurred at the leaf's vigorous growth stage. The optimal PFD of photosynthesis for single leaf was approximately $1290 \mu mol/m^2/s^1$ but $1950 \mu mol/m^2/s^1$ for canopy. Nevertheless, CO_2 concentration for single leaves and CPn of ginger was similar, approximately $1500 \mu l$–$1 l$. Moreover, the optimal temperature and soil water content for photosynthesis of ginger were 25–30°C and 75%, respectively (Xu et al., 2004).

Stomatal conductance (gsc) increased rapidly as irradiance increased and was saturated at relatively low values of light intensity $(400 \mu mol/m^2/s^1)$. At different leaf temperatures, gsc peaked at 29°C, but tr increased with increasing irradiance and temperature. At soil moisture contents of 2–3.5 g/g, gsc increased approximately in linear with increasing tr. Fluorescence (Fv/Fm) decreased with increasing PFD. In leaves exposed to high PFD and different temperatures, Fv/Fm was lowest at <15°C, and highest at >25°C. In leaves exposed to low PFD, Fv/Fm remained at a similar value over all temperatures tested (Yun et al. 1988).

Stomatal conductance and stomatal frequency decreased with increasing shade levels, and were highest in plants grown under open conditions. In contrast, stomatal resistance increased with increasing shade levels and was highest $(77.68 m^2/s^1/mol)$ with 80% shading. Photosynthetic and tr decreased with increasing shade levels and were highest $(7.76 \mu mol \ CO_2/m^2/s^1$ and $2.27 mol \ H_2O/m^2/s^1$, respectively) in plants grown under open conditions. LAI and DM yield were highest with 20% shading (8.18 and 25.32 g/plant, respectively) and decreased with higher shade levels (Sreekala and Jayachandran, 2001).

Light and Physiological Processes

The chlorophyll and carotenoid contents of ginger leaves covered with a green, red, or blue film at the seedling stage were markedly higher than those of the control during the whole growth period. At the seedling stage, leaves covered with a blue or green film had the highest concentrations of chlorophyll and carotenoid, followed by leaves covered with a red film. At the vigorous growth stage, the chlorophyll and carotenoid contents were highest in leaves covered with a green film, followed by leaves covered with red and blue film. Pn was highest in ginger leaves covered with a green film, followed by leaves covered with either red or blue film. This suggests that covering ginger leaves with colored plastic films alleviated photosynthetic depression at noon without changing the diurnal variation in photosynthesis (RuiHua et al., 2007).

The diurnal variation of chlorophyll fluorescence in leaves of ginger covered with different colored films was similar. But the photochemical efficiency (Fv/Fm), the efficiency of excitation energy capture by open PSII reaction centers (Fv2/Fm2), quantum of PSII (OPSII), photochemical quenching coefficient (qP), and photochemical reflectance index (PRI) of leaves treated with green film were the highest, followed by blue, white, or red film. The diurnal change in net Pn was highest with green film, followed by white, red, or blue film and that of Pr was reverse. Increasing the ratio of green light in light quality could reduce the photoinhibition, make the correspondence of transferring electron between PSI and PSII well, decrease heat dissipation of excitation energy, and enhance the light energy utilization efficiency (RuiHua et al., 2008). White or red light induced the greatest growth, chlorophyll content, and net Pn. The maximal photochemical efficiency (Fv/Fm) and efficiency of excitation energy capture by open PS U reaction centers (Fv2/Fm2), as well as qP, decreased markedly under midday strong light (RuiHua et al., 2008).

Ginger plants grown as an intercrop were significantly taller than those under open conditions (sole crop), when measured 200 DAP and had significantly lower number of functional leaves and tillers per clump. Interception of photosynthetically active radiation (PAR) by ginger was maximum at 110 DAP, both in open conditions (1.088 ly/min) and in the intercrop (0.788 ly/min). Percentage of PAR intercepted by ginger out of total PAR was lowest at 170 DAP in both open (74.4%) and under arecanut shade (56.41%). Mean duration of ginger crop grown in open conditions was 184.4 days, while it was 198.5 days when grown as intercrop. Per plant yield of ginger under arecanut plantation was significantly higher (154.5 g) when compared with open conditions (118.8 g/plant) (Hegde et al., 2000). Nonvolatile ether extract (NVEE) concentration was low when ginger was grown under low light (Vastrad et al., 2001). Under 0–80% shade, thickness, cell size, chloroplast size, and number of leaves decreased with increased shading, but the number of starch grains increased with increased shading (Zhen Xian et al. 1996).

On sunny days, the photochemical efficiency of PSII (Fv/Fm) and apparent quantum yield of ginger leaves declined gradually in the morning and increased continuously after 12 h. The MDA content increased while the Pn declined under midday high light stress. SOD activity increased gradually before 14 h, followed by a subsequent decrease. In plants which were 60% shaded during the seedling stage, Fv/Fm and AQY of ginger leaves increased while MDA content, SOD activity, and Pn decreased (Xi Zhen et al., 2000). Photoinhibition was observed under high light stress during midday. The extent of photoinhibition reduced as the growth phase progressed and PFD decreased. After shading, AQY and photochemical efficiency of photosystem PSII (Fv/Fm) increased and the degree of photinhibition declined markedly. Greater shading resulted in reduced photoinhibition but Pn declined (ZhenXian et al., 2000).

Light intensity in ginger has a role on the activity of major enzymes, which respond to stress signals. MDA content remained constant and SOD activity increased slightly during the first 40 days of leaf development. Subsequently, MDA content increased rapidly and SOD activity declined. Peroxidase and catalase activities increased steadily during the first 60 days of development, but the rate of increase declined after 40 days (Xi Zhen et al., 1998). A full length (2000 bp) cDNA encoding violaxanthin de-epoxidase (GVDE) was cloned from ginger using RT-PCR and 5′, 3′ rapid amplification of cDNA ends (RACE). The expression patterns of GVDE in response to light were characterized. Chlorophyll fluorescence measurements showed that transgenic plants had lower values of nonphotochemical quenching (NPQ) and the maximum efficiency of PSII photochemistry (Fv/Fm) compared with the untransformed controls under high light intensity. Results showed that violaxanthin de-epoxidase (VDE) plays a major role in alleviating photoinhibition (Jin et al., 2007).

Water Stress and Mulching

Soil water stress decreased the light saturation point, the rate and diurnal changes of photosynthesis, and enhanced midday depression of photosynthesis. Under high temperatures and light conditions, soil water stress decreased the SOD, peroxidase, and catalase activities (Kun and Sheng, 2000). The number of sprouts, sprout frequency, tiller height, and finger size of ginger were better in the mulched plots when compared with the unmulched (control) plots (Vanlalhluna and Sahoo, 2008). The Pn in unmulched fields was lower than that in mulched and shaded fields. Mulching increased the ability of plants to adapt to strong light (Kun and Guo, 2000).

Growth Regulators

Cycocel (CCC) increased the volatile oil content when applied at the rate of 100 ppm. There was an undesirable increase in crude fiber with increase in CCC concentration. Etherel at lower concentrations increased the contents of volatile oil and NVEE as well as crude fiber. An application rate of 200 ppm of CCC significantly reduced rhizome yield although the volatile oil content rose. Kinetin at the rate of 50 ppm increased the contents of volatile oil, NVEE, and starch, but did not affect crude fiber content. At 75 ppm, there was a maximum yield of NVEE (10.6%) and of rhizome yield (15.03 kg/plot) (Jayachandran and Sethumadhavan, 1988). GA inhibited flowering and shoot emergence, while etherel had no effect on flowering but enhanced shoot emergence. LA, leaf development rate (LDR), stem elongation rate (SER), vigor index (VI), and rhizome yield were increased with CCC and decreased with GA and etherel. Application of CCC at 250 ppm significantly improved rhizome yield by 36.4% (Obasi and Atanu, 2004). While starch and crude fiber contents showed significant positive correlation with yield, the crude protein content was negatively correlated (Shankar and Muthuswami, 1985). Phogat and Singh (1987) reported the highest rhizome yields in 2 years of experimentation, from plants treated with 200 ppm etherel.

References

Anderson, T., 1991. Growth phases of the ginger plant. Inlightingsbulletin. Navorsingsinstituut vir Sitrus en Subtropiese Vrugte 225, 17–19.
De Wan, Z., Zhen Xian, Z., Xian Chang, Y., Kun, X., Xi Zhen, A., 1995. Study on canopy photosynthetic characteristics of ginger. Acta Hortic. Sin. 22 (4), 359–362.
Hegde, N.K., Sulikere, G.S., Rethinum, B.P., 2000. Distribution of Photosynthetically Active Radiation (PAR) and performance of ginger under arecanut shade Spices and aromatic

plants: challenges and opportunities in the new century. Contributory papers. Centennial conference on spices and aromatic plants. Indian Institute of Spices Research, Calicut, Kerala State, India. pp. 107–112

Jayachandran, B.K., Sethumadhavan, P., 1988. Effect of CCC, etherel and kinetin on quality of ginger (*Zingiber officinale* R.). Agric. Res. J. Kerala 26 (2), 277–279.

Jin, Li H., Cheng, L.L., Zhen Xian, Z., 2007. Molecular cloning and characterization of violaxanthin de-epoxidase (VDE) in *Zingiber officinale*. Plant Sci. 172 (2), 228–235.

Kandiannan, K., Utpala, P., Krishnamurthy, K.S., Thangamani, C.K., Srinivasan, V., 2009. Modeling individual leaf area of ginger (*Zingiber officinale* Roscoe) using leaf length and width. Sci. Hortic. 120 (4), 532–537.

Kun, X., Guo, M., 2000. Effects of mulching with straw on the photosynthetic characteristics of ginger. China Vegetables 2, 18–20.

Kun, X., Sheng, G., 2000. Effects of soil water stress on photosynthesis and protective enzyme activity of ginger. Acta Hortic. Sin.

Obasi, M.O., Atanu, S.O., 2004. Effect of growth regulators on growth, flowering and rhizome yield of ginger (*Zingiber officinale* Rosc.). Nigerian J. Hortic. Sci. 9, 69–73.

Pandey, Y.R., Sagwansupyakorn, C., Sahavacharin, O., Thaveechai, N., 1996. Influence of photoperiods on dormancy and rhizome formation of ginger (*Zingiber officinale* Roscoe). Kasetsart J. Nat. Sci. 30 (3), 386–391.

Phogat, K.P.S., Singh, O.P., 1987. Effect of cycocel and etherel on growth and yield of ginger. Prog. Hortic. 19 (3–4), 223–226.

RuiHua, Z., KunYou, Z., Kun, X., 2007. Effects of covering with colored plastic films on the pigment content and photosynthesis in ginger leaves. Acta Hortic. Sin. 34 (6), 1465–1470.

RuiHua, Z., Kun, X., Can Xing, D., Ying, L.Y., Jie, Lu, 2008. Effects of light quality on growth and light utilization characteristics in ginger. Acta Hortic. 35 (5), 673–680.

Shankar, C.R., Muthuswami, S., 1985. Influence of CCC on the quality of ginger (*Zingiber officinale* Roscoe.) rhizome. S. Indian Hortic. 33 (4), 271–275.

Shankar, C.R., Swamy, S.M., 1988. Influence of light and temperature on leaf area index, chlorophyll content and yield of ginger. J. Maharashtra Agric. Univ. 13 (2), 216–217.

Sreekala, G.S., Jayachandran, B.K., 2001. Photosynthetic rate and related parameters of ginger under different shade levels. J. Plantation Crops 29 (3), 50–52.

Vanlalhluna, P.C., Sahoo, U.K., 2008. Effect of different mulches on soil moisture conservation and productivity of rainfed ginger in an agroforestry system of Mizoram. Range Manag. Agroforestry 29 (2), 109–114.

Vastrad, N.V., Sulikeri, G.S., Hegde, R.V., 2001. Effect of light intensity and vermicompost on the quality of ginger. Karnataka J. Agric. Sci. 14 (4), 1143–1144.

Xi Zhen, A., Zhen Xian, Z., Zhi Feng, C., LiPing, C., 1998. Changes in photosynthetic rate, MDA content and the activities of protective enzymes during development of ginger leaves. Acta Hortic. Sin. 25 (3), 294–296.

Xi Zhen, A., Zhen Xian, Z., Shao Hui, W., Zhi Feng, C., 2000. The role of SOD in protecting ginger leaves from photoinhibition damage under high light stress. Acta Hortic. Sin. 27 (3), 198–201.

Xian Chang, Y., Kun, X., Xi Zheng, A., Li Ping, C., Zhen Xian, Z., 1996. Studies on the relationship between population, canopy photosynthesis and yield formation in ginger. J. Shandong Agric. Univ. 27 (1), 83–86.

Xu, K., Guo, Y.Y., Wang, X.F., 2004. Studies on the photosynthetic characteristics of ginger. In: Craker Simon, L.E., Jatisatienr, J.E., Lewinsohn, A. (Eds.), The Future for Medicinal and Aromatic Plants. Proceedings of the XXVI International Horticultural Congress Acta Horticulturae, Toronto, Canada, pp. 347–353.

Xu, K., Guo, Y.Y., Wang, X.F., 2004a. Transportation and distribution of carbon and nitrogen nutrition in ginger. In: Craker, L.E., Simon, J.E., Jatisatienr, A., Lewinsohn, E. (Eds.), The Future for Medicinal and Aromatic Plants. Proceedings of the XXVI International Horticultural Congress Acta Horticulturae, Toronto, Canada, pp. 341–346.

Xu, K., Kang, L.M., Zhen, Y.Q., Su, H., 2008. Changes of carbon and nitrogen nutrition during growth of ginger. In: Gardner, G., Craker, L.E. (Eds.), Proceedings of the International Symposium on Plants as Food and Medicine: The Utilization and Development of Horticultural Plants for Human Health, IHC 2006 Acta Horticulturae, Seoul, Korea, pp. 263–268.

Yun, H.D., Ki Young, K., In Lok, C., Dong, K.S., Moon Soo, P., 1988. Change of stomatal behavior and chlorophyll fluorescence to environmental conditions in ginger (*Zingiber officinale* Rosc.). J. Korean Soc. Hortic. Sci. 39 (2), 145–148.

Zhao, D.W., Liu, S.L., Chen, L.P., 1987. A study on the characteristics of translocation and distribution of ^{14}C–labelled assimilates in Lai-wu ginger. Acta Hortic. Sin. 14 (2), 119–124.

Zhao, D.W., Xu, K., Chen, L.P., 1991. A study of the photosynthetic characteristics of ginger. Acta Hortic. Sin. 18 (1), 55–60.

ZhenXian, Z., XiZhen, A., Qi, Z., ShiJie, 2000. Studies on the diurnal changes of photosynthetic efficiency of ginger. Acta Hortic. Sin. 27 (2) 107–111.

Zhen Xian, Z., Yankui, G., Qi, Z., 1996. Effects of shading on ultrastructure of chloroplast and microstructure of ginger leaves. Acta Hortic. Sin. 26 (2), 96–100.

18 Cropping Zones and Production Technology

Climatic Requirements

Ginger requires either a tropical or a subtropical climate, preferring a warm, humid climate and cannot withstand very low temperature. It comes up well up to an altitude of 1500 m above MSL, the optimum being 300–900 m. The base temperature requirement is 13°C with an optimum range of 19–28°C for good growth and crop performance (Hackett and Carolane, 1982). Seed rhizome sprouts better at a soil temperature of 25–26°C and for optimum growth and development of the crop it is 27.5°C (Eveson et al., 1978). It was noted that air temperature of 22–28°C at seedling and early growth stages and 25°C at rhizome enlargement stage are ideal, and the crop stopped growing below 15°C in China (Xizhen et al., 1998). In Kerala, an important ginger-growing state in India, the crop is being cultivated in the air temperature range of 28–35°C. Warm sunny days are preferred for good growth. Extremely high temperature causes sunburn to the plant, while low temperature leads to dormancy. In Australia, it was observed that soil temperature at the top 10 cm layer takes several months to reach optimum after the normal planting season, and hence, full emergence is delayed up to 6 weeks (Weiss, 1997). The crop requires short or long day length for its growth and development. With the increase in day length from 10 to 16 h, the vegetative growth is enhanced, while the decrease in day length from 16 to 10 h promoted rhizome swelling (Adaniya et al., 1989). Kandiannan et al. (2010) reported that heat sum (growing degree days (GDD)) of 278, 753, and 1660 degree days were required for the emergence of new shoot after planting and emergence of first and last suckers, respectively. Good sunshine, heavy rainfall, and high RH are necessary to get good yield (Ridley, 1912). A well-distributed annual rainfall of 1500–3000 mm over a span of 5–7 months is ideal for a rainfed ginger crop. If rainfall is low, water is supplemented with irrigation. The crop is sensitive to waterlogging, frost, and salinity, and tolerant to wind and drought (Hackett and Carolane, 1982). Ginger requires partial shade for better rhizome yield. Frost is injurious to the foliage and rhizome of ginger. Hence, in areas where frost is common, as in hilly regions of north and northeastern India, harvesting is finished before the onset of frost.

Soil Requirements

Ginger can be grown in a wide variety of soils, such as sandy loams, clay loams, alluvial, and lateritic soils. However, it is mainly grown in red and laterite soils of Kerala,

The Agronomy and Economy of Turmeric and Ginger. DOI: http://dx.doi.org/10.1016/B978-0-12-394801-4.00018-1

Karnataka, Odisha, West Bengal, Maharashtra, and northeastern states of India. Well-drained soil, with at least 30 cm depth, loose and friable in texture is preferred for ginger cultivation. By adopting cultural practices such as raised beds and surface mulching, shallow soil can be utilized. Alluvial soils and drained paddy fields or well-drained marshy areas can also be utilized for ginger cultivation as is commonly practiced in China, Taiwan, and Japan (Weiss, 1997). Deep soils with rich organic matter content and nutrient availability are more suitable for ginger cultivation (Cho et al., 1987). However, the crop performs best on medium-textured loams having enough quantity of humus. Even tough virgin forest soils after deforestation are found ideal (Paulose, 1973). Ridley (1912) quoted about the possibility of growing ginger for longer years in the same patch of land, and it is therefore quite unnecessary to destroy forests of great value. Compact clay soils characterized by waterlogging or coarse sands with poor water holding capacity, gravelly soils, or soils with hardpan are unsuitable (Lawrence, 1984). Deep slopes in hilly areas are not advisable for ginger cultivation, as it leads to soil erosion during heavy rainfall. It is grown on volcanic soils in Mauritius and other Indian Ocean Islands and on silty clay loams in Madagascar. In West Africa, forest soils are preferred. In Jamaica, ginger is grown mainly on hilly terrains having clay loam over limestone or conglomerates. In Australia, moderate to heavy soils are utilized.

Growth, development, and maturity of ginger are also affected by soil type. For instance, oil content was highest in cultivar Rio de Janeiro at 180–210 days after planting, when it was grown in calcareous soils in northeast India, compared to those in clayey soils of southern India. Soil hardiness less than 15.7 mm is optimum. The optimum soil pH preferred for ginger is in the range of 5–7 and if the pH is more than 8, growth is retarded. Maximum rhizome yield can be achieved in sandy loam soil having minimum BD (1.20 g/cc), moderately acidic (pH 5.7), high in organic matter content and available potassium, and the yield increased with increasing clay content in soil and decrease in pH (Sahu and Mitra, 1982).

The soil with low BD promotes the growth of ginger and improves yield (Wen et al., 2006). The plant height, stem diameter, branch and leaf number in soil with BD of 1.20 g/cm^3 were higher than the control (BD of 1.49 g/cm^3 by 17.9%, 14.0%, 35.2%, and 36.1%, respectively). The fresh weight of root, stem, leaf, and rhizome were higher by 43.3%, 24.4%, 28.1%, and 35.2%, respectively. The yield, dry recovery, starch, and volatilized oil increased with the increase in soil BD, while the protein and amino acid contents decreased. The change in soil BD had little effect on the fiber and sugar contents. It was found that the soil quality influences the elemental distribution within the ginger rhizome; however, the plant has the inherent ability to control the amount of each element entering the rhizome (Govender et al., 2009). Saline soil was converted for cultivation of ginger with amendment (vermicompost) in Pakistan (Ahmad et al., 2009). The soil should be relatively free of root knot nematode infection and soil-borne diseases causing rhizome rot and bacterial wilt.

Cropping Zones

The site suitability for ginger cultivation in India was investigated with the help of GIS using DIVA-GIS by Utpala et al. (2006). An Ecocrop map for ginger was prepared based on temperature, rainfall, and growth parameters (Hijmans et al., 2005). The study showed that Kerala, parts of Mizoram, Manipur, and Assam in India are most suitable for ginger production. West Bengal and Odisha fall into the category "very suitable." Similarly, the eastern and western parts of Meghalaya are also very suitable. Northeastern states in India with high-yielding local varieties coupled with favorable climate makes them better suited for ginger production. But in states like Tamil Nadu and Gujarat, the climate is only marginally suitable and the area under cultivation is low. The site suitability map shows that Assam, Mizoram, Tripura, and the western part of Meghalaya are all very suitable areas for ginger production. West Bengal, Odisha, Kerala, and the Western Ghat region of Karnataka and Maharashtra states have the highest suitability for the cultivation of ginger in India. The area under the crop has increased from 62,000 ha in 2003 compared to 17,000 ha in 1950–1951. The study further indicated that though the secondary data of area, production, and productivity of ginger have consistently shown an increasing trend during the last three decades, increase in area is not always in proportion to the increase in production.

The area and production of ginger in Mizoram increased over the years with high productivity. Increase in area is less compared to the increase in productivity. Three high-yielding varieties, *Maran, Thingpuri*, and *Thingria* are popular in Mizoram (Utpala et al., 2006) and the state is classified as "Excellent" for ginger production. Here, the impact of intensive agriculture has been negligible in the past decades.

Odisha state is also very well suited for ginger cultivation. But the productivity level is low (about 2000 kg/ha). Misra reported that the tribal farmers, who are mostly economically backward, cultivate ginger. It is grown as a rainfed crop with their local varieties. This is the principal reason for low productivity in the state. As the suitability of ginger cultivation in the state is high, farmers obtain good returns and of late the area under ginger cultivation is increasing. With appropriate technology transfer, Odisha could become one of the highest ginger producing states of India.

Utpala et al. (2008) studied the impact of climate change on the area under ginger cultivation. If the temperature increase is about 1.5–2°C, the suitability of Odisha and West Bengal as good states for ginger production will decline, from the current "high suitability" to "marginal suitability," which establishes the adverse impact of climate change.

Production Technology

Planting Material

Seed Size and Number of Seeds

Ginger is propagated through vegetative seed rhizome called "piece" or "set" or "knob" or "cutting," and the length and weight of pieces vary from place to place and variety to variety. In general, bigger-sized seed material results in quicker emergence and, consequently, better growth and yield, leading to more economic returns. Rhizome pieces weighing 14–56 g with two or three sprouts are sufficient for increased yield and better economic returns (Meenakshi, 1959). Larger seed pieces have produced higher yield and profit (Whiley, 1990). In India, seed pieces weighing 15 g up to 20–25 g are used (Kannan and Nair, 1965; Korla et al., 1989). In Australia, seed sizes vary from 28 to 56 g (BAE, 1971) to 60 g (Smith et al., 2004). In Bangladesh, it is 21–30 g (Ahmed et al., 1988), while in China it is 75 g (Wang et al., 2003). In Puerto Rico, it varies from 86 to 114 g (Beale et al., 2006). In Ethiopia, it is 32 g (Girma and Kindie, 2008), while in Nigeria, it varies from 20 to 25 g (Attoe and Osodeke, 2009). In Nepal, it is 60 g (NARC, 2008).

Yield increase of 33%, 51%, and 80% by increasing the weight of ginger rhizome from 10 to 20, 30, and 40 g, respectively, has been reported (Sengupta et al., 1986), and recorded yield of 50.18 t/ha using 40 g seed set. Whiley (1990) has recorded increased knob size (12%) and yield (26%) by using 85.5 g seed piece compared to 42.5 g seed piece. Nizam and Jayachandran (2001) have compared 5, 10, and 15 g seed size under shade and open conditions and found that mini sets weighing 10 g were more profitable. Wang et al. (2003) from China have tested four seed sizes, namely 100 ± 5 g, 75 ± 5 g, 50 ± 5 g, and 25 ± 5 g and found that the use of large seed size resulted in early emergence and vigorous growth at the early stage. However, the largest and smallest sizes were not beneficial for rhizome growth and yield and concluded that 75 ± 5 g seed size was better than all the other sizes tested. Beale et al. (2006) have compared six types of seed pieces, namely 18–20 g, 29–43 g, 44–57 g, 58–85 g, 86–114 g, and 115–128 g, and concluded 86–114 g to be economically feasible in East Central Puerto Rico. Girma and Kindie (2008) from Ethiopia have tested four types, namely small (4 g), medium (8 g), large (16 g), and very large (32 g), and found that 32 g sets were better for higher yield and higher returns. The International Union for the Protection of New Varieties of Plants, Geneva, has formulated Distinctness' Uniformity and Stability (DUS) guidelines for testing ginger. Their specification for seed piece weight is 80–100 g for raising test crop (UPOV, 1996), whereas the Protection of Plant Varieties and Farmers' Rights Authority, Government of India, New Delhi, suggested 25–30 g seed piece for DUS guidelines for ginger.

Seed pieces with two or three constrictions along their length resulted in better growth than with four constrictions (Lee, 1974). In general, bigger seed pieces have a greater number of buds. Randhawa et al. (1972) have used two types of seed pieces, namely 60 g with two buds and 150 g with 4–6 buds and larger piece with 18.3 buds of 32 g. Xizhen et al. (2005) have provided good information on ginger seed bud and suggested that one or two strong and short buds are sufficient on each piece and the rest removed to pool the whole nutrition on the buds which are left. In most places, planting starts before sprouts emerge from the seed. There may be one or more apical buds on seed rhizome piece, but only one bud becomes active and emerges after planting. When large pieces are used, more than one bud may develop into shoot simultaneously (Ravindran et al., 2005). Irrespective of seed size, it should have one or two viable buds to obtain better emergence and performance. Ginger seeds planted in the soil give birth to new plants and remain attached with daughter rhizomes and are viable without any damage, unless otherwise infected with pests and diseases. Okwuowulu (1988) has recycled the same seed planted in the first year to the following year and the buds remain viable even after 1 year. Studying the effect of set size and weight on plant growth and yield response are of interest and so far only partially understood (Okwuowulu, 2005). Rhizomes after harvest are cleaned, clumps separated into convenient sizes for handling, and good quality rhizomes stored for the next season's planting. At the commencement of the following planting season, the stored rhizomes are taken out, broken into small pieces (that usually contain one or two viable buds), and used for planting. This is done manually by an experienced person. Kandiannan et al. (2010) have observed that 27% of broken seed piece belongs to the seed size category of 21–30 g.

Seed Rate

Seed size determines the seed rate in ginger, in addition to other factors, such as spacing used for planting, soil fertility conditions, and the variety used. Bigger the seed size, the more the seed rate needed. In India, seed rate varies from 1500 to 2500 kg/ha depending on seed size and spacing. Seed rate of 1250 kg/ha with each seed rhizome weighing 30 g was suggested by Randhawa and Nandpuri (1970). Table 18.1 indicates seed size and rates used for different regions within India and other countries where ginger is grown.

Seed rate for plains and lower altitudes is 1500–1800 kg and for higher altitudes (>1000 m above MSL) is 2000–2500 kg/ha (Aiyadurai, 1966). In Queensland, seed rates of 8–10 t/ha are required on fully mechanized farms, but farmers invariably use 4–6 t/ha (Whiley, 1974). Seed rate of 6 t/ha, corresponding to 1,400,000 plant population/ha, was adopted by Lee et al. (1981). In China, seed rate of 5.25–7.5 t/ha with seed sizes of 50–75 g are used. However, under less intensive agricultural conditions, as in India and Sri Lanka, 1.5–4 t/ha seed rate is used (Weiss, 1997). In India, in the state of Kerala, a seed rate of 800–1100 kg/ha is adopted (Mirchandani, 1971), whereas in Himachal Pradesh, a seed rate of 2.3–3.5 t/ha is used. In the northeastern states of India and adjoining areas, farmers plant whole rhizomes and unearth the mother rhizome when the crop reaches 30–35 cm height after 3 months of planting

Table 18.1 Planting Time, Seed Rate, and Spacing Recommended at Different Places

Country/State	Planting Time	Seeding Rate (kg/ha)	Seed Size (g)	Spacing (cm)
India				
Kerala State	April–June	1500–1800	15	20–25 × 20–25
Odisha	April	1800	15–20	25 × 20
Himachal Pradesh	March–April	2500–3500	50–100	30 × 20
Sikkim	February–April	3000–6000	75–150	45–60 × 30
Bihar	April–June	1800	18–20	25 × 20
Meghalaya	March–April	2500–3500	25–50	30 × 30
Andhra Pradesh	April–June	1700	20–25	30 × 20
Other Countries				
Australia	September	8000–10000	50–80	40–60 × 15
China	January–April	5250–7250	75–100	60–65 × 20; 50–55 × 20
Jamaica	May–June	4000–6000	45–50	15–20 × 15–20

Source: Srinivasan et al. (2008).

and 94.6% of seed is recovered (germinability) (Singh, 1982). By using this method, farmers get back 60–75% of the seed cost. Thus, detaching and recycling the sets in the same season or subsequently provides a means of achieving rapid seed ginger multiplication and higher aggregate yield (Okwuowulu, 2005). It is not followed in intensive irrigated production systems and other places where agricultural laborers are unavailable in sufficient number.

Seed Treatment

Seed treatment induces early germination and prevents seed-borne pathogens and pests. Seed rhizomes have to be treated before planting to control rhizome rot and other seed-borne diseases. Park (1937) reported the beneficial effect of seed treatment against *Sclerotium rolfsii*. Seed treatment by farmers in Kerala is done by dipping seed rhizome in cow dung emulsion (Mirchandani, 1971) and another way is to smoke the seed rhizomes once or twice before storage (Mirchandani, 1971). Hot water treatment of seed rhizomes at 48°C for 20 min before planting was also an effective seed treatment (Colbran and Davies, 1969). The cut end of seeds is dipped for 10 min in benomyl 0.25% to prevent entry of pathogens (Whiley, 1974). Treating the seed with formulations such as Agallol 0.5% for 3 min or wettable Ceresan 0.1% for 30 min or Coppersan 0.3% for 60 min (CSIR, 1976), wettable Ceresan 0.25% (Kannan and Nair, 1965), Dithane M 45 (Mohanty et al., 1990) are recommended as seed treatments. Treating the rhizome in ethrel increased the growth and development of ginger (Islam et al., 1978). A pre-plant soak application of Ethephon at 750 ppm with warm (51°C) water for 10 min has increased shoot number of ginger cultivars (Furutani et al., 1985). Ra et al. (1989) have observed that the low temperature (5°C) treatment of seed rhizomes decreased plant weight and rhizome yield. Treating by steeping seed rhizomes in spore suspension of *T. viride* or *T. hamatum* was found effective against the pathogen *P. aphanidermatum* causing rot (Bharadwaj et al., 1988). *T. harzianum*, *T. aureoviride*, *G. virens*, and *T. viride* treatments reduced rhizome rot caused by *F. solani* or *P. myriotylum* and significantly increased growth, development, and yield of ginger (Ram et al., 2000). The cut rhizome pieces are dipped in fresh and clean plant ash to seal off the wounds (Xizhen et al., 2005).

Land Preparation

Fields put to ginger cultivation must have soil which is loose and friable. The preparation of the field depends on climate, soil type, farm size, and irrigation methods. In arriving at the right decision as to how to go about ginger cultivation, the farmer must exercise his or her discretion and judgment. The soil should be thoroughly broken up, pulverized with a hoe or plow, and if possible, harrowed subsequently. Without such improvement in the tilth, the crop would fail to produce well-shaped rhizomes. Well-shaped rhizomes are desirable to fetch a good market and amenable for ease of postharvest curing methods. In India, the land plowed several times or dug thoroughly with the receipt of early summer showers help in obtaining good tilth. In Jamaica, clayey soils are forked and left to dry for 3 months prior to planting, and again a month before planting the soils are forked and drained by means of trenches (Graham, 1936), and for planting, the ridge and furrow method is used. In the furrow method, the seed rhizomes are planted in furrows 22.5 cm deep and 22.5 cm apart. When the ridge method is used, the rhizome pieces are planted 30 cm apart, but only a few centimeters below the surface in prepared ridges, which is about 30 in. high (Lawrence, 1984). In Queensland, after initial plowing during November–December, a cover crop of maize is planted during summer to build up organic matter or the land is kept fallow. Poultry manure or press mud (sugarcane factory waste) provides additional organic matter and root knot nematode infestation is controlled by soil fumigation (Whiley, 1974).

Plowing six times brings soil to good tilth. In West Bengal, after the first rains, during March–April, 12–13 plowings are done and the soil is then leveled and water channels are formed to irrigate the crop. The water channels are made 60–80 ft apart and connected by smaller ones running at right angles to the main channels, which is about 8 ft apart (Ridley, 1912). However, there is no added advantage in plowing the land over the minimum requirement of 3–5 plowings (Aiyadurai, 1966). Solarization of beds for 40 days by using transparent polythene sheets with *Trichoderma* application is recommended for areas prone to rhizome rot and nematode infestation. Tillage starts after the harvest of previous crop during autumn or winter in China. Plowing 25–30 cm deep promotes root penetration and enlarges root foraging area. After harrowing once or twice, 75–120 t/ha of good quality farmyard manure is applied and the land is leveled (Xizhen et al., 2005). Plowing followed by harrowing several times at a 5- to 7-day interval to kill weed growth and weed seed germination and rotovating is common in African region (Okwuowulu, 2005). In hill slopes, beds are laid along contours to reduce soil erosion. Ridge and furrow with 40 cm spacing apart for irrigated crop is practiced. Aiyadurai (1966) suggested that the flatbed system for sandy loam soil and raised beds for clay loam soil were the most suited for successful ginger cultivation.

Two distinct methods of cultivation, namely Malabar (Kerala State) and South Kanara (Karnataka State) systems, are prevalent in India. In the Malabar system, beds of 3 m × 1 m in size is formed 30–45 cm apart with small shallow pits on beds for planting the sets at required spacing and a handful of cattle manure is applied to each of these pits. In the South Kanara system, there are no beds, instead, a mixture of manure and burnt earth is applied in the form of a small 5 cm thick ridge in between the rows 100–200 cm apart from each other and the seed rhizomes are placed in the rows and earthed to make the ridges 15–20 cm high (CSIR, 1976). Ginger is planted on a raised bed to facilitate drainage in China. Planting ginger in raised beds, and irrigating the crop, gave a higher yield compared to the crop planted in ridges, furrows, and flat ground in field research trials. Raising ginger in flatbeds in sandy loam soil and on raised beds in clay loam soil, followed by earthing up, with the application of fertilizers, is most suited for successful cultivation of ginger (Singh et al., 2003). Shaikh et al. (2006a) compared three systems of planting, namely flatbed, ridges and furrows, and raised beds, and observed that planting in flatbeds resulted in the highest yield of fresh and dry rhizomes (153.8 and 30.35 g/plant, respectively), and green rhizomes yield (20.34 t/ha).

Regions away from environmental contamination sources with good drainage are eminently suitable for organic ginger cultivation. A buffer zone with suitable physical/tree barriers or buffer crops all around the organic plantation to avoid contamination by drift from neighboring conventional plantations is absolutely necessary. Ginger grown in the isolation zone as a buffer crop is not sold as organic. The area of buffer zone may vary with the landscape and will be specified and certified by the certification agencies based on the local agro situations (Srinivasan et al., 2008). A large number of tribal farmers in northeastern India still practice the traditional methods of cultivation, wherein ginger is cultivated in *Jhum* lands, *Buns*, *Zabo* lands, terraced lands, and in the plains. In the traditional methods of cultivation, farmers rely on organic inputs, local resources, and practices (Rahman et al., 2009).

Planting

The time of planting is very important in ginger cultivation as it affects both yield and quality. The main factors considered while planting are season, date of planting, spacing, and depth of seed placement (Parthasarathy et al., 2003). The planting time depends on the onset of monsoon. In India, generally planting is done during March–June (Table 18.1) (Phogat and Pandey, 1988; Randhawa et al., 1972). Studies have shown that planting during April gives better growth and development of rhizomes and fewer incidences of diseases. In Australia and Fiji, planting is done during September (Whiley, 1974), while in West Indies it is during March–April (Ridley, 1912). It is in the mid-April in southeastern Nigeria (Okwuowulu et al., 1989), May–June in Jamaica (Prentice, 1959), February–April in Taiwan (Lawrence, 1984), April–May in Sierra Leone and Hawaii (Furutani et al., 1985), June in Ghana (Lawrence, 1984), and January–April in China (Xizhen et al., 2005). However, the irrigated crop can be planted any time of the year, the optimum time being mid-February in India. Normally, ginger planting commences with the onset of SW monsoon in India, which normally breaks in the first week of June in the Kerala State. Aiyadurai (1966) observed a yield increase of 200% by planting ginger during the first week of April, with the receipt of summer showers, than in the normal planting time of May–June. Early planting of ginger is beneficial as the crop can grow sufficiently well to withstand heavy rains, by July–August (Nair and Varma, 1970). Better yield was obtained when planting was done on May 15, which was found on par with planting on April 30 (Nimkar and Korla, 2009). The cost–benefit ratio of ginger cultivation indicated that planting on April 30 with mulching and application of FYM was quite remunerative in the mid-hilly regions of Himachal Pradesh, India. Generally, ginger production is not by transplantation, but in the hilly regions of Himachal Pradesh this practice is followed. The highest yield of 1.08 kg/plant was recorded transplanting 90-day-old seedlings and the lowest yield of 0.48 kg/plant was obtained transplanting 30-day-old seedlings (Kumar and Korla, 1998).

Spacing

Plant population in a unit area of land is an important factor, which decides the yield and it depends on spacing and varies with soil type, variety, climate, and management practices. Closer spacing gives higher yield with maximum plant density (Aiyadurai,

1966; Nair, 1982). Different spacings are 15–45 cm × 15–45 cm, which is recommended (Table 18.1), 40 cm × 15 cm (Lee et al., 1981), and 60 cm × 11.8 cm in Australia (Whiley, 1981), 38.1 cm × 38.1 cm in Mauritius, and 45 cm × 40 cm in Trinidad, West Indies (Wilson and Ovid, 1993). In China, 60–65 cm × 19–20 cm spacing with a planting density of 1,05,000–1,12,500 plants/ha for *Lai Wu* slice ginger gave better yield (Xizhen et al., 2005). In Africa, spacing of 20 cm × 20 cm for ware—ginger production and 10–15 cm × 10–15 cm for seed ginger production through mini set technique is followed (Okwuowulu, 2005). By adopting a correct spacing of 20 cm × 20 cm and by timely planting, weed density could be reduced to the bare minimum. Whiley (1990) observed that higher planting densities with intrarow spacings of 11.2 and 17.0 cm increased rhizome yield up to 43% and reduced the time to final first harvest maturity (35% fiber-free rhizome) by about 10 days. Zaman et al. (2008) reported that spacing of 40 cm × 20 cm significantly gave the highest among all yield attributes, leading to the highest (30.81 t/ha) yield. The combination of 40 cm × 20 cm spacing and planting at 9 cm depth gave the maximum yield (40.0 t/ha) and cost–benefit ratio of 1.00:3.41.

Depth of Planting

Optimum depth is essential for early germination/emergence, and planting depth may vary depending on seed size, soil type, and soil moisture conditions at the time of planting. In general, planting is deeper for bolder rhizomes and shallower for smaller ones. Based on field trials, planting on raised beds at a spacing of 20–25 cm × 20–25 cm and a depth of 4–10 cm with the viable bud facing upward is recommended (Kannan and Nair, 1965; Lee et al., 1981). Okwuowulu (1992) has reported that planting at 5 cm depth predisposes the rhizome to significant desiccation loss under delayed harvest; hence, the author recommended deeper seed placement at a depth of 10 cm. Zaman et al. (2008) have found that rhizome placement at 9.0 cm depth produced the highest yield (31.58 t/ha). Lighter irrigation after planting is beneficial for better germination. Ginger takes 10–15 days after planting to sprout under ideal conditions, but sprouting might prolong up to 2 months. Kandiannan et al. (2010) have reported that around 400 growing degree days (GDD) or heat sums/heat units are required for proper emergence.

Mulching

The purpose of mulching, in rainfed areas, is to primarily conserve soil moisture and check weed growth. Additionally, it also checks surface run-off, when there is excessive downpour, as can happen in Indian conditions, and help conserve soil. It also provides shade to the sprouting rhizome. In subtropical and temperate conditions, mulching can lead to raising of soil temperature. Other advantages are: it enhances germination, increases water infiltration, conserves soil moisture, regulates soil temperature, decreases surface evaporation, enhances microbial activity, and improves soil fertility by adding organic matter. Mulching could change the physical and chemical properties of the soil, resulting in increased availability of certain nutrients like phosphorus and potassium (Muralidharan, 1973). In Fiji, it was found that mulching led to a decrease in nematode infestation (Haynes et al., 1973). The quantity of mulch applied varies with the availability of the material used for mulching. In general, 10–30 t/ha or even more, is applied twice or thrice during crop growth, the first at planting time, the second 45 days after planting, and the third 90 days after planting. Mulching the beds with green leaves at 15 t/ha after planting followed by two mulchings (at 7.5 t/ha) at 45 days and 90 days after planting is an essential practice (Nybe and Mini Raj, 2005).

Research findings have shown that straw mulching increased yield by 12.2% over unmulched conditions (Joachim and Pieris, 1934). Application of forest leaves at 20 t/ha in two equal splits, the first at planting time and the second 45 days after planting, increased ginger yield by 200% (Kannan and Nair, 1965). Application of 35 t/ha FYM as mulch increased yield by 65% (Aiyadurai, 1966). Application of mulch at the rate of 12.5, 5.0, and 5.0 t/ha at the first, second, and third time, respectively is optimum (Randhawa and Nandpuri, 1970). Kingra and Gupta (1977) used dry grass and forest leaves as mulch, at a rate of 15 t/ha, whereas Mohanty and Sarma (1979) used 15 t/ha green leaves at planting and 7.5 t/ha each at 45 days and 90 days after planting. Owadally et al. (1981) stated that mulching with sugarcane trash and rice straw was beneficial. Performance of different live mulches (like a standing crop of sunhemp, *Crotalaria juncea*, in the ginger field) was similar but superior to unmulched conditions. Korla et al. (1990) found that FYM mixed with grass, pine needles, and paddy straw was effective mulch and led to increased yield. Mulching three times with leaves and growing an intercrop like soybean as live mulch was equally effective (AICRPS, 1992a). Polythene as mulch material gave fresh yield of 19.9 t/ha compared to 12.0 t/ha obtained in unmulched plots (Mohanty et al., 1990).

Commonly used mulch materials are green and dry forest leaves, residues such as sugarcane trash, paddy, wheat, finger millet, barley straw and coconut leaves, banana leaves, dry sal leaves, and vegetation from the locality. Coconut leaves (Aclan and Quisumbing, 1976), banana leaves (Mohanty, 1977), dry sal leaves (AICSCIP, 1985), *shisham* (Jha et al., 1986), and green forest leaves (Roy and Wamanan, 1989) performed well as good mulches and increased ginger yield. Mishra and Mishra (1982) have reported that mulching with dry leaves suppressed the weed growth and increased seedling emergence, growth, and yield. Ahmad et al. (1983) have tested three different types of mulches, namely saw dust, sand, and saw dust + sand, and found that saw dust has proved the best of all as there was positive association between this material and the yield components. Das (1999) has showed that mulching with neem leaves (*Melia azadirachta*) reduced the rhizome rot caused by *Pythium* infection. FYM and compost are also used as mulches. If the quantity of the above-mentioned materials are in short supply, live mulches like sunhemp, green gram, horse gram, black gram, niger, sesbania, cluster bean, French bean, soybean, cowpea, dhaincha, and red gram can also be grown as intercrops and mulched *in situ* between 45 and 60 days after planting of ginger (Kandiannan et al., 1996). Growing green manure crops like *Sesbania rostrata*, *S. aculeate*, *S. speciosa*, or fodder cowpea among ginger crops and using as a second or third mulch provided biomass of 4 kg green leaves per 1.5 m^2 (Valsala et al., 1990), which reduced weed growth and increased ginger yield (Kurian et al., 1997). In Sikkim, ginger beds are usually covered with leaves and twigs of various forest trees after planting. In Meghalaya, application of locally available organic mulches like paddy straw and dry leaf mulch of *Schima wallichii* at a rate of 16 t/ha increased the dry yield of ginger. One-fourth of the recommended rate (30 t/ha) of green mulch could be saved if ginger is inter-cropped under coconut plantations with 25% shade (Babu and Jayachandran, 1997). Ginger when intercropped with green manure crops, like daincha and sunhemp, the latter are plowed into the soil at the time of their flowering and used as mulch.

In African countries, green leaves of guinea grass (*Panicum maximum*), Elizabeth weed (*Chromolaena odoratum*), and rubber at the rate of 10 t/ha, decomposed saw dust (with low C:N ratio) at 30 t/ha, or well-rotted deep litter poultry manure or cow dung at 20 t/ha are commonly used for mulching (Okwuowulu, 2005). In China, mulching with a layer of 3–5 cm thick wheat straw, applied at the rate of 3–4.5 t/ha decreased soil tem-perature, maintained soil moisture, and improved the field microclimate. Plastic film mulching, stretching a double layer of black plastic film tightly over the ridge, is also popular among ginger growers, as it is a better material than straw, which led to about 8–30% increase in yield and economic benefit (Xianchang et al., 1996). Polythene as mulch material gave 19.9 t/ha of fresh rhizome compared to 12 t/ha in unmulched plots (Mohanty et al., 1990). Chukwu and Onyekwere (1996) have reported that inorganic fertilizer + mulch or mulch alone increased the productivity and sustainable production of irrigated ginger on soils of moderate fertility in Nigeria.

Three mulches, namely paddy straw, *Gliricidia maculata*, and *Lantana* (a perni-cious weed in red laterite soils), were compared as mulches in rainfed ginger. Paddy straw retained the highest soil moisture. However, *Gliricidia* recorded the highest average rhizome yield (7.29 t/ha), net returns (Rs 69,095/ha approximately US$ 1200 at current rate of exchange) and a benefit–cost ratio of 1.76 in India. Highest rainfall water use efficiency was also obtained with *Gliricidia* mulch (7.15 kg/ha/mm) (Dass et al., 2006). Vanlalhluna and Sahoo (2008) compared three mulches, namely rice straw, subabul leaf, and common weeds, at 6, 8, and 10 t/ha, respectively. Soil moisture retention in the plots was in the following order: straw > subabul leaf > weeds and the rates used were classified as follows: 10 > 8 > 6 t/ha. The number of sprout, sprout frequency, tiller height, and finger size of ginger were better in the mulched plots when compared with the unmulched (control) plots. Ginger yield in the mulched plots found to be in the following order: subabul leaf > rice straw > weeds. This clearly shows that though straw retained more soil moisture, it was subabul leaf that led to higher ginger yield. The slower decomposition of rice straw (because of wide C:N ratio) caused greater retention of moisture, while quick decomposition of subabul leaf caused increased addition of nitrogen to soil thereby increased ginger yield. Ghosh (2008) has reported that paddy straw mulch produced the highest yield (28.70 t/ha) followed by composted coir pith (27.60 t/ha) and water hyacinth (27.40 t/ha), as compared to the lowest yield (15.3 t/ha) in the control treatment. The net return from intercrop was high-est in the case of straw mulch at Rs 132,400/ha (approximately US$ 2300 at the current exchange rate) with the maximum benefit–cost ratio of 1.36. Application of FYM at 10 t/ha as mulch resulted in the maximum rhizome yield, compared to other mulches (Mishra and Pandey, 2009). Mulched plants matured earlier than unmulched ones (Ning et al., 2009). Sengupta et al. (2009) have reported that dry leaves mulch at the rate of 5 t/ha led to maximum height of the ginger plants (78.05 cm), number of pseudostem per clump (4.26), leaves per clump (62.65), and the highest yield (52.17 t/ha). Mulching is especially beneficial when ginger is grown under rainfed conditions.

Weed Management

Initial slow growth of the ginger plant facilitates rapid weed growth. Weeds compete for moisture, nutrition, space, and sunlight. Mulching suppresses weed growth and increases the crop emergence, growth, and, ultimately, yield (Mishra and Mishra, 1982). In general, 2–3 weedings are done depending on weed growth (Mohanty et al., 1990). The first weeding is done just before second mulching (at 45 days after planting) and the second weeding is done 120–135 days after planting (Kannan and Nair, 1965). Weed flora varies from place to place. Melifonwu and Orkwor (1990) have recorded some of the commonly occurring weeds in ginger fields in Nigeria. *Galinsoga parviflora*, *Cyperus* sp., *Euphorbia* sp., and *Setaria glauca* (*S. pumila*) are the major weeds in Palampur, Himachal Pradesh, India (Sinha, 2002). Manual weeding consists of either just pulling out the weeds, chipping with a hoe, or just cutting the roots with a knife (Purseglove et al., 1981). The preemergent herbicide diuron (4.5 kg a.i./ha) is used to control weeds in Queensland, Australia, but its action might vary depending on the soil in which it is used (Whiley, 1974). Paraquat is used as a postemergence herbicide in the early stages of plant growth, applied between plant rows, and in later stages limited to spot spraying between the beds. Preemergence application of 2,4-D at the rate of 1 kg a.i./ha (Mishra and Mishra, 1982) or atrazine at 1.5 kg a.i./ha was also found effective in controlling weeds (Rethinam et al., 1994). Preemergence applications of mixtures of alachlor + chloramben or fluometuron at the rate of 0.75 + 0.75 kg a.i./ha has provided effective control of some weed species, though not all, and has led to higher rhizome yield (Melifonwu and Orkwor, 1990). Kurian et al. (1997) have reported that *in situ* green manuring reduced the weed problem in ginger. Herbicide usage reduced labor requirement in Nigeria (Aliyu and Lagoke, 2001). Black polythene mulch was also effective in weed suppression. But, the cost–benefit ratio was higher in the case of grass + FYM mulching (Sinha, 2002). Emulsifiable concentrate (EC) of 40% pendimethalin + acetochlor as a preemergence herbicide was found to effectively check weed growth than in the case of separate applications (Yang et al., 2004). Barooah and Saikia (2006a) have reported that mulching after planting + hoeing 40 days after planting + grubber at 60 days after planting + hand weeding at 90 days after planting + mulching resulted in significant increase in yield attributes, dry ginger yield, oleoresin, and volatile oil contents. Weed management is an important cultural operation in ginger production, but traditional manual weeding is no more possible because of labor paucity. An important offshoot of excessive use of chemical herbicides is the adverse environmental fallout.

Earthing Up (Hilling)

Soil stirring and earthing up are essential to break the soil hardpan formed by rain or irrigation, and it also helps in checking weed growth, conserving soil moisture, mixing applied manure thoroughly with the soil, enlargement of daughter rhizomes, and provides adequate aeration for root expansion and protects the rhizomes from the ravages of scale insects (Panigrahi and Patro, 1985). Earthing up may be combined with hand hoeing (weeding) and mulching. Shaikh et al. (2006b) have found that earthing up 4 months after planting led to higher rhizome yield (19.10 t/ha) compared to earthing up 3 and 5 months after planting (14.33 and 15.94 t/ha, respectively).

Irrigation

Ginger is principally cultivated as a rainfed crop. Where rainfall is less, the crop needs regular/supplemental irrigation. According to the Queensland Irrigation and Water Supply Commission, ginger requires 1320–1520 mm of water during its complete crop cycle. In India, irrigation at fortnightly intervals, usually during the middle of September to the middle of November increased the yield by 56% and resulted in better quality of the product (Aiyadurai, 1966). In Queensland, Australia, irrigation is provided to ensure that the crop receives approximately 10 mm of water every second day from mid-January until early March, when the most rapid growth takes place (Whiley, 1974). Overhead sprinkling to protect against sunburn is extremely important in late October, November, and December. Australia has a polar climate and the summer is intense from November onward until the end of December. At this time, irrigation is usually provided between 10 AM and 3 PM daily. This establishes a cool microclimate for the crop. In South Africa (also where the climate is polar), when the ambient temperature reaches 26°C during October–November and 30°C during December–March, impulse sprinkler irrigation creates a favorable microclimate for the crop (Anderson et al., 1990). This results in better plant growth and enhances yield by about 25%. Ginger planted in the first week of May in Odisha State, India, needs two or four initial pot watering at weekly interval depending on soil type. Following this the rainfall breaks, which lasts until the end of September. Subsequently, the crop has to be provided pot watering commencing from mid-October to the end of December at fortnightly intervals (Panigrahi and Patro, 1985). The critical stages for irrigation are germination stage, rhizome initiation stage (90 days after planting), and rhizome development stage (135 days after planting). Vaidya et al. (1972) has recommended that the first irrigation is at immediately after planting and subsequent irrigations are given at intervals of 10 days with a total water usage of 90–100 cm in 16–18 irrigations. A fortnightly irrigation during the drier part of monsoon months contributed to increased yield and quality of the produce (Gupta, 1974). Increased water supply has increased the yield of rhizomes and essential oil content (Lawrence, 1984). Scheduling of irrigation at 60 mm cumulative pan evaporation (CPE) (Gavande, 1986) and Irrigation Water (IW)/CPE ratio of I.O (KAU, 1994) produced the maximum rhizome yield. Irrigation once in a 4- to 6-day interval during summer or drought months is practiced in China to keep the soil moisture at about 70–80% (Xizhen et al., 2005).

Shade

Ginger prefers light shade for good growth (CSIR, 1976) and shade requirement depends on location. Shading is helpful in reducing water loss and general cooling of the plant (Lawrence, 1984). Instead of shade, overhead sprinkler irrigation protects the crop from sunburn by evaporative cooling (Whiley, 1974). Aclan and Quisumbing (1976) have recorded shorter plants with fewer leaves per tiller when ginger was grown under full sunlight in the Philippines, whereas Wilson and Ovid (1993) have observed increased height and tiller number under shade. Shade also enhanced the net assimilation rate (NAR) and chlorophyll content (KAU, 1992). Under low to medium shade, higher dry matter production, nutrient uptake (KAU, 1992), yield (Aclan and Quisumbing, 1976), and quality (Ancy and Jayachandran, 1993) were obtained. Heavier shade (beyond 50%) decreased tiller and leaf number (KAU, 1992) and yield (Aclan and Quisumbing, 1976). Ginger grown in shade shrinks and hence, Graham (1936) recommended that ginger be grown in the open in Jamaica. Growing ginger completely exposed to sun resulted in higher yield (Aiyadurai, 1966) and oleoresin content (KAU, 1992), than when the crop was grown in shade. But, Ravisankar and Muthusamy (1986) have reported that shade had no adverse effect on the quality of the rhizome.

Variety *Himachal* was found to perform well both in the open and in low light intensity (75% shade) with respect to dry matter accumulation and uptake of N, P, and K (George et al., 1998a). Ginger grown under low to medium shade (25%) recorded good quality parameters (volatile oil and starch contents, less crude fiber) (Ajithkumar and Jayachandran, 2003), compared to those grown under open and heavy shade in the climatic conditions prevailing in Kerala State, India. Under low and medium shade, nutrient uptake and rhizome yield of ginger were more comparable with that grown in high shade or in the open (Ancy and Jayachandran, 1996). The cultivars Jamaica, Valluvanad, Jorhat, and Rio de Janeiro are suitable for intercropping in plantations with 50% or more shade levels. The incidence of *Phyllosticta* leaf spot is also much less under shade. Hazarika et al. (2009) have found that intercropping ginger in *som* (*Persea bombycina*) plantations planted with a spacing of 3 m × 3 m showed maximum plant height, leaf number per plant, finger number per rhizome, length and weight of single finger, and single rhizome weight compared to that grown in the open. Shade requirement depends on location and differential response of varieties also. The farmer should be the best judge to arrive at the right decision considering various aspects, detailed above, to get the best possible results.

Cropping Systems

Generally, ginger is grown as a sole crop, and often it can also be grown as an intercrop under other perennial crops, such as arecanut, coconut, cocoa, and rubber, for the farmer to get additional income. It can also be intercropped with vegetables, such as cabbage, tomato, chili (green), French bean, and lady's finger, pulses (pigeon pea, black gram, and horse gram), cereals (maize, finger millet), oilseeds (castor, soybean, sunflower, and niger), and other crops such as sesamia, tobacco, and pineapple. Intercropping with soybean (AICRPS, 1992a), lady's finger (Chowdhury, 1988), pineapple (Lee, 1974), and maize (Kandiannan et al., 1999) was found advantageous. Ginger can also be grown as a mixed crop with castor, red gram, maize, and finger millet. As ginger requires partial shade, it can also be grown as an under crop in coconut, arecanut, rubber, orange, stone fruit, litchi, guava, mango, papaya, loquat, peach, coffee, and poplar plantations. Singh et al. (1991) indicated that ginger is the most favored crop component in agroforestry.

In coffee-based spices (black pepper, cardamom, etc.), a multistoried cropping system, local cultivars of ginger grown as an intercrop fetched a gross income of Rs 79,800/ha (approximately US$ 1400 at the current rate of exchange) during the fourth year from the date of the commencement of the multistoried cropping system (Korikanthimath et al., 1995). Ginger is a very common intercrop in the coconut-based cropping system. As an intercrop in an arecanut plantation, ginger showed a higher yield coupled with bigger rhizomes and taller plants, but the duration of the crop was lengthier (199 days) than in the case where the crop is grown solely in the open (Hegde et al., 2000).

It is possible to raise green manure crops like *dhaincha* (*Sesbania aculeata*) in the interspaces of beds (of the mai crop), which could add 50% of the green leaves required for mulching, suppressing weed growth, and reducing cost of production. Ginger is grown as a mixed crop with castor, red gram, finger millet, and maize. Sharma and Bajaj (1998) have recommended intercropping ginger with bell pepper (*Capsicum annum*) for reducing nematode infestation. This system reduced the yellow ginger mosaic virus disease by 76% (Dohroo and Sharma, 1997). In Bangladesh, ginger produced high yield (4.33 t/ha) and benefit–cost ratio (1:6.2) when grown under juvenile mango trees (8–10 years old) (Haque et al., 2004). Growing the fodder tree *Quercus leucotrichophora* with ginger is found to be the most ideal and remunerative silvi–horticultural combination in the hills of Uttar Pradesh in India (Bisht et al., 2000). Nutrient use efficiencies were higher under *Ailanthus*–ginger mixed cropping (Thomas et al., 1998). Ginger in the interspaces of 5-year-old *Ailanthus* trees with 52% mean daily PAR flux or 73% midday PAR flux exhibited better growth and maximum rhizome yield compared to sole crop and also *Ailanthus* + ginger combinations, which also improved the soil nutrient status (Kumar et al., 2001). Ginger yield reduced when it was intercropped under closely planted poplar tree (*Populus deltoides* "G-3" March) when compared to the situation where the poplar trees were planted at a spacing of 5 m × 4 m (Jaswal et al., 1993). In Eastern China, ginger was found to be an ideal shade crop in *Paulownia elongata* plantations (Newman et al., 1997).

Yield of ginger was found to be higher in the open field (14.7 ha) compared to that obtained in plots of agroforestry (Ghosh et al., 2006). In an agroforestry system, the highest ginger yield (12.3 t/ha) was obtained when ginger was grown in the alleys of *P. deltoides* (poplar tree) and the lowest (3.3 t/ha) when grown in the alleys of *Casuarania equisetifolia*. There exists a significant positive correlation between yield of ginger and PAR and atmospheric and soil temperature. However, yield showed a negative correlation with RH. Sanwal et al. (2006) found that different intercropping systems affected some of the growth characteristics and yield of ginger in a negative manner except when legumes are grown as intercrop. Obviously, the soil fertility enriching characteristics of the legumes must have been the dominant source of influence in this positive effect. The highest net monetary return recorded was in the ginger + cowpea combination followed by ginger + French bean combination. When ginger was intercropped with nonleguminous crops, the total N and K uptake differed significantly. Land equivalent ratio (LER) values were always more than one in intercropping systems. Overall, ginger intercropped with cowpea and

French bean was found to be the most suitable and an economically viable combination in the mid-hill agroclimatic conditions of northeast hill regions of India, such as Assam, Meghalaya, and Manipur. Investigations of Prajapati et al. (2007) showed that the growth and biomass of ginger were higher in unpruned *Ceiba pentandra* (L.) Gaertn compared to pruned condition. Kumar et al. (2010) reported that interception of PAR by ginger crop 150 days after planting as intercrop in tamarind plantation (6 m × 6 m spacing) was 25,229 lx compared with 31,643 lx in open space. Higher number of rhizomes and yield (173.89 g/plant) were recorded in the intercropped area compared to sole crop grown in open space.

Crop Rotation

Crop rotation is essential, as ginger depletes a lot of soil nutrients and monoculture with ginger can lead to soil infestation with pathogens, primarily *Pythium*, which causes the dreaded rhizome rot. Ginger is rotated with tapioca, chilies, sesame, little millet, and dry paddy in rainfed conditions, while finger millet, groundnut, maize, and vegetables are rotated with ginger where the crop is grown in irrigated conditions. Pegg et al. (1969) recommended beans, cucurbits, and strawberries as suitable crops for rotation as cover crops to minimize nematode infestation. Crop rotation with tomato, potato, eggplant, and peanut should be avoided as these plants are hosts of the wilt-causing pathogen *Ralstonia solanacearum*. Ridley (1912) reported of an instance of a farmer who continuously grew ginger in the same field year after year without any yield reduction or any of the associated problems mentioned above. Obviously, this was a long time ago, when farming was still uncomplicated and without the trappings of the chemically driven so-called "green revolution" that started in the early 1960s when high-input technology took over. Along with it, a host of soil and environmental-related problems also began to prevail. There are a number of instances now to prove this.

Harvesting

The harvest time depends on the product for which the rhizomes are eventually used, prevalent price trend in the market, and climatic conditions. When the rhizome is used as a vegetable or for preparation of ginger preserve, candy, soft drinks, pickles, and alcoholic beverages, harvesting should be done 4–5 months after planting, whereas when it is used as dried ginger, and also for the preparation of value-added products like ginger oil, oleoresin, dehydrated and bleached ginger, harvesting should be done between 8 and 10 months. Fiber, volatile oil contents, and pungency levels are the most important criteria to assess the suitability of ginger rhizomes for processing (Purseglove et al., 1981). And relative abundance of these components depends on the stage of maturity at harvest (Natarajan et al., 1972). Oleoresin and oil contents rose sharply after 5½–6 months, beyond which there was a decline and fiber development was extremely rare between 6 and 7 months of growth (Winterton and Richardson, 1965). Although there is fiber in the rhizome which develops from time to time, the amount is insignificant in the initial stages. Aiyadurai (1966) has reported that crude fiber content increased beyond 260 days after planting. The diameter and strength of fiber increase with the physiological age of the rhizome. Fibrous ginger is unacceptable in the manufacture of processed confectionary because of its reduced palatability. In Australia, harvesting of confectionary grade ginger begins when 40–50% by weight of the rhizome is free of commercial fiber, which continues down to a 35% level (Whiley, 1990). Increase in crude fiber and decrease in fat and protein contents of rhizome was noticed after 6½ months after planting (Jogi et al., 1972). Oleoresin and oil contents of different cultivars reached their maximum 265 DAP (Nybe et al., 1982). Ratnambal et al. (1989) have observed that dry recovery, starch, and crude fiber were positively correlated with maturity, whereas essential oil, oleoresin, and protein contents were negatively correlated with maturity.

In India, early harvest at 200–215 DAP gave higher yield than at late harvest at 230–245 DAP (Aiyadurai, 1966). In Australia, early harvest yielded 50 t/ha and late harvest 90–100 t/ha (Lee et al., 1981). However, Nair and Varma (1970) and Pawar and Patil (1987) have observed no differences in yield when ginger was harvested 215–275 DAP. Harvesting of ginger is done using a spade, hoe, or digging fork in India, while mechanical diggers are used in Australia. It is extremely important that care is taken to minimize physical damage to the rhizomes while harvesting. The soil, roots, and tops are separated from the rhizomes and washed in clean running water. In India, harvest is generally carried out in the months of January–April and the date may vary with the locations. Irrigation is completely stopped a month prior to harvest allowing the plant tops to dry. Leaves and stem are cut to the ground and the rhizomes are dug out by hoeing or plowing. In the plains where the average temperature fluctuates between 30°C and 35°C, the crop is ready for harvest in about 8 months after planting, when the plants turn yellow and start drying up gradually. In Nepal, Sikkim, and Darjeeling in India, removal of seed (mother) rhizome after 2–3 months of planting is a local practice. By this practice, farmers get back their investment on seed rhizome. Late harvest is practiced in areas where there is no damage to the rhizome due to heat or pests and diseases. The fresh rhizome yield varies with variety and time of harvest; in general, yield may vary between 20 and 50 tons of fresh rhizomes/ha. The dry recovery also varies as a consequence of dry yield. Organic farming experiments on ginger produced a yield (14.3 t/ha) which was 22.8% and 25% lower compared to chemical and integrated farming in the first year of experimentation (IISR, 2005). But the practical experience with organic farming is that in the first 3 years after a switch from intensive chemical farming to organic farming, there is a yield depression, thereafter yield stabilizes and then gradually increases.

Seed Storage

Seed storage in ginger is a very important operation to ensure good germination when the sets from stored seed are used for subsequent planting. Duration between first harvest and the next planting is approximately 120–150 days, sometimes even longer. As ginger is vegetatively propagated, the rhizomes should be stored safely during the off-season, but they are highly perishable and susceptible to soil-borne fungal and insecticidal attacks. The seed rhizomes should be appropriately stored so that rotting, shriveling, dehydration, and sprouting are avoided until the following season. Farmers adopt different methods of storage of seed ginger (Kannan and Nair, 1965). Storage losses can often be as high as 50%. Recovery of seed rhizomes at planting was as high as 96.05% and dipping in Dithane M 45 0.3% solution for 30 min and drying under shade and then storing in pits (wherever bacterial wilt is a problem, the seeds should be treated with streptocycline 200 ppm) (NRCS, 1989b). Elpo et al. (2008) reported that paucity of good storage facilities is an important constraint in ginger production.

Maintaining a storage temperature of 22–25°C make the growing buds fat and strong and temperature higher than 28°C in the long run make the buds thin and weak. If the storage humidity is too low, the ginger rhizome epidermis may also lose water and then wrinkle, consequently adversely affecting the rhizome quality and speed of sprouting (Xizhen et al., 2005). Ginger can be stored in pits (1 m × 1 m × 1 m size), with the inner walls lined with stones/bricks. The bottom of these pits is filled with 10 cm thick dry sand. Disease-free bold rhizomes are selected after harvest, cleaned, and treatment with Bordeaux mixture (1:1:100) for 20 min is recommended for seed rhizomes (Xizhen et al., 2005). Prestorage steeping of rhizomes in *T. hamatum* or *T. viride* culture also showed inhibition against *Fusarium equiseti* infection (Bharadwaj et al., 1988). Treated rhizomes are placed in pits leaving 10–15 cm space on the top, covered with wooden plank to have space for aeration and plastered with cow dung. Covering the seed material with a layer of *Glycosmix pentaphylla* leaves is also beneficial (Nybe and Mini Raj, 2005). In China, cellars, holes dug out of the earth in shady or covered places, are used commonly for storage of seed rhizomes. A zero energy cool chamber (ZECC) is found ideal for storing fresh ginger. The loss in weight of the rhizomes was only 23% after storing for 4 months in this chamber, while the ginger stored in open conditions was shrunken in 4 months (IISR, 2002). Studies on storage of "seed pieces" of ginger showed that, in general, the number of days to germinate decreased with length of storage period, while percentage germination and yield increased from 0 to 42 days of storage. However, germination and yield were consistently lower after 35 days of storage. This anomalous response may be due to secondary dormancy during which the seed pieces lose their dormancy up to 21 days of storage, but regain or enter into secondary dormancy at 35 days and again lose dormancy after 42 days (Timpo and Oduro, 1977). In Nigeria, delayed harvesting with grass mulching up to 15 cm thick in the field itself or storage of harvested rhizomes in shade on layers of sand or grass cover are some alternate methods of seed ginger storage (Anonymous, 1970b).

Mature ginger rhizomes can be stored at 12–14°C with 85–90% RH for 60–90 days (Paull et al., 1988). Storage at 13°C with 65% RH leads to extensive dehydration and a wilted appearance (Akamine, 1962). Superficial mold growth can occur if condensation occurs on rhizomes. Chhata et al. (2007) suggested that ginger rhizomes might be stored at 10°C and less than 30% RH to reduce storage rot. The stored rhizomes are examined each month and rotten rhizomes are taken out to preserve the rest in a good healthy condition, free from disease and pest infestation.

Alternate Method of Ginger Production

Soil-borne diseases and nematode infestation are high in ginger fields. Aeroponic cultivation of ginger can provide high-quality rhizomes which are free from pesticides and nematodes, which are produced in mild-winter temperature greenhouses. The hydroponic system produced more yield and better quality rhizomes. Preliminary observations in South Florida showed that cost of production was lower in a hydroponic system due to reduced maintenance costs, arising from disease, pest, and weed infestations (Rafie et al., 2003). Hayden et al. (2004) in Arizona also tried soil-less aeroponic cultivation of ginger to obtain high-quality rhizomes, in mild-winter temperature greenhouses, which are pesticide free and without nematode infestation. The unique aeroponic system of ginger cultivation has growing units, incorporating "rhizome compartments" separated and elevated above an aeroponic spray chamber. Plants received bottom heat on perlite medium which showed accelerated growth and faster maturity. A noncirculating hydroponic method was used in Hawaii to produce disease-free ginger for seed (Hepperly et al., 2004). The disadvantages of this system are (i) initial heavy capital investment for erecting greenhouses or shelter structures, plastic composite benches, an efficient irrigation system, pots, and clean potting medium; (ii) a reliable source of clean water, preferably from a reliable "piped-in" source; (iii) availability of wilt-free starters; and (iv) assurance of strict sanitation measures in the greenhouse to preempt onset of disease and pest attack.

Ratooning is not common in ginger and is not encouraged or preferred, as the end product contains more fiber. However, Ridley (1912) mentioned that a few farmers in Jamaica practiced ratooning. Extracts from ratoon ginger had a higher "fiery taste" and less flavor than from those obtained from normal planting (Purseglove et al., 1981).

References

Aclan, F., Quisumbing, F.C., 1976. Fertilizer requirement, mulch and light attenuation on the yield and quality of ginger. Philipp. Agric. 60, 183–191.

Adaniya, S., Shoda, M., Fujiada, R., 1989. Effect of day length on flowering and rhizome swelling in ginger. J. Jpn. Soc. Hortic. Sci. 58, 49–56.

Ahmad, S.N., Asghar, K.A., Mohar, T.A., 1983. Ginger cultivation as affected by various mulches in the Punjab plains. Pak. J. Agirc. Res. 4 (1), 34–36.

Ahmad, R., Azeem, M., Ahmad, N., 2009. Productivity of ginger (*Zingiber officinale*) by amendment of vermicompost and biogas slurry in saline soils. Pak. J. Bot. 41 (6), 3107–3116.

Ahmed, N.U., Rahman, M.M., Hoque, M.M., Hossain, A.K.M., 1988. Effect of seed size and spacing on the yield of ginger. Bangladesh Hort. 16 (2), 50–52.

AICRPS, 1992a. All India Coordinate Research Project on Spices, Annual Report 1991–1992. National Research Centre for Spices, Calicut, Kerala State.

AICSCIP, 1985. All India Coordinated Spices and Cashewnut Improvement Project, Annual Report 1983–1984. Central Plantation Crops Research Institute, Kasaragod, Kerala State.

Aiyadurai, S.G., 1966. A Review of Research on Spices and Cashewnut in India Regional Office (Spices and Cashewnut). Indian Council of Agricultural Research, Ernakulam, Kerala State, pp. 228.

Ajithkumar, K., Jayachandran, B.K., 2003. Influence of shade regimes on yield and quality of ginger (*Zingiber officinale* Rosc.). J. Spices Aromat. Crops 12, 29–33.

Akamine, E.K., 1962. Storage of fresh ginger rhizomes. Bull. Hawaii Agric. Exp. Sta. No. 130, 23.

Aliyu, L., Lagoke, S.T.O., 2001. Profitability of chemical weed control in ginger (*Zingiber officinale* Roscoe.) production in northern Nigeria. Crop Protection 20 (3), 237–240.

Ancy, J., Jayachandran, B.K., 1993. Effect of shade and fertilizers on the quality of ginger (*Zingiber officinale* R.). S. Indian Hortic. 41 (4), 219–222.

Ancy, J., Jayachandran, B.K., 1996. Nutrient requirement of ginger under shade. Indian Cocoa Arecanut Spices J. 20 (4), 115–116.

Anderson, T., du Pleissis, S.F., Niemand, T.R., Scholtz, A., 1990. Evaporative cooling of ginger (*Zingiber officinale* R.). Acta Hortic. 275 (1), 173–179.

Anonymous, 1970b. Guide to Production of Ginger Extension guide No 7. Aerals, ABU, Nigeria.

Attoe, E.E., Osodeke, V.E., 2009. Effects of NPK on growth and yield of ginger (*Zingiber officinale* Roscoe.) in soils of contrasting parent materials of cross river state. Electronic J. Environ. Agric. Food Chem. 8 (11), 1261–1268.

Babu, P., Jayachandran, B.K., 1997. Mulch requirement of ginger (*Zingiber officinale* Rosc.) undershade. J. Spices Aromat. Crops 6 (2), 141–143.

BAE, 1971. Bureau of Agricultural Economics. The Australian Ginger Growing Industry. Bureau of Agricultural Economics, Canberra.

Barooah, L., Saikia, S., 2006a. Effect of integrated weed control measures on yield of rhizome, volatile oil and oleoresin in ginger cv. *Nadia*. J. Asian Hortic. 2 (4), 317–319.

Beale, A.J., Ramirez, L., Diaz, M., Munoz, M.A., Flores, C., 2006. Effect of seed set weight of ginger (*Zingiber officinale*) on yield. In: 42nd Annual Meeting of Caribbean Food Crops Society, 9–14 July, San Juan, Puerto Rico. Available from: <http://cfcs.eea.uprm.edu/Presentations>.

Bharadwaj, S.S., Gupta, P.K., Dohroo, N.P., Shyam, K.R., 1988. Biological control of rhizome rot of ginger in storage. Indian J. Plant Pathol. 6 (10), 56–58.

Bisht, J.K., Chandra, S., Chauhan, V.S., Singh, R.D., 2000. Performance of ginger (*Zingiber officinale*) and turmeric (*Curcuma longa*) with fodder tree based silvi–horti system in hills. Indian J. Agric. Sci. 70 (7), 431–433.

Chhata, L.K., Jeewa, R., Thakore, B.B.L., 2007. Effect of meteorological parameters on severity of storage rot of ginger. J. Agrometeorol. 9 (1), 125–127.

Cho, G.H., Yaoo, C.H., Choi, J.W., Park, K.H., Hari, S.S., Kim, S.J., 1987. Research report. Rural development administration, plant environment mycology and farm products utilisation. Korea Republic 29 (2), 30–42.

Chowdhury, P.C., 1988. Intercropping short duration summer crops with ginger in Darjeeling hills. Indian Farming 37 (11), 45.

Chukwu, G.O., Onyekwere, P.S.N., 1996. Complementary effect of NPK fertilizer and mulch on the growth and yield of irrigated ginger (*Zingiber officinale* Rosc.). Afr. J. Root Tuber Crops 1 (2), 26–29.

Colbran, R.C., Davies, J.J., 1969. Studies on hot water treatment and soil fumigation for control of root-knot in ginger. Queensland J. Agric. Animal Sci. 26, 439–445.

CSIR, 1976. The Wealth of India, Raw Materials, vol. II. Council of Scientific and Industrial Research, New Delhi.

Das, A., Sudhisri, S., Paikaray, N.K., 2006. Planting techniques and alternate mulches for enhancing soil moisture retention and productivity of rainfed ginger (*Zingiber officinale* Rosc.). Int. J. Agric. Sci. 2 (1), 211–215.

Das, N., 1999. Effect of organic mulching on root-knot nematode population, rhizome rot incidence and yield of ginger. Ann. Plant Protect. Sci. 7 (1), 112–114.

Dohroo, N.P., Sharma, S., 1997. Effect of organic amendments and intercropping on spore density of VAM fungi and yellows of ginger. Plant Dis. Res. 12 (1), 46–48.

Elpo, E.R.S., Negrelle, R.R.B., Rucker, N.G. de A., 2008. Ginger production in Morrets Town, Parana State. Sci. Agraria 9 (2), 211–217.

Eveson, J.P., Bryant, P.J., Asher, O.J., 1978. Germination and early growth of ginger (*Zingiber officinale* Rosc.) effect of constant and fluctuation in soil temperature. Trop. Agric. Trinidad 55, 1–7.

Furutani, S.C., Villanueva, J., Tanable, M.J., 1985. Effect of ethephon and heat on the growth and yield of edible ginger. Hortic. Sci. 20, 392–393.

Gavande, S.S., 1986. Studies on the effect of scheduling of irrigation and nitrogen levels on growth and yield of ginger (*Zingiber officinale* Rosc.) M.Sc(Ag) Thesis, Mahatma Phule Agricultural University, Rahuri, India.

George, B.E., Sreedevi, P., Vikraman Nair, R., 1998a. Effect of shade on plant characters and net assimilation rate of ginger cultivars. J. Trop. Agric. 36 (1–2), 65–68.

Ghosh, D.K., 2008. Performance of ginger as intercrop in coconut plantation with organic mulches and the effect of intercropping on the main crop. Indian Coconut J. 51 (4), 20–24.

Ghosh, K.K., Vinoy, T., Saha, D., 2006. Performance of ginger under agroforestry system. J. Interacademicia 10 (3), 337–341.

Girma, H., Kindie, T., 2008. The effects of seed rhizome on the growth, yield and economic return of ginger (*Zingiber officinale* Rosc.). Asian J. Plant Sci. 7 (2), 213–217.

Govender, A., Kinness, A., Jonnalgadda, S.B., 2009. Impact of soil quality on elemental uptake by *Zingiber officinale* (ginger rhizome). Int. J. Environ. Anal. Chem. 89 (5), 367–382.

Graham, J.A., 1936. Methods of ginger cultivation in Jamaica. J. Jamaica Agric. Soc. 40, 231–232.

Gupta, R., 1974. Process ginger for new products. Indian Fmg. 23 (1), 79, 14.

Hackett, C., Carolane, J., 1982. Edible horticultural crops A Compendium of Information on Fruit, Vegetable, Spice and Nut Species. Part I: Introduction and Crop Profiles. Academic Press, London.

Haque, M.E., Roy, A.K., Sikdar, B., 2004. Performance of ginger, turmeric and mukhi kachu under shade of mango orchard. Hortic. J. 17, 101–107.

Hayden, A.L., Brigham, L.A., Giacomelli, GA, 2004. Aeroponic cultivation of ginger (*Zingiber officinale*) rhizomes. Acta Hortic. 659, 397–402.

Haynes, P.H., Patridge, I.J., Sivan, P., 1973. Ginger production in Fiji. Fiji Agric. Res. J. 35, 51–56.

Hazarika, U., Dutta, R.K., Chakravorty, R., 2009. Morphology and yield attributing features of ginger and turmeric under natural shade of som (*Persea bombycina* Kost) plants. Adv. Plant Sci. 22 (1), 115–117.

Hegde, N.K., Sulikere, G.S., Rethinam, B.P., 2000. Distribution of photosynthetically active radiation (PAR) and performance of ginger under arecanut shade. In: Ramana, K.V., Santhosh, J., Eapen, J., Nirmal Babu, K., Krishnamurthy, K.S., Kumar, A. (Eds.), Centennial Conference on Spices and Aromatic Plants Indian Society of Spices, Calicut, Kerala State, pp. 107–112.

Hepperly, P., Zee, F.T., Kai, R.M., Arakawa, C.N., Meisner, M., Kraky, B., et al., 2004. Producing bacterial wilt-free ginger in greenhouse culture Extension Service Bulletins P.6. University of Hawaii at Manoa., Available from: <http://www.ctahr.hawaii.edu/oc/freepubs/pdf/scm-8.pdf> (accessed 29.02.2008).

Hijmans, R.J., Guarino, L., Jarvis, A., O'Bien, R., Mahtur, P., Bussink, C., et al., 2005. DIVA-GIS Version 5.2 Manual. International Potato Center and International Plant Genetic Resources Research Institute, Lima, Peru.

IISR, 2002. Indian Institute of Spices Research, Calicut, Kerala State, India, p. 17.

IISR, 2005. Annual Report 2004–2005. Indian Institute of Spices Research, Calicut, Kerala State, India, pp. 38–39.

Islam, A.K.M.S., Asher, C.J., Edwards, D.G., Evenson, J.P., 1978. Germination and early growth of ginger (*Zingiber officinale* Rosc.) II. Effects of 2-chloroethyl phosphonic acid or elevated temperature pretreatments. Trop. Agric. 55, 127–134.

Jaswal, S.C., Mishra, V.K., Verma, K.S., 1993. Inercropping ginger and turmeric with poplar (*Populus deltoides* "G-3" Marsh.). Agroforestry Syst. 22 (2), 111–117.

Jha, R.C., Maurya, K.R., Pandey, R.P., 1986. Influence of mulches on the yield of ginger in Bihar. Indian Cocoa Arecanut Spices J. 9 (4), 87–90.

Joachim, A.W.R., Pieris, H.A., 1934. Ginger manurial and cultural experiments. Trop. Agric. 82, 340–353.

Jogi, B.S., Singh, I.P., Dua, H.S., Sukhija, P.S., 1972. Changes in crude fibre, fat and protein content of ginger at different stages of ripening. Indian J. Agric. Sci. 42, 1011–1015.

Kandiannan, K., Sivaraman, K., Thankamani, C.K., 1996. Agronomy of ginger (*Zingiber officinale* Rosc.)—a review. J. Spices Aromat. Crops 5, 1–27.

Kandiannan, K., Koya, K.M.A., Peter, K.V., 1999. Mixed cropping of ginger (*Zingiber officinale* Rosc.) and maize. J. Spices Aromat. Crops 5, 1–27.

Kandiannan, K., Parthasarathy, U., Krishnamurthy, K.S., Srinivasan, V., 2010. Ginger phenology and growing degree days—a tool to schedule cultural operations Proceedings of the National Seminar on Soil, Water and Crop Management for Higher Productivity of Spices, 11–12 February 2010. Centre for Water Resources Development and Management (CWRDM), Calicut, Kerala State, pp. 172–180.

Kannan, K., Nair, K.P.V., 1965. Ginger (*Zingiber officinale* Rosc.) in Kerala. Madras Agric. J. 52, 168–176.

KAU, 1992. Final Research Report of ICAR Ad Hoc Scheme Shade Studies on Coconut Based Intercropping Situations. Kerala Agricultural University, Vellanikkara, Thrissur District, Kerala State.

KAU, 1994. Scheduling irrigation to ginger under varying nitrogen levels Sixteenth Zonal Research and Extension Advisory Council Meeting. Regional Agricultural Research Station, Kerala Agricultural University, Pattambi, Kerala State.

Kingra, I.S., Gupta, M.L., 1977. Ginger cultivation in Himachal. Kurukshetra 25 (19), 14–15.

Korikanthimath, V.S., Hegde, R., Sivaraman, K., 1995. Integrated input management in coffee based spices multistoried cropping system. Indian Coffee 59 (2), 3–6.

Korla, B.N., Rattan, R.S., Dohroo, N.P., 1989. Effect of seed rhizome size on growth and yield in ginger. Indian Cocoa Arecanut Spices J. 13, 47–48.

Korla, B.N., Rattan, R.S., Dohroo, N.P., 1990. Effect of mulches on rhizome growth and yield of ginger. S. Indian Hortic. 38, 163–164.

Kumar, B.M., Thomas, J., Fisher, R.F., 2001. *Ailanthus triphysa* at different density and fertilizer levels in Kerala, India: tree growth, light transmittance and understorey ginger yield. Agroforestry Syst. 52 (2), 133–144.

Kumar, M., Korla, B.N., 1998. A note on age of transplants, mulches and Ethrel treatments on yield and quality of ginger. Vegetable Sci. 25 (1), 100–101.

Kumar, R.D., Sreenivasulu, G.B., Prashanth, S.J., Jayaprakashnarayayan, R.P., Nataraj, S.K., Hegde, N.K., 2010. Performance of ginger in tamarind plantation (as intercrop) compared to sole cropping (ginger). Int. J. Agric. Sci. 6 (1), 193–195.

Kurian, A., Valsala, P.A., Nair, G.S., 1997. *In situ* green manure production as mulch material for ginger. J. Trop. Agric. 35 (1/2), 56–58.

Lawrence, B.M., 1984. Major tropical spices—ginger (*Zingiber officinale* Rosc.). Perfum. Flavorist 9 (5), 1–40.

Lee, M.T., Edward, D.G., Asher, C.J., 1981. Nitrogen nutrition of ginger (*Zingiber officinale* Rosc.) II. Establishment of leaf analysis test. Field Crops Res. 4 (1), 69–81.

Lee, S.A., 1974. Effect of rhizome size and fungicide treatment on the sprouting and early growth of ginger. Malaysian Agric. J. 48, 480–491.

Meenakshi, S., 1959. Ginger earns good income for the west coast. Indian Farming 9 (4), 12–13.

Melifonwu, A.A., Orkwor, G.C., 1990. Chemical weed control in ginger production from mini-setts. Nigerian J. Weed Sci. 3, 43–50.

Mirchandani, T.J., 1971. Investigations into Methods and Practices of Farming in Various States. Indian Council of Agricultural Research, New Delhi.

Mishra, A.C., Pandey, V.K., 2009. Effect of mulching on plant growth and rhizome yield in ginger (*Zingiber officinale* Rosc.) in rainfed plateaus of Jharkhand. Int. J. Plant Sci. (Muzaffarnagar) 4 (2), 385–387.

Mishra, S., Mishra, S.S., 1982. Effect of mulching and weedicides on growth and fresh rhizome yield of ginger. In: Abstracts of Papers Annual Conference of Indian Society of Weed Science, Hissar, India.

Mohanty, D.C., 1977. Studies on the effect of different mulch materials on the performance of ginger in the hills of Pottangi. Orissa J. Hortic. 5, 11–17.

Mohanty, D.C., Sarma, Y.N., 1979. Performance of ginger in tribal areas of Orissa, India, as influenced by method of planting, seed treatment, manuring and mulching. J. Plantation Crops 6, 14–16.

Mohanty, D.C., Naik, B.S., Panda, B.S., 1990. Ginger research in Orissa with reference to its varietal and cultural improvement. Indian Cocoa Arecanut Spices J. 14 (2), 61–65.

Muralidharan, A., 1973. Effect of graded doses of NPK on the yield of ginger (*Zingiber officinale* R.) var Rio de Janeiro. Madras Agric. J. 60, 664–666.

Nair, P.C.S., 1982. Agronomy of ginger and turmeric. In: Nair, M.K., Premkumar, T., Ravindran, P.N., Sarma, Y.R. (Eds.), Proceedings of the National Seminar on Ginger and Turmeric Central Plantation Crops Research Institute, Kasaragod, Kerala State, pp. 63–68.

Nair, P.C.S., Varma, A.S., 1970. Ginger in Kerala: steps towards increased production. Indian Farming 20 (3), 37–39.

NARC, 2008. Ginger—Crop Management—Seed Size and Spacing Available from: <http://www.narc.gov.np/NARHome/HIGHLIGHTS/AnnReport9798/Ginger.htm> (accessed 29.02.2008).

Natarajan, C.P., Bai, R.P., Krishnamurthy, M.N., Raghavan, B., Sankaracharya, N.B., Kuppuswamy, S., et al., 1972. Chemical composition of ginger varieties and dehydration studies of ginger. J. Food Sci. Technol. 9, 120–124.

Newman, S.M., Bennet, K., Wu, Y., 1997. Performance of maize, beans and ginger as intercrops in Paulownia plantations in China. Agroforestry Syst. 39 (1), 23–30.

Nimkar, S.A., Korla, B.N., 2009. Yield and economics of ginger (*Zingiber officinale* Rosc.) cultivation. Green Farming 13 (1), 948–949. (Special).

Ning, T.G., Shu, L.L., Jing, J., Wang, H.H., 2009. Early growing technique of deep sowing ginger in level row with plastic mulch. China Vegetables 2, 65–67.

Nizam, S.A., Jayachandran, B.K., 2001. Performance of mini seed rhizome as planting material in ginger (*Zingiber officinale* Rosc.) Proceedings of the 13th Kerala Science Congress. STEC, Trivandrum, Kerala, pp. 457–458.

NRCS, 1989b. Storage Method of Ginger Seed Rhizomes. National Research Centre for Spices, Calicut, Kerala State.

Nybe, E.V., Mini Raj, N., 2005. Ginger Production in India and Other South Asian Countries I In: Ravindran, P.N. Nirmal Babu, K. (Eds.), Ginger: The Genus *Zingiber* Medicinal and Aromatic Plants—Industrial Profiles, 41 CRC Press, Washington, DC, pp. 211–240.

Nybe, E.V., Nair, P.C.S., Mohankumar, N., 1982. Assessment of yield and quality components in ginger. In: Nair, M.K., Premkumar, T., Ravindran, P.N., Sarma, Y.R. (Eds.), Proceedings National Seminar on Ginger and Turmeric Central Plantation Crops Research Institute, Kasaragod, Kerala State, India, pp. 24–29.

Okwuowulu, P.A., 1988. Effect of seed ginger weight on flowering and rhizome yield of field grown edible ginger (*Zingiber officinale* Rosc.) in Nigeria. Trop. Sci. 28, 171–176.

Okwuowulu, P.A., 1992. Effects of depth of seed placement and age at harvest on the tuberous stem yield and primary losses in edible ginger (*Zingiber officinale* Rosc.) in south eastern Nigeria. Trop. Landwirtsch Vet. Med. 2, 163–168.

Okwuowulu, P.A., 2005. Ginger in Africa and the Pacific Ocean islands In: Ravindran, P.N. Nirmal Babu, K. (Eds.), Ginger: The Genus *Zingiber* Medicinal and Aromatic Plants—Industrial Profiles, 41 CRC Press, Washington, DC, pp. 279–303.

Okwuowulu, P.A., Ene, L.S.O., Odurukwe, S.O., Ojinaka, T., 1989. Effect of time of planting and age at harvest on yield of stem tuber and shoots in ginger (*Zingiber officinale* Rosc.) in the rain forest zone of southeastern Nigeria. Exp. Agric. 26, 209–212.

Owadally, A.L., Ramtohul, M., Heasing, J.M., 1981. Ginger production and research in its cultivation. Rev. Agric. sucr. Maurice 60, 131–148.

Panigrahi, U.C., Patro, G.K., 1985. Ginger cultivation in Orissa. Indian Farming 35 (5), 3–4.

Park, M., 1937. Seed treatment of ginger. Trop. Agric. 89, 3–7.

Parthasarathy, V.A., Kandiannan, K., Anandaraj, M., Devasahayam, S., 2003. Technical advances for improving production of ginger. In: Singh, H.P., Tamil Selvan, M. (Eds.), Indian Ginger—Production and Utilization Directorate of Arecanut and Spices Development, Calicut, Kerala State, pp. 24–40.

Paull, R.E., Chen, N.J., Goo, T.T.C., 1988. Control of weight loss and sprouting of ginger rhizome in storage. Hortic. Sci. 23, 734–736.

Paulose, T.T., 1973. Ginger cultivation in India Proceedings of the Conference on Spices. Tropical Products Institute, London, pp. 117–121.

Pawar, H.K., Patil, B.R., 1987. Effects of application of NPK through FYM and fertilizers and time of harvesting on yield of ginger. J. Maharashtra Agric. Univ. 12, 350–354.

Pegg, K.G., Moffett, M.L., Colbran, R.C., 1969. Diseases of Ginger in Queensland. The Buderim Ginger Growers Cooperative Association Limited, Australia.

Phogat, K.P.S., Pandey, D., 1988. Influence of the time of planting on the growth and yield of ginger var Rio de Janeiro. Prog. Hortic. 20, 40–68.

Prajapati, R.K., Nongrum, K., Singh, L., 2007. Growth and productivity of ginger (*Zingiber officinale* Rosc.) under kapok (*Ceiba pentandra* (L.) Gaertn) based agri-silviculture system. Indian J. Agroforestry 9 (1), 12–19.

Prentice, A., 1959. Ginger in Jamaica. World Crops 11, 25–26.

Purseglove, J.W., Brown, E.G., Green, C.L., Robbins, S.R.J., 1981. Spices, vol. 2. Longman Group Limited, London.

Ra, S.W., Shin, D.K., Shin, B.W., Lee, E.M., Roh, T.H., 1989. A study of cold injury to seed ginger research reports of the rural development administration. Horticulture 31 (3), 39–42.

Rafie, A.R., Teresa, O., Guerrero, W., 2003. Hydroponic production of fresh ginger roots (*Zingiber officinale*) as an alternative method for South Florida. Proc. Florida State Hortic. Soc. 116, 51–52.

Rahman, H., Karuppaiyan, R., Kishore, K., Denzongpa, R., 2009. Traditional practices of ginger cultivation in Northeast India (Special Issue: indigenous knowledge of the ethnic people of Northeast India in bioresources management). Indian J. Trad. Knowledge 8 (1), 23–28.

Ram, D., Mathur, K., Lodha, B.C., Webster, J., 2000. Evaluation of resident biocontrol agents as seed treatments against ginger rhizome rot. Indian Phytopathol 53 (4), 450–454.

Randhawa, K.S., Nandpuri, K.S., 1970. Ginger (*Zingiber officinale* Rosc.) in India—a review. Punjab Hortic. J. 10, 111–112.

Randhawa, K.S., Nandpuri, K.S., Bajwa, M.S., 1972. The growth and yield of ginger (*Zingiber officinale* Rosc.) as influenced by different dates of sowing. J. Res. (PAU Ludhiana) 9, 32–34.

Ratnambal, M.J., Gopalm, A., Nair, M.K., 1989. Quality evaluation in ginger (*Zingiber officinale* Rosc.) in relation to maturity. J. Plantation Crops 15, 108–117.

Ravindran, P.N., Nirmal Babu, K., Shiva, K.N., 2005. Botany and crop improvement of ginger. In: Ravindran, P.N., Nimal Babu, K. (Eds.), Ginger: The Genus *Zingiber* Medicinal and Aromatic Plants—Industrial Profiles CRC Press, Washington, DC.

Ravisankar, C., Muthusamy, S., 1986. Dry matter production and recovery of dry ginger in relation to light intensity. Indian Cocoa Arecanut Spices J. 10, 4–6.

Rethinam, P., Sivaraman, K., Sushama, P.K., 1994. Nutrition of turmeric. In: Chadha, K.L., Rethinam, P. (Eds.), Plantation and Spice Crops. Part 1. Advances in Horticulture Malhotra Publishing House, New Delhi, pp. 477–489.

Ridley, H.N., 1912. Ginger Spices. Macmillan Co, London, pp. 389–421.

Roy, A.R., Wamanan, P.P., 1989. Effect of seed-rhizome size on growth and yield of ginger. Adv. Plant Sci. 2, 62–66.

Sahu, S.K., Mitra, G.N., 1982. Influence of physicochemical properties of soil on yield of ginger and turmeric. Fert. News 37 (10), 59–63.

Sanwal, S.K., Yadav, R.K., Yadav, D.S., Rai, N., Singh, P.K., 2006. Ginger-based intercropping: highly profitable and sustainable in mild hill agroclimatic conditions of North East hill region. Vegetable Sci. 33 (2), 160–163.

Sengupta, D.K., Maity, T.K., Dasgupta, B., 2009. Effect of mulching on ginger (*Zingiber officinale* Rosc.) in the hilly region of Darjeeling district. J. Crop Weed 5 (1), 206–208.

Sengupta, D.K., Maity, T.K., Som, M.G., Bose, T.K., 1986. Effect of different rhizome size on the growth and yield of ginger (*Zingiber officinale* Rosc.). Indian Agric. 30, 201–204.

Shaikh, A.A., Ghadage, H.L., Jawale, S.M., 2006a. Effect of planting layouts and times of earthing up on growth contributing characters and rhizome yield in ginger. J. Maharashtra Agric. Univ. 31 (3), 275–278.

Shaikh, A.A., Ghadage, H.L., Jawale, S.M., 2006b. Influence of planting methods and time of earthing up on yield, nutrient uptake and balance in ginger (*Zingiber officinale*). Indian J. Agric. Sci. 76 (4), 246–248.

Sharma, G.C., Bajaj, B.K., 1998. Effect of intercropping bell pepper with ginger on plant parasitic nematode populations and crop yields. Ann. Appl. Bio. 133 (2), 199–205.

Singh, B.S., 1982. Cultivation of ginger and turmeric in Meghalaya. In: Nair, M.K., Premkumar, T., Ravindran, P.N., Sarma, Y.R. (Eds.), Proceedings National Seminar on Ginger and Turmeric Central Plantation Crops Research Institute, Kasaragod, Kerala State, India, pp. 224–226.

Singh, A., Singh, B., Singh, A., Singh, P., 2003. Response of ginger (*Zingiber officinale*) to methods of planting and levels of phosphorus in a rehabilitated forest developed in sodic land. J. Spices Aromat. Crops 12, 63–66.

Singh, K.A., Rai, R.N., Pradhan, I.P., 1991. Agroforestry systems in Sikkim hills. India Farming 41 (3), 7–10.

Sinha, B.N., 2002. Effect of different weed management practices on growth, yield and quality parameters of ginger (*Zingiber officinale* Rosc.). Himachal J. Agric. Res. 28 (1/2), 30–33.

Smith, M.K., Hamill, S.D., Gogel, B.J., Seven-Ellis, A.A., 2004. Ginger (*Zingiber officinale*) autotetraploids with improved processing quality produced by an *in vitro* colchines treatment. Australian J. Exp. Agric. 44, 1065–1072.

Srinivasan, V., Shiva, K.N., Kumar, A., 2008. Ginger Organic Spices. New India Publishing Agency, New Delhi, pp. 335–386.

Thomas, J., Kumar, B.M., Wahid, P.A., Kamalam, N.V., Fisher, R.F., 1998. Root competition for phosphorus between ginger and *Ailanthus triphysa* in Kerala. India Agroforestry Syst. 41 (3), 293–305.

Timpo, G.M., Oduro, T.A., 1977. The effect of storage on growth and yield of ginger (*Zingiber officinale* Rosc). Acta Hortic. 53, 337–340.

UPOV, 1996. Guidelines for the Conduct of Tests for Distinctness, Uniformity and Stability for Ginger (*Zingiber officinale* Rosc.). International Union for the Protection of new Varieties of Plants, Geneva.

Utpala, P., Johny, A.K., Parthasarathy, V.A., Jayarajan, K., Madan, M.S., 2006. Diversity of ginger cultivation in India—a GIS study. J. Spices Aromat. Crops 15 (2), 93–99.

Utpala, P., Jayarajan, K., Johny, A.K., Parthasarathy, V.A., 2008. Identification of suitable areas and effect of climate change on Ginger—a GIS study. J. Spices Aromat. Crops 17 (2), 25–29.

Vaidya, V.G., Sahasrabuddhe, K.R., Khuspe, V.S., 1972. Crop Production and Field Experimentation. Continental Prakashan, Poona, India.

Valsala, P.A., Amma, S.P., Sudhadevi, P.K., 1990. Feasibility of growing daincha in the inter-spaces of ginger beds. Indian Cocoa Arecanut Spices J. 14 (2), 65–66.

Vanlalhluna, P.C., Sahoo, U.K., 2008. Effect of different mulches on soil moisture conservation and productivity of rainfed ginger in an agroforestry system of Mizoram. Range Manag. Agroforestry 29 (2), 109–114.

Wang, G., Xu-Kun Zhang, Y., 2003. Effect of seed ginger size on growth and yield of ginger (*Zingiber officinale*). China Vegetables 1, 13–15.

Weiss, E.A., 1997. Essential Oil Crops. CAB International, Wallingford, pp. 539–567.

Wen, S.Q., Kong, X.K., Bo, X., 2006. Effect of soil bulk density on the growth, yield and quality of ginger. China Vegetables 11, 18–20.

Whiley, A.W., 1974. Ginger growing in Queensland. Queensland Agric. J. 91, 100.

Whiley, A.W., 1981. Effect of plant density on time of first harvest, maturity, knob size and yield in two cultivars of ginger (*Zingiber officinale* Rosc.) grown in Southeast Queensland. Trop. Agric. Trinidad 8, 245–251.

Whiley, A.W., 1990. Effect of "seed piece" size and planting density on harvested "knob" size and yield in two cultivars of ginger (*Zingiber officinale* Rosc.) grown in Southeast Queensland. Acta Hortic. 275, 167–172.

Wilson, H., Ovid, A., 1993. Growth and yield responses of ginger (*Zingiber officinale* Rosc.) as affected by shade and fertilizer applications. J. Plant Nutr. 16, 1539–1546.

Winterton, D., Richardson, K., 1965. An investigation into the chemical constituents of Queensland grown ginger. Queensland J. Agric. Anim. Sci. 22, 205.

Xianchang, Y., Zhifend, C., Xiheng, A., 1996. Effect of plastic mulch on growth and yield of ginger production. China Vegetables 6, 15–16.

Xizhen, A., Jinfeng, S., Xia, X., 2005. Ginger production in SouthEast Asia In: Ravindran, P.N. Nirmal Babu, K. (Eds.), Ginger: The Genus *Zingiber* Medicinal and Aromatic Plants–Industrial Profiles, 41 CRC Press, Washington, pp. 241–278.

Xizhen, A., Zhenxian, Z., Shaohui, W., 1998. Effect of temperature on photosynthetic characteristics of ginger leaves. China Vegetables 3, 1–3.

Yang, Z., Hua, Z.L., Long, Z.H., You, W.T., Yu, D.Y., Qiao, Z., 2004. Control effect of pendimethalin-acetochlor emulsifiable concentrate on weeds in ginger field. Weed Sci. (China) 1, 46–48.

Zaman, M.M., Sultana, N.A., Rahman, M.M., 2008. Yield of ginger as influenced by plant spacing and planting depth. Int. J. Sustain. Agric. Tech. 4 (2), 54–58.

19 The Biotechnology of Ginger

Tissue Culture

Micropropagation

In vitro culture is one of the key tools of plant biotechnology that exploits the totipotency nature of plant cells, a concept proposed by Haberlandt (1902) and unquestionably demonstrated, for the first time, by Steward et al. (1958). Tissue culture is alternatively called cell, tissue, and organ culture through *in vitro* conditions (Debergh and Read, 1991). The cell and tissue culture techniques have immense advantages in ginger, a vegetatively propagated crop, mainly because conventional breeding programs are hampered due to lack of fertility and natural seed set. Protocols for micropropagation and somaclonal variation, plant regeneration, meristem culture, *in vitro* pollination, microrhizome induction, protoplast culture, synseeds, and cryopreservation are optimized in ginger. Exploitation of somaclonal variation can help in inducing variability, leading to high-yielding, high-quality, and disease-resistant lines. Meristem culture can help in isolating disease-free plants.

Even though micropropagation technology is more expensive than conventional propagation methods, and unit cost per plant reaches unaffordable levels, compelling adoption of alternative strategies other than micropropagation to minimize production cost and to lower cost per plant and its dependability has been the reason for its preferential choice and widespread adoption by many researchers. Clonal multiplication of ginger from vegetative buds has been reported by many investigators Babu et al. (1997) and Nadgauda et al. (1980). Shooting was observed in the presence of BAP, IAA, IBA, Kinetin, and NAA. NAA was found to increase shoot length in both solid and liquid media (Arimura et al., 2000a). However, absence of BAP increased shoot number in liquid media (Arimura et al., 2000b). Some authors report maximum shoot production in MS media (Murasighe and Skoog, 1962) supplemented with ordinary sugar and no growth regulators (Sharma and Singh, 1995a).

Among the different explants, shoot tips gave the best results (Rajani and Patil, 2009) and emerged as suitable explants for ginger culture establishment. Behera and Sahoo (2009) tried liquid culture with vegetative buds and obtained good multiplication, rooting, and establishment of plants in the field. Production of etiolated adventitious shoots is also reported in ginger (Arimura et al., 2000a). On account of diseases of ginger being spread through infected seed rhizomes, meristem culture was optimized in ginger with a view to produce disease-free planting materials (Rout et al., 2001). Meristems of developed shoots were cultured on MS medium supplemented with coconut water initially for rapid screening of bacterial blight infection. The bacteria-free shoots were multiplied on MS medium supplemented with BAP or 2iP, and ginger plantlets were transferred *ex vitro*, which produced bacteria-free rhizomes showing vigorous growth,

The Agronomy and Economy of Turmeric and Ginger. DOI: http://dx.doi.org/10.1016/B978-0-12-394801-4.00019-3

high survival percentage, and high yield (Kirdmanee et al., 2004). Minas (2010) reports a protocol for micropropagation from meristem tips on MS with BAP ascorbic acid.

An efficient method for micropropagation of ginger that can increase plantlet production from a single shoot tip by 19-fold after every 12 weeks was developed. Among the various growth regulators used, a combination of BAP, IBA, Kinetin, and/ or IAA, with half- and three-quarter-strength MS medium was superior for morphogenesis. The liquid phase enhanced shoot multiplication due to agitation. All regenerated shoots produced functional roots in the same media (Tauqeer et al., 2007). *In vitro* shoot clusters of ginger were initiated from the meristem culture in a bioreactor in the presence of BAP and NAA. Increased levels of sucrose promoted both proliferation and growth of *in vitro* shoots without hyperhydricity (Ahn et al., 2007). Field performance of micropropagated (MP) ginger was evaluated in earthen pots with different compositions of sand, soil, cow dung (a common source of organic manure in India), and chemical fertilizer. Rooting performance was good in MS basal liquid media, while field establishment was better in garden soil:sand (1:1) mixture, which resulted in maximum height of the plants. Jasarai et al. (2000) reported good establishment in the mixture of garden soil, sand, and compost (40:20:40). A low-cost glass fermenting device for massive micropropagation of ginger in liquid media was developed by Hernandez-Soto et al. (2008).

Rooting was induced best with IBA for direct regeneration (Rajani and Patil, 2009) and also callus-regenerated (Sultana et al., 2009) plantlets from different explants. Kavyashree (2009) reports rooting on LSBM fortified with BAP, where simultaneous shooting and rooting was observed. Dogra et al. (1994) and Arimura et al. (2000b) reported the use of NAA for rooting. Kambaska and Santilata (2009) reported rooting on half-strength MS. Root number and length were enhanced in liquid medium (Arimura et al., 2000b) regardless of plant growth regulator treatment, while Devi et al. (1999) and Arimura et al. (2000a) reported establishment of roots in a growth regulator-free medium. The longest roots were obtained on potato extract and dextrose medium (Sharma and Singh, 1995a).

Cha-um et al. (2005) studied *ex vitro* survival and growth of ginger plantlets cultured photoautotrophically after *in vitro* acclimatization and found that the plantlets acclimatized under high RH with CO_2-enrichment conditions showed the highest adaptive abilities, resulting in the highest survival percentage (90–100%) after transplantation to *ex vitro*.

The regenerated plantlets were successfully established in the field with 86% survival frequency after a few days of indoor acclimatization (Kavyashree, 2009). The potential of beneficial bacteria to improve the growth of MP plants of ginger using reduced fertilizer inputs for improved growth of ginger tissue culture plants, where rhizome fresh weight, shoot, and root dry weights increased significantly compared to controls. Protocols for micropropagation of an economically and medicinally important *Zingiber* species viz *Z. cassumunar* (Roxb.) was optimized by Chirangini and Sharma (2005). They reported *in vitro* propagation on MS media fortified with NAA and BAP. Prathanturarug et al. (2004) reported *in vitro* propagation of *Z. petiolatum*, a rare plant from Thailand using seedling explants on MS medium containing 6-BAP alone or in combination with NAA. Rooting was spontaneously achieved in MS medium without growth regulators.

Direct Regeneration for Aerial Stem

Direct organogenesis from the aerial stem in the form of shoots and roots was observed in all combinations of BAP and NAA, leading to multiple shoots. Eighty-five percent of the plantlets could be established in the field, which after 8 months yielded approximately 100 g fresh rhizome/plant (Lincy, 2007). Lincy et al. (2009) reported indirect and direct somatic embryogenesis from aerial stem explants of ginger. Somatic embryos were induced in a medium containing BAP. The mature, club-shaped somatic embryos germinated in the presence of BAP and NAA in different concentrations. Direct somatic embryogenesis was observed from the plants in aerial stem and leaf base explants in the presence of TDA and IBA.

Kackar et al. (1993) successfully induced embryogenic callus in ginger on MS medium from young leaf segments of *in vitro* shoot cultures. Dicamba was the most effective in inducing and maintaining embryogenic cultures. Efficient plant regeneration was achieved on transfer to BAP-supplemented media. Histological studies revealed various stages of somatic embryogenesis characteristic of the monocot system. The *in vitro*-raised plants were successfully established in soil.

Anther Culture

Production of haploids through anther culture can be used for inducing mutations, for transformation, for cell line selection, and in the production of homozygous lines for further breeding programs. Plant regeneration from anther callus was reported from diploid and tetraploid ginger (Babu, 1997). Ginger anthers collected at the uninucleate microspore stage were subjected to a cold treatment (0°C) for 7 days and induced to develop profuse callus on MS medium supplemented with 2,4-D. Plantlets could be regenerated from these calli on MS medium supplemented with BAP and 2,4-D. The regenerated plantlets could be established in soil with 85% success. This protocol can be used for the possible development of androgenic haploids and dihaploids in ginger. Callus formation, development of roots, and rhizome-like structures were reported earlier from excised ginger anthers cultured on MS medium containing 2,4-D and coconut milk (Ramachandran and Chandrashekaran, 1992). Kim et al. (2000) induced compact embryogenic calli from anther explants on N6 medium with NAA. Regeneration occurred after 40 days of transfer to media in the presence of BA.

Inflorescence Culture and *In Vivo* Development of Fruit

Immature inflorescences form ideal explants for micropropagation of many crop species mainly due to the fact that the regenerated plants derived immature inflorescence, which usually maintain the genetic characteristics of the parent plants. In ginger, immature floral buds were converted into vegetative buds which subsequently developed into complete plants. Vegetative shoots were produced in 70% of the explants when 1-week-old inflorescences were cultured on MS medium supplemented with 10 mg/l BAP and 0.2 mg/l 2,4-D. These shoots grew into complete plantlets in 7–8 weeks' time (Babu et al., 1992).

It is already known that ginger does not set fruit. However, by supplying required plant nutrients to young flowers, *in vitro* pollination could be effected to develop "fruit" and subsequently plants could be recovered from these fruits (Babu et al., 1992). Development of seeds by *in vitro* pollination in ginger was also reported by Valsala et al. (1996). Dark-colored arillate mature seeds obtained after pollination germinated on half-strength MS medium with 2,4-D, BA, and NAA. The *in vitro*-produced seeds grew rapidly during the initial period (20 DAP), but there was little subsequent growth. The maturation period of seeds was 2–3 months under *in vitro* conditions as observed from the change in seed color.

In vitro pollination was successfully attempted by Nazeem et al. (1996) to overcome the prefertilization barriers like spiny stigma, long style, and coiling of pollen tube that interfered with natural seed set in ginger. They also developed a viable protocol for rapid multiplication of tiny seedlings which emerged from *in vitro*-raised seeds. This opens up new possibilities of sexual reproduction and development of seed-derived progenies of ginger, bringing in new possibilities in the improvement of the ginger crop.

Microrhizomes

Low efficiency of vegetative propagation, susceptibility of rhizomes used for vegetative propagation to diseases, and degeneration of rhizomes on long-term storage, coupled with poor flowering and seed set, have adversely affected ginger cultivation and breeding (Zheng et al., 2008). These can all be easily overcome through microrhizome technology. *In vitro* induction of microrhizomes in ginger was reported by many investigators (Babu, 1997; Babu et al., 2003; Sakamura et al., 1986).

Microrhizomes resemble the normal rhizomes in all respects, except in their smaller size. The microrhizomes consist of 2–4 nodes and 1–6 buds. They also have the aromatic flavor of ginger, and they resemble the normal rhizome in anatomical features by the presence of well-developed oil cells, fibers, and starch grains.

The microrhizome-derived plants have more tillers, but the plant height is smaller. *In vitro*-formed rhizomes are genetically more stable compared to MP plants. Seed rhizomes weigh about 2–8 g as compared to 20–30 g in the case of conventionally propagated plants. Microrhizomes gave very high recovery through lesser yield per bed. Microrhizomes also were genetically stable. This, coupled with its disease-free nature, will make microrhizomes an ideal source of planting material suitable for germplasm exchange, transportation, and conservation (Babu et al., 2003).

Microrhizomes were produced on medium BA, IAA (Rout et al., 2001), while Sharma and Singh (1995b) obtained microrhizome induction in MS supplemented with only sucrose. Zheng et al. (2008) reports a major role of sucrose and larger effect of GA compared to Kinetin and NAA on microrhizome induction. They also reported that increasing macro- and microelement content in MS medium can increase the weight of each microrhizome and the induction of microrhizomes;

however, when macroelements were above 2.0, toxic effects on ginger plantlets occurred *in vitro*, resulting in etiolation and ultimately plant death. Tyagi et al. (2006) reported no significant effect of maleic hydrazide on microrhizome production.

Tyagi et al. (2006) reported a significant effect of light treatments on survival of cultures but not on rhizome formation. Zheng et al. (2008) reported a photoperiod of 24L:0D (light/dark periods). At 16/8 h (day/night) photoperiod, light intensity of 40% of full sunlight and air temperature of 22–30°C resulted in optimum rhizome growth and photosynthetic ability as per Hyun et al. (1997). Cho et al. (1997) reported plant acclimatization under 85 μE/m/s light for 7 days before transplantation to be advantageous.

In vitro-grown rhizomes of ginger grew well on the carbonized rice husk:peat medium (5:1) ratio) (Cho et al., 1997). The percentage of established plants and sanitary conditions tended to be better in the presence of sand only or sand in combination with other media (Him-de-Freitez and Paez-de-Casares, 2004). The microrhizomes sprouted in a soil mixture within 2 weeks of planting (Rout et al., 2001). The sprouted plantlets survived under field conditions with normal growth. Sharma and Singh (1995b) observed 80% survival on storage in moist sand at room temperature for 2 months. Geetha (2002) reports 85% success on direct field planting without any hardening, and Zheng et al. (2008) reports a survival rate of 100%. A multiplication rate of 70,000 plants-rhizome/year was observed (Him-de-Freitez and Paez-de-Casares, 2004).

Quality analysis of *in vitro*-developed rhizomes indicated that they contain the same constituents as the original rhizome but with quantitative differences. The composition of basal medium seems to affect the composition of oil (Sakamura et al., 1986). Chirangini and Sharma (2005) reported development of microrhizomes induction in *Z. cassumunar* (Roxb.) when *in vitro*-derived shoots were cultured on MS media supplemented with sucrose within 8 weeks of incubation. Average fresh weight of 0.81 g was observed.

Plant Regeneration from Callus Culture

Plant regeneration from calli is possible by *de novo* organogenesis or somatic embryogenesis. Callus cultures also facilitate the amplification of limiting plant material. In addition, plant regeneration from calli permits the isolation of rare somaclonal variants which result either from an existing genetic variability in somatic cells or from the induction of mutations, chromosome aberrations, and epigenetic changes by the *in vitro*-applied environmental stimuli, including growth factors added to the cultured cells. Regeneration through callus phase via both organogenesis and embryogenesis was reported in ginger from leaf, vegetative bud, ovary, anthers, and shoot tips (Nadgauda et al., 1980). Callus from ovary tissues showed both organogenesis and embryogenesis while those from other explants showed plantlet regeneration on media supplemented with BAP and 2,4-D. NAA was found to enhance secondary embryoid production and formation of plantlets and rooting (Babu et al., 1992). Palai et al. (2000) reported plantlet regeneration from callus in four varieties of ginger in MS basal medium supplemented with BAP, IAA, and adenine sulfate.

Jamil et al. (2007) reported best callus induction on media supplemented with IAA and BAP. Xuan et al. (2004) reported 2,4-D and Kinetin for better callusing and Kinetin and NAA for differentiation. Callusing was usually found in the presence of IAA. However, dicamba induced callusing; Sultana et al. (2009) reported that leaf explants gave best results over shoot tip and root explants. Basal leaf sections of *in vitro* plants were found to be good for callus induction as per Him-de-Freitez and Paez-de-Casares (2004). Dicamba-promoted callusing (Sultana et al., 2009), however, showed a tendency to develop rhizogenic callus-like tissue in media containing only 2,4-D (Him-de-Freitez and Paez-de-Casares, 2004). Kinetin and BAP was best for regeneration (Sultana et al., 2009). Him-de-Freitez and Paez-de-Casares (2004) maintained cultures under dark conditions for 60 days.

Organogenic callus was found to have high carbohydrate and protein content than nonorganogenic callus. Six specific polypeptides accumulated during organogenesis and five expressed during rooting. Peroxidase and catalase activities were higher during rooting. Acid phosphatase activities were high during organogenesis and declined during root induction.

Ishida and Adachi (1997) studied the effects of phytohormones on the regeneration of plantlets in two morphological types of callus developed by culturing shoot meristem domes of three cultivars (Indo, Taiwan, and Ooshouga) on MS medium containing 2,4-D. Two types of callus, Type-I, which is compact with parenchyma-like cells and large vacuoles that developed green spots and became covered in structures like root hairs and Type-II, a sticky callus composed of small, square to rectangular cells with dense cytoplasms, and small or no vacuoles were found. Adventitious shoots and plantlet production was obtained only from sticky callus on MS medium containing NAA and BA.

Jamil et al. (2007) studied the effect of CO_2 enrichment on the differentiation and plant regeneration in ginger plant from callus culture through organogenesis. Callus initiated from shoot tips on exposure to elevated (5400–4000 ppm) atmospheric CO_2 leads to an increase in adventitious bud and shoot primordial. But the growth of adventitious bud and shoot primordium were reduced at 800 ppm.

Guo and Zhang (2005) reported establishment of somatic embryogenic cell suspension cultures of four ginger cultivars on MS agar medium supplemented with 2,4-D, Kinetin, and NH_4NO_3. The somatic embryos produced shoots and roots and developed into complete plantlets on MS medium supplemented with BA and NAA. The suspension cultures retained their embryonic potential even after subculture for 8 months. Palai et al. (2000) reported rooting on half-strength basal MS medium supplemented with either IAA or IBA within 7–8 days. Sultana et al. (2009) reported that IBA gave the best results for rooting.

Suspension Culture

When a callus tissue is dispersed into a liquid medium and agitated, it is found to disperse throughout the medium and form a cell suspension culture. These cells are capable of synthesizing any of the compounds normally associated with the plant.

They provide an ideal system when rapid cell division and uniform conditions are required. They also can facilitate rapid production of variant cell lines through selection procedures that can help in production of new plant varieties and secondary metabolites (Gurel et al., 2002).

Suspension cultures have been used to study the metabolic activities as well as to screen cell lines for resistance to diseases. The accumulation of (6)-gingerol and (6)-shogaol was much higher in culture systems of ginger where morphological differentiation was apparent (Zarate and Yeoman, 1996). Cultures grown on a callus-inducing medium also accumulated these metabolites but to a lesser extent. There is a positive relationship between product accumulation and morphological differentiation, although unorganized callus tissue also seems to possess the necessary biochemical machinery to produce and accumulate some phenolic pungent principles. There was no positive correlation between the amount of (6)-gingerol accumulated and the number of pigmented cells in either of the culture systems investigated (Zarate and Yeoman, 1996).

Protoplast Culture

Cultured protoplasts can be used not only for somatic cell fusions but also for taking up foreign DNA, cell organelles, bacteria, and virus particles. Using single cells such as protoplasts as starting material, one can also overcome the problems associated with multicellular tissues or organs like induction of chimeric polyploids. Protoplasts could be isolated successfully from leaf tissues as well as from cell suspension cultures in ginger on digestion in a mixture of macroenzyme R10, hemicellulase, and cellulose Onozuka R10. Seventy-two percent of the protoplasts were viable with a size of 0.39 mm. These viable protoplasts could be successfully plated on culture media to develop microcalli (Babu, 1997; Geetha, 2002).

Guo et al. (2007) described a procedure to regenerate plants from embryogenic suspension-derived protoplasts of ginger from shoot tips in the presence of NH_4NO_3, 2,4-D, and Kinetin. Protoplasts were isolated from embryogenic suspensions using a mixture of cellulose, maceroenzyme, and pectolyase in the presence of 2-(N-morpholino) ethane sulfonic acid. The protoplasts were cultured initially in liquid MS medium with 2,4-D and Kinetin and protoplast-derived calli were transferred to 2,4-D and BA to produce white somatic embryos that later developed into shoots and complete plants on plain MS medium lacking growth regulator and solid MS medium supplemented with BA and NAA. PVP and DTT could improve callus formation, shoot differentiation, growth, and rooting.

In Vitro Selection/Induction of Systemic Resistance

Kulkarni et al. (1987) reported isolation of *Pythium*-tolerant ginger by using culture filtrate (CF) as the selecting agent. *In vitro* selection for resistant types to soft rot and bacterial wilt using CFs of the pathogen or pathotoxin as the selecting agent was attempted by Dake et al. (1997).

Tilad et al. (2002) treated ginger callus cultures with salicylic acid (SA) prior to selection with CF of *F. oxysporum* f. sp. *Zingiberi* that increased the callus survival against the pathogen. Exogenous application of SA resulted in increased activity of peroxides and β-1,3-glucanase enzymes in the callus cultures. Two new protein bands of approximately 97 and 38 kDa were observed in SDS-PAGE analysis of soluble proteins from SA-treated calli and the 38 kDa protein cross-reacted with PR-1 monoclonal antibody used in immunodetection. *In vitro* antifungal activity of protein extract of SA-treated calli showed significant reduction in spore germination and germ tube elongation. The induction of resistance to fungus may be due to increased activity of peroxides, β-1,3-glucanase and antifungal PR-proteins.

Gosh and Purkayastha (2003) induced systemic protection against *P. aphanidermatum* in ginger (cultivar Suprabha) by soaking rhizome seeds separately in selected synthetic chemicals or herbal extracts for 1 h prior to sowing. Jasmonic acid (JA, 5 mM) and 10% leaf extract of *Acalypha indica* reduced disease significantly, with concomitant increase of defense-related proteins.

Somaclonal Variation

Coupling somaclonal variation methods with strategic and efficient *in vitro* selection pressures can lead to generation of variants. This is especially useful in ginger, where the lack of sexual reproduction hampers crop-improvement programs. Somaclones could also be used for screening against biotic and abiotic stresses after screening and confirmation using standard reliable methods. Field evaluation of somaclones indicated variability with regard to various agronomic characters and other yield parameters (Babu, 1997; Samsudeen, 1996).

Tashiro et al. (1995) treated excised shoot tips from three ginger cultivars with 5 mM *N*-methyl-*N*-nitrosourea (MNU) for 5–20 min and cultured on MS medium supplemented with NAA and BA. About 25% of MNU-treated plants differed in the basic isoenzyme patterns of GOT, 6-PGDH, PGM, or SKDH. All of them showed morphological mutations, such as multiple shoot formation, dwarfing, and abnormal leaves. The results indicated that treating cultured shoot tips with MNU effectively induces physiological and morphological mutations in ginger plants.

Production of Secondary Metabolites

Plant cells cultured *in vitro* produce a wide range of primary and secondary metabolites of economic value. Hence, plant tissue culture can be used as an alternative to whole plants as a biological source of potentially useful metabolites and biologically active compounds (Yeoman, 1987). Production of phytochemicals from plant cell cultures has been used for pharmaceutical products. Production of flavor components and secondary metabolites *in vitro* using immobilized cells may be tried in some species.

Successful establishment of cell suspension cultures in ginger was reported by Babu (1997). These cultures are maintained by weekly transfers to the fresh nutrient medium containing MS basal salts and 2,4-D for over 2 years and are in continuous growth and multiplication. The cells were heterogenous and during the process of subcultures, some of the cells differentiated into a few oil-producing cells. Callus and cell cultures and cell immobilization techniques for production of essential oils have been standardized for ginger (Ilahi and Jabeen, 1992). Production of volatile constituents in ginger cell cultures was reported earlier by Sakamura et al. (1986). Ilahi and Jabeen (1987) also reported preliminary studies on alkaloid biosynthesis in callus cultures of ginger, while Charlwood et al. (1988) have reported the accumulation of flavor compounds.

In Vitro Polyploidy

Ginger displays high sterility as a result of chromosome aberrations such as translocations and inversions. Hence, other breeding methods such as mutation and polyploidy breeding are required to obtain genetically improved plants. *In vivo* induction of tetraploid is a difficult process and *in vitro* colchicine treatment produces polyploids with higher efficiency than *in vivo* treatments (Adaniya and Shirai, 2001).

Adaniya and Shirai (2001) reported *in vitro* induction of tetraploid ginger on MS medium containing BA, NAA, and colchines. Induced tetraploid had higher pollen fertility and germinability. In the tetraploid strains, pollen fertility and germination rates were very high.

Smith et al. (2004) reported development of autotetraploid ginger with improved processing quality through *in vitro* colchicine treatment. Shoot tips immersed in colchicines and dimethyl sulfoxide on culture produced stable autotetraploid lines with significantly wider, greener leaves and fewer but thicker shoots. One superior line with improved aroma/flavor profile, fiber content, and consistently good rhizomes yield was released as "Buderim Gold." More importantly, it produced large rhizome sections, resulting in a higher recovery of premium grade confectionery ginger and a more attractive fresh market product. Babu et al. (1993) also reported selection of a tetraploid line with extra-bold rhizomes from ginger somaclones.

Wang et al. (2010) compared using shoot cultures and the doubling effects of somatic cell chromosome under *in vitro* conditions with different colchines, concentration, and treating time period. Shoots when treated in liquid culture medium of 0.2% colchicines for 8 days showed the best inducement and survival rates of 42.86% and 70.0%, respectively. The chromosome number of tetraploid was $2n = 4x = 44$, while that of the control diploid was $2n = 2x = 22$. The tetraploids were taller with a thicker stem, larger, broader, and thicker leaf blades, fewer stomata per unit area, larger stomata, and more chloroplasts in stomatal guard cell. The density of stomata and chloroplast number in guard cells could be important characteristics distinguishing tetraploid from diploid. Studies on pollen germination *in vitro* of a tetraploid ginger showed that the optimal pollination environment of the tetraploid ginger was 17–20°C under 100% RH (Adaniya, 2001).

Field Evaluation of Tissue-Cultured Plants

Field evaluation of tissue-cultured plants indicated that MP plants require at least two crop seasons to develop rhizomes of normal size that can be used as seed rhizomes for commercial cultivation (Babu et al., 1998). Similar results were obtained by Smith and Hamill (1996) wherein the first generation *ex vitro*, MP plants had significantly reduced the rhizome yield with smaller knobs and more roots. MP plants have a greater shoot–root (rhizome) ratio than seed-derived plants. Shoots from MP plants were also significantly smaller with a greater number of shoots per plant. They suggested that factors that can improve rhizome yield at low production costs need to be identified before MP plants can be recommended for use as a source of pest-free planting material.

Lincy et al. (2008) studied the relationship between vegetative and rhizome characters and rhizome yield in MP ginger plants over two generations. Correlation and path analysis for yield and yield contributing characters in plantlets directly regenerated from aerial stem explants and plantlets regenerated from aerial stem-derived callus over two generations revealed a high positive correlation of *in vitro*-derived plants for yield to rhizome characters and negative correlation for yield to tiller number in the first generation. In the second generation of the aerial stem-regenerated plants, tiller number, number of nodes per cormlets, circumference of cormlets, number of cormlets, and plant height exhibited high positive correlation and maximum direct effect with rhizome yield. However, the callus-regenerated plants showed the same trend in both generations.

Tissue-cultured ginger plantlets developed a functional photosynthetic apparatus and antioxidant enzymatic protective system during acclimation. The development of newly formed leaves after plant transplanting is crucial to the plant photosynthesis and growth (Guan et al., 2008). Based on the high survival rate of acclimatized plants and the higher levels of total soluble sugars and starch in early stages of cultivation, acclimation is recommended to ensure higher plant survival and reserve allocation (Giradi and Pescador, 2010). The growth and performance of plants derived from tissue culture is as good as plants propagated by seed after second generation *ex vitro* (Smith and Hamill, 1996).

Seedlings from *in vitro* cultivation were assessed for the efficacy of coconut-based substrates in ginger flower acclimation for survival rate, aerial part height, fresh matter, number of seedlings, root dry matter, and substrate pH. Sand, powder coconut + sand, and defibered coconut + sand showed the best results for all variables analyzed and was recommended for ginger flower (*Etlingera elatior*) acclimatization (Assiss et al., 2009).

Ginger plants were propagated *in vitro* and acclimated under different photosynthetic photon flux densities. *In vitro* plantlets exhibited higher chlorophyll (Chl) content and Chla/b ratio and exhibited low photosynthesis. The new leaves formed during acclimation in both treatments showed higher photosynthetic capacity than the leaves formed *in vitro*. Also, activities of antioxidant enzymes of MP ginger plantlets changed during acclimation (Guan et al., 2008). Ginger rhizome buds were induced to produce multiple shoots on MS with BA and IBA and successfully field planted with high survival percentage (Chan and Thong, 2004).

Metabolic profiling using GC/MS and LC-ESI-MS revealed a chemical difference between the different lines and tissues (such as rhizome, root, leaf, and shoot) tested. However, conventional greenhouse-grown and *in vitro*-derived plants did not show any significant difference on analysis of gingerols and gingerol-related compounds, other diarylheptanoids, and methyl ether-derivatives of these compounds, as well as major mono- and sesquiterpenoids identified.

The field performance of MP and conventionally grown plants (CP) of ginger were compared qualitatively and quantitatively and it was observed that the MP plants took 2 months longer to yield as much oleoresin and gingerols as CP plants. Periodic fluctuations in the yield and rhizome composition of various compounds were observed (Bhagyalakshmi and Singh, 1994).

Screening and inoculation of arbuscular mycorrhizal fungi in ginger plants is a feasible procedure to increase the oleoresin production of *Z. officinale* and consequently increase the aggregate value in ginger rhizome production (Silva et al., 2008). Field establishment of *in vitro* plantlets was better in a combination of garden soil:sand (1:1 ratio, Ipsita et al., 2010).

Germplasm Conservation

In Vitro Conservation and Cryopreservation

Being a vegetatively propagated crop, ginger is seriously affected by an accumulation of pathogens. Establishing *in vitro* germplasm collection is a process that cleans the plants from all diseases but viruses. It gives a good control on the preservation of genetic resources and facilitates international exchange of healthy plant material. Two kinds of *in vitro* germplasm preservation were considered: slow growth condition culture for mid-term preservation and cryopreservation using the encapsulation/dehydration technique for long-term preservation. *In vitro* conservation of germplasm of gingers was reported by Babu et al. (1999) and Geetha (2002). Minimal growth was induced and ginger plantlets could be successfully conserved for an extended period over 12 months on half-strength MS basal medium supplemented with a high concentration of sucrose and mannitol, sealed with aluminum foil, and maintained at $22 \pm 2°C$ (Babu et al., 1999; Geetha, 2002).

Balachandran et al. (1990) reported that ginger cultures could be maintained up to 7 months without subculture by using polypropylene caps. Dekkers et al. (1991) reported that ginger shoots could be maintained for over 1 year at ambient temperatures (24–29°C) in mannitol overlayed with mineral oil.

Use of *in vitro* rhizomes for germplasm conservation revealed that they remain healthy and viable up to 22 months at 25°C. Presence of light-enhanced survival of cultures (Tyagi et al., 2006). An optimum subculture interval of 8–24 months was reported for different *Zingiber* species. Tyagi et al. (1998) and Yamuna et al. (2007) developed efficient cryopreservation technique for *in vitro*-grown shoots of ginger based on encapsulation–dehydration, encapsulation–vitrification, and vitrification procedures. Plants could be successfully regenerated from cryopreserved shoots of ginger (Ravindran et al., 2004; Yamuna et al., 2007). Vitrification procedure resulted in higher regrowth than in other methods (Yamuna, 2007). DMSO and glycerol were used as cryoprotectants, and glycerol and sucrose were used as osmoprotectants. The genetic stability of cryopreserved ginger shoot buds was confirmed using IISR and RAPD profiling.

Synthetic Seeds

Germplasm of ginger conserved in field gene banks is often infected by soil-borne pathogens and the exchange of germplasm as rhizomes is problematic. Synseed technology is an important asset for (i) micropropagation, possessing the ability of regrowth and development into plantlets (conversion) for *in vitro* and *in vivo* use and (ii) for exchange of germplasm with potential storability, ease of handling, limited quarantine restrictions, and low cost of production. Furthermore, they can be used for cryopreservation through encapsulation–dehydration and encapsulation–vitrification methods. Fabre and Dereuddre reported encapsulation–dehydration, a new approach in cryopreservation of *Solanum* shoot tips. Thus, the synthetic seed technology is designed to combine the advantages of clonal propagation with those of seed propagation and storage. Reports on synthetic seeds are available in ginger. Sharma et al. (1994) successfully encapsulated ginger shoot buds in 4% sodium alginate gel for production of disease-free encapsulated buds, which were germinated *in vitro* to form roots and shoots.

Sajina et al. (1997) and others reported development of synthetic seeds in ginger by encapsulating the somatic embryos and *in vitro*-regenerated shoot buds in calcium alginate. These synseeds are viable up to 9 months at room temperature of $22°C \pm 2°C$ and germinated into normal plants on MS medium supplemented with BAP and IBA (Geetha, 2002; Sajina et al., 1997).

Synseeds of ginger were produced using aseptically proliferated 2-week-old microshoots. Shoot/synseed production was enhanced significantly higher in the presence of BAP. For short-term storage, sucrose-dehydrated synseeds were found to be better than air-dehydrated or fresh synseeds. Plantlets obtained from stored synseeds established successfully *ex vitro* and exhibited genetic fidelity. This synseed protocol could be useful for short-term storage and exchange of germplasm of ginger between national as well as international laboratories (Sundararaj et al., 2010).

Molecular Markers and Diversity Studies

Markers are of interest as a source of genetic information in a particular crop and for use in indirect selection of marker-linked traits. The use of molecular markers is very common these days and facilitates molecular breeding which involves primarily "gene tagging," followed by "marker-assisted selection" of desired genes or genomes. Gene tagging refers to the identification of existing DNA or the introduction of new DNA that can function as a tag or label for the gene of interest.

Molecular markers fall into two broad classes: those based on gene product (biochemical markers like isozymes) and those relying on a DNA assay (molecular markers) (Semagn et al., 2006). The two types of markers were extensively used in the improvement of this obligatory asexual crop. The most significant advancement is the development and utilization of DNA markers to detect and exploit genetic polymorphism in germplasms, in genetic fidelity testing of *in vitro*-propagated plantlets, barcoding and phylogeny construction, in checking adulteration in traded commodities and drugs, besides gene tagging and marker-assisted selection.

Isoenzyme studies have been reported in ginger. He et al. (1995) compared 28 ginger cultivars for peroxidase isoenzyme patterns by fuzzy cluster analysis and found differences in zonal number, activity intensity, and isoenzyme pattern, which were related to rhizome size, growth intensity, and blast (*P. solanacearum*) resistance. Isoenzymes of glutamate dehydrogenase (GDH), glutamate-oxaloacetate transaminase (aspartate aminotransferase) (GOT), malate dehydrogenase (MDH), 6-phosphogluconate dehydrogenase (6-PGDH), phosphoglucoisomerase (glucose-6-phosphate isomerase) (PGI), phosphoglucomutase (PGM), and shikimate dehydrogenase (SKDH) were analyzed in ginger cultivars and its relatives by electrophoresis (Tashiro et al., 1995). There was no variation between the three ginger cultivars in any of the isoenzyme profiles. However, isoenzyme profiles varied between the species were investigated.

Excised shoot tips from 5 mM *N*-methyl-*N*-nitrosourea were treated, and the regenerants on analysis showed variation in characteristic isoenzyme profiles of GOT, 6-PGDH, PGN, or SKDH in 25% of plantlets. All of these mutants also showed morphological mutations, such as multiple shoot formation, dwarfing, and abnormal leaves (Tashiro et al., 1995).

Babu (1997) and Sumathi (2007) morphologically characterized platelets obtained through micropropagation, callus regeneration, and microrhizome pathways and observed certain amount of variations. The callus-regenerated plants show maximum variation in plant height, number of tillers per plant, number of leaves per plant, leaf length, rhizome yield, and, primary and secondary fingers. However, plant height, number of leaves, yield, primary and secondary fingers, number of nodes of mother rhizome, intermodal length of primary and secondary fingers did not show any variation.

Biochemical characterization of ginger somaclones indicated significant variations in percentage of oil, oleoresin, starch, and fiber content. Negligible variation was observed between MP microrhizome and CP with regard to biochemical characters. However, callus regenerants showed significant variation and a few promising lines having important useful traits were selected (Babu, 1997; Sumathi, 2007).

Chromosome indexing of selected somaclones showed deviations in chromosome numbers and structural aberrations in the case of callus-regenerated plants. The numerical variations were mainly aneuploids with lesser number of chromosomes ($2n = 53, 55, 57$) than the normal number except one plant with higher number of chromosomes ($2n = 65$) (Sumathi, 2007). A few promising high-yielding somaclones tolerant to rhizome rot were identified (Babu et al., 1997). RAPD characterization of these somaclones showed profile variations indicating genetic differences

(Sumathi, 2007). A comprehensive metabolic fingerprinting from the leaves of three MP ginger cultivars was done to detect chemical variations including the volatile and nonvolatile oleoresins. It was apparent that the chemical variations were due to genetic effects rather than environmental and intrinsic factors.

Molecular characterization of MP plants by Rout et al. (1998) indicated that RAPD profiles did not indicate any polymorphism among the MP plants. However, Babu et al. (2003) reported RAPD profile differences among MP ginger also. Variations among regenerated plants were reported in *Kaempferia galangal* (Ajithkumar and Seeni, 1995) using RAPD markers. Isozymes and RAPD markers have been used in this direction (Tashiro et al., 1995).

Jatoi et al. (2006) tested cross-amplification potential of microsatellite markers among taxa to identify a larger number of genetic markers. Eight to twelve rice SSR markers amplified fragments in 14 genotypes from three genera—*Zingiber*, *Curcuma*, and *Alpinia*. Though the sequence analysis of these bands confirmed the absence of target repeat motif, amplification of a large number of polymorphic bands provided a basis for genetic diversity analysis. Lee et al. (2007) reported isolation and characterization of eight polymorphic microsatellite markers for ginger. These were used to detect a total of 34 alleles across the 20 accessions with an average of 4.3 alleles per locus. The data generated indicate the existence of moderate level of genetic diversity among the ginger accessions genotyped with eight markers.

Molecular Phylogeny of *Zingiber*

Chase (2004) investigated the phylogeny and relationships in monocots based on DNA analysis sequence data of seven genes representing all the three genomes of Zingiberales. A phylogenetic analysis of the tribe Zingibereae (Zingiberaceae) was performed by Ngamriabsakul et al. (2003) using nuclear ribosomal DNA (ITSI, 5.8S, and ITS2) and chloroplast DNA (trnL (UAA) 5′ exon to trnF (GAA)) and suggested that the tribe Zingiberaceae as well as the genus *Curcuma* are monophyletic. Kress et al. (2002) studied the phylogeny of the gingers (Zingiberaceae) based on DNA sequences of the nuclear ITS and plastid matK regions and proposed a new classification of the Zingiberaceae. Gao et al. (2006) used SRAP markers to analyze phylogenetic relationships of 22 species of Chinese *Hedychium* which grouped into three clusters on phylogenetic analysis.

Limited research on characterization at the DNA level has hindered the improvement of cultivated ginger through molecular approaches. Ninety-six accessions of ginger were analyzed using RAPD profiling and interrelationships studied. The polymorphism detected is moderate to low in ginger (Sasikumar and Zachariah, 2003). Wahyuni et al. (2003) studied genetic relationships among ginger accessions based on AFLP markers. Kavitha and Thomas (2008) reported *Z. zerumbet* (L) Smith, a wild species, as a potential resistance donor against soft rot disease in ginger caused by *P. aphanidermatum* (Edson) Fitzp. They studied the genetic diversity and *P. aphanidermatum* resistance of *Z. zerumbet* accessions in 15 populations. AFLP analysis of *Z. zerumbet* accessions revealed a high genetic diversity in

Z. zerumbet, unexpected for a clonal species. In the UPGMA dendrogram, accessions were clustered mostly according to their geographical origin. Though good variability for pathogen resistance among *Z. zerumbet* accessions was observed, no clear correspondence was observed between the clustering pattern of accessions and their responses to *P. aphanidermatum*.

Wahyuni et al. (2003) studied genetic relationships among ginger accessions based on AFLP markers. Intraspecific genetic variation was assessed in cultivated ginger and its three wild congeners: *Z. neesanum*, *Z. nimmonii*, and *Z. zerumbet* from Western Ghats in southern India using 169 AFLP markers generated by six primer combinations. The very low level of genetic diversity recorded in ginger complies with its obligatory asexual breeding behavior (Kavitha et al., 2010). The genetic variation among three *Z. officinale* cultivars was studied using 16 arbitrary RAPD primers to determine genetic differences between the cultivars (Muda et al., 2004).

The existing variation among 16 promising cultivars as observed through differential rhizome yield (181.9–477.3 g) was proved to have genetic basis using different genetic markers such as karyotype, 4C nuclear DNA content, and RAPD. The karyotypic analysis revealed a differential distribution of A, B, C, D, and E type of chromosomes among different cultivars as represented by different karyotype formulae. A significant variation of 4C DNA content was recorded at intraspecific level. RAPD analysis revealed a different polymorphism of DNA showing a number of polymorphic bands ranging from 26 to 70 among 16 cultivars. The RAPD primers OPCO2, OPAO2, OPD20, and OPNO6 showing strong resolving power were able to distinguish all of the 16 cultivars. Genetic diversity analyses revealed a conspicuous genetic diversity among different cultivars investigated (Nayak et al., 2005).

The RAPD-based method was optimized for the identification and differentiation of the commercially important *Z. officinale* Roscoe from the closely related species *Z. zerumbet*. RAPD was used to identify putative species-specific amplicons for *Z. officinale* that were cloned and sequenced to develop SCAR markers. The developed SCAR markers were tested in several non-*Zingiber* species commonly used in ginger-containing formulations. One of the markers, P3, was found to be specific for *Z. officinale* and was successfully applied to detect *Z. officinale* from *Trikatu*, a multicomponent formulation.

The genetic relatedness among 51 accessions, 14 species of the genus *Zingiber*, and the genetic variability of a clonally propagated species, *Z. montanum* (Koenig) Link ex Dietr. from Thailand, was studied by using RAPD method. Jaccard's coefficient of similarity varied from 0.119 to 0.970, indicative of distant genetic relatedness among the genotype investigated. UPGMA clustering indicated eight distinct clusters of *Zingiber*, with a high cophenetic correlation (r=1.00) value. Genetic variability in *Z. montanum* was exhibited by the collections from six regions of Thailand. High molecular variance (87%) within the collection regions of *Z. montanum* accessions was displayed by ANOVA data and also explained by the significant divergence among the sample from six collection regions. The results indicate that RAPD technique is useful to detect the genetic relatedness within and among species of *Zingiber* and that high diversity exists in the clonally propagated species *Z. montanum* (Bua-in and Paisooksantivatana, 2010).

Diversity of 21 cultivars and 3 wild accessions of *Z. officinale*, 84 accessions of 5 wild species of *Zingiber*, and 1 accession of *Hedychium* were examined using AFLP markers (Kavitha et al., 2007). No polymorphism was detected at more than 500 loci screened in ginger cultivars, showing an extreme case of genetic narrowness in a species. The study points out that monotypic nature of ginger cultivars may hinder the possibility of their identification based on their nucleic acid profiles. The present study also suggests the application of genomic tools to access the soft rot resistance trait in *Z. zerumbet* to develop transgenic ginger tolerant to this disease.

Ahmad et al. (2009) examined genetic variation in 22 accessions belonging to 11 species in 4 genera of Zingiberaceae, mainly from Myanmar by PCR-RFLP to investigate their relationships within this family. Two out of ten chloroplast gene regions (trnS-trnfM and trnK2-trnQr) showed differential PCR amplification across the taxa. The study showed that *C. zedoaria* and *C. xanthorrhiza* appeared to be identical, supporting their recent classification as synonyms.

Nucleotide diversity of a defense gene, *PR5*, and a nondefense gene, *methionine synthase*, were compared in *Zingiber* species with contrasting breeding system to reveal higher levels of nucleotide diversity in *PR5* genes as compared to *methionine synthase*, and evidence for transient episodes of positive selection experienced by this gene in the past. Contrary to expectations, the nucleotide diversity was higher in obligatory asexual *Z. officinale*, suggesting the possibility of accumulation of deleterious mutations. Results indicate accelerated evolution of *PR5* genes in *Zingiber* species; perhaps to coevolve variants against evolving pathogen populations (Thomas et al., 2010).

Application of Molecular Markers

Detection of Adulteration in Traded Ginger

A simple and rapid method to isolate good quality DNA with fairly good yields from mature rhizome tissues of turmeric and ginger has been perfected. Isolated DNA was amenable to restriction digestion and PCR amplification. Genetic profiling of traded ginger from India and China using 20 RAPD primers and 15 ISSR primers gave consistent amplification pattern. Significant variation was observed between the produce from the two countries.

The 46 accessions were characterized using two types of molecular markers, RAPD and ISSR. UPGMA dendrograms constructed based on three similarity coefficients, i.e., Jaccard's, Sorensen-Dice, and Simple Matching using the combined RAPD and ISSR markers placed the accessions in four similar clusters in all three dendrograms revealing the congruence of clustering patterns among the similarity coefficients and a rather less genetic distance among the accessions. Improved varieties/cultivars are grouped together with primitive types. Moreover, in the clustering pattern it was evident that germplasm collected from nearby locations, especially with vernacular identity, may not be generally distinct. The clustering of accessions was largely independent on its agronomic features (Jaleel and Sasikumar, 2010).

Jiang et al. (2006) used metabolic profiling and phylogenetic analysis for authentication of ginger species and closely related species in the genus *Zingiber*.

Z. *officinale* samples from different geographical origins were genetically indistinguishable, while other *Zingiber* species were significantly divergent, allowing all species to be clearly distinguishable. The metabolic profiling revealed that ginger from different origins showed no qualitative differences in major volatile compounds, but some quantitative differences in nonvolatile composition, regarding the contents of (6)-, (8)-, and (10)-gingerols were observed. Comparative DNA sequence/chemotaxonomic phylogenetic trees showed that the chemical characters of the investigated species generated essentially the same phylogenetic relationships as the DNA sequences. This supports the contention that chemical characters can also be used effectively to identify relationships between plant species. However, identification and quantification of the actual bioactive compounds are required to guarantee the bioactivity of a particular *Zingiber* sample even after performing authentication by molecular and/or chemical markers.

Molecular Markers in Genetic Fidelity Testing

Studies on RAPD profiling within the replicates of *in vitro*-conserved and cryopreserved lines of ginger using Operon Random Primers (RAPD) did not detect any polymorphism between the conserved lines in any of the primers tested, indicating genetic stability (Geetha, 2002). Rout et al. (1998) reported that molecular characterization of MP plants by RAPD revealed no variation. Babu et al. (2003) used RAPD profiles as an index to estimate genetic fidelity of selected "variants" among MP and callus-generated plants. It was observed that micropropagation even without callus phase induced variations. Similar differences were noticed among callus-regenerated plants (Sumathi, 2007). In general, this indicates that variability exists at genetic level among the MP and callus-regenerated plants of ginger, probably due to the accumulated natural mutations in ginger, which has eventually led to many varieties and cultivars even without sexual reproduction.

Tagging Genes of Interest Using Markers

Ginger being sterile, no sexual progenies are available and hence no mapping population could be developed. Thus, other approaches need to be used to tag genes of interest. Gao et al. (2008) reported development of a mapping population in a related genera, *Hedychium* using two species—one is *Hedychium coronarium* J. Konig, a cultivated species with sweet fragrance and imbricate bracts and the other is *Hedychium forrestii* Diels., a wild species with light aroma and tubular brachts, as parents. Using the same technique, medium-density maps of two species used as parents were constructed. The maternal parent contained 139 loci and spanned 917.1 cM. These loci were distributed in 18 linkage groups. But the data was insufficient to join these two maps.

Isolating Candidate Genes for Resistance

In the absence of sexual reproduction, using information from other sources with comparisons at heterologous genomes or genes will give us the necessary leads for tagging. Candidate genes responsible for pathogenesis can also be identified from sequence information available in databases. This approach using degenerate primes and functional genomics is more suitable for ginger improvement. Information available from the *Arabidopsis* genome on resistance (Laurent et al., 1998) can be used for tagging and isolating genes for biotic and abiotic resistance in ginger.

In ginger, Aswati Nair and Thomas (2006) reported isolation, characterization, and expression of resistance gene candidates (RGCs) using degenerate primers based on conserved motifs from the NBS domains of plant resistance from cultivated and wild *Zingiber* species. They also reported evaluation of resistance gene (R-gene) specific primer sets and characterization of RGCs in ginger. They designed 14 oligonucleotide primers, specific to the conserved regions of 4 classes of cloned R-genes. Clones derived from 17 amplicons generated by 12 successful primers were sequence characterized. Clones derived from three primers showed strong homology to cloned R-genes or RGCs from other plants and conserved motifs characteristic of non-TIR subclass of NBS-LRR R-gene superfamily. Phylogenetic analysis separated ginger RGCs into two distinct subclasses corresponding to clades 3 and 4 of non-TIR-NBS sequences described in plants. This study provides a base for future RGC mining in ginger and valuable insights into the characteristics and phylogenetic affinities of non-TIR-NBS-LR R-gene subclass in ginger genome.

Priya and Subramanian (2008) reported isolation and molecular analysis of R-gene in resistant ginger varieties against *F. oxysporum*. They observed that the R-gene is present only in resistant ginger varieties. These cloned R-genes provide a new resource of molecular markers for marker-assisted selection (MAS) and rapid identification of *Fusarium* yellow-resistant ginger varieties.

Kavitha and Thomas (2006) reported *Z. zerumbet*, a close relative of ginger, as a potential donor for *P. aphanidermatum* resistance in ginger. They employed AFLP markers and mRNA differential display to identify genes whose expression was altered in a resistant accession of *Z. zerumbet* before and after inoculation. A few differentially expressed transcript-derived fragments (TDFs) were isolated, cloned, and sequenced. Homology searches and functional categorization of some of these clones revealed the presence of defense/stress/signaling groups, which are homologous to genes known to be actively involved in various pathogenesis-related functions in other plant species. They found that *Z. zerumbet* showed adequate variability both at the DNA level and in response to *Pythium* (Kavitha and Thomas, 2006). Aswati Nair et al. (2010) identified a member of the pathogenesis-related protein group 5 (*PR5*) gene family in *Z. zerumbet* that is expressed constitutively but is unregulated in response to infection by *P. aphanidermatum*. Isolation of R-genes from such related genera will help in ginger improvement via transgenic approaches.

Transient knockdown of gene expression for downregulating phytoene desaturase was achieved in the culinary ginger in relation to infection by barley stripe mosaic virus infecting two species within Zingiberaceae. The results suggest the extension of BSMV-VIGS to monocots other than cereals, which has the potential for directed genetic analyses of many important temperate and tropical crop species (Renner et al., 2009).

Isolating Candidate Genes for Other Agronomically Important Traits

Chen et al. (2005) reported for the first time isolation, cloning, and characterization of a mannose-binding lectin from cDNA derived from ginger rhizomes. The full-length cDNA (746 bp) of *Z. officinale* agglutinin (ZOA) was cloned by rapid amplification of cDNA ends (RACE) and this contained a 510 bp open reading frame (ORF) encoding a lectin precursor of 169 amino acids with a signal peptide. ZOA have three typical mannose-binding sites (QDNY). Semiquantitative RT-PCR analysis revealed that ZOA expressed in all the tested tissues of *Z. officinale*, including leaf, root, and rhizome, suggesting it to be a constitutively expressing one. ZOA protein was successfully expressed in *E. coli* with the molecular weight as expected.

Yua et al. (2008) reported isolation and functional characterization of β-eudesmol synthase from *Z. zerumbet* Smith. They identified a new sesquiterpene synthase gene (ZSS2) from *Z. zerumbet* Smith. Functional expression of ZSS2 in *E. coli* and *in vitro* enzyme assay showed that the encoded enzyme catalyzed the formation of β-eudesmol and five additional by-products. Quantitative RT-PCR analysis revealed that ZSS2 transcript accumulation in rhizomes has strong seasonal variations. They introduced a gene cluster encoding six enzymes of the mevalonate pathway into *E. coli* and coexpressed it with ZSS2 to further confirm the enzyme activity of ZSS2 and to assess the potential for metabolic engineering of β-eudesmol production. When supplemented with mevalonate, the engineered *E. coli* produced a similar sesquiterpene profile to that produced in the *in vivo* enzyme assay, and the yield of β-eudesmol reached 100 mg/l.

Huang et al. (2007) reported molecular cloning and characterization of violaxanthin de-epoxidase (VDE) in ginger. A full-length (2000 bp) cDNA encoding violaxanthin de-epoxidase (GVDE) (AY876286) was cloned from ginger using RT-PCR and RACE. The expression patterns of GVDE in response to light were characterized. GVDE has a 1431 bp ORF and the predicted polypeptide contains 476 amino acids with the molecular mass of 53.7 kDa. Northern blot analysis showed that the GVDE was mainly expressed in leaves. GVDE mRNA level increased as the illumination time prolonged under high light. To determine the GVDE function, its antisense sequence was inserted into tobacco plants via EHA105. PCR-Southern blot analysis confirmed the integration of antisense GVDE in the tobacco genome. Chlorophyll fluorescence measurements showed that, transgenic plants had lower values of non-photochemical quenching (NPQ) and the maximum efficiency of PSII photochemistry (Fv/Fm) compared with the untransformed controls under high light. The size and the ratio of xanthophylls cycle pigment pool were lower in T-VDE tobacco plants than in control, indicating that GVDE was suppressed in antisense T-VDE tobacco. These results showed that VDE plays a major role in alleviating photoinhibition.

Genetic Transformation

Genetic transformation is one of the most promising strategies to introduce new resistance factors. Gene transfer compatible regeneration protocols have been optimized in ginger for full exploitation of this technique to introduce novel genes, especially those for disease resistance/tolerance. Transient expression of GUS was successful in ginger. Embryogenic callus was bombarded with plasmid vector pAHC 25 and promoter Ubi-1 (maize ubiquitin) using a gene gun (Babu, 1997). The pAHC 25 vector harboring GUS (β-glucuronidase) and BAR (phosphinothricin-acetyl transferase) genes were used (Christensen and Quail, 1996).

Agrobacterium tumefaciens strain EHA105/p35SGUSInt, effective in expressing β-glucuronidase activity, was used to standardize the transformation protocol in ginger, which was confirmed by histochemical GUS assay and PCR (Suma et al., 2008). The transcripts of the (S)-α-bisabolene synthase gene was detected in young rhizomes of ginger. The cDNA, designated ZoTps1, potentially encoded a protein that comprised 550 amino acid residues and exhibited 49–53% identity with those of the sesquiterpene synthases already isolated from the genus *Zingiber* (Fugisawa et al., 2010).

References

Adaniya, S., 2001. Optimal pollination environment of tetraploid ginger (*Zingiber officinale* Roscoe) evaluated by *in vitro* pollen germination and pollen tube growth in styles. Sci. Hortic. 90, 219–226.

Adaniya, S., Shirai, D., 2001. *In vitro* induction of tetraploid ginger (*Zingiber officinale* Roscoe) and its pollen fertility and germinability. J. Hortic. 88, 277–287.

Ahmad, D., Kikuchi, A., Jatoi, S.A., Mimura, M., Watanabe, K.N., 2009. Genetic variation of chloroplast DNA in Zingiberaceae taxa from Myanmar assessed by PCR-restriction fragment length polymorphism analysis. Ann. Appl. Biol. 155, 91–101.

Ahn, J.H., Lee, J.J., Kim, T.S., Kim, H.S., Lee, S.Y., 2007. Effects of growth regulators and sucrose concentrations on proliferation of *in vitro* shoot using bioreactor culture in *Zingiber officinale*. Hortic. Environ. Biotechnol. 48, 354–358.

Ajithkumar, P., Seeni, S., 1995. Isolation of somaclonal variants though rhizome explants cultures of *Icampferia galanga* L. Proceedings of All India Symposium on Recent Advances in Biotechnological Applications of Plant Tissue Cell Culture (RABAPTCCAS). CFTRI, Mysore, pp. 43.

Arimura, C.T., Finger, F.L., Casali, V.W.D., 2000a. A fast method for *in vitro* propagation of ginger. Trop. Sci. 40, 86–91.

Arimura, C.T., Finger, F.L., Casali, V.W.D., 2000b. Effect of NAA and BAP on ginger (*Zingiber officinale* Roscoe) sprouting in solid and liquid medium. Rev. Bras. Plant. Med. 2, 23–26.

Assiss, A.M., de Faria, R.T., Unemoto, L.K. de Colombo, L.A., Lone, A.B., 2009. Ginger flower (*Etilingera elatior*) acclimation in coconut-based substrates. Acta Sci. 31, 43–47.

Aswati Nair, R., Thomas, G., 2006. Isolation, characterization and expression studies of resistance gene candidates (RGCs) from *Zingiber* spp. Theor. Appl. Genet. 116, 123–134.

Aswati Nair, R., Kiran, A.G., Sivakumar, K.C., Thomas, G., 2010. Molecular characterization of an oomycete-responsive PR-5 protein gene from *Zingiber zerumbet*. Plant Mol. Biol. Rep. 28, 128–135.

Babu, K.N., 1997. *In vitro* studies in *Zingiber officinale* Rosc. Ph.D. Thesis, Calicut University, Calicut, Kerala State.

Babu, K.N., Samsudeen, K., Ratnambal, M., 1992. *In vitro* plant regeneration from leaf derived callus in ginger, *Zingiber officinale* Rosc. Plant Cell Tissue Organ Culture 29, 71–74.

Babu, K.N., Sasikumar, B., Ratnambal, M.J., George, K.J., Ravindran, P.N., 1993. Genetic variability in turmeric (*Curcuma longa* L.). Indian J. Genet Plant Breed. 53, 91–93.

Babu, K.N., Ravindran, P.N., Peter, K.V., 1997. Protocols for Micropropagation in Spices and Aromatic Crops. Indian Institute of Spices Research, Calicut, Kerala State, p. 35.

Babu, K.N., Minoo, D., Geetha, S.P., Samsudeen, K., Rema, J., Ravindran, P.N., et al., 1998. Plant biotechnology–its role in improvement of spices. Indian J. Agric. Sci. 68, 533–547.

Babu, K.N., Geetha, S.P., Minoo, D., Ravindran, P.N., Peter, K.V., 1999. *In vitro* conservation of cardamom (*Elettaria cardamomum*) germplasm. Plant Genet. Resour. Newslet. 119, 41–45.

Babu, K.N., Ravindran, P.N., Sasikumar, B., 2003. Field evaluation of tissue cultured plants of spices and assessment of their genetic stability using molecular markers Final Report Submitted to Department of Biotechnology, Government of India, p. 94.

Balachandran, S.M., Bhat, S.R., Cahndel, K.P.S., 1990. *In vitro* multiplication of turmeric (*Curcuma* sp.) and ginger (*Zingber officinale* Rosc.). Plant. Cell. Rep. 8, 521–524.

Behera, K.K., Sahoo, S., 2009. An efficient method of micropropagation of ginger (*Zingiber officinale* Rosc. cv. *Suprava* and *Suruchi*) *in vitro* response of different explants' types on shoot and root development of ginger. Indian J. Plant Physiol. 14, 162–168.

Bhagyalakshmi, N.S., Singh, N.S., 1994. The yield and quality of ginger produced by micropropagated plants as compared with conventionally propagated plants. J. Hortic. Sci. 69, 321–327.

Bua-in, S., Paisooksantivatana, Y., 2010. Study of clonally propagated cassumunar ginger (*Zingiber montanum* (Koenig) Link ex Dietr.) and its relation to wild *Zingiber* species from Thailand revealed by RAPD markers. Genetic Res. Crop Evol. 57, 405–414.

Chan, L.K., Thong, W.H., 2004. *In vitro* propagation of Zingiberaceae species with medicinal properties. J. Plant Biotechnol. 6, 181–188.

Charlwood, K.A., Brown, S., Charlwood, B.V., 1988. The accumulation of flavor compounds by cultivars of *Zingiber officinale*. In: Robins, R.J., Rhoades, M.J.C. (Eds.), Manipulating Secondary Metabolites in Culture AFRC Institute of Food Research, Norwich, pp. 195–200.

Cha-um, S., Tuan, N.M., Phimmakong Kirdmanee, C., 2005. The *ex vitro* survival and growth of ginger (*Zingiber officinale* Rosc.) influence by *in vitro* acclimatization under high relative humidity and CO_2 enrichment conditions. Asian J. Plant Sci. 4, 109–116.

Chase, M.W., 2004. Monocot relationships: an overview. Am. J. Bot. 91, 1645–1655.

Chen, Z.H., Kai, G.Y., Liu, X.J., Lin, J., Sun, X.F., Tang, K.X., 2005. cDNA cloning and characterization of a mannose-binding lectin from *Zingiber officinale* Roscoe (ginger) rhizomes. J. Biosci. 30, 213–220.

Chirangini, P., Sharma, G.H., 2005. *In vitro* propagation and microrhizome induction in *Zingiber cassumunar* (Roxb.)—an antioxidant-rich medicinal plant. J. Food Agric. Environ. 3, 139–142.

Cho, S.K., Roh, K.H., Hyun, D.Y., Choi, I.L., Kim, K.Y., Kim, S.D., et al., 1997. Mass production of rhizome induced by tissue culture on ginger 2. Selection of the optimal nutrient solution and media in hydroponics. RDA J. Indian Crop Sci. 39, 16–21.

Christensen, A.H., Quail, P.H., 1996. Ubiquitin promoter-based vectors for high-level expression of selectable and/or screenable marker genes in monocotyledonous plants. Transgenic. Res. 5, 213–218.

Dake, G.N., Babu, K.N., Rao, T.G.N., Leela, N.K., 1997. *In vitro* selection for resistance to soft rot and bacterial wilt in ginger International Conference on Integrated Plant

Disease Management for Sustainable Agriculture, 10–15 November 1997. Indian Phytopathological Society, New Delhi, pp. 339.

Debergh, P.C., Read, P.E., 1991. Micropropagation. In: Debergh, P.C., Zimmerman, R.H. (Eds.), Micropropagation: Technology and Application Kluwer Academic Publishers, Dordrecht, the Netherlands.

Dekkers, A.J., Rao, A., Goh, C.J., 1991. *In vitro* storage of multiple shoot cultures on gingers at ambient temperature of 24–29°C. Sci. Hortic. 47, 157–167.

Devi, S., Taylor, M.B., Powaseu, I., Thorpe, P., 1999. Micropropagation of ginger PRAP report-pacific regional. Agric. Prog. 7, 11–12.

Dogra, S.P., Korla, B.N., Sharma, P.P., 1994. *In vitro* clonal propagation of ginger (*Zingiber officinale* Rosc.). Hortic. J. 7, 45–50.

Fugisawa, M., Harada, H., Kenmoku, H., Mizutani, S., Misawa, N., 2010. Cloning and characterization of a novel gene that encodes (S)-beta-bisabolene synthase from ginger, *Zingiber officinale*. Planta 232, 121–130.

Gao, D.M., Liu, Z.W., Fan, S.J., 2006. RAPD analysis of genetic diversity among *Zingiber officinale* cultivars. J. Agrl. Biotechnol. 14, 245–249.

Gao, L., Liu, N., Huang, B., Hu, X., 2008. Phylogenetic analysis and genetic mapping of Chinese *Hedychium* using SRAP markers. Sci. Horti. 117, 369–377.

Geetha, S.P., 2002. *In Vitro* Technology for Genetic Conservation of Some Genera of Zingiberaceae. Ph.D. Thesis, Calicut University, Calicut, Kerala State.

Giradi, C.G., Pescador, R., 2010. Aclimatacao de gengibre (*Zingiber officinale* Roscoe) e a relacao com carboidratos endogenos. Rev. Bras. Plant Med. Botucatu. 12, 62–72.

Gosh, R., Purkayastha, R.P., 2003. Molecular diagnosis and induced systemic protection against rhizome rot disease of ginger caused by *Pythium aphanidermatum*. J. Phytopathol. 85, 1782–1787.

Guan, Q.Z., Guo, Y.H., Sui, X.L., Li, W., Zhang, Z.X., 2008. Changes in photosynthetic capacity and antioxidant enzymatic systems in micropropagated *Zingiber officinale* plantlets during their acclimation. Photosynthetica 46, 193–201.

Guo, Y., Bai, J., Zhang, Z., 2007. Plant regeneration from embrogenic suspension-derived protoplasts of ginger (*Zingiber officinale* Rosc.). Plant. Cell. Tiss. Organ. Cult. 89, 151–157.

Guo, Y., Zhang, Z., 2005. Establishment and plant regeneration of somatic embryogenic cell suspension cultures of the *Zingiber officinale*. Rosc. J. Sci. Hortic. 107, 90–96.

Gurel, S., Ekrem, G., Zeki, K., 2002. Establishment of cell suspension cultures and plant regeneration in sugar beet (*Beta vulgaris* L.). Turk. J. Bot. 26, 197–205.

Haberlandt, G., 1902. Experiments on the culture of isolated plant cells. Bot. Rev. 35, 68–85.

He, C.K., Li, J.S., Guo, S.Z., Zheng, M., Chen, W.S., 1995. The relationship between geographic distribution and the genetic difference of peroxidase isozyme of ginger germplasm. Fujian Acta Hortic. 402, 125–132.

Hernandez-Soto, A., Gatica-Arias, A., Alvarenga-Venutolo, S., 2008. Low cost glass fermentor design for mass micropropagation of ginger. Agron. Mesoam. 19, 87–92.

Him de Freitez, Y.Y., Paez de Casares, J., 2004. Anatomia foliar comparada de plantas de Jengibre (*Zingiber officinale* Roscoe) cultivadas en tres ambientes de crecimiento. Bioagro. Ene. 16, 27–30.

Huang, J.L., Cheng, L.L., Zhang, Z.X., 2007. Molecular cloning and characterization of violaxanthin depoxidase (VDE) in *Zingiber officinale*. Plant Sci. 172, 228–235.

Hyun, D.Y., Cho, S.K., Roh, K.H., Kim, K.Y., Choi, I.L., Kim, S.D., Park, M.S., Choi, K.G., 1997. Mass production of rhizome induced by tissue culture on ginger 1. Environmental factor related to the increasing rhizome RDA. J. Ind. Crop. Sci. 39, 10–15.

Ilahi, I., Jabeen, M., 1987. Micropropagation of *Z. officinale* Rosc. Pak. J. Bot. 19, 61–65.

Ilahi, I., Jabeen, M., 1992. Tissue culture studies for micropropagation and extraction of essential oils from *Zingiber officinale*. Rosc. Pak. J. Bot. 24 (1), 54–59.

Ipsita, R., Nuruzzaman, M., Habiba, S.U., Uddin, A.F.M.J., 2010. Evaluation of micropropagated ginger plantlets in different soil composition of pot culture. Int. J. Sustain. Agric. Tech., 18–21.

Ishida, M., Adachi, T., 1997. Plant regeneration from two callus types of ginger (*Zingiber officinale* Rosc.). SABRAO J. 29, 53–60.

Jaleel, K., Sasikumar, B., 2010. Genetic diversity analysis of ginger (*Zingiber officinale* Rosc.) germplasm based on RAPD and ISSR markers. Sci. Hortic. 125, 73–76.

Jamil, M., Kim, J.K., Akram, Z., Ajmal, S.U., Rha, S.S., 2007. Regeneration of ginger plant from callus culture through organogenesis and effect of CO_2 enrichment on the differentiation of regenerated plant. Biotechnology 6, 101–104.

Jasarai, Y.T., Patel, K.G., George, M.M., 2000. Micropropagation of *Zingiber officinale* Rosc. and *Curcuma amada* Roxb. In: Ramana, K.V. (Ed.), Spices and Aromatic Plants: Challenges and Opportunities in the New Century India Indian Society for Spices, pp. 52–54. Centennial Conference on Spices and Aromatic Plants, 20–23 September, Calicut, Kerala State.

Jatoi, S.A., Akira, K., San, S.Y., Khaw, W.N., Shinsuke, Y., Watanabe, J.A., et al., 2006. Use of RICE SSR markers as RAPD markers for genetic diversity analysis in *Zingiberaceae*. Breed. Sci. 56, 107–111.

Jiang, H., Xie, Z., Koo, H.J., McLaughlin, S.P., Timmermann, B.N., Gang, D.R., 2006. Metabolic profiling and phylogenetic analysis of medicinal *Zingiber* species: tools for authentication of ginger (*Zingiber officinale* Rosc.). Phytochemistry 67, 1673–1685.

Kackar, A., Bhat, S.R., Chandel, K.P.S., Malik, S.K., 1993. Plant regeneration via somatic embryogenesis in ginger. Plant Cell Tissue Organ Culture 32, 289–292.

Kambaska, K.B., Santilata, S., 2009. Effect of plant growth regulator on micropropagation of ginger (*Zingiber officinale* Rosc.) cv. *Suprava* and *Suruchi*. J. Agric. Tech. 5, 271–280.

Kavitha, P.G., Thomas, G., 2006. *Zingiber zerumbet*, A potential Donor for Soft Rot Resistance in Ginger: Genetic Structure and Functional Genomics, Extended Abstract XVIII, Kerala Science Congress, pp. 169–171.

Kavitha, P.G., Thomas, G., 2008. Defence transcriptome profiling of *Zingiber zerumbet* (L) Smith by mRNA differential display. J. Biosci. 33, 81–90.

Kavitha, P.G., Pratibha, N., Aswati Nair, R., Jayachandran, B.K., Sabu, M., Thomas, G., 2007. AFLP polymorphism and *Pythium* response in *Zingiber* species. In: Raghunath, K. (Ed.), Recent Trends in Horticultural Biotechnology New India Publishing Agency, New Delhi, India, pp. 497–503.

Kavitha, P.G., Kiran, A.G., Dinesh Raj, R., Sahu, M., Thomas, G., 2010. Amplified fragment length polymorphism analyses unravel a striking difference in the intraspecific genetic diversity of four species of genus *Zingiber* Boehm. From the Western Ghats. S. India Curr. Sci. 98, 242–247.

Kavyashree, R., 2009. An efficient *in vitro* protocol for clonal multiplication of Ginger var *Varada*. Indian J. Biotech. 8, 328–331.

Kim, T., Choi, I., Kim, H., Kim, S., Park, M., Ko, J., 2000. Investigation of floral structure and plant regeneration through anther culture in ginger. Korean J. Crop. Sci. 45, 207–210.

Kirdmanee, C., Mosaleeyanon, K., Tanticharoen, M., Craker, L.E., Simon, J.E., Jatisatienr, A., et al., 2004. A novel approach of bacteria-free rhizome production of ginger through biotechnology. Acta Hortic. 629, 457–461.

Kress, W.J., Prince, L.M., Williams, K.J., 2002. The phylogeny and a new classification of the gingers (Zingiberaceae): evidence from molecular data. Am. J. Bot. 89, 1682–1696.

Kulkarni, D.D., Khupse, S.S., Mascarenhas, A.F., 1987. Isolation of *Pythium* tolerant ginger by tissue culture. In: Potty, S.N. (Ed.), Proceedings of VI Symposium on Plantation Crops, Calicut, Kerala State, India, pp. 3–13.

Laurent, D., Frederic, P., Laurence, L., Sylvie, C., Canan, C., Kevin, W., et al., 1998. Genetic characterization of RRS1, a recessive locus in *Arabidopsis thaliana* that confers resistance to the bacterial soil borne pathogen *Ralstonia solanacearum*. MPMI 11, 659–667.

Lee, S.Y., Fai, K.W., Zakaria, M., Ibrahim, H., Othman, Y.R., Gwag, G.J., et al., 2007. Characterization of polymorphic microsatellite markers, isolated from ginger (*Zingiber officinale* Rosc.). Mol. Ecol. Notes 7, 1009–1011.

Lincy, A.K., 2007. Investigation on Direct *In Vitro* Shoot Regeneration from Aerial Stem Explants of Ginger (*Zingiber officinale* Rosc.) and Its Field Evaluation, Ph.D. Thesis, Calicut University, Calicut, Kerala State.

Lincy, A.K., Jayarajan, K., Sasikumar, B., 2008. Relationship between vegetative and rhizome characters and final rhizome yield in micropropagated ginger plants (*Zingiber officinale* Rosc.) over two generations. Sci. Hortic. 118, 70–73.

Lincy, A.K., Remashree, A.B., Sasikumar, B., 2009. Indirect and direct somatic embryogenesis from aerial stem explants of ginger (*Zingiber officinale* Rosc.). Acta Bot. Croat. 68, 93–103.

Minas, G.J., 2010. Ginger (*Zingiber officinale* Rosc.) sanitation and micropropagation *in vitro*. Acta Hortic. 853, 93–98.

Muda, A.M., Ibrahim, H., Norzulaani, N., 2004. Differentiation of three varieties of *Zingiber officinale* Rosc. by RAPD finger printing. Malaysian J. Sci. 23, 135–139.

Murasighe, T., Skoog, F., 1962. A revised medium for rapid growth and bioassays with tobacco tissue cultures. Physiol. Plant 15, 473–497.

Nadgauda, R.S., Kulkarni, D.B., Mascarenhas, A.F., 1980. Development of plantlets from tissue cultures of ginger. In: Proceedings of the Annual Symposium on Plantation Crops, Calicut, Kerala State, India, pp. 143–147.

Nayak, S., Naik, P.K., Laxmikanta, A., Mukherjee, A.K., Panda, P.C., Premananda, D., 2005. Assessment of genetic diversity among 16 promising cultivars of ginger using cytological and molecular markers. Z. Naturforsch. C 60, 485–492.

Nazeem, P.A., Joseph, L., Rani, T.G., Valsala, P.A., Philip, S., Nair, G.S., 1996. Tissue culture system for *in vitro* pollination and regeneration of plantlets from *in vitro* raised seeds of ginger—*Zingiber officinale* Rosc. International Symposium on Medicinal and Aromatic Plants, ISHS, Acta Hortic, 426, 10–15.

Ngamriabsakul, C., Newman, M.F., Cronk, Q.C.B., 2003. The phylogeny of tribe Zingiberaceae (*Zingiberaceae*) based on its (nrDNA) and trnl-f (cDNA) sequences. Edinb. J. Bot. 60, 483–507.

Palai, S.K., Rout, G.R., Samantaray, S., Das, P., 2000. Biochemical changes during *in vitro* organogenesis in *Zingiber officinale* Rosc. J. Plant Biol. 27, 153–160.

Prathanturarug, S., Angsumalee, D., Pongsiri, N., Suwacharangoon, S., Jenjittikul, T., 2004. *In vitro* propagation of *Zingiber petiolatum* (Holttum) I. Theilade, A rare Zingiberaceous plant from Thailand. J. In Vitro. Cell Dev. Biol. Plant. 40, 317–320.

Priya, R.S., Subramanian, R.B., 2008. Isolation and molecular analysis of R-gene in resistant *Zingiber officinale* (ginger) varieties against *Fusarium oxysporum* f.sp. *zingiberi*. Bioresour. Tech. 99, 4540–4543.

Rajani, H., Patil, S.S., 2009. *In vitro* responses of different explants types on shoot and root development of ginger. Acta Hortic. 829, 349–353.

Ramachandran, K., Chandrashekaran, P.N., 1992. *In vitro* roots and rhizomes from anther explants of ginger. J. Spices Aromat. Crops 1 (1), 72–74.

Ravindran, P.N., Babu, K.N., Saji, K.V., Geetha, S.P., Praveen, K., Yamuna, G., 2004. Conservation of spices genetic resources *in vitro* gene banks ICAR Project Report. Indian Institute of Spices Research, Calicut, Kerala State, pp. 81.

Renner, T., Bragg, J., Driscoll, H.E., Cho, J., Jackson, A.O., Specht, C.D., 2009. Virus-induced gene silencing in the culinary ginger (*Zingiber officinale*): an effective mechanism for down-regulating gene expression in tropical monocots. Mol. Plant 2, 1084–1094.

Rout, G.R., Das, P., Goel, S., Raina, S.N., 1998. Determination of genetic stability of micropropagated plants of ginger using random amplified polymorphic DNA (RAPD) markers. Bot. Bull. Acad. Sin. 389 (1), 23–29.

Rout, G.R., Palai, S.K., Samantaray, S., Das, P., 2001. Effect of growth regulator and culture conditions on shoot multiplication and rhizome formation in ginger (*Zingiber officinale* Rosc.) *in vitro*. In Vitro Cell. Dev. Biol. Plant 37 (6), 814–819.

Sajina, A., Mini, P.M., John, C.Z., Babu, K.N., Ravindran, P.N., Peter, K.V., 1997. Micropropagation of large cardamom (*Amomum subulatum* Roxb.). J. Spices Aromat. Crops 6, 145–148.

Sakamura, F., Oghihara, K., Suga, T., Taniguchi, K., Tanaka, R., 1986. Volatile constituents of *Zingiber officinale* rhizome produced by *in vitro* shoot tip culture. Phytochemical 25 (6), 1333–1335.

Samsudeen, K., 1996. Studies on Somaclonal Variation Produced by *In Vitro* Culture in *Zingiber officinale* Rosc., Ph.D. Thesis, University of Calicut, Calicut, Kerala State.

Sasikumar, B., Zachariah, J., 2003. Organization of ginger and turmeric germplasm based on molecular characterization. In: Final Report ICAR Ad Hoc Project IISR, Calicut, Kerala State.

Semagn, K., Bjornstad, A., Ndjiondjop, M.N., 2006. An overview of molecular marker methods for plants. Afr. J. Biotechnol. 5, 2540–2568.

Sharma, T.R., Singh, B.M., 1995a. Simple and cost-effective medium for propagation of ginger (*Zingiber officinale*). Indian J. Agric. Sci. 65, 506–508.

Sharma, T.R., Singh, B.M., 1995b. *In vitro* micro rhizome production in *Zingiber officinale* Roscoe. Plant Cell Rep. 15, 274–277.

Sharma, T.R., Singh, B.M., Chauhan, R.S., 1994. Production of encapsulated buds of *Zingiber officinale* Roscoe. Plant Cell Rep. 13, 300–302.

Silva, M.F., da Pescador, R., Rebelo, R.A., Stumer, S.L., 2008. The affect of arbuscular mycorrhizal fungal isolates on the development and oleoresin production of micropropagated *zingiber officinale*. Brazil. J. Plant Physiol. 20, 119–130.

Smith, M.K., Hamill, S.D., 1996. Field evaluation of micropropagated and conventionally propagated ginger in subtropical Queensland. Aus. J. Exp. Agric. 36, 347–354.

Smith, M.K., Hamill, S.D., Gogel, B.J., Severn-Ellis, A.A., 2004. Ginger (*Zingiber officinale*) autotetraploid with improved processing quality produced by an *in vitro* colchicines treatment. Australian J. Exp. Agric. 44, 1065–1072.

Steward, F.C., Mapes, M.O., Mears, J.S., 1958. Growth and organized development of cultured cells. II Organization in cultures grown from freely suspended cells. Am. J. Bot. 45, 705–708.

Sultana, A., Hassan, L., Ahmad, S.D., Shah, A.H., Batool, F., Rahman, R., et al., 2009. *In vitro* regeneration of ginger using leaf, shoot tip and root explants. Pak. J. Bot. 41, 1667–1676.

Suma, B., Keshavachandran, R., Nybe, E.V., 2008. *Agrobacterium tumefaciens* mediated transformation and regeneration of ginger (*Zingiber officinale* Rosc.). J. Trop. Agric. 46, 38–44.

Sumathi, V., 2007. Studies on Somaclonal Variation in *Zingiberaceous* Crops, Ph.D. Thesis, Calicut University, Calicut, Kerala State, p. 227.

Sundararaj, G., Anuradha, A., Rishi, K.T., 2010. Encapsulation for *in vitro* short-term storage and exchange of ginger (*Zingiber officinale* Rosc.) germplasm. Sci. Hortic. 125, 761–766.

Tashiro, Y., Onimaru, H., Shigyo, M., Isshiki, S., Miyazaki, S., 1995. Isozyme mutations induced by treatment of cultured shoot tips with alkylating agent in ginger cultivars (*Zingiber officinale* Rosc.). Bull. Fac. Agric. Saga. Univ. 79, 29–35.

Tauqeer, A., Nazreen, Z., Khan, N.H., 2007. Enhanced *Zingiber officinale* shoot multiplication in liquid culture. Pak. J. Sci. Ind. Res. 50, 145–148.

Thomas, E.G., Aswati Nair, R., Sabu, M., George, T., 2010. Molecular evolution of a PR-5 protein gene in *Zingiber* species with contrasting breeding systems. In: Proceedings of International Symposium on Biocomputing No 45.

Tilad, P., Sharma, R., Singh, B.M., 2002. Salicylic acid induced insensitivity to culture filtrate of *Fusarium oxysporum* f. sp. *Zingiberi* in the calli of *Zingiber officinale* Roscoe. Eur. J. Plant Pathol. 108, 31–39.

Tyagi, R.K., Bhat, S.R., Chandel, K.P.S., 1998. *In vitro* conservation strategies for species crop germplasm: *Zingiber*, *Curcuma* and *Piper* species. In: Mathew, N.M., Jacob, C.K. (Eds.), Developments in Plantation Crop Research Allied Publishers Limited, New Delhi, pp. 77–82.

Tyagi, R.K., Agarwal, A., Yusuf, A., 2006. Conservation of *Zingiber* germplasm through *in vitro* rhizome formation. Sci. Hortic. 108, 210–219.

Valsala, P.A., Nair, G.S., Nazeem, P.A., 1996. Seed set in ginger (*Zingiber officinale* Rosc.) through *in vitro* pollination. J. Trop. Agric. 34, 81–84.

Wahyuni, S., Xu, D.H., Bermawie, N., Tsunematsu, H., Ban, T., 2003. Genetic relationships among ginger accessions based on AFLP marker. J. Bioteknologi Pertanian 8, 60–68.

Wang, M., Niu, Y., Song, M., Tang, Q.L., 2010. Tetraploid of *Zingiberofficinale* Roscoe. *in vitro* inducement and its morphology analysis. China Vegetables DOI: CNKI: SUN: ZGSC. O.2010-04-013.

Xuan, P., Guo, Y., Yue, C., Yin, C., 2004. Study on tissue culture and rapid propagation of ginger (*Zingiber officinale*). Southwest China J. Agric. 17, 484–486.

Yamuna, G., 2007. Studies on Cryopreservation of Spices Genetic Resources, Ph.D. Thesis, Calicut University, Calicut, Kerala State.

Yamuna, G., Sumathi, V., Geetha, S.P., Praveen, K., Swapna, N., Babu, K.N., 2007. Cryopreservation of *in vitro* grown shoot of ginger (*Zingiber officinale* Rosc). Cryo. Lett. 28, 241–252.

Yeoman, M.M., 1987. Bypassing the plant. Ann. Bot. 60, 175–188.

Yua, F., Haradab, H., Yamasakia, K., Okamotoa, S., Hirasec, S., Tanakac, Y., et al., 2008. Isolation and functional characterization of a β-eudesmol synthase, a new sesquiterpene synthase from *Zingiber zerumbet* Smith. FEBS Lett. 582, 565–572.

Zarate, R., Yeoman, M.M., 1996. Changes in the amounts of (6) gingerol and derivatives during a culture cycle of ginger, *Zingiber officinale*. Plant Sci. (Limerick) 121, 115–122.

Zheng, Y., Liu, Y., Ma, M., Xu, K., 2008. Increasing *in vitro* microrhizome production in ginger (*Zingiber officinale* Roscoe.). Acta Physiol. Plant 1, 519–523.

20 Ginger Nutrition

Uptake and Requirement

Ginger is a very nutrient exhaustive crop, and so supplemental application of chemical fertilizers and/or organic manures is absolutely essential. Ginger rhizomes are mainly N and K exhausting, intermediate in P and Mg removal, and least in Ca removal (Nagarajan and Pillai, 1979). Haag et al. (1990) reported an accumulation of macronutrients in the following decreasing order: N < K < Ca < Mg < S, and P and micronutrients in this order: Fe < Mn < Zn < B, and Cu. However, nutrient uptake varies considerably with soil type, climatic conditions, levels of nutrients in the soil, and variety/cultivar cultivated. Govender et al. (2009) found that the soil quality influenced the elemental distribution within the ginger rhizome and the plant has the inherent capacity to control the amount of elements absorbed. The ginger "flesh" tends to accumulate Mn and Mg. A synergistic relationship between Cr and Mn and an antagonistic relationship between Fe and Cu and Fe and Cr is observed. The development of ginger with respect to the development of the aerial tissues can be classified into three distinct growth phases, namely (i) active growth (90–120 DAP); (ii) slow vegetative growth (120–180 DAP); and (iii) senescence (180 DAP), in which the rhizome development continues until harvest. The uptake of N, P, and K in leaf and pseudostem increases until harvest (Johnson, 1978). Xu et al. (2004) observed that ginger shoots and leaves are the centers of growth at the seedling stage, and 80.7% of the carbon (C) assimilated is transferred to these parts. Subsequently, the distribution rate for shoots and leaves decreased gradually with the growth of the plant, whereas the distribution rate into the rhizome lessened. During the vigorous growth stage of the rhizome, C assimilation is mainly transported from the leaves to the rhizomes, thus rhizome becomes the growth center. The absorption and utilization of N were the same as for C assimilates. About 48.41% of the N absorbed from the applied fertilizer at seedling stage was distributed to the shoots and leaves. While 65.43% of the N derived from the applied fertilizer at various stages of growth of the rhizomes was distributed to the rhizomes, and only 32.04% was distributed to the shoots and leaves. The results indicated that the rate of fertilizer–N utilization increased with delayed application. The exchange of C and N nutrition between the aboveground parts and underground seed of ginger and their transportation and distribution during the growth of the ginger plant ensures that the seed ginger will not shrivel and dry.

An average yield of 40 t/ha dry ginger rhizomes removes 70 kg N, 17 kg P_2O_5, 117 kg K_2O, 8.6 kg Ca, 9.1 kg Mg, 1.8 kg Fe, 500 g Mn, 130 g Zn, and 40 g Cu/ha. For healthy growth of ginger, very low external Ca is required. A heavy ginger crop removes 35–50 kg P/ha. The leaves of healthy ginger plants contain 1.1–1.3% of Ca

The Agronomy and Economy of Turmeric and Ginger. DOI: http://dx.doi.org/10.1016/B978-0-12-394801-4.00020-X

and concentrations as low as 2 ppm is sufficient to achieve 90% of maximum yield. In order to calculate the nutrient requirement of the crop, Johnson (1978) recommended the fifth pair of leaves during the 90–120 DAP stage for foliar diagnosis of N, P, and K.

Organic and Integrated Nutrient Management

The requirement of organic matter for ginger may be met from various sources, such as green/organic manures and mulches. This was very much evidenced by the good performance of the crop shown in high fertility conditions of Waynad, Kerala State, India, which supplied with 10 t of organic manure and 15 t of green leaf mulch per hectare, without any application of chemical fertilizers (Thomas, 1965). Ginger performs well with the supply of humus and organic matter, which showed a positive correlation with yield (Cho et al., 1987). FYM, poultry manure, green leaf, compost, press mud, oil cakes, and biofertilizers are used as sources of organics. The quantity of organics applied may vary with availability of the material and generally it varies between 5 and 30 t/ha. Organic manures are mostly applied as basal doses and in certain places it is also applied after the emergence of the crop, as a mulch. However, farmers in Maharashtra apply heavy doses of FYM, about 40–50 t/ha. In Kerala, the recommended dose of organic manure is 25–30 t/ha of FYM and 30 t/ha of green leaf mulch applied in three split applications. Application of FYM up to 48 t/ha resulted in the highest yield and benefit–cost ratio in Kerala (Chengat, 1997).

Rhizome yield increases with the increase in the level of FYM applied. Application of FYM at 6 t/ha was found to be superior to achieve a high yield (3.2 t/ha) and additional net profit of Rs 6820/ha in Madhya Pradesh (Khandkar and Nigam, 1996). Application of organic cakes increased nutrient availability, improved physical conditions of the soil, and increased yield and oleoresin extracted from the rhizomes. Maximum availability of exchangeable Ca in the soil was recorded by the application of FYM at a rate of 10 t/ha (Sadanandan and Iyer, 1986). Application of coconut cake (0.3% w/w, Rajan and Singh, 1973), neem cake at 2 t/ha (Sadanandan and Iyer, 1986), or groundnut cake at 1 t/ha (Sadananadan and Hamza, 1998b) improved yield and quality parameters and reduced soft rhizome rot incidence. Neem cake application significantly increased the N availability and the highest available N was recorded in soils, which received half the dose of urea along with neem cake at 2 t/ha. In addition to neem cake, phosphobacteria and P as rock phosphate increased the availability of P, Ca, Mg, Zn, and Mn (Annual Report IISR, 2004). Incorporation of inorganic N with neem cake (50% N), poultry manure (25% N), and groundnut cake (50% N) was found to be the best treatment to increase ginger yield up to 29 t/ha. Also, application of neem cake at 2 t/ha along with NPK at 75:50:50 kg/ha increased the yield of ginger and reduced the rhizome rot incidence in Kerala, India. Among other organic sources, application of coir compost (Terracare) at 2.5 kg/ha increased the yield by 37.5% over the control treatment (Srinivasan et al., 2000a). Gradual availability of nutrients through decomposition of organics throughout the growth phase may be the probable cause for better growth and development of the plant and ultimate yield, when inorganics were substituted with organics at different levels of application (Roy and Hore, 2007).

Trials on different management systems on ginger at the IISR, Calicut, Kerala State, India, showed that higher soil nutrient build up with the highest organic carbon content (2.33%) was in the organic system, which was on par with the integrated system of nutrient management and among the different systems of nutrient management. Organic management system recorded the highest availability of soil P, Ca, Mg, Zn, and Cu. The effect of different cropping systems on microbial population in soil also showed that the population of *Pseudomonas fluorescence, Azospirillum*, and phosphobacteria was highest in the organic system of nutrient management. The activities of enzymes, such as dehydrogenase, acid phosphatase, alkaline phosphatase, cellulase, and urease, were significantly higher under the organic system of nutrient management as compared to the exclusive inorganic system or integrated system of nutrient management. However, during the initial years, 15–20% reduction in yield under the organic system of nutrient management was encountered (Srinivasan et al., 2008). The results also indicated that certain crop quality parameters registered an increase in organically managed plots as compared to integrated or exclusively inorganic nutrient management systems. In ginger, oil content did not vary significantly among the treatments, as shown in Table 20.1.

However, the fiber content was significantly reduced in the organic system of nutrient management. Interestingly, both oleoresin and starch contents were the maximum in the organic system of nutrient management, and in both cases, there were statistically significant differences among the three systems of nutrient management. The maximum yield and oleoresin content was obtained with the application of 10t/ha of FYM + 1.25t/ha of compost + 20kg/ha of *Azospirillum*, which also showed higher nutrient uptake (Srinivasan et al., 2000a).

Gas chromatography in conjunction with mass spectrometry analysis of organically grown samples of fresh Chinese white and Japanese yellow varieties of ginger showed the presence of 20 hitherto unknown natural products and 31 compounds of previously reported ginger constituents (Jolad et al., 2004). These include paradols, dihydroparadols, gingerols, acetyl derivatives of gingerols, shogaols, 3-dihydroshogaols, gingerdiols, mono- and diacetyl derivatives of gingerdiols, 1-dehydrogingerdiones, diarylheptanoids, and methyl ether derivatives of some of these compounds.

Among the different combinations of composts/manures tried, to supplement the nutrient requirement of ginger, the highest (statistically significant) fresh yield of

Table 20.1 Effect of Different Nutrient Management Systems on the Quality of Ginger

Management System	Oil Content (%)	Oleoresin Content (%)	Starch (%)	Fiber (%)
Organic	1.20	3.96	70.07	1.69
Inorganic	1.20	3.15	62.21	1.90
Integrated	1.25	3.36	55.82	1.90
LSD (95%)	NS	0.23	9.89	0.08

LSD: Least significant difference; NS: not significant.

rhizome of ginger (14 t/ha) was obtained in FYM (10 t/ha) + neem cake (2 t/ha) + vermicompost (4 t/ha) combination followed by 10 t/ha coir compost + 8 t/ha of vermicompost. These combinations also resulted in higher oleoresin and lower fiber contents in ginger rhizomes. Field trials conducted through the AICRP on spices at various locations in India revealed that combined application of different organic sources such as FYM + pongamia oil cake + neem oil cake + stera meal + rock phosphate + wood ash yielded on par with the conventional practice. The experiment on the effects of organic manure on the yield components of ginger conducted in Nigeria showed that organic manures gave significantly higher rhizome yield, between 114.3% and 250.6% relative to the control treatment. The highest yield of 11.42 t/ha was obtained in plots where 30 t/ha cow dung and 10 t/ha of poultry litter were applied. The cow dung and poultry litter application increased the soil pH, organic matter, total N content, available P and K contents, and exchangeable K, Ca, and Mg contents relative to the control treatment (Ayuba et al., 2005). Organic manures, namely FYM, vermicompost, neem cake, green leaf manures, and microbial inoculants, namely AMF and *Trichoderma* and their combinations, were tried for the organic production of ginger grown as an intercrop in coconut gardens. Among the different combinations tried, FYM (30 t/ha) + neem cake + AMF + *Trichoderma* and FYM + AMF produced significantly higher yield (Sreekala and Jayachandran, 2006), compared to other treatments.

Mulches add organic matter, check soil erosion, conserve soil moisture, reduce soil temperature, improve soil physical properties, and suppress weed growth. Mulching can alter the physical and chemical properties of the soil and increase the availability of P and K. In general, 10–30 t/ha mulch is applied twice or thrice, the first time at planting, the second time at 45 DAP, and the third time at 90 DAP. Some of the commonly used mulch materials are green and dry forest leaves, coconut leaves, banana leaves, dry sal leaves, *shisam* residues like sugarcane trash, paddy, wheat, finger millet, little millet, and barley straw, and also weeds and other vegetation are used as mulch. FYM and compost are also used as mulch. If these materials are not available in sufficient quantity, live mulches such as sunhemp, green gram, horse gram, black gram, niger, sesbania, cluster bean, French bean, soybean, cowpea, dhaincha, and red gram can be grown as an intercrop and used for *in situ* mulching from 45 to 60 DAP (Kandiannan et al., 1996; Valsala et al., 1990). Fodder cowpea and dhaincha are suitable green manure crops for a self-sustainable system to provide mulch material of 89% and 69%, respectively, of the total requirement during the first sowing itself. Fodder cowpea was also found to be ideal for a third mulching (Kurian et al., 1997).

Application of green leaf mulch at 2 t/ha in two equal splits, one at planting and a second at 45 DAP resulted in 200% increase in yield under high fertility conditions of Waynad, Kerala State, India (Kannan and Nair, 1965). Polythene mulch also gave a yield of 19.9 t/ha (Mohanty et al., 1990). Ginger beds were covered by leaves and twigs of various forest trees after sowing in Sikkim. In East Khasi hills, traditionally "*jhum* cultivation" (slash-and-burn farming) is practiced, where dry grasses (30–45 t/ha) on the beds are burnt and ginger is planted (Chandra, 1996). *Jhum* cultivation not only improves the soil fertility but also controls the weeds. One-fourth (7.5 t/ha) of the recommended dose

(30 t/ha) of green mulch could be saved, resulting in a yield of 5246 kg/ha of dry ginger, if grown under 25% shade (Babu and Jayachandran, 1997). In Meghalaya, application of locally available organic mulches, namely paddy straw and *Schima wallichii* dry leaf mulches at 16 t/ha increased ginger yield by 43.6% and 39.7%, respectively. Mulching three times with leaves and growing soybean intercrops as live mulch was found to be equally effective. Mulching at 12.5, 5.0, and 5.0 t/ha for the first, second, and third mulching, respectively, are considered to be optimum for ginger cultivation (Randhawa and Nandpuri, 1970). The CPCRI in Kasaragod, Kerala State, India, recommended mulching with dry sal leaves. Intercropping ginger under *Ceiba pentadra* obtained a higher yield and income under unpruned conditions with an application of FYM at 30 t/ha, compared with 25% pruning of the principal crop. As ginger is a shade-loving crop, it gives good yield with organic matter application when grown under optimum shade. Both photosynthetic activity and translocation of photosynthates to the roots were enhanced due to FYM application due to the initiation of more stems and consequently more dry matter production (rhizome yield) going up to 14.4 t/ha (Prajapati et al., 2007).

Investigations at the Regional Research Station of Kerala Agricultural University at Ambalavayal, Kerala State, India, showed that organic amendments such as coconut oil cake, sesamum cake, and cashew shell when applied to the soil to supply 50 kg N/ha were found to reduce soft rot incidence caused by *Pythium aphanidermatum*. Lime application is practiced in acidic soils to correct soil acidity, which leads to nutrient availability imbalance in soil. In Kerala, where soils are acidic, application of rock phosphate along with FYM enhanced P availability and overcame P fixation. One of the branded rock phosphates, Rajphos, was found to be quite efficient among those tried. However, inasmuch as phosphate recovery, agronomic efficiency of the applied rock phosphate, and percentage increase in yield were concerned, another branded rock phosphate, Gafsaphos, was found to be superior to Rajphos and was applied with FYM (Srinivasan et al., 2000b).

Both the nature and level of organic manures markedly influence the growth and yield attributes of ginger. Roy and Hore (2007) reported that the maximum plant height, tiller number, leaf number, clump weight and length, and per plot yield (13.76 kg/3 m^2) and projected yield (29.72 t/ha) was recorded when one-fourth of N was applied through poultry manure and three-fourths of N as urea. Plants raised when groundnut cake (one-half of N) + urea (the other half) were applied had the highest number of primary fingers compared to the lowest number when one-fourth of N was applied as mustard cake + three-fourths of N as urea. A field experiment on ginger (cultivar Nagaland) conducted on an Alfisol in Medziphema, Nagaland, India, revealed that the highest number of tillers per plant (10.1), highest fresh rhizome yield (29.4 t/ha), highest oil content (0.47 ml %), and highest oleoresin content (5.57 ml %) were observed with the application of pig manure, and this treatment was significantly superior to the treatment where either FYM or poultry manure was applied. Singh and Singh (2007) also observed the highest uptake of N (136.7 kg/ha) and K (73.9 kg/ha) with the application of pig manure.

Azospirillum has a prominent role in increasing the productivity and quality of ginger, while reducing rhizome rot. In some situations, application of inorganic N rate can be reduced up to 30% in the presence of biofertilizers and economic returns can be obtained. *Azospirillum* has a marked positive influence on ginger production. The influence of *Azospirillum*, FYM, and their combinations were investigated at Pottangi in Odisha State, India. The highest fresh rhizome yield (1.87 t/ha), lowest rhizome rot (11%), oleoresin content (5.82%), and a benefit–cost ratio of 2.4 was obtained with 100% recommended rates of fertilizers, along with *Azospirillum* application at a rate of 10 kg/ha combined with FYM at 10 t/ha (Dash et al., 2008). Earlier, Jana (2006) reported that soil application of *Azospirillum* at 5 kg/ha, FYM at 15 t/ha, and 50% of recommended rate of N fertilizer registered the highest fresh rhizome yield (15.9 t/ha), highest essential oil content (1.50%), highest oleoresin content (6.65%), and a cost–benefit ratio of 1.59 compared to the recommended rates of fertilizers only. Besides, *Azospirillum* inoculation with VAM and *Glomus mosseae* led to better growth of the ginger plant and final yield in Himachal Pradesh, India (Sharma et al., 1997). Soil application of Gigaspora at the time of planting (2.5 g/rhizome) was also found to increase the yield as in the case of pine needle organic amendment and seed treatment with *T. harzianum*. Also, the effects of humic acid fertilizer on soil urease activity and available N content, N uptake, and rhizome yield were reported (Mei et al., 2009). The beneficial effects of slow-releasing humic acid fertilizer in increasing soil urease activity, available N, and N uptake and increase in rhizome yield by 9.17% have been noted. Time and again, a combination of inorganic and organic fertilizers have proved very beneficial in ginger production. Field studies on ginger cultivar Nadia, conducted at Umiam, Meghalaya, India, in a sandy loam soil revealed that N at 100 kg/ha + FYM was the best treatment to obtain the highest yield and good crop quality (Majumdar et al., 2003).

References

Annual Report, 2004. Indian Institute of Spices Research (ICAR), Calicut, Kerala State, India pp. 44–45.

Ayuba, S.A., John, C., Obasi, M.O., 2005. Effects of organic manure on soil chemical properties and yield of ginger—research note. Nigerian J. Soil Sci. 15, 136–138.

Babu, P., Jayachandran, B.K., 1997. Mulch requirement of ginger (*Zingiber officinale* Rosc.) under shade. J. Spices Aromat. Crops 6, 141–143.

Chandra, R., 1996. Growth and yield of ginger as influenced by method of planting, seed rate and removal of old rhizomes. J. Plant. Crops 24, 116–119.

Chengat, T., 1997. Influence of organic manures and *Azospirillum* on growth, yield and quality of ginger (*Zingiber officinale*), M.Sc. (Hort.) Thesis, Kerala Agricultural University, Kerala, India.

Cho, G.H., Yoo, C.H., Choi, J.W., Park, K.H., Hari, S.S., Kim, S.J., 1987. Research report rural development administration, plant environment mycology and farm products utilisation. Korea Republic 29, 30–42.

Dash, D.K., Mishra, N.C., Sahoo, B.K., 2008. Influence of nitrogen, *Azospirillum* sp. and farm yard manure on the yield, rhizome rot and quality of ginger (*Zingiber officinale* Rosc.). J. Spices Aromat. Crops 10, 177–179. Special issue on Proceedings of the National Symposium on Spices and Aromatic Crops: Threats and solutions to spices and aromatic crops industry.

Govender, A., Kindness, A., Jonnalagadda, S.B., 2009. Impact of soil quality on elemental uptake by *Zingiber officinale* (ginger rhizome). Int. J. Environ. Anal. Chem. 89, 367–382.

Haag, H.P., Saito, S., Dechen, A.R., Carmello, Q.A.C., 1990. Anais da Escola Superior de agriculture. Luiz de Queiroz 47, 435–457.

Jana, J.C., 2006. Effect of *Azospirillum* and graded levels of nitrogenous fertilizer on growth, yield and quality of ginger (*Zingiber officinale* Rosc.). Environ. Ecol. 24, 551–553.

Johnson, P.T., 1978. Foliar Diagnosis, Yield and Quality of Ginger in Relation to N, P and K, M.Sc. (Agric.) Thesis, Kerala Agricultural University, Kerala, India.

Jolad, S.D., Lantz, R.C., Solyom, A.M., Chen, G.J., Bates, R.B., Timmermann, B.N., 2004. Fresh organically grown ginger (*Zingiber officinale*): composition and effects on LPS-induced PGE2 production. Phytochemistry 65, 1937–1954.

Kandiannan, K., Sivaraman, K., Thankamani, C.K., 1996. Agronomy of ginger (*Zingiber officinale* Rosc.)—a review. J. Spices Aromat. Crops 5, 1–27.

Kannan, K., Nair, K.P.V., 1965. Ginger (*Zingiber officinale* R.) in Kerala. Madras Agric. J. 52, 168–176.

Khandkar, U.R., Nigam, K.B., 1996. Effect of farmyard manure and fertility level on growth and yield of ginger (*Zingiber officinale*). Indian J. Agric. Sci. 66, 549–550.

Kurian, A., Valsala, P.A., Nair, G.S., 1997. *In situ* green manure production as mulch material in ginger. J. Trop. Agric. 35, 50–58.

Majumdar, B., Venkatesh, M.S., Kumar, K., Patiram, 2003. Effect of N levels, FYM and mother rhizome removal on yield, nutrient uptake and quality of ginger (*Zingiber officinale* Rosc.) and different forms of N build up in an acidic Alfisol of Meghalaya. Crop Res. (Hissar) 25, 478–483.

Mei, L.Z., Lan, L.L., Yu, S.C., Guang, C.X., Chao, Z., Lan, Y.H., 2009. Effects of humic acid fertilizer on urease activity in ginger growing soil and nitrogen absorption of ginger. China Vegetables 4, 44–47.

Mohanty, D.C., Naik, B.S., Panda, B.S., 1990. Ginger research in Orissa with reference to its varietal and cultural improvement. Indian Cocoa Arecanut Spices J. 14, 61–65.

Nagarajan, M., Pillai, N.G., 1979. Note on nutrient removal by ginger and turmeric rhizomes. Madras Agric. J. 66, 56–59.

Prajapati, R.K., Nongrum, K., Singh, L., 2007. Growth and productivity of ginger (*Zingiber officinale* Rosc.) under Kapok (*Ceiba pentadra* L. Gaertn) based agri–silviculture system. Indian J. Agroforestry 9, 12–19.

Place of Publication: Calicut, Kerala State, India , Editor SN Potty

Rajan, K.M., Singh, R.S., 1973. Effect of organic amendment of soil on plant growth yield and incidence of soft rot of ginger. In: Potty, S.N. (Ed.), Proceedings of the National Symposium on Plantation Crops, Kerala, 1. Journal of Plantation Crops, Calicut, Kerala State, India, pp. 102–106.

Randhawa, K.S., Nandpuri, K.S., 1970. Ginger (*Zingiber officinale* Rosc.) in India—a review. Punjab Hortic. J. 10, 111–112.

Roy, S.S., Hore, J.K., 2007. Influence of organic manures on growth and yield of ginger. J. Plantation Crops 35, 52–55.

Sadananadan, A.K., Hamza, S., 1998b. Effect of organic farming on nutrient uptake, yield and quality of ginger (*Zingiber officinale* R). In: A.K. Sadanandan, K.S. Krishnamurthy, K. Kandiannan, V.S. Korikanthimath (Eds.). Proceedings of the Water and Nutrient Management for Sustainable Production and Quality of Spices, Madikeri, Karnataka State, India, pp. 89–94.

Sadanandan, N., Iyer, R., 1986. Effect of organic amendments on rhizome rot of ginger Indian Cocoa Arecanut Spices J. 9, 94–95.

Sharma, S., Dohroo, N.P., Korla, B.N., 1997. Effect of VAM inoculation and other field practices on growth parameters of ginger. J. Hill Res. 10, 74–76.

Singh, V.B., Singh, A.K., 2007. Effect of types of organic manure and methods of nitrogen application on growth, yield and quality of ginger. Environ. Ecol. 25, 103–105.

Sreekala, G.S., Jayachandran, B.K., 2006. Effect of organic manures and microbial inoculants on nutrient uptake, yield and nutrient status of soil in ginger intercropped coconut garden. J. Plantation Crops 34, 25–31.

Srinivasan, V., Sadanandan, A.K., Hamza, S., 2000a. An IPNM approach in spices with special emphasis on coir compost Proceedings of the International Conference on Managing Natural Resources for Sustainable Agricultural Production in the 21st Century, 14–21 February. Indian Agricultural Research Institute, New Delhi, pp. 1363–1365.

Srinivasan, V., Sadananadan, A.K., Hamza, S., 2000b. Efficiency of rock phosphate sources on ginger and turmeric in an Ustic Humitropept. J. Indian Soc. Soil Sci. 48, 532–536.

Srinivasan, V., Shiva, K.N., Kumar, A., 2008. Ginger. In: Parthasarathy, V.A., Kandiannan, K., Srinivasan, V. (Eds.), Organic Spices, New India Publishing Agency, New Delhi, pp. 335–386.

Thomas, K.M., 1965. Influence of N and P_2O_5 on the yield of ginger. Madras Agric. J. 52, 512–515.

Valsala, P.A., Prasannakumari, A.S., Sudhadevi, P.K., 1990. Glycosidically bound aroma compounds in ginger (*Zingiber officinale* Roscoe.). J. Agric. Food Chem. 38, 1553–1556.

Xu, K., Guo, Y.Y., Wang, X.F., 2004. Transportation and distribution of carbon and nitrogen nutrition in ginger. Acta Hortic. 629, 347–353.

21 The Diseases of Ginger

Ginger is grown by small and marginal farmers in the states of Assam, Himachal Pradesh, Karnataka, Kerala, Meghalaya, Odisha, Sikkim, and other northeastern regions of India, as well as Southeast Asian countries, Africa, and Hawaii (USA). The ginger crop is affected by many pests and diseases. Of these, soft rot, bacterial wilt, yellows, *Phyllosticta* leaf spot, and several types of storage rots are major diseases which cause severe economic losses. Bacterial wilt and soft rot are prevalent in all major ginger-growing areas across the ginger-growing countries. *Pythium* spp., *Fusarium oxysporum*, *Ralstonia solanacearum*, and *Pratylenchus coffeae* are potent pathogens causing soft rot, yellows, bacterial wilt, and dry rot in the field. Table 21.1 details the disease, the causal pathogen, and the distribution within India and overseas.

Diseases Caused by Oomycetes and True Fungi

Soft Rot (Pythium *spp.*)

Soft rot is also referred to as rhizome rot or *Pythium* rot in scientific literature. The disease is prevalent in all the ginger-growing countries across the world, such as India, Japan, China, Nigeria, Fiji, Taiwan, Australia, Hawaii, Sri Lanka, and Korea (Lin et al., 1971). Several species of *Pythium*, such as *Pythium aphanidermatum*, *P. myriotylum*, *P. vexans*, *P. ultimum*, *P. splendans*, and *P. deliense* have been recorded as causal agents of soft rot of ginger across the world. Among the diseases that affect ginger crop in the world, soft rot is the most destructive in all stages of its growth. Soft rot disease is both soil and seed borne (McRae, 1911; Mundkar, 1949). The disease occurs during the months of July and September in India, the time coinciding with the onset of the southwest monsoon in India. High soil moisture and low ambient temperature (25–28°C) in this period are highly conducive for the spread of the disease. *Pythium* is a large genus of the class Oomycetes, including more than 120 described species (Dick, 1990). *Pythium* species often infect immature and undifferentiated parts of the plant. Roots, rhizomes, emerging sprouts, and the pseudostem of ginger are all prone to infection, depending on the stage of their maturity. About six *Pythium* species, namely, *P. aphanidermatum*, *P. butleri*, *P. deliense*, *P. myriotylum*, *P. pleroticum*, *P. ultimum*, and *P. vexans* have been reported to cause soft rot in different parts of India (Dohroo, 1987). The most commonly encountered are *P. aphanidermatum* and *P. myriotylum* (Sarma, 1994). Using conational, taxonomical tools/keys, the species are separated primarily by differences in oogonial diameter and number of antheridia per oogonium (Van der Plaats-Niterink, 1981). Identification and characterization of *Pythium* species is often based on the morphological characteristics,

The Agronomy and Economy of Turmeric and Ginger. DOI: http://dx.doi.org/10.1016/B978-0-12-394801-4.00021-1

Table 21.1 Diseases of Ginger and Their Causal Organisms

Disease	Causal Pathogen	Distribution in India	Global Distribution
Soft rot	*Pythium* species	Andhra Pradesh, Assam	India, Japan, China, Nigeria, Fiji,
	P. aphanidermatum	Bihar, Gujarat, Kerala	Taiwan, Australia,
	P. gracile, P. deliense	Karnataka, Maharashtra	Hawaii, Sri Lanka, Korea
	P. myriotylum	Madhya Pradesh	
	P. ultimum	West Bengal, Madhya Pradesh	
	P. pleroticum	Himachal Pradesh	
	P. vexans	Kerala, Maharashtra	
	P. splendons	Rajasthan	
	F. solani	Himachal Pradesh	
Bacterial wilt	*R. solanacearum* Biovar III and IV	All over India	India, Japan, China, Nigeria, Fiji, Taiwan, Australia, Hawaii, Sri Lanka, Korea
Yellows	*Fusarium* spp.	Himachal Pradesh	India, Australia,
	F. oxysporum f. sp. *zingiberi*	Rajasthan	Hawaii
	F. moniliforme		
	F. graminearum		
	F. equiseti		
Leaf spots	*Phyllosticta zingiberis*	Kerala, Karnataka	India
	Helminthosporium maydis	Himachal Pradesh	
	Colletotrichum zingiberis	Bihar	
	Pyricularia zingiberis	Andhra Pradesh	
	Leptosphaeria zingiberis	Assam	
	Coniothyrium zingiberis	Meghalaya	
	Curvularia lunata	Assam	
	Vermicularia zingiberis	Bihar	
	Septoria zingiberis	Andhra Pradesh	
Nematodes	*Meloidogyne* spp., *Radopholus similis, Pratylenchus coffeae*	Kerala, Sikkim	India, Fiji, Australia
Leaf blight/Dry rot	*Rhizoctonia solani*	Himachal Pradesh	India
	R. bataticola	Haryana, Kerala	
Thread blight	*Pellicularia filamentosa*	Kerala	India
Basal rot	*Sclerotium rolfsii* (*Corticum rolfsii*)	Maharashtra	India
Sheath rot	*Fusarium* spp.	Maharashtra	India
Virus diseases	Cucumber mosaic virus (CMV), Chlorotic Fleck virus (GCFV), Chirke virus	Kerala, Assam	India, Malaysia, Mauritius, Australia

but complications might arise due to the absence of sexual structures and the failure to induce zoosporogenesis. In addition, environmental factors such as ambient temperature, media type, and age of the culture can affect the morphological and physiological characteristics and thus hinder the identification process (Hendrix and Papa, 1974).

Molecular techniques are now widely used to study fungal taxonomy and phylogeny (Taylor, 1986). The technique is useful to determine the relatedness between the isolates within a species and to distinguish species from one another. Ribosomal DNA (rDNA) of eukaryotes is arranged in tandem repeats in specific chromosomes and is subjected to concerted evolution. The ITS region of rDNA is used as a target for species-specific detection of fungi and is variable between species, but is largely conserved within species (Chen, 1992). The ITS region consists of noncoding variable regions that are located within two rDNA repeats, the highly conserved small subunit rRNA genes. The ITS region is a particularly useful area for molecular characterization investigations in fungi (Sreenivasaprasad, 1996).

Symptoms of the Disease

The disease is predisposed to waterlogged conditions and is caused by several species of *Pythium*. The plant is susceptible to *Pythium* infection during all the stages of growth. All the underground parts like roots, stem, and emerging sprouts are susceptible to this disease. The buds, roots, developing underground stem, the rhizome, and collar regions are the portals of infection. When the seed rhizomes are infected, they fail to sprout due to the rotting of young buds. After sprouting, the infection takes place through the root or through the collar region, finally reaching the rhizome. Symptoms appear initially as water-soaked patches at the collar region. These patches enlarge and the collar region becomes soft and watery and then rots. *P. aphanidermatum* was found pathogenic to germinating buds and mature rhizomes of ginger and infection was more severe when the tissues were wounded (Indrasenan and Paily, 1973). In mature plants, collar infection leads to yellowing of leaves. This yellowing starts from the leaf tip and spreads downward, mainly along the margins of the leaves resulting in the death of the leaves. The dead leaves droop and hang down the pseudostem until the entire shoot becomes dry. The basal portion of the plant exhibits a pale translucent coloration. This area subsequently becomes water-soaked and soft to such an extent that the whole shoot either topples or can easily be pulled out. Rhizomes first turn brown and gradually decompose, forming a watery mass of putrefying tissue enclosed by the tough skin of the rhizome. The fibrovascular strands are not affected and remain isolated within the decaying mass. Roots arising from the affected regions of the rhizome become soft and rotten. The rotten parts emit a foul smell. Rotting attracts opportunistic fungi, bacteria, and insects, and in particular, scavengers. Rhizome rot of ginger is the eventual status of the ginger rhizomes incited by primary pathogens such as *Pythium* or *Ralstonia*. The term rhizome rot is loosely used to describe any ultimate state of a rhizome due to any of the above-mentioned organisms. Though each of the above-mentioned organisms causes its own disease in ginger, the final state of the rhizome is complete rotting or partial rotting, hence the term rhizome rot.

Genetic Diversity of Pythium

Twenty-nine isolates of *Pythium*, isolated from the major ginger-growing locations of India, were characterized by adopting certain phenotypic and molecular methods. PCXR-RFLP by using ITS primers revealed five clusters among the isolates, which were morphologically identified as *P. myriotylum*, *P. ultimum* var. *sporangiiferum*, *P. ultimum* var. *ultimum*, and *P. deliense* (Jooju, 2005). The pathogenic potential of the isolates varied among the isolates collected from the different geographical locations (Kumar et al., 2007). The isolates obtained from high-altitude regions like Sikkim, India, showed less aggressive infective ability on ginger than when assayed in the plains, at lower altitudes. Isolates required 2–10 weeks to induce soft rot in ginger, and their rhizome rot potential also varied among the isolates. *P. myriotylum* and *P. deliense* were among the most virulent species as they were found to cause over 85% reduction in rhizome yield. The majority of the isolates collected from throughout Kerala State in India belonged to *P. myriotylum* (Kumar et al., 2008).

PCR-Based Identification of Pythium

The species of *Pythium*, which causes soft rot of ginger in the states of Kerala, Karnataka, Uttar Pradesh, and Sikkim, India, were identified as *P. myriotylum*. A PCR-based method was found suitable for identification of *P. myriotylum*. Primers specific for *P. myriotylum* were found to amplify 150 bhp sequences in the genomic DNA of *P. myriotylum* (Kumar et al., 2008). Among the 29 *Pythium* isolates, 14 isolates from the States of Assam, Sikkim, Uttar Pradesh, Kerala, and Karnataka were identified as *P. myriotylum*, based on the size of the species-specific amplicon (150 bp) using the oligo primers Pmy5 and ITS2. It was found that the isolates from different geographical regions were *P. myriotylum*. The suitability of the primer combination of Pmy5 (5'-GTCGCTGTTATGGCGGAG-3') and ITS2 (5'-GCTGCGTTCTTCATCGATGC-3') (Wang et al., 2003) at the species level identification of *P. myriotylum* is validated among Indian isolates (Kumar et al., 2008). Six pathogenic *Pythium* isolates obtained from diseased ginger (*Z. officinale*) rhizomes were identified as *P. myriotylum* based on various morphological and physiological characteristics. The isolates showed strong virulence on buds, crowns, rhizomes, and roots, as well as on leaves and stems. The maximum, optimum, and minimum growth temperatures for *P. myriotylum* were 39–45°C, 33–37°C, and 5–7°C, respectively. The optimum pH for growth is 6–7. Mycelial linear growth was most rapid on V-8 juice agar, but aerial mycelia were most abundant on PDA and cornmeal agar. Zoosporangial and oogonial formation was greatest on V-8 juice agar. Optimum temperatures for the production of zoosporangia and oogonia were 20–35°C and 15°C, respectively (Kim et al., 1997).

Management of Soft Rot

1. Seed rhizome selection and treatment: Infected rhizomes are the primary source of infection and spread of soft rot in the field. The best method to manage the disease is by the use of disease-free rhizomes for planting. Use apparently good-looking and healthy rhizomes for planting.

2. Chemical treatment: Treat the seed rhizomes for 30 min with Mancozeb (0.3%) or carbendazim (0.3%) in the case of soft rot prior to storing and planting. Carbendazim alone or in combination with Mancozeb is also used to prevent the seed-borne inoculums of *Pythium* and *Fusarium*.
3. Cultural methods: One of the predisposing factors for soft rot spread of ginger is ill-drained fields in continuous wet weather. Proper drainage in sandy loam soil for cultivation ensures a healthy crop.
4. Soil solarization: Soil solarization is a soil disinfection practice achieved by covering moist soil with transparent polythene film during the period of prevalent high temperatures and intense solar radiation. Wherever possible, this practice can be adopted.
5. Soil drenching: Soil drenching with Mancozeb (0.3%) or cheshunt compound to control soft rot or metalaxyl at the rate of 500 ppm as a soil drench was found to reduce the soft rot incidence. Metalaxyl in combination with copper or biocontrol organisms has been used successfully to reduce crop loss.
6. Biological control: Antagonistic fungi, namely *Trichoderma harzianum*, *T. hamatum*, *T. virens*, and bacterial isolates *Bacillus* and *Pseudomonas fluorescens* have been reported to suppress soil-borne pathogens of ginger.

Fusarium *Yellows or Dry Rot (*Fusarium oxysporum *f. sp.* zingiberi*)*

This disease is caused by *F. oxysporum* f. sp. *zingiberi*, predisposed by nematode infestation by *Pratylenchus coffeae*. Ginger yellows was originally reported from Queensland and subsequently from Hawaii (USA) and India (Haware and Joshi, 1974; Trujillo, 1963).

Fusarium rhizome rot: A very common and serious fungal disease which is specific to ginger. Infected plants are stunted and yellow, lower leaves dry out and turn brown. Eventually, all of the above-ground shoots dry out completely. The plant's collapse is very slow, which can last for several weeks, compared with the rapid collapse associated with bacterial wilt infection. Diseased rhizomes show a brown internal discoloration, are normally shriveled in appearance, and eventually decay, leaving the outer shell intact with fibrous internal tissue remaining. Increased nematode infestations are usually associated with *Fusarium* rhizome rot, accentuating yield losses. *Fusarium* is also responsible for serious loss of planting pieces and poor germination.

Symptoms of the disease: On leaves, symptoms appear as yellowing of the leaf margins of the lower leaves, which gradually spread over the entire leaf. Older leaves dry up first, followed by the younger leaves. Plants may also show premature drooping, wilting, and drying in patches in the field or in the whole bed. Plants generally do not lodge on the ground as noticed in the case of soft rot or bacterial wilt. In rhizomes, a cream-to-brown discoloration accompanied by shriveling is commonly seen. Vascular rot is also prominent. In the final stages of infection, only the fibrous tissues remain within the rhizomes. A white cottony fungal growth may develop on the surface of stored rhizomes. This disease, along with nematode infestation, severely reduces the marketability of the rhizomes. When used as seed rhizome, the disease may affect the germination of the rhizomes. To ensure that there is no incidence of this disease, collect the planting materials from areas or regions known to be free from the disease. This is a very important precautionary measure to preempt the disease incidence. When cutting up the rhizome, the pieces showing shriveling or discoloration symptoms must be discarded, and the cutting knife must be regularly dipped in methyl alcohol or in any other commercial disinfectant. As soon as possible after preparation, the seed pieces should be dipped in benomyl solution for a minute. Avoid areas with heavy nematode infestation. Crop rotation for at least two to five years between ginger crops will help reduce *Fusarium* infestation. The maximum germination of ginger sets occurred with a treatment consisting of pine needles, organic amendment, and a seed treatment consisting of a mixture of Mancozeb + thiophanate-methyl at 0.25% concentration and carbendazim at 0.1% concentration for 1 h. Pine needle amendment alone and in combination with a fungicidal seed treatment and a combination of *T. harzianum* and *Gigaspora margarita* gave the best control of ginger yellows caused by *F. oxysporum* f. sp. *zingiberi*, resulting in the highest yield of fresh ginger. Inhibition of *Meloidogyne* spp. and *Pratylenchus* spp. was obtained best with pine needle organic amendment in combination with *T. harzianum* seed and soil application.

1. Healthy rhizome selection: As the disease spreads through contaminated rhizomes, selection of healthy rhizomes has been found to be an effective preventive measure for disease control.
2. Rhizome treatment with hot water: This is done with hot water (51°C) for 10 min, and is particularly recommended in places where the disease is endemic.
3. Chemical control: Mancozeb (0.3%) and carbendazim (0.05%) are found to reduce the disease incidence.
4. Biological control: Biochemical agents such as *T. harzianum*, *T. hamatum*, and *T. virens* as seed treatment and soil application were found to control the disease.

Phyllosticta *Leaf Spot (*Phyllosticta zingiberi*)

This disease is widespread in most ginger-growing countries, including India. Leaf spot disease of ginger was reported for the first time in the Godavari and Malabar districts in Andhra Pradesh and Kerala, India, respectively, by Ramakrishnan (1942).

Symptoms: A phyllosphere fungus, *Phyllosticta zingiberi*, causes this disease. Small, spindle to oval or elongated spots appear on younger leaves. The spots have white papery centers and dark brown margins surrounded by yellowish halos. The spots later increase in size and coalesce to form large spots, which eventually decrease the effective photosynthetic area on the leaf surface. As the plants put forth fresh leaves, they subsequently become infected. Such infected areas often dry up at the center, forming holes. In the case of a severe attack, the entire leaf dries up. As a result of infection, the crop develops a grayish, disheveled look.

Management of the Disease

1. Seed rhizome selection: The seed rhizomes should be selected from disease-free areas, as the disease can spread through rhizomes which appear to be normal.
2. Rhizome treatment: Seed rhizomes can be treated with a carbendazim + Mancozeb combination or carbendazim (0.25%) before planting. Prochloraz, Tebuconazole, Chlorothalonil, Mancozeb, Captan, and Chlorothalonil + Copper gave the best control and increased ginger yield in Brazil (De Nazareno, 1995).
3. The natural way to control leaf spot is to grow ginger under shade trees, such as coconut trees.

Thread Blight (Pellicularia filamentosa)

Sundram (1954) reported this disease for the first time in the Malabar district of Kerala State, India. The disease is not of much significance and occurs very rarely during heavy rainfall. This disease is caused by *Pellicularia filamentosa*.

Symptoms: Small water-soaked lesions appear on the leaf margins or other parts of the leaf during the initial stages of this disease. Subsequently, the infected leaves lose their turgidity, wilt, and may become detached from the sheath. Fine hyphal threads spread over the infected parts, and small brown sclerotia are present on the lower surface. The infected portion of the leaf turns white and papery on drying. Protective spraying with Bordeaux mixture (1%) before the start of heavy rains and an application of carbendazim (0.2%) as a spray is found to reduce the incidence of the disease.

Leaf Spot

Dry rot caused by *Macrophomina phaseolina* (Tassi) Goid, basal sheath rot by *Aphelenchus* (nematode), and a *Fusarium* sp. basal rot by *Sclerotium rolfsii* (*Corticum rolfsii*), violet rot pathogen, *Helicobasidium mompa* Tanaka, black rot by *Rosellinia zingiberis* Stevens, leaf spot by *Leptosphaeria zingiberis*, *Coniothyrium zingiberis*, and *Cercoseptoria* sp., *Curvularia lunata* (*Cochliobolus lunatus*), *Vermcularia zingiberae*, *Pyricularia zingiberis*, *Colletotrichum zingiberis*, *Septoria zingiberis*, *Helminthosporium* sp. were all found to be associated with ginger leaves or rhizomes. *Rhizoctonia* (*Corticium*) *solani* causing pseudostem rot and *Rhizoctonia bataticola* (*Macrophomina*) causing leaf blight are other diseases affecting ginger crops on a limited scale.

Storage Diseases

As ginger undergoes three months of dormancy in storage during the months of April and May, it is important to protect it from various storage losses due to micro-organisms and insect pests apart from abiotic stress such as heat build up. Under storage, different fungi and bacteria have been found to be associated with ginger rhizomes which results in rotting and decaying of the rhizomes. In storage, the fungi, such as *F. oxysporum, P. deliense, P. myriotylum, Geotrichum candidum, Aspergillus flavus, Cladosporium lennissimum, Gliocladium roseum, Graphium album, Mucor racemosus, Stachybotrys sansevieriae, Thanatephorus cucumeris*, and *Verticillium chlamydosporium* are known to affect the ginger rhizomes. The fungus *A. flavus* in association with ginger rhizomes was implicated in the production of carcinogenic aflatoxin. Root patterns were grouped into four different types: yellow soft rot, brown rot, localized ring rot, and water-soaked rot. Water-soaked rot was the most frequent (40%) and ring rot the least frequent (14%). Causal pathogens were identified as *Erwinia carotovora* and *Pseudomonas aeruginosa* (yellow soft rot), *F. solani* and *Pseudomonas aeruginosa* (brown rot), *F. solani* (localized ring rot), and *P. spinosum* and *P. ultimum* (water-soaked rot). *P. myriotylum*, the causal pathogen of *Z. officinale* rhizome rot which occurs severely in the fields, was rarely detected from storage seed-rhizomes, suggesting its minor involvement with storage rot. Pathogenic *Pythium* isolates were frequently obtained from both the rhizome surface and the inner tissues of rotten rhizomes (Kim et al., 1998).

Diseases Caused by Bacterial Pathogens

Bacterial Wilt of Ginger *(R.* solanacearum*)*

Bacterial wilt disease of ginger blast is one of the most important production constraints in tropical, subtropical, and warm temperate regions of the world. Bacterial wilt of ginger inflicts serious economic losses to small and marginal farmers who depend on this crop for their sustenance. Bacterial wilt of ginger is an important production constraint and it is widely distributed in most of the tropical and subtropical regions of the world (Kumar and Sarma, 2004). The causative organism, *R. solanacearum (Pseudomonas solanacearum* Smith, 1896) is a soil- and plant-inhabiting bacterium. *R. solanacearum* affects monocotyledonous and dicotyledonous plants. *R. solanacearum* is widely distributed in tropical, subtropical, and temperate regions of the world. The bacterial wilt is characterized by the entry of the bacterium into the host followed by its multiplication and movement through the xylem vessels of the host plant. In the process, the bacteria interfere with the translocation of water and nutrients, which in turn results in drooping, wilting, and death of the above-ground plant parts. In the case of the ginger plant, the first noticeable symptom of bacterial wilt is downward curling of leaves due to loss of turgidity and within three to four days, the leaves dry up. The affected rhizomes start to rot and putrefy due to the attack of saprophytic soil microorganisms. The rotted rhizomes emit a foul smell and the affected plants wither and die within two to three weeks.

Geographical distribution of the pathogen has expanded in the last few years, because of inadvertent transmission of the bacterium along with contaminated planting material (infected rhizomes) (Kumar and Hayward, 2005; Kumar et al., 2004). Bacterial wilt infection spreads when environmental conditions such as heavy rainfall and cool weather prevail, making them predisposing conditions for the spread of the disease. Rhizome-borne inoculum is primarily responsible for the initiation of the disease in the field, which further spreads horizontally across the field due to incessant rain. It is speculated that the rhizomes collected from previously diseased fields carry the inoculum to new locations, as well as to the next season (Kumar and Hayward, 2005). A PCR-based method has been described for the detection of the bacterial wilt pathogen in soil and in planting material such as potato tubers (Kumar et al., 2002). The detection method for the bacterial wilt pathogen on symptomless tuber has been reported (Janse, 1988). Several advances have been made for the detection of the bacterial wilt pathogen in environmental samples (Kumar and Anandaraj, 2006; Schaad et al., 1995). Primers specific for PCR-based detection of *Rhizoctonia solanacearum* in plant and soil samples have been reported (Ito et al., 1998). This bacterial pathogen survives in soil and makes it unsuitable for ginger cultivation for long periods once introduced through infected planting material. Once introduced in an area, the soil becomes unsuitable for further cultivation of ginger. The severity of the disease is evident from its rapid spread in the field when environmental conditions (heavy and incessant rainfall, alternating with warm weather) are favorable for the onset and spread of the disease. Each of the infected plants is capable of releasing hundreds of thousands of bacterial cells in the form of bacterial ooze.

Preplant detection of the bacterium in seed rhizomes and soil assumes significance to avoid the disease epidemic. Serological methods, such as the indirect ELISA method, has been indicated for the detection of the disease in the soil (Prioru et al., 1999), in addition to environmental methods, such as isolation on semiselective medium (Englebrecht, 1994) or bioassays using indicator host plants. Conventional methods are unsuitable to detect the pathogen, as it survives at a very low population density in the soil. However, these methods, particularly the serological ones, are not universal, as they are known to yield false-positive results when adopted in new host–pathogen systems. Another potential alternative approach would be DNA-based methods such as PCR using pathogen-specific probes or oligo primers to detect the pathogen.

The Bacterial Wilt Pathogen

Bacterial wilt of ginger is caused by the bacterium *R. solanacearum* biovar III (Smith) Yabuuchi, which is one of the important rhizome-borne diseases affecting ginger in the field. In India, biovar III causes rapid wilt in ginger within five to seven days after infection under artificial stem inoculation and seven to ten days under soil inoculation of the pathogen (Kumar and Sarma, 2004). Traditionally, ginger is cultivated in previously fallowed soil or on virgin soil. Incidence of bacterial wilt noticed in such fields is one of the indirect evidences of the rhizome-borne nature of *R. solanacearum* in ginger. Being a vascular pathogen, it is presumed that the pathogen *R. solanacearum* can survive in ginger plants at a very low level of inoculum without adversely affecting the normal state of the plant growth. Bacterial wilt caused by *R. solancearum* (Smith) Yabuuchi is one of the important production constraints in ginger production in India and other parts of the world. In India, this disease has been found in all major ginger-growing states and is particularly severe in hot and humid southern states (ambient temperature varying between 28°C and 30°C), as well as in the cold high-altitude Eastern Himalayan state of Sikkim (ambient temperature is 7–22°C), where ginger farming in the Northern and Eastern districts has been severely affected by bacterial wilt during the last decade. These geographical, micro-, and macroclimatic variations and differences in the method of ginger production in these locations did not deter the severity of bacterial wilt in the Indian subcontinent. Genetic comparison was attempted between these two populations of strains causing bacterial wilt of ginger from these geographically well-isolated locations. Initially, the bacterial wilt pathogen was isolated from wilted ginger plants from these geographical locations. Ten isolates were obtained from wilted ginger plants from the North and the East Sikkim districts of the Eastern Himalayan regions, at an altitude of over 5500 m above mean sea level (msl). These isolates were phenotypically and genotypically compared with 13 other strains isolated from Kerala and Karnataka, in the southern states of India. The strains were isolated on CPG agar and identified by PCR-based assay using universal Rs-specific primers which produced a single 280 bp amplicon specific for *R. solanacearum*. Phenotypic characterization revealed the occurrence and dominance of biovar III over IV among the collections. Interestingly, biovar IV was rarely encountered in both the locations compared with biovar III. The biovar III strains were highly aggressive on the ginger plant, causing wilt in 5–7 days of soil inoculation, whereas the biovar IV strains took 3–4 weeks to wilt the ginger plants. However, in places like Hawaii (USA), biovar III is of little significance and biovar IV is responsible for a very rapid spread, leading to wilting of the plant and causing heavy losses to the crop. The genetic diversity of *R. solanacearum* strains isolated from ginger growing on the Hawaiian island was determined by analysis of AFLPs, which revealed that *R. solanacearum* strains obtained from ginger grown in Hawaii are genetically distinct from the local strains from tomato (Race 1) and *Heliconia* (Race 2) (Yu et al., 2003). A weakly pathogenic strain of *R. solanacearum* isolated from ginger was shown to differ from a local tomato strain in cross-inoculation studies. Infected plants become stunted and yellow and the lower leaves dry out over a prolonged period before the plants finally wilt and die (Lum, 1973).

Diversity of Bacterial Wilt Pathogen

Isolates of *R. solanacearum* causing bacterial wilt of ginger in the northeastern States of Sikkim and Kerala were found to be 100% similar in REP, ERIC, and BOX PCR profiles. A very close similarity coefficient between these two geographically well-separated locations clearly indicated the strain migration from one location to another. They belong to biovar III or biovar IV and caused wilt within five to seven days of inoculation. Biovar III and biovar IV could be differentiated in REP-PCR-based fingerprinting (Kumar et al., 2004).

Close similarity among the isolates from these geographically well-separated locations indicated that the strains had migrated from one place to another, most likely through the rhizomes, as ginger rhizome exchanged among small and marginal farmers in India is an important activity during peak planting season. This result further confirmed the role of apparently good-looking but contaminated/latently infected rhizomes in the spread and distribution of bacterial wilt in India. This emphasized the need for an effective internal quarantine measure to regulate the unorganized movement of rhizomes from the affected areas to prevent the possible spread and outbreak of bacterial wilt in newer locations in India.

Resistance

Over the years, the Indian Institute of Spices Research in Calicut, Kerala State, India, has collected more than 700 accessions of ginger through a germplasm exploration program. The systematic selection program has resulted in the release of three high-yielding ginger varieties. However, none of these varieties are resistant (or tolerant) to any of the economically important diseases of ginger, in particular, *Pythium* rot and *Ralstonia* wilt. This is due to lack of genetic variability among the accessions for genetically important traits, such as pest and disease resistance, which is one of the bottlenecks in ginger genetic improvement. Mutation breeding, though it appeared promising initially, did not subsequently yield any desirable results, until now. In recent years, the search for resistance has been extended to another genus in the family Zingiberaceae. Among the Zingiberaceae members, such as *C. amada, C. longa, C. zedoria, C. aromatica, Kaempferia galanga, Elettaria cardamomum, Z. officinale,* and *Z. zerumbet,* evaluated for their reaction to ginger strain *R. solanacearum* biovar III and *Pythium* species, *C. amada* Roxb., the Indian mango ginger (rhizomes have an aroma of green mango), was found to resist infection by both pathogens. The pathogen *R. solanacearum* could be detected in soil, root, and on the surface of the rhizome of mango ginger in PCR-mediated assay using *R. solanacearum*-specific primers. The survival of *R. solanacearum* in soils planted with *C. amada* was confirmed by bioassay where the ginger plants transplanted in soil in the vicinity of surviving *C. amada* was found to wilt within five to seven days of infection. This clearly indicated that the pathogens were unable to infect the plant when inoculated in soil. Interestingly, the *C. amada* plants succumbed to wilt when the pathogen was directly delivered into the pseudostem through pin pricking. This further confirms that the plant *per se* is not antagonistic to the pathogen in soil. It could be due to nonrecognition of the plant by *R. solanacearum*. The *C. amada* plants hold promise for developing bacterial wilt-resistant-ginger plants if the precise mechanism of resistance is unraveled. The literature survey did not indicate any resistant host in Zingiberaceae for bacterial wilt. A thorough genetic analysis would unravel the factors (genes) governing the resistance in *C. amada* against *R. solanacearum* and *Pythium* species.

PCR-Based Identification/Detection of R. solanacearum

A PCR-based method for identification of bacterial wilt pathogen was optimized for unambiguous identification of *R. solanacearum*. The bacterium produced 280 bp amplicon in a PCR performed with a primer (primer sequence). An efficient DNA isolation protocol and PCR-based detection of bacterial pathogen in soil is standardized. The DNA isolation method and PCR-based approach using universal *R. solanacearum*-specific primer to detect the bacterium in the soil offer rapid methods for unambiguous detection of this pathogen in soil which can be employed to monitor soil samples for this globally important plant pathogen. The PCR-based assay could detect the pathogen concentration at a concentration of 10^3–10^4 cells per gram of soil (Kumar and Anandaraj, 2006). The rhizome-borne nature of the bacterial wilt pathogen *R. solanacearum* in ginger was confirmed using PCR-based detection assay. Surviving rhizomes collected from previous bacterial wilt-infected fields, which appeared outwardly healthy, were found to be infected with *R. solanacearum*. The soil collected from the vicinity of the healthy rhizomes in the bacterial wilt-affected fields also tested positive for *R. solanacearum*. The results underscore the potential threat of using apparently "healthy" rhizomes from such latently infected fields, as well as the "avoidance" of visibly healthy-looking rhizomes for planting purpose, from such locations. The study further emphasizes the need for strict rules in restricting the movement of such rhizomes from endemic locations to nontraditional areas (Kumar and Abraham, 2008). Healthy and disease-free planting material is the foundation for good ginger production.

Management of Bacterial Wilt

Various control measures have been evaluated to combat the disease only with limited success. Bacterial wilt is a major problem in the production of ginger and other vegetable crops, owing to the wider host range and genetic variability that it exhibits. Besides, the pathogen is endowed with multiple modes of survival and rapid transmission capability within and between crop fields. The strategies for the wilt disease management are:

1. Selection of healthy rhizome material from disease-free area,
2. Selection of field with no history of bacterial wilt in the past,
3. Preplant rhizome treatment by heat or rhizome solarization (Kumar et al., 2005),
4. Strict phytosanitation in the field including restrictions on movement of farm workers and irrigation water across the fields,
5. Clean cultivation and minimum tillage,
6. Crop rotation with nonhost plants such as paddy and maize,
7. Insect pests and nematode control in the field,
8. Soil amendments including biological control agents.

Preplant rhizome treatment: Rhizome heat treatment aided by solar radiation is called "rhizome solarization." This method has been standardized in the case of bulk rhizome disinfection. Rhizome temperature over 45°C was achieved when rhizomes were subjected to sunlight, either directly or after blanketing them with polythene sheets (100 or 200 μm). Initially the rhizome heating was optimized using 1 kg of seed rhizomes in polythene bags of 200-μm quality, which was found to be cumbersome and labor intensive. To ease this, the method of rhizome treatment was modified to suit the bulk requirement where a large quantity of rhizome could be treated with minimal labor in a short span of time. Soon after harvest and before sprouting, the rhizomes were spread on a long polythene sheet followed by wrapping them with another sheet and exposing them to direct sunlight during a bright sunny day to achieve the required temperature of 47°C in the vascular region. The treatment can be done at any time in a day when the uninterrupted sunlight intensity is 2100 mmoles/m^2/s. The light intensity and concentration of CO_2 inside the polythene blanket was found to be 1600–1700 mmoles/m^2/s and 1800–1850 ppm (four- to fivefold increase compared to the ambient concentration), respectively. The elevated rhizome temperature was observed especially in the vascular region, where the pathogen is reported to survive. One of the major sources of variability in rhizome heat build-up *vis-à-vis* the survival of *R. solanacearum* as well as the viability of the rhizome is the variation due to the varying size and shape of the rhizome. Larger rhizomes (100 g) recorded 1–3°C higher temperatures than smaller ones (10 g). Many ideally heated rhizomes were found that were optimally solarized, besides a certain proportion of under- or over-heated rhizomes, as the variation in the sizes and shapes of the rhizome is inherent. Postenrichment DAS-ELISA for *R. solanacearum* and microbiological plate assay with solarized (heat-treated) rhizomes further confirmed that *R. solanacearum* and other microorganisms could not survive in solarized rhizomes. The assay clearly indicated that the rhizome solarization was capable of disinfecting the rhizomes infected by the pathogen *R. solanacearum*, as indicated by the low A405 values recorded in DAS-ELISA. Greenhouse trials using naturally and artificially infected rhizomes after rhizome solarization produced healthy plantlets. It was found that the freshly harvested rhizomes were highly amenable to heat treatment, as these rhizomes are intact without any juvenile and heat-sensitive sprouts. Besides, interruption by premonsoon clouds seldom occurred when the treatment was done before storage of the rhizomes in storage cabinets. The other effects observed due to "rhizome solarization" were early breaking of rhizome dormancy and obtaining a significantly high number of good sprouts. The partially shrunk rhizomes become shriveled or completely rotten after solarization. Thus, the method of rhizome solarization was proven to be an efficient method for selection of good and pathogen-free rhizomes for planting purposes. A greenhouse-based culture system to produce ginger rhizomes free of bacterial wilt has been developed by a group of Hawaiian scientists working on bacterial wilt of ginger (Hepperly et al., 2004).

Diseases Caused by Viruses

Mosaic Disease of Ginger

The symptoms of this disease appear as a yellow and dark green mosaic pattern on leaves. The affected plants show stunting. The virus causing mosaic disease in ginger has spherical particles with a diameter of 23–38 µm. It shows a positive serological reaction with antiserum to cucumber mosaic virus (CMV). The virus is known to be transmitted by the plant sap to different plants known to be hosts to CMV.

Chlorotic Fleck Disease

The geographical distribution of the virus is uncertain but is thought to occur in India, Malaysia, and Mauritius. The virus is mechanically transmitted by *Myzus persicae*, *Pentalonia nigronervosa*, *Rhopalosiphum maidis*, or *R. padi* (Thomas, 1986).

Big Bud

The tomato big bud organism causes this disease in ginger. The affected plants cease to grow and leaves become bunched at the top of the stem. As the disease advances, plants turn yellow and die. The pathogen has a wide range of hosts, and the disease is transmitted by leafhoppers. In seed-production areas, affected plants are removed and destroyed carefully.

Chirke Virus

Raychaudhary and Ganguly (1965) reported Chirke virus on ginger, which is a known disease in the large cardamom crop in India.

Diseases Caused by Nematodes

Several plant parasitic nematodes infect ginger and among them *Meloidogyne* spp., *Radopholus similis*, and *Pratylenchus* spp. have been reported to be the major ones of economic importance as they cause significant damage to ginger plants. *Meloidogyne incognita* was found to cause damage to *Z. officinale* in Kerala State, India (Mammen, 1973). Marketability of the ginger rhizome was found to be severely affected if the rhizomes are infested by nematodes, in particular by *Pratylenchus coffeae*. The nematode is found to reduce the germination of the rhizomes when used as seed rhizomes and further aggravated the infection by *Fusarium*. Linear growth and hyphal thickness of *F. oxysporum* f. sp. *zingiberi* were greater when exposed to an extract of ginger root infected with *Meloidogyne incognita* than when exposed to an extract of healthy root (Agrawal et al., 1974). Butler and Vilsoni (1975) reported that *Radopholus similis* was found on rhizomes of *Z. officinale* and transmission was mainly through planting infested vegetative "seeds" in Fiji. Histopathological studies demonstrate that the nematodes enter the rhizomes and penetrate the tissue intracellularly, large infestations resulting in the destruction of tissues and the formation of channels and galleries within the rhizomes. Secondary organisms eventually rot the entire rhizome. The symptoms shown by infested ginger plants include stunting, chlorosis, and failure to tiller profusely (Vilsoni et al., 1976).

Management of the Nematode Infection

Soil application of carbofuran at 3 kg a.i./ha three weeks after planting of the ginger crop was found to decrease yield losses due to the infestation of *Meloidogyne*. Preplanting application of neem (*Azadirachta indica*) cake at 1 t/ha followed by postplanting application of carbofuran at 1 kg a.i./ha 45 days after planting (DAP) is recommended to control *Meloidogyne incognita* infestation (Mohanty et al., 1995). The highest level of control was achieved by sawdust mulching combined with postplanting treatment with branded nematicides, Nemacur or Oxamyl. Postplant treatment with Nemacur or Oxamyl reduced the incidence of *Fusarium*.

References

Agrawal, P.S., Joshi, L.K., Haware, M.P., 1974. Effect of root knot extract of ginger on *Fusarium oxysporum* f. *zingiberi* Trujillo causing yellows disease. Curr. Sci. 43, 23–52.

Butler, L.D., Vilsoni, F., 1975. Potential hosts of burrowing nematode in Fiji. Fiji Agric. J. 37, 38–39.

Chen, W., 1992. Restriction fragment length polymorphisms in enzymatically amplified ribosomal DNAs of three heterothallic *Pythium* species. Phytopathology 82, 1467–1472.

De Nazareno, N.R.X., 1995. Control of yellow leaf spot (*Phyllosticta* sp.) of ginger with commercial fungicides. Horticultura-Brasileira 13, 142–146.

Dick, M.W., 1990. Keys to *Pythium*. University of Reading, UK, pp. 214–216.

Dohroo, N.P., 1987. *Pythium ultimum* on *Zingiber officinale*. Indian Phytopathol. 40 (2), 275.

Englebrecht, M.C., 1994. Modification of a semi-selective medium for the isolation and quantification of *Pseudomonas solanacearum* In: Hayward, A.C. (Ed.), Bacterial Wilt Newsletter, 10 Australian Centre for International Agricultural Research, Canberra, Australia, pp. 3–5.

Haware, M.P., Joshi, L.K., 1974. Studies on soft rot of ginger from Madhya Pradesh. Indian Phytopathol. 27, 158–161.

Hendrix, F.J., Papa, K.E., 1974. Taxonomy and genetics of *Pythium*. Proc. Am. Phytopathol. Soc. 1, 200–207.

Hepperly, P., Zee, F.T., Kai, R.M., Arakawa, C.N., Meisner, M., Krarky, B., et al., 2004. Producing bacterial wilt-free ginger in green house culture. Exten. Service Bull., 10, 6.

Indrasenan, G., Paily, P.V., 1973. Studies on the soft rot of ginger (*Zingiber officinale* Roscoe) caused by *Pythium aphanidermatum* (Edison) Fitz. Agric. Res. J. Kerala 11 (1), 53–56.

Ito, S., Ushijima, Y., Fuji, T., Tanaka, S., Kameya-Iwaki, M., Yoshiwara, S., et al., 1998. Detection of viable cells of *Ralstonia solanacearum* on soil using semi selective medium and a PCR technique. J. Phytopathol. 146, 379–384.

Janse, J.D., 1988. A detection method for *Pseudomonas solanacearum* in symptomless potato tubers and some data on its sensitivity and specificity. Bull. OEPP 18, 343–351.

Jooju, B., 2005. Evaluation of Genetic Diversity of *Pythium* spp. Causing Soft Rot of Ginger Using Phenotypic and Molecular Methods, M. Phil. Thesis, Bharathidasan University, Trichy, Tamil Nadu, India, p. 72.

Kim, C.-H., Yang-Sung, S., Park-Kyong, S., Kim, C.H., Yang, S.S., Park, K.S., 1997. Pathogenicity and mycological characteristics of *Pythium myriotylum* causing rhizome rot of ginger. Korean J. Plant Pathol. 13 (3), 152–159.

Kim, C.-H., Yang-Jong, M., Yang-Sung, S., Kim, C.H., Yang, J., Yang, S.S., 1998. Identification and pathogenicity of microorganisms associated with seed-rhizome rot of ginger in underground storage caves. Korean J. Plant Pathol. 14 (5), 484–490.

Kumar, A., Abraham, S., 2008. PCR based detection of bacterial wilt pathogen *Ralstonia solanacearum* in ginger rhizomes and soil collected from bacterial wilt affected field. J. Spices Aromat. Crops 17 (2), 109–113.

Kumar, A., Anandaraj, M., 2006. Method for isolation of soil DNA and PCR based detection of ginger wilt pathogen, *Ralstonia solanacearum*. Indian Phytopathol. 59 (2), 154–160.

Kumar, A., Anandaraj, M., Sarma, Y.R., 2005. Rhizome solarization and microwave treatment: ecofriendly methods for disinfecting ginger rhizomes. In: Prior, P., Allen, C., Hayward, A.C. (Eds.), Bacterial Wilt and *Ralstonia Solanacearum* Spices Complex, American Phytopathological Society Press, pp. 185–196.

Kumar, A., Hayward, A.C., 2005. Bacterial diseases of ginger and their control. In: Ravindran, P.N., Babu, K.N. (Eds.), Monograph on Ginger CRC Press, Boca Raton, FL, pp. 341–366.

Kumar, A., Sarma, Y.R., 2004. Characterization of *Ralstonia solanacearum* causing bacterial wilt of ginger in India. Indian Phytopathol. 57, 12–17.

Kumar, A., Sarma, Y.R., Priou, S., 2002. Detection of *Ralstonia solanacearum* ginger rhizomes using post-enrichment NCM-ELISA. J. Spices Aromat. Crops 51, 35–40.

Kumar, A., Sarma, Y.R., Anandaraj, M., 2004. Evaluation of genetic diversity of *Ralstonia solanacearum* causing bacterial wilt of ginger using Rep-PCR and RFLP-PCR. Curr. Sci. 87 (11), 1555–1561.

Kumar A., Thomas R.S., Jooju B., Suseelabhai R., Shiva K.N., 2007. PCR based identification of *Pythium myriotylum* causing soft rot of ginger. In: Proceedings of the Nineteenth Kerala Science Congress, 29–31 January 2007, Kannur, Kerala State, India, pp. 700–702.

Kumar, A., Reeja, S.T., Suseela Bhai, R., Shiva, K.N., 2008. Distribution of *Pythium myriotylum* Dreschsler causing soft rot of ginger. J. Spices Aromat. Crops 17 (1), 5–10.

Lin, L.N., Cheong, S.S., Leu, L.S., 1971. Soft rot of ginger. Plant Prot. Bull. 13, 54–67.

Lum, K.Y., 1973. Cross inoculation studies of *Pseudomonas solanacearum* from ginger. MARDI-Research Bull. 1 (1), 15–21.

Mammen, K.V., 1973. Root gall nematodes as a serious pest of ginger in Kerala. Curr. Sci. 42, 15–549.

McRae, W., 1911. Soft rot of ginger in Rangpur district of East Bengal (E. Pakistan). Agric. J. India 6, 139–146.

Mohanty, K.C., Ray, S., Mohapatra, S.N., Patnaik, P.R., Ray, P., 1995. Integrated management of root knot nematode in ginger (*Zingiber officinale* Rosc.). J. Spices Aromat. Crops 4, 70–73.

Mundkar, B.B., 1949. Fungi and Plant Disease. Macmillan and Co., London, p. 246.

Prioru, S., Gutarra, L., Fernandez, H., Alley, P., 1999. Sensitive detection of *Ralstonia solanacearum* in latently infected potato tubers and soil by post enrichment ELISA CIP programme report 1997–98. International Potato Centre, Lima, Peru. pp. 111–121.

Ramakrishnan, T.S., 1942. A leaf spot disease of *Zingiber officinale* caused by *Phyllosticta zingiberi* sp. Proc. Indian Acad. Sci. Sect. B 20 (4), 167–171.

Raychaudhary, S.P., Ganguly, B., 1965. Further studies on Chirke disease of large cardamom by aphid species. Indian Phytopath. 18, 373–377.

Sarma, Y.R., 1994. Rhizome rot disease of ginger and turmeric. Adv. Horticult. 10, 1134–1136.

Schaad, N.W., Cheong, S.S., Tanaka, S., Hatziloukas, E., Panpaulos, N.J., 1995. A combined biological and enzymatic amplification (BIO-PCR) technique to detect *Pseudomonas syrigiae* pv. *Phaseolicola* in bean seed extracts. Phytopathology 85, 243–248.

Sreenivasaprasad, S., 1996. Phylogeny and systematics. Genome 39, 499–512.

Sundram, S.V., 1954. Thread blight of ginger. Indian Phytopathol. 6, 80–85.

Taylor, J.W., 1986. Fungal evolutionary biology and mitochondrial DNA. Exp. Mycol. 10, 259–269.

Trujillo, E.E., 1963. *Fusarium* yellows and rhizome rot of common ginger. Phytopathology 53, 1370–1371.

Thomas, J.E., 1986. Purification and properties of ginger chlorotic fleck. Ann. Appl. Bio. 108 (1), 43–50.

Van der Plaats-Niterink, A.J., 1981. Monographs of the genus *Pythium*. Stud. Mycol. 21, 1–242. Centraalbureau voor Schimmelcultures, Baam.

Vilsoni, F., McCure, M.A., Butler, L.D., 1976. Occurrence, host range and histopathology of *Radopholus similis* in ginger (*Zingiber officinale*). Plant Dis. Rep. 60 (5), 417–420.

Wang, P.H., Chung, C.Y., Lin, Y.S., Yeh, Y., 2003. Use of polymerase chain reaction to detect the soft rot pathogen, *Pythium myriotylum* infected ginger rhizomes. Lett. Appl. Microbiol. 36, 116–120.

Yu, Q., Alvarez, A.M., Moore, P.H., Zee, F., Kim, M.S., de Silva, A., et al., 2003. Molecular diversity of *Ralstonia solanacearum* isolated from ginger in Hawaii. Phytopathology 93 (9), 1124–1130.

22 The Insect Pests of Ginger and Their Control

Shoot Borer (*Conogethes punctiferalis* Guen.)

The shoot borer (*C. punctiferalis*) (Pyralidae) is the most serious insect pest of ginger and turmeric in India. The larvae bore into the pseudostems and feed on the internal shoot, resulting in yellowing and drying of infested pseudostems. The presence of a borehole on the pseudostem through which frass is extruded and the withered central shoot is the characteristic symptom of the pest infestation (Devasahayam and Koya, 2005). The adults are medium-sized moths with a wingspan of 18–24 mm; the wings and body are pale straw yellow with minute black spots. There are five larval instars and fully grown larvae are light brown with sparse hairs and measure 16–26 mm in length. Adult females lay 30–60 eggs during their life span and six to seven generations are completed during a crop season in the field. The pest is observed in the field throughout the crop season and its population is higher during September–October in Kerala State, India. The shoot borer is highly polyphagous and has been recorded on more than 35 host plants, including several economically important plants in India (Devasahayam and Koya, 2005; Jacob, 1981).

Management of the Pest

Spraying malathion 0.1% solution at monthly intervals during July–October is effective for the management of the shoot borer on ginger. A sequential sampling strategy to monitor the pest infestation in the field as a guide to take control measures has also been formulated (Koya et al., 1986). An integrated strategy involving pruning and destroying freshly infested shoots during June–August and spraying insecticide such as malathion 0.1% solution during September–October has also been suggested to control the pest infestation on ginger (IISR, 2001a). In Nagaland, mulching with *mahaneem* (*Melia dubia* Cav.) leaves (Lalnuntluanga and Singh, 2008) or spraying quinalphos 0.05% + Ozoneem 1500 ppm (3 ml/l) (Mhonchumo et al., 2010) has been suggested for the management of the pest. On turmeric, spraying of malathion 0.1% solution during July–October at 21-day intervals is effective to control the pest infestation (IISR, 2001b).

Various natural enemies including the mermithid nematode (*Hexamermis* sp.), hymenopterous parasitoids (*Xanthopimpla australis* Kr. Ichneumonidae), *Apanteles taragamae* Viereck, and *Myosoma* sp. (Braconidae) and general predators such as spiders, earwigs, and asilid flies have been recorded on shoot borers infesting ginger and turmeric. Conservation of natural enemies plays a significant role in reducing the population of the pest in the field (Devasahayam, 1996).

The Agronomy and Economy of Turmeric and Ginger. DOI: http://dx.doi.org/10.1016/B978-0-12-394801-4.00022-3

Rhizome Scale (*Aspidiella hartii* Ckll.)

The rhizome scale (*A. hartii*) (Diaspididae) infests rhizomes of ginger and turmeric both in the field and in storage in India. In the field, the pest infestation is generally seen during the dry post monsoon season and severely infested plants wither and dry. In storage, the pest infestation results in shriveling of buds and rhizomes; when the infestation is severe, it adversely affects the sprouting rhizome (Devasahayam and Koya, 2005). The adult females are minute in size, circular, and light brown to gray in color, measuring about 1.5 mm in diameter. Females are ovoviviparous and also reproduce parthenogenetically. The rhizome scale also infects yams, tannia, and taro (Devasahayam and Koya, 2005).

Management of the Pest

Timely harvest and discarding severely infested rhizomes during storage reduces further spread of the pest infestation in storage. Dipping of seed rhizomes in quinalphos 0.075% solution after harvest and storage in dry leaves of *Strychnos nux-vomica* L.+ sawdust in a 1:1 ratio is effective in controlling rhizome scale infestation on ginger and turmeric (IISR, 2004–2005).

White Grubs (*Holotrichia* spp.)

White grubs (*Holotrichia* spp.) (Melolonthidae) often cause serious damage to ginger plants in certain regions of northeastern India. The grubs feed on roots and on newly formed rhizomes. The pest infestation leads to yellowing of leaves, and in severe cases of infestation, the pseudostem must be cut at the base to stop further infestation. The entire crop may be lost in severely infested plantations. The adults of *Holotrichia* spp., commonly occurring in Sikkim, India, are dark brown beetles measuring about 2.5 cm × 3.5 cm in size. The grubs are creamy white, occurring in the soil. The adults emerge in large numbers with the summer showers during the months of April and May (Varadarasan et al., 2000).

Management of the Pest

Mechanical collection and destruction of adults during their peak periods of manifestation and application of the entomophagous fungus *Metarhizium anisopliae* mixed with fine cow dung is effective in managing the white grubs. However, in severely affected areas, drenching with chlorpyriphos 0.075% solution may be necessary, along with mechanical collection and destruction of beetles (IISR, 2001c).

Minor Insect Pests

Leaf/Shoot-Feeding Caterpillars

The larvae of the leaf roller (*Udaspes folus* Cram.) (Pieridae) cut and fold leaves of ginger and turmeric and feed from within, especially during the monsoon season in India. A spray with carbaryl 0.1% solution or 0.05% solution of dimethoate may be done if the infestation is severe (IISR, 2001b, 2001c). Larvae of cutworms of *Heliothis* sp. (Noctuidae) feed on the young shoots at the ground level in Australia. Spraying of a branded insecticide is recommended for the control of this minor pest (Broadley, 2010).

African Black Beetle (Heteronychus *sp.*)

The adults of the African black beetle (*Heteronychus* sp.) (Scarabaeidae) feed on ginger pseudostems below the ground level in Australia, causing the plant to collapse. The adults are dark brown to black in color and measure about 12 mm in length. The larvae are soil dwelling and white in color with a prominent brown head, measuring about 25 mm in length. Spraying with a branded insecticide, when the first symptoms of infestation emerge, so as to wet the soil up to 5 cm depth, is recommended to manage the pest effectively (Broadley, 2010).

Nematode Pests of Ginger

Among the various nematode species, which infest ginger and turmeric in India, the root-knot nematode (*Meloidogyne* spp.) (Meloidogynidae), burrowing nematode (*Radopholus similis*), lesion nematode (*Pratylenchus coffeae*) (Pratylenchidae), and reniform nematode (*Rotylenchulus reniformis* Linford and Oliveira, 1940) (Rotylenchulidae) are important (Koshy et al., 2005). *Meloidogyne* spp. and *R. similis* are major nematode pests of ginger in Australia (Pegg et al., 1974) and Fiji (Butler and Vilsoni, 1975).

Root-knot nematodes cause galling and rotting of roots and rhizomes of both ginger and turmeric. Extensive internal lesions are formed in the fleshy roots and rhizomes. The infested rhizomes have brown, water-soaked areas on the outer tissues, and the nematodes continue to develop after the crop gets matured and is harvested. Heavily infested plants are stunted with very few tillers and have chlorotic leaves with marginal necrosis, which die prematurely leaving a poor crop stand on the field at harvest. The incidence of rhizome rot in ginger caused by the fungus *P. aphanidermatum* (Edson) Fitzp. 1923 is reported to be severe when rhizomes are also infested with nematodes, such as infestation by *M. incognita* and *P. coffeae* (Dohroo et al., 1987; Ramana and Eapen, 1998). Root-knot nematode infestation is reported to cause yield loss from 46.4% to 57.0% in India (Charles, 1978) and Australia (Pegg et al., 1974), respectively.

Ginger and turmeric plants infected by *R. similis* exhibit stunting, loss of vigor, and production of tillers. The topmost leaves become chlorotic with scorched tips. The affected plants tend to mature and dry out faster than the healthy ones. Incipient infections of the rhizomes are evidenced by small, shallow, sunken, and water-soaked lesions (Sundarraju et al., 1979). *P. coffeae* is reported to cause "ginger yellows" disease and is more prevalent in Himachal Pradesh. The nematode infestation causes yellowing of leaves and dry rot-like symptoms on rhizomes. Dark brown necrotic lesions can be observed within the infected rhizomes (Kaur and Sharma, 1988).

Table 22.1 Major Insect and Nematode Pests of Ginger

Crop	Order/Family
Shoot borer (*C. punctiferalis* Guen.)	Lepidoptera: Pyralidae
Rhizome scale (*A. hartii* Ckll.)	Homoptera: Diaspididae
White grub (*Holotrichia* spp.)	Coleoptera: Melolonthidae
Root-knot nematode (*M. incognita*)	Tylenchida: Meloidogynidae
(Kofoid and White, 1919), (Chitwood, 1949)	
Burrowing nematode (*R. similis*) (Cobb, 1893),	Tylenchida: Pratylenchidae
(Thorne, 1949)	
Lesion nematode (*P. coffeae*) (Zimmerman, 1898),	
(Filipjev and Stekhoven, 1941)	
Reniform nematode (*R. reniformis* Linford and	Tylenchida: Rotylenchulidae
Oliveira, 1940)	

Management of Nematode Infestation

1. Cultural Control: Since the seed material generally harbors nematodes, selection of seed rhizomes is critical for the management of nematode infestation. Nematode-free planting material should be selected from fields with a known history. Disinfestation of ginger rhizomes can also be achieved by hot water or steam treatment (Vadhera et al., 1998). Crop rotation with forage sorghum (*Sorghum bicolr* (L.) Moench cv. Jumbo) or lablab (*Lablab purpureus* (L.) Sweet cv. Highworth) along with application of poultry manure and/or sawdust resulted in negligible nematode damage (Stirling and Nikulin, 2008).

2. Organic Amendments: Mulching with *mahaneem* leaves or sawdust or applying well-decomposed cattle manure, poultry manure, compost, or neem oil cake reduces nematode buildup in ginger (Kaur, 1987; Stirling, 1999).

3. Biological Control: Application of fungal antagonists such as *Verticillium chlamydosporium* Goddard, *Purpureocillium lilacinus*, *Fusarium* sp., *Aspergillus nidulans* (Eidam) G.Winter, and *Scopulariopsis* sp., along with organic material, suppressed nematode populations (Eapen and Ramana, 1996).

4. Host Resistance: "IISR Mahima," an improved variety of ginger, is reported to be resistant to *M. incognita* (Sasikumar et al., 2003). The high-yielding varieties PCT-8, PCT-10, Suguna, and Sudharshana are generally free from *M. incognita* infestation in Andhra Pradesh, India (Rao et al., 1994).

5. Chemical Control: Application of carbofuran or phorate at 1 kg a.i./ha suppresses *M. incognita* infestations in ginger and turmeric (Gunasekharan et al., 1987) (Table 22.1).

References

Broadley, R., 2010. Ginger in Queensland. Commercial Production. Department of Employment, Economic Development and Innovation, Queensland Government. Accessed from: <http://www.deedi.qld.gov.au/> 20 July 2010.

Butler, E.J., Vilsoni, F., 1975. Potential hosts of burrowing nematode in Fiji. Fiji Agric. J. 37, 38–39.

Charles, J.S., 1978. Studies on the Nematode Diseases of Ginger, M.Sc. Thesis, Kerala Agricultural University, Vellayani, Kerala State.

Devasahayam, S., 1996. Biological control of insect pests of spices. In: Anandaraj, M., Peter, K.V. (Eds.), Biological Control in Spices, Indian Institute of Spices Research, Calicut, Kerala State, pp. 33–45.

Devasahayam, S., Koya, K.M.A., 2005. Insect pests of ginger. In: Ravindran, P.N., Babu, K.N. (Eds.), Ginger. The Genus *Zingiber*. CRC Press, Washington, DC, pp. 367–389.

Dohroo, N.P., Shyam, K.R., Bharadwaj, S.S., 1987. Distribution, diagnosis and incidence of rhizome rot complex of ginger in Himachal Pradesh. Indian J. Plant Pathol. 5, 24–25.

Eapen, S.J., Ramana, K.V., 1996. Biological control of plant parasitic nematodes of spices. In: Anandaraj, M., Peter, K.V. (Eds.), Biological Control in Spices: Indian Institute of Spices Research, Calicut, Kerala, India, pp. 20–32.

Gunasekharan, C.R., Vadivelu, S., Jayaraj, S., 1987. Experiments on nematodes of turmeric—a review. Proceedings of the Third Group Discussion on the Nematological Problems of Plantation Crops, 29–30 October 1987. Sugarcane Breeding Institute, Coimbatore, Tamil Nadu, pp. 45–46.

IISR, 2001a. Annual Report. Indian Institute of Spices Research, Calicut, Kerala State.

IISR, 2001b. Annual Report. Turmeric (Extension Pamphlet). Indian Institute of Spices Research, Calicut, Kerala State.

IISR, 2001c. Annual Report. Ginger (Extension Pamphlet). Indian Institute of Spices Research, Calicut, Kerala State.

IISR, 2004–2005. Annual Report. Indian Institute of Spices Research, Calicut, Kerala State.

Jacob, S.A., 1981. Biology of *Dichocrocis punctiferalis* Guen. on turmeric. J. Plantation Crops 9, 119–123.

Kaur, D.J., Sharma, N.K., 1988. Occurrence and pathogenicity of *Meloidogyne arenaria* on ginger. Indian Phytopathol. 41, 467–468.

Kaur, K.J., 1987. Studies on Nematodes Associated with Ginger (*Zingiber officinale* Rosc.) in Himachal Pradesh, Thesis submitted to the Himachal Pradesh University, Shimla, India.

Koshy, P.K., Eapen, S.J., Pandey, R., 2005. Nematode parasites of spices, condiments and medicinal plants. In: Luc, M., Sikora, R.A., Bridge, J. (Eds.), Plant Parasitic Nematodes in Subtropical and Tropical Agriculture, Second ed. CABI Publishing, Wallingford, pp. 751–791.

Koya, K.M.A., Balakrishnan, R., Devasahayam, S., Banerjee, S.K., 1986. A sequential sampling strategy for the control of shoot borer (*Dichocrocis punctiferalis* Guen.) on ginger (*Zingiber officinale* Rosc.) in India. Trop. Pest Manag. 32, 343–346.

Lalnuntluanga, J., Singh, H.K., 2008. Performance of certain chemicals and neem formulations against ginger shoot borer (*Dichocrocis punctiferalis* Guen.). Indian J. Entomol. 70, 183–186.

Mhonchumo, Neog, P., Singh, H.K., 2010. Eco-friendly management of shoot borer and rhizome fly of ginger in Nagaland. In: Sema, V., Srinivasan, V., Shitri, M. (Eds.), Proceedings and Recommendations, National Symposium on Spices and Aromatic Crops (SYMSAC–V), 30–31 October 2009. Medziphema. Central Institute of Horticulture, pp. 117–123.

Pegg, K.C., Moffett, M.L., Colbran, R.C., 1974. Diseases of ginger in Queensland. Queensland Agric. J. 100, 611–618.

Ramana, K.V., Eapen, S.J., 1998. Plant parasitic nematodes associated with spices and condiments. In: Trivedi, P.C. (Ed.), Nematode Diseases in Plants. CBS Publishers and Distributors, New Delhi, India, pp. 217–251.

Rao, P.S., Krishna, M.R., Srinivas, C., Meenakumari, K., Rao, A.M., 1994. Short duration, disease-resistant turmerics for northern Telangana. Indian Hortic. 39, 55–56.

Sasikumar, B., Saji, K.V., Antony, A., George, K.G., Zachariah, T.J., Eapen, S.J., 2003. IISR Mahima and IISR Rejatha – two high yielding and high quality ginger (*Zingiber officinale* Rosc.) cultivars. J. Spices Arom. Plants 12, 34–37.

Stirling, G.R., 1999. Organic amendments for control of root-knot nematode (*Meloidogyne incognita*) on ginger. Aust. Plant Pathol. 18, 39–44.

Stirling, G.R., Nikulin, A., 2008. Crop rotation, organic amendments and nematicides for control of root-knot nematode (*Meloidogyne incognita*) on ginger. Aust. Plant Pathol. 27, 234–243.

Sundarraju, P., Sosamma, V.K., Koshy, P.K., 1979. Pathogenicity of *Radopholus similis* on ginger. Indian J. Nematol. 9, 91–94.

Vadhera, I., Tiwari, S.P., Dave, G.S., 1998. Plant parasitic nematodes associated with ginger (*Zingiber officinale*) in Madhya Pradesh and denematization of infested rhizome by thermotherapy for management. Indian J. Agric. Sci. 68, 367–370.

Varadarasan, S., Singh, J., Pradhan, L.N., Gurung, N., Gupta, S.R., 2000. Bioecology and management of white grub *Holotrichia seticollis* Mosher (*Melolonthinae*: Coleoptera), a major pest on ginger in Sikkim. In: Muraleedharan, N., Kumar, R.R. (Eds.), Recent Advances in Plantation Crops Research. Allied Publishers Limited, New Delhi, pp. 323–326.

23 The Postharvest and Industrial Processing of Ginger

Harvest Maturity

Ginger is used both as a fresh vegetable and as a dried spice. The crop is ready for harvest in about eight months after planting, when leaves turn yellow and start drying up gradually. The clumps are lifted carefully with spade or digging fork and the rhizomes are separated from the dried leaves, roots, and adhering soil. To prepare vegetable ginger harvesting is done from the sixth month onward after planting. The most important criteria in assessing the suitability of ginger rhizomes for specific processing purposes are the fiber and volatile oil contents and the pungency level. The relative abundance of these three components in the fresh rhizome is governed by its state of maturity at harvest. Tender rhizomes lifted at the beginning of the harvesting season, about five to seven months after planting, are preferred for the production of preserved ginger as the fiber content is negligible and the pungency is mild. As the season progresses, the relative abundance of the volatile oil, the pungent constituents, and the fiber content increase. At about nine months after planting, the volatile oil and pungent principle contents reach the maximum and thereafter their relative abundance falls as the fiber content continues to increase. In India, the volatile oil content of ginger has been reported to be at maximum between 215 and 260 days after planting (Purseglove et al., 1981).

The fresh ginger after harvest is sometimes subjected to washing, which is performed to remove soil dirt, spray residues, and other foreign materials. Manual cleaning by rubbing under the thumb is commonly followed for small-scale cleaning in the farm. In large plantations, a high-pressure jet is used for washing. For this type of cleaning, the ginger rhizomes are soaked in still water overnight and the following day, a high-pressure water spray jet is directed on the rhizomes, which forcefully removes the firmly attached dirt. A drum-type washer is also quoted as a commercial device (Kachru and Srivastava, 1988) for cleaning ginger rhizomes. However, the use of such a washer has not been reported earlier. The washing of ginger is essentially followed when the oil is extracted from green ginger without drying. Sreekumar et al. (2002) have reported the use of washed, cleaned fresh ginger for extraction of essential oil in the processing facility set up at Manipur in India.

Processing

Peeling

Peeling serves to remove the scaly epidermis and to facilitate drying. The outer skin of ginger is scraped off with a bamboo splinter or wooden knife with a pointed end.

The Agronomy and Economy of Turmeric and Ginger. DOI: http://dx.doi.org/10.1016/B978-0-12-394801-4.00023-5

An iron knife is not recommended as it may leave black stains on the peeled surface, adversely affecting the appearance of the rhizome, or it may lead also to fading the rhizome's color. During peeling, it should be ensured that the cortical parenchyma, which is rich in essential oil-bearing cells, are not removed or cut, as it would cause the loss of volatile oil, and, thereby, decrease the aroma of the peeled rhizome. Since scraping of ginger is a laborious process, attempts have been made for chemical and mechanical peeling of ginger.

Chemical Peeling

Chemical peeling using sodium hydroxide (NaOH), widely known as lye peeling, is one of the most common and the oldest methods for peeling fruits and vegetables. Theoretically, lye peeling is a complex process involving diffusion and chemical reactions. Once the caustic solution of NaOH comes in contact with the surface of the fruit, it dissolves the epicuticular waxes, penetrates the epidermis, and diffuses through the skin into the fruit (Floros et al., 1987).

Randhawa and Nandpuri (1970) reported that peeling of ginger was made easier by dipping it in boiling lye, followed by washing and steeping in acid solution. The process consisted of putting the ginger to be peeled in a wire-gauze cage and dipped into hot boiling lye for the required period. The lye solution causes the separation of the skin of the rhizome from the flesh beneath the epidermal layer. Dipping for 5½ min in a boiling lye solution of 20%, 25%, and 50% concentrations, respectively, removed the peel. The ginger was then washed in running water and finally kept in a 4% citric acid solution for 2 h. However, this process did not result in quality dried product and is not commercially practiced. Trials have been carried out by dipping the ginger rhizomes in boiling water for a short time prior to peeling (Lawrence, 1984; Natarajan et al., 1972). Ginger treated with boiling water gives a dark final product, and hence, this treatment is not recommended.

The effect of lye pretreatment on the peeling efficiency and ginger meat loss before mechanical peeling was investigated by Charan et al. (1993). Lye treatment of ginger in a 7.5% solution for 5 min before machine peeling indicated that the peeling efficiency of the machine could be increased and the meat loss decreased. Peeling efficiency of the machine operating in three and fourpasses increased from 73% to 83% and from 75% to 86% respectively by lye pretreatment, whereas the corresponding meat loss was reduced from 3.3% to 1.9% and from 3.8% to 2.4% respectively. However, the application of this unit in large scale needs to be investigated further and then optimized.

Mechanical Peeling

Ginger rhizomes have an irregular shape for a machine to handle, especially when only a thin layer of its skin must be removed. Natarajan et al. (1972) have reported about some commercial undertakings that attempted to use machines fitted with abrasive rollers, but this met with little success. As the time of abrasive peeling increased, more and more of the outer skin and tissue layers were lost, resulting in a progressive decrease in oil content recovered.

A mechanical brush-type ginger peeling machine was developed by Agrawal et al. (1983). It essentially consisted of two continuous brush belts being driven in opposite directions with a downward relative velocity by a variable speed motor. The movement of the brush belts in opposite directions provides the abrasive action on the ginger rhizomes passing in between, while the downward relative velocity provided the downward flow of ginger. The spacing between the belts and the belt velocity could be varied. The machine was reported to function satisfactorily during the limited tests performed on stored ginger.

Evaluating the machine parameters was essential because peeling of the rhizome's skin was associated with the loss of ginger meat from underneath the skin. The epidermal cells in ginger contain most of the essential oil which imparts the ginger's characteristic aroma and is perhaps the most important factor in determining its market price (Jaiswal, 1980). Therefore, the loss of ginger meat from underneath the skin would result not only in the reduction of ginger weight but also in the heavy economic loss of the value of the ginger.

Agrawal et al. (1987) optimized the operational parameters of the abrasive brush-type ginger peeling machine for maximum peeling efficiency with minimum loss of ginger meat. The parameters optimized were brush belt spacing (1 cm) and belt speed (65 rpm) of the driving brush belt, resulting in the belt relative velocity of 199 cm/s. Number of passes required was four to five and the capacity of the machine at the recommended parameter values with five passes was 20 kg/h. When operated at full capacity, the machine had a peeling efficiency of 71% with ginger meat loss of 1.6%.

The performance of the abrasive brush-type ginger peeling machine could be improved by redesigning the peeling unit (Ali et al., 1990). Experiments were conducted to study the effect of various combinations of brush spacing, height, and recommended optimum machine operational parameters. It was recommended that brush spacing of 1.9 cm and brush height of 2 cm and four or five passes with brush tip spacing of 1 cm at the relative velocity of 199 cm/s in a downward direction were the best parameters to peel 2 cm thick ginger. The average peeling efficiency and material loss were found to be 87.92% and 8.22%, respectively.

The final prototype of the abrasive brush-type ginger peeling machine was reported by Ali et al. (1991). The machine essentially consists of two continuous vertical abrasive belts with the brush of 32 SWG thick steel wires, 2 cm long, with a spacing of 1.90 cm. The peeling zone of the ginger peeling machine was increased from 15 cm × 90 cm to 30 cm × 150 cm. Thus, the number of the passes was reduced from five to three. The gear reduction device was replaced by the jack pulley for power transmission. The peeling efficiency and material loss were 83.46% and 4.33%, respectively. The capacity of the machine at the recommended parameters was 200 kg/h.

Charan et al. (1993) developed a small manually operated ginger peeling machine for application by farmers. The machine was fabricated using locally available materials. The moving abrasive surface was made of coconut fiber brushes (30 mm length) mounted on two endless canvas belts that were 40 mm wide and 5 mm thick. The stationary abrasive surface was also developed with the same abrasive brushes arranged side by side on a wooden plank of 780 mm × 240 mm × 15 mm in size. The capacity of the machine to peel untreated ginger is 24 kg/h. The peeling efficiency and loss of ginger meat were 71% and 1.3% respectively, showing that the machine is quite efficacious in doing its intended job.

Drying

Peeled rhizomes are washed and dried uniformly under the sun for 7–10 days. Ginger should be dried on a clean surface to ensure that no extraneous matter contaminates the product. Care should be taken to avoid growth of mold on the rhizomes while drying. During the first few days of drying, each rhizome is turned at least once a day to ensure that it dries uniformly. In order to get rid of the last bit of debris sticking to the rhizomes or to the peeled skin, the rhizomes are rubbed together during drying. Rhizomes must be dried to a moisture content of 10% and stored properly to avoid infestation by storage pests. Improperly or inadequately dried ginger is susceptible to microbial and fungal growth, which in turn will vastly reduce its market value.

Traditionally, ginger is sun dried in a single layer in the open farmyard. The dried ginger presents a brown, irregular wrinkled surface and when broken shows a dark brownish color. The drying of ginger usually leads to the loss of volatile oil by evapo-ration. Mathew et al. (1973) have shown that this loss may be as high as 20%. The extent of cleaning the rhizomes prior to drying has a considerable influence on the volatile oil content of the end product. Removal of cork skin not only reduces the fiber content but also enhances the loss of volatile oil through rupture of the oil cells, which are present near the epidermis. Jamaican ginger, which is cleanly peeled, has some-what lower quantities of volatile oil and fiber content than the commercially dried gin-gers which are partially peeled or unpeeled (Purseglove et al., 1981).

Mani et al. (2000) studied different drying methods for dry ginger. The drying methods reported were sun drying, vacuum oven drying at 60°C under 500 mmHg vacuum pressure, and hot-air oven drying at 60°C. Sun drying took 104 h for com-plete drying. Vacuum oven drying, though it took only 48 h to completely dry, caused a growth of black fungi all over the surface of the rhizome. In a hot-air oven, drying continued even after 88 h. However, the ginger dried after cutting the rhizomes into slices of 1.27 cm thick, without peeling, and dried under the sun, helped to recover higher amounts of volatile oil.

Kachru and Srivastava (1988) reported about a tray-type solar dryer, designed and developed by Sukhadia University, Udaipur, Rajasthan State, India, for drying gin-ger. The total time required to dry ginger from initial moisture content of 455% to 10% (dry basis) or less by use of a solar cabinet dryer was 30 h compared to 45–60 h required when the rhizomes are dried on a concrete surface in the open.

Drying of ginger in an integral-type natural convection solar dryer coupled with a biomass stove was reported by Prasad and Vijay (2005). It was found that 18 kg of fresh ginger with an initial moisture content of 319.74% (dry basis) was dried to a final moisture content of 11.8% (dry basis) within 33 h. Drying of the product was also investigated under "solar only" conditions and in the open in the same cli-matic conditions, and the results indicated that drying was faster in the hybrid dryer. It took only 33 h in the hybrid dryer, as against 72 h in the "solar only" operation of the same dryer and 192 h in the open sun. The developed dryer is simple, can be manufactured locally, and can be used to dry other agricultural products.

Charan (1995) reported mechanical drying of peeled ginger in two stages: drying up to 50% moisture content (wet basis) at 85°C and then to the required moisture at 65°C gave the best organoleptic and biochemical qualities.

A tray-type dryer to dry ginger was reported by Philip et al. (1996). The main parts were a drying chamber, plenum chamber, and a chimney with butterfly valve. Trays of wire mesh were provided in the drying chamber to keep the materials to be dried. The plenum chamber encloses the burning-cum-heat exchanging unit. Preliminary tests showed that 10 h of drying at 60°C reduced the moisture content of ginger from 90% to 11%. The product obtained was also of high quality.

Polishing and Storage

Polishing of dried ginger is done to remove the wrinkles developed during the drying process. It is generally done by rubbing the dried rhizomes against a hard surface, such as a granite stone/slab. Polishing of dry ginger is also carried out by putting the dried rhizomes in a gunnysack and swirling or swaying the sack between two persons standing in opposite directions. Hand- or power-operated mechanical polishers are also employed for this purpose. Kachru and Srivastava (1988) reported a mechanical polisher developed at Sukhadia University, Udaipur, Rajasthan, India, as in the case of a tray-type solar cabinet dryer. It has a capacity to dry 15–20 kg/h and could give 5–7% polish. As far as storage of ginger is concerned, it could be stored in the dried rhizome form without any significant change in biochemical constituents for over a year.

Cleaning and Grading

Once the ginger rhizomes are dry, they are sorted and graded. Grading of ginger takes into consideration the size of the rhizome, its color, shape, extraneous matter, the presence of light pieces, and the extent of residual lime it carries in the case of bleached ginger. The specifications of various grades of Indian ginger and ginger powder under the Indian AGMARK specifications are given Tables 23.1–23.5. AGMARK certification is currently not mandatory for export trade in ginger. However, it is still valued as a mark of quality. Indian Standard Specifications (ISS) for ginger are almost in line with AGMARK specifications. The minimum size of rhizomes for domestic trade as per ISS is 20 mm (IS 1908, 1993).

The American Spice Trade Association (ASTA) specifications for cleanliness of ginger are given in the following table (Table 23.6).

The tolerance levels of pesticide residues in ginger under the US regulations are given in the following table (Table 23.7).

The Incidence of Aflatoxin

An important issue in the quality of spice is the presence of aflatoxins. Aflatoxins are a group of secondary metabolites of the fungi *Aspergillus flavus* and *Aspergillus parasiticus* and are rated as potent carcinogens. Inadequate and unhygienic drying of the ginger rhizomes leads to the growth of these fungi on ginger. Aflatoxins in

Table 23.1 AGMARK Grade Designations of Garbled Nonbleached Ginger (Whole)

Grade/Designation	Size of Rhizome (length in mm) Min	Organic Extraneous Matter % (m/m), Max	Inorganic Extraneous Matter % (m/m) Max	Moisture(%) (m/m) Max	Total Ash (%) (m/m) Max	Calcium (as CaO %) Max	Volatile Oil (%) (ml/100 g) (m/m), Max Min
Special	20.0	1.5	0.5	12.0	8.0	1.1	1.5
Standard	15.0	1.5	0.5	13.0	8.0	1.1	1.0

Source: Spices Board (2011).

Table 23.2 AGMARK Grade Designations and Quality of Ungarbled and Nonbleached Ginger (Whole)

Grade/Designation	Size of Rhizome (length in mm) Min	Organic Extraneous Matter % (m/m) Max	Inorganic Extraneous Matter % (m/m) Max	Very Light Pieces % (m/m) Max	Moisture % (m/m) Max	Total Ash % (m/m) Max	Calcium as Cao % (m/m), Max	Volatile oil, % (ml/100g) Min
Special	20.0	1.5	0.5	4.0	12.0	8.0	1.1	1.5
Standard	15.0	1.5	0.5	6.0	13.0	8.0	1.1	1.0

Source: Spices Board (2011).

Table 23.3 AGMARK Grade Designations of Garbled Bleached Ginger (Whole)

Grade/Designation	Size of Rhizomes (length in mm) Min	Organic Extraneous Matter, % (m/m) Max	Inorganic Extraneous Matter % (m/m) Max	Moisture (%) (m/m), Max	Total Ash (%) (m/m) Max	Calcium (Cao %) (m/m) Max	Volatile oil (%) (ml/100g) Min
Special	20.0	1.5	0.5	12.0	12.0	2.5	1.5
Standard	15.0	1.5	0.5	13.0	12.0	4.0	1.0

Pieces of rhizomes smaller than 15 mm can be graded with the marking "Garbled Bleached Ginger (Pieces)." It may be marked as "Garbled Bleached Calicut" (BGK) or "Garbled Bleached Cochin" (BGC) depending on its place of origin.
Source: Spices Board (2011).

Table 23.4 AGMARK Grade Designations of Ungarbled Bleached Ginger (Whole)

Grade/Designation	Size of Rhizomes (length in mm) Min (m/m)	Extraneous Matter (%) Max (m/m)	Very Light Pieces (%) Max	Moisture (%) (m/m) Max	Total Ash (%) (m/m) Max	Calcium (Cao %) (m/m) Max	Volatile Oil (%) (ml/100g) Min
Special	20.0	2.0	4.0	12.0	12.0	2.5	1.5
Standard	15.0	2.0	5.0	13.0	12.0	4.0	1.0

Pieces of rhizomes smaller than 15 mm can be graded with the marking "Garbled Nonbleached Ginger Pieces." It may be marked as "Ungarbled Bleached Calicut" (BUGK) or "Ungarbled Bleached Cochin" (BUGC) depending on its place of origin.
Source: Spices Board (2011).

Table 23.5 AGMARK Grade Designations and Quality of Ginger Powder

Grade/Designation	Moisture (%) (m/m) Max	Total Ash (%) (m/m) Max	Acid Insoluble Ash (%) (m/m) Max.	Water Soluble Ash (%) ash (%) Min	Calcium Soluble(as CaO) (m/m) Max	Alcohol Soluble Extract (%)	Cold Water Extract (%) (m/m) Min	Volatile Oil (%) (ml/100 g) Min
Special	20.0	8.0	1.0	1.9	1.1	5.1	11.4	1.5
Standard	13.0	8.0	1.0	1.7	4.0	4.5	10.0	1.0

Source: Spices Board (2011).

Table 23.6 American Spice Trade Association (ASTA)
Cleanliness Specifications for Ginger

Parameter	Upper Limit
Whole insects, dead (by count)	4
Excreta, mammalian (mg/lb)	3
Excreta, other (mg/lb)	3
Mold (% by weight)[*]	3
Insect defiled/infested (% by weight)[*]	3
Extraneous foreign matter (% by weight)	

[*]Moldy pieces and/or insect-infested pieces by weight.

Table 23.7 Tolerance Levels for Pesticide Residues in Ginger under
U.S. Regulations

Pesticides	Tolerance Limit (ppm)
Lindane	0.50
BHC	0.05
Heptachlor	0.01
Heptachlor epoxide	0.01
Trifluralin	0.05
Ethylene oxide	50.0
Propylene oxide	300.0
Diquat	0.02
Dichlorvos	0.50
Dalapon	0.20
Aluminum phosphide	0.10
2,4-D	0.10
Glyphosate	0.20
Methyl bromide	100

spices are generally classified into four categories: B_1, B_2, G_1, and G_2. B_1 and B_2 are produced by *A. flavus*, whereas G_1 and G_2 are produced by *A. parasiticus*. Of these, B_1 is the most virulent carcinogen and has received the most attention. The tolerance limits for aflatoxins under German law (Spices Board, 2002) are given in Table 23.8.

The maximum permissible limits for trace metals in ginger powder under Japanese specifications are given in the following table (Table 23.9).

The nutritional data for 100 g dry ginger is given in the following table (Table 23.10)

Chemical Composition of Ginger

Ginger rhizomes contain volatile oil, pungent compounds, starch and other saccharides, proteins, crude fiber, waxes, coloring matter, and trace elements. The presence of vitamins and amino acids also has been reported. The relative percentages of these

Table 23.8 Tolerance Limits for Aflatoxins in Spices

S. No.		Aflatoxin	Tolerance Limit (ppb Maximum)
1	German Law	$B_1+B_2+G_1+G_2$	4
		B_1	2
2	European Commission Regulations	$B_1+B_2+G_1+G_2$	10
		B_1	5

ppb, parts per billion.

Table 23.9 Maximum Permissible Limits for Trace Metals in Ginger Powder Under Japanese Specifications

Metal	Upper Limit (ppm)
Magnesium	2000
Zinc	33
Copper	3.7
Aluminum	42
Arsenic	0
Boron	1.5
Barium	15
Beryllium	0.038
Chromium	0.4
Manganese	270
Molybdenum	0.47
Nickel	0.97
Antimony	2.2
Selenium	0.14
Silicon	21
Tin	4.4
Lithium	0
Strontium	1.2
Titanium	0
Bismuth	0
Cadmium	0
Gallium	0
Lead	0
Tellurium	0

components vary with the ginger cultivar, soil in which grown, and climate. Starch is the most abundant of the constituents, comprising 40–60% of the weight of the dry rhizome (Lawrence, 1984). Crude protein, total lipids, and crude fiber have been reported to vary between 6.2% and 19.8%, 5.7–14.5%, and 1.1–7.0% respectively in different cultivars (Jogi et al., 1972). Minute glands containing essential oil and resin are scattered throughout the rhizome but are particularly numerous in the epidermal tissues (Guenther, 1952).

Table 23.10 Nutritional Data for 100 g Dry
Ginger

Details	Content
Water	9.4 g
Food energy	347 kcal
Protein	9.1 g
Fat	6.0 g
Total carbohydrate	70.8 g
Fiber	5.9 g
Ash	4.8 g
Calcium	116 mg
Iron	12 mg
Magnesium	184 mg
Phosphorus	148 mg
Potassium	1342 mg
Sodium	32 mg
Zinc	5 mg
Niacin	5 mg
Vitamin A	147 IU
Other vitamins	Negligible

Ginger rhizome also contains a number of amino acids. The presence of aspartic acid, threonine, serine, glycine, cysteine, valine, isoleucine, leucine, and arginine has been reported (Takahashi et al., 1982).

Ginger Powder

Ginger powder is made by pulverizing the dry ginger to a mesh size of 50–60 (Natarajan and Lewis, 1980). Ginger powder forms an important component in curry powder. It also finds direct application in a variety of food products.

Distillation of the Volatile Oil

Essential oils are the volatile organic constituents of fragrant plant matter. They are generally composed of a number of compounds, including some which are the solids at normal temperature, possessing different chemical and physical properties. The aroma profile of the oil is a cumulative contribution from the individual compounds. The boiling points of most of these compounds range from 150°C to 300°C at atmospheric pressure. If heated to this temperature, labile substances would be destroyed and strong resinification would occur. Hydrodistillation permits the safe recovery of these heat-sensitive compounds from the plant matter. Hydrodistillation involves the use of water or steam to recover volatile principles from plant materials.

The fundamental feature of the hydrodistillation process is that it enables a compound or mixture of compounds to be distilled and subsequently recovered at a temperature substantially below that of the boiling point of the individual constituents.

Water and Steam Distillation

Here the plant material is supported on a perforated grid inside the still. The lower part of the still is filled with water to a level below the grid. The water is heated to generate steam. The steam, usually wet and at low temperature, rises through the charge carrying the essential oil. The advantage of this method over water distillation is that the raw material is not in contact with boiling water. The exhausted plant material can be handled easily as it does not form a slurry with water.

Steam Distillation

This is the most widely used industrial method for the isolation of essential oil from plant material. Here the steam is produced outside the still, usually in a steam boiler. Steam at optimum pressure is introduced into the still below the charge through a perforated coil or jets. Steam distillation is relatively rapid and is capable of greater control by the operator. The steam pressure inside the still could be progressively increased as distillation proceeds for complete recovery of high-boiling constituents. The still can be emptied and recharged quickly. With the immediate reintroduction of steam, there is no unnecessary delay in the commencement of the distillation process. Oils produced by this method are of more acceptable quality than those produced by other methods.

Ginger oil is commercially produced by steam distillation of the comminuted dried spice. The spice is powdered and charged in a stainless steel still of optimum dimensions. The still is attached to a heat exc hanger (condenser) and a separator. Direct steam is admitted from the bottom of the still. The steam, which rises through the charge, carries along with it the vapors of the volatile oil. The oil vapor–steam mixture is cooled in the condenser. The oil is separated from water in the separator and collected in glass or stainless bottles. The oil is thoroughly dried and stored airtight in full containers in a cool dry place protected from light.

Composition of ginger oil

Essential oils, in general, contain volatile compounds of many classes of organic substances. Guenther (1972) has classified the essential oil components into four main groups:

1. Terpenes, related to isoprene or isopentene.
2. Straight-chain compounds, not containing any branches.
3. Benzene derivatives.
4. Miscellaneous (compounds other than those belonging to the first three groups shown above, specific for a few species).

The most characteristic group of compounds present in essential oils are the terpenes, which comprise of hydrocarbons $(C_5H_8)_n$ and their oxygenated derivatives. Even though the chemistry of the volatile components of ginger oil has been investigated exhaustively, for a long time, modern analytical techniques have served to decipher the composition quite exhaustively. The main constituent of the ginger oil is the sesquiterpene zingiberene $(C_{15}H_{24})$.

Lawrence (1984) has carried out extensive investigations on the composition of ginger oil using a combination of distillation, CC, GC, NMR spectroscopy, IR spectroscopy, and MS. The results showed that the oil, which was found to be very complex in nature, contained about 83% hydrocarbons and 10% oxygenated compounds, with the rest of the constituents unidentified.

Organoleptic Properties

The specific gravity, refractive index, and optical rotation are the primary quality determinants of ginger oil, occasionally supplemented by gas chromatographic composition of the constituents. Even if the physical properties and the level of major components as disclosed by the chromatogram exhibit similarity, variation in the minor components can influence the organoleptic profile of the oil. For applications that cannot accommodate any flavor variations, repeated supplies of the oil must possess consistent flavor quality. Sensory analysis of the oil by a well-trained panel of technicians remains the ultimate method for validating flavor quality of in such cases.

Oil of Green (Fresh) Ginger

Essential oil can also be distilled from fresh ginger. When ginger is dried, numerous complex chemical changes take place within the tissues and the freshness of the spice is lost. Oil from fresh ginger retains the true aroma of the fresh spice and finds application in delicate flavor and perfumery formulations. An analysis of oil prepared from green ginger is given in Table 23.11.

Ginger Oil from Scrapings

The ginger scrapings, normally thrown away by the farmers, also can be utilized for the recovery of oil (Moudgill, 1928). This author obtained 0.9% and 0.8% oil respectively from air-dried scrapings of Cochin ginger. The properties of these oils are given in Table 23.12. The oil contains all major components present in normal ginger oil. However, the odor of the oil is heavy and earthy and the color is darker.

Table 23.11 Properties of Oil from Green
Ginger

Details	Value
Specific gravity (25°C)	0.8702
Refractory index (25°C)	1.4895
Optical rotation	(−) 40°
Acid number	1.41
Ester number	6.20

Source: Natarajan et al. (1970).

Table 23.12 Properties of Oil from Air-Dried Scrapings of Cochin Ginger
Varier (1945)

Property	Moudgill	Varier
Specific gravity	0.8816 (27°C)	0.8905 (30°C)
Optical rotation	−9.85° (30°C)	−5.2°
Refractive index	1.4862 (25°C)	1.4859 (30°C)
Acid number	1.0	0.90
Ester number	10.0	6.10
Ester number after acetylation	103.0	72.2

Characteristics of Ginger Oil

Ginger oil possesses the following characteristics that make it a better substitute for
the raw spice in a number of applications (Balakrishnan, 1991):

1. Represents the true aroma of the spice.
2. Does not impart color to the end product.
3. Has uniform flavor quality.
4. Is free from enzymes and tannins.

However, the oil lacks the nonvolatile principles which contribute to the taste
characteristics of the spice.

Oleoresin Ginger

The essential oil of ginger derived by steam distillation represents only the aromatic
odorous constituents of the spice; it does not contain the nonvolatile pungent principles for which ginger is highly in demand. The oleoresin, obtained by extraction of
the spice with volatile solvents, contains the aroma as well as the taste principles of
ginger in highly concentrated form.

The oleoresin represents the wholesome flavor of the spice—a cumulative
effect of the sensation of smell and taste. It consists of the volatile essential oil and

nonvolatile resinuous fraction comprising taste components, fixatives, antioxidants, pigments, and fixed oils naturally present in the spice. The oleoresin is, therefore, designated as "true essence" of the spice and can replace spice powders in food products without altering the flavor profile.

Green Ginger Oleoresin

The oleoresin extracted from fresh ginger retains the fresh aroma and wholesome flavor that closely matches the parent spice. The oleoresin of fresh ginger is termed green ginger oleoresin. Green ginger oleoresin finds application in flavor formulation where the fresh note of the spice is the prime quality determinant.

Modified Oleoresin

Sometimes a straight extracted oleoresin may require modification to suit specific applications. It can be fortified with distilled essential oil to achieve a balance between pungency and aroma. The strength of the oleoresin can be adjusted to the required level by dilution with permitted diluents. Diluents also improve the flow properties of the product.

The oleoresin may be rendered water soluble using permitted emulsifiers or converted to powder form by dispersing on dry carriers such as flour, salt, dextrose, or rusk powder. These plated products impart the strength of good-quality freshly ground spices and can be easily incorporated in food.

Microencapsulated Ginger Oil and Oleoresin

Microencapsulated extracts are microfine particles of oils and oleoresins coated with an envelope of an edible medium such as starch, maltodextrin, or natural gums so that the flavor is locked within the tiny capsule. The encapsulated product is usually prepared by spray-drying technique. When incorporated in food, the outer coating dissolves off, thereby releasing the flavor. Encapsulated oleoresins can be designed to contain a predetermined level of the core material.

Ginger oil as well as oleoresin can be microencapsulated/spray dried to convert to powder form with an improved shelf life and application convenience. Microencapsulation serves the following purposes:

1. Controls the release of the core material.
2. Locks in the flavor to ensure against loss on storage.
3. Offers convenience in handling by converting liquids and semisolids into free-flowing powder.
4. Provides uniform dispersibility in the food matrix.

Preserved Ginger

Traditional methods of preserving ginger by immersion in brine or syrups consisting of a mixture of dissolved sugars have been practiced for centuries (Brown, 1969a).

The quality requirements of ginger for making preserved ginger are different from those for dried ginger. The ginger for the preserve should not be very hot or fibrous and hence should be harvested at an earlier stage than that for drying and further processing.

Ginger in Syrup

In commercial sugar syruping, the peeled rhizomes are normally held in a preserving solution prior to treatment to prevent possible mold formation (Ingleton, 1966). The preserving solution is usually brine at a concentration range of 14–17%. The preserving solution is drained out on a loose mesh sieve. The salt must be washed off the rhizomes before the sugar treatment. Sometimes the rhizomes are also given a boiling treatment prior to syruping to soften it (Brown, 1969a). The ginger is then diced into pieces of required size and shape and immersed in sugar syrup. Alternatively, the sugar syrup may be circulated through the ginger pieces held in vats. The syrup concentration is gradually increased to minimize shrinkage. Process conditions are so selected to ensure optimum sugar absorption (Brown, 1969a). The invert sugar:sucrose ratio requires close control to prevent crystallizing of the syrup. It has been found that the optimum reducing sugar concentration is 25–33% of the total sugars. The optimum pH for syruping has been reported to be 4.3 for the desirable flavor. The ginger, after attaining the desirable level of absorption, is packed in its own syrup in small glass jars. The addition of honey up to 15% of the weight of the syrup produced a distinctive flavor in ginger in syrup.

Ginger in Brine

Tender rhizomes are also preserved in brine. This is used to make sauces and pickles and can also form the raw material for syruping after the salt has been removed by boiling in water (Sills, 1959).

Crystallized Ginger

Another version of preserved ginger is the crystallized or candied ginger produced by taking the ginger-in-syrup process to a stage further. To produce crystallized ginger, the ginger is further dipped in syrups of progressively increasing concentration (Ingleton, 1966). Optimum process conditions are followed for satisfactory sugar absorption, weight gain, desirable color, and desirable texture. The ginger is then removed from the syrup, rolled in castor sugar in a rotating drum, and dried in air-draft dehydrators (Leverington, 1969). The product is then cooled and packed in sealed polythene bags. In crystallized ginger, the pieces are small but of regular size and uniformity of cut. Different additives may also be incorporated to give a noticeable improvement in crispness of texture as well as to modify the flavor (Natarajan et al., 1970). Solutions of gelatin of varying strengths and temperatures, as well as hot pectin solutions, were found to be suitable adhesives to hold the sugar crystals onto the ginger (Leverington, 1969).

These candies are perfect to settle the stomach and soothe the throat during long travel by road or rail in India. They may be popped into the mouth and chewed to enjoy the flavor and taste the sweet heat. They also help to relieve the heaviness in the stomach after a full meal. Candied ginger can be chopped and used as a topping on ice cream or added in cookies or cakes.

Ginger Puree and Ginger Paste

Ginger puree consists of fresh ginger that has been peeled, washed, sanitized, cooked briefly, and ground. The puree may be stored frozen or permitted preservatives may be added to keep the quality. Ginger paste, on the other hand, is salted and seasoned. Commercial packs contain salt, oil, acetic acid/vinegar, permitted preservatives, and occasionally, spices.

Smooth and instant ginger in these forms adds convenience to cooking. Ginger puree and paste are used in stir fries, soups, sauces, cakes, breads, puddings, chutneys, creams, filings, and marinades.

Uses of Ginger and Ginger Products

Ginger is perhaps the most widely used spice, both for flavoring and for medicinal purposes.

Flavoring Applications

Culinary Purposes

Fresh ginger is an essential ingredient in the preparation of Oriental dishes, both sweet and savory, from entrees to desserts. It finds application in almost all meat, poultry, seafood, and vegetable preparations. Ginger contributes a freshness to foods that other spices do not (Farrell, 1985). A tenderizing effect has been observed when meat is cooked with slices of fresh ginger (Lawrence, 1984). Tender rhizomes are used in pickling. Ginger is used extensively in the preparation of different types of condiments and to flavor breads, cakes, biscuits, cookies, candy, jelly, toffees, and beverages.

Gingerbread

Gingerbread is prepared by incorporating finely grated fresh ginger, cold fresh ginger juice, or ginger powder in the dough. Occasionally, this is supplemented with other spices such as garlic, cinnamon, and cloves.

Ginger Biscuits, Cookies and Cakes

These traditional family favorites are unique in their warm spicy flavor. Typical dosage is one teaspoon of ginger powder for four cups of flour. Sometimes small quantities of cinnamon and nutmeg are also added.

Ginger Drinks

Ginger drinks are cool, refreshing beverages and provide health benefits to the consumer. The most popular ginger drinks are ginger ale and ginger beer, which are carbonated ginger-flavored soft drinks. Even though the two terms are used interchangeably, often the term "ginger beer" is associated with the spicier ginger ale. Ginger ale can be conveniently prepared at home. Even though a vast number of recipes are available, the essential steps for a typical preparation may be summarized as follows:

To 50 g grated or crushed ginger, add 5 l of boiling water. Add 500 g sugar and stir to dissolve. Leave to cool. When the contents cool to lukewarm, strain and add 15 g yeast. Flavoring agents like lime or lemon juice, vanilla essence, or other spices may also be added at this stage to suit the taste. Cover the container opening with a clean cloth and leave in a warm place overnight. Remove any scum from the top of the mixture the following day, strain, and transfer the clear liquid into sterilized bottles. Seal the bottles with screw caps and leave for two days at room temperature. Leaving the bottles at room temperature too long will cause over-carbonation and the drink will taste too yeasty. Refrigerate to finish the aging process. Add sparkling water or club soda and serve. The quantities of ingredients may be altered to suit the taste. Drinking ginger ale or ginger beer has all the benefits of consuming ginger.

Ginger Wine

Ginger wine is a popular beverage that is very warming in the winter weather. It is brewed by the fermentation of grapes or raisins with sugar, ginger, and yeast. Fifty grams of ginger would be required for making 1 l of wine. The wine is usually stored for three to four months to age. It may be served as is, or blended with other alcoholic drinks.

Ginger Tea

Ginger tea is a standard remedy for sore throat, colds, and influenza. This is an infusion prepared by steeping grated fresh ginger in boiling water for 5–0 min. The liquid is then strained and mixed with honey or sugar to taste. Some lemon juice also may be added. This soothing tea may be taken hot in winter and iced in summer. This tea is also good after a meal to aid good digestion. Powdered dry ginger can also replace fresh ginger for making ginger tea.

Ginger Syrup

Ginger syrup can make perfectly spiced sweet ginger drinks. To make ginger syrup, boil 50 g of finely chopped ginger in sugar syrup containing one cup of sugar in two cups of water. Simmer and cook for 1 h. Strain the liquid and add vanilla or lemon essence to taste. Cover and refrigerate. This syrup will keep for several days. The concentrated syrup may be extended with carbonated or plain water. The syrup could also be drizzled over ice cream. The syrup from candied/crystallized ginger processing can also be used in the same manner.

Table 23.13 Approximate Dosages of Ginger Oil and Oleoresin for Typical Applications

Details of Applications	Ginger Oil (ppm)	Ginger Oleoresin (ppm)
Nonalcoholic beverages	17	79
Ice cream, ices, and so on	20	36–65
Candy	14	27
Baked goods	47	52
Condiments	13	10–1000
Meats	12	30–250

Ginger Coffee

Ginger coffee is a blend of roasted coffee powder and ginger powder. A hot beverage is prepared by boiling the powder in water. Milk and sugar are optional. Ginger coffee is a cordial and beneficial beverage in cold weather. It is a remedy for colds, cough, and influenza.

Ginger is also the major component, along with turmeric, in the preparation of Indian *masala*, the base ingredient for the famous Indian curry. Ginger is also an important ingredient in the *masala tea*, a spicy tea which is very popular in the North Indian States.

Uses of Ginger Oil and Oleoresin

Ginger oil and oleoresin can replace raw ginger in all flavoring and medicinal applications. These concentrates, which can be standardized to the required level of aroma and taste, overcome all the disadvantages associated with the raw spice, especially with respect to flavor consistency and shelf life. They are used extensively in the processed food industry for formulating seasonings for meat, poultry, seafood, and a variety of vegetable preparations. They also find applications in flavoring baked goods, confectionery, beverages, cordials, liqueurs, spicy table sauces, and in pharmaceutical preparations for cough syrups and creams for the relief of joint pain. Ginger oil finds a limited use in perfumery, where it imparts an individual note to compositions of the Oriental type. It is also recognized as a masking agent for mouth odor in dentifrices and oral hygiene products. Toothpaste flavored with ginger oil has a unique, refreshing taste.

The approximate dosages of ginger oil and oleoresin for typical applications are given in Table 23.13 (Fenaroli, 1975). The figures are only indicative and vary with regional preferences. Exact dosage levels may be determined through application trials.

References

Agrawal, Y.C., Singhvi, A., Sodhi, R.S., 1983. Ginger peeling—development of an abrasive brush type ginger peeling machine. J. Agric. Eng. 20 (3&4), 179–182.

Agrawal, Y.C., Hiran, A., Galundia, A.S., 1987. Ginger peeling machine parameters. Agric. Mech. Asia Af. 18 (2), 59–62.

Ali, Y., Kandelwal, N.K., Sharma, P., Agrawal, Y.C., 1990. Standardization of peeling unit of an abrasive brush type ginger peeling machine. J. Agric. Eng. 27 (1–4), 85–90.

Ali, Y., Jain, G.C., Kapdi, S.S., Agrawal, Y.C., Bhatnagar, S., 1991. Development of an abrasive brush type ginger peeling machine. Agric. Mech. Asia Af. 22 (2), 71–73.

Balakrishnan, K.V., 1991. An insight into spice extractives. Indian Spices 28 (2), 22–26.

Brown, B.I., 1969a. Processing and preservation of ginger by syruping under atmospheric conditions I. Preliminary investigations of vat systems. Food Technol. 23, 87–91.

Charan, R., 1995. Developments in ginger processing. Agric. Mech. Asia Af. 26 (4), 49–51.

Charan, R., Agrawal, Y.C., Bhatnagar, S., Mehta, A.K., 1993. Application of abrasive and lye peeling of ginger at individual farmer's level. Agric. Mech. Asia Af. 24 (2), 61–64.

Farrell, K.T., 1985. Spices, Condiments and Seasonings. The AVI Publishing, Westport, CT, pp. 121–126.

Fenaroli, G., (Ed.), 1975. Ginger, second ed. Fenaroli's Handbook of Flavor Ingredients, vol. 1 CRC Press Inc., Boca Raton, FL, pp. 364–365.

Floros, J.D., Wetzstein, H.Y., Chinnan, M.S., 1987. Chemical (NaOH) peeling as viewed by scanning electron microscopy: pimiento peppers as a case study. J. Food Sci. 52 (5), 1312–1320.

Guenther, E., 1952. The Essential Oils, vol. 5. Van Nostrand Reinhold, New York, NY, pp. 106–120.

Guenther, E., 1972. The Essential Oils, vol. 1. Robert K. Krieger Publishing, New York, NY, pp. 18, 88–104.

Ingleton, J.F., 1966. Preserved and crystallised ginger. Confect. Prod. 32, 527–528.

IS 1908, 1993. Spices and Condiments – Ginger, Whole, in Pieces or Ground–Specification (second rev.), Bureau of Indian Standards, New Delhi, India.

Jaiswal, P.L., 1980. Handbook of Agriculture. Indian Council of Agricultural Research, New Delhi, India, pp. 1179–1184.

Jogi, B.S., Singh, I.P., Dua, H.D., Sukhija, P.S., 1972. Changes in crude fibre, fat and protein content in ginger (Zingiber officinale Rosc.) at different stages of ripening. Indian J. Agric. Sci. 42, 1011–1015.

Kachru, R.P., Srivastava, P.K., 1988. Post harvest technology of ginger. Cardamom 21 (5), 49–57.

Lawrence, B.M., 1984. Major tropical spices—ginger (Zingiber officinale Rosc.). Perfum. Flavor. 9, 1–40.

Leverington, R.E., 1969. Ginger processing investigations improving the quality of processed ginger. Queensland J. Agric. Anim. Sci. 26, 264–270.

Mani, B., Paikada, J., Varma, P., 2000. Different drying methods of ginger (Zingiber officinale) a comparative study. Spice India 13 (6), 13–15.

Mathew, A.G., Krishnamurthy, N., Namboodiri, E.S., Lewis, Y.S., 1973. Oil in ginger. Flavour India 3, 78–81.

Moudgill, K.L., 1928. Essential oils of Travancore. Part VII. From the rhizomes of ginger, Zingiber officinale. J. Indian Chem. Soc. 5, 251–259.

Natarajan, C.P., Lewis, Y.S., 1980. Technology of ginger and turmeric. Status papers and Abstracts National Seminar on Ginger and Turmeric. 8–9 April 1980, Central Plantation Crops Research Institute, Calicut, Kerala State, India.

Natarajan, C.P., Kuppuswamy, S., Sankaracharya, N.B., Padma Bai, R., Raghavan, B., Krishnamurthy, M.N., et al., 1970. Product development of ginger. Indian Spices 7 (4), 8–12. 24–28.

Natarajan, C.P., Panda Bai, R., Krishnamurthy, M.N., Raghavan, B., Sankaracharya, N.B., Kuppuswamy Govindarajan, V.S., et al., 1972. Chemical composition of ginger varieties and dehydration studies on ginger. J. Food Sci. Technol. 93 (3), 120–124.

Philip, G.J., Bastin, A., Devi, Manuel, 1996. Ginger drier. Spice India 9 (10), 4.

Prasad, J., Vijay, V.K., 2005. Experimental studies on drying of *Zingiber officinale, Curcuma longa* and *Tinospora cordifolia* in solar-biomass hybrid drier. Renew. Energ. 30 (14), 2097–2109.

Purseglove, J.W., Brown, E.G., Green, C.L., Robin, S.R.J., 1981. Spices—II. Longman, New York, NY.

Randhawa, K.S., Nandpuri, K.S., 1970. Ginger (*Zingiber officinale* Roscoe.) in India—review. Punjab Hort. J. 10, 111–122.

Sills, V.E., 1959. Ginger products. Colon. Fiji. Agric. J. 29, 13–16.

Spices Board, 2002. In: Sivadasan, C.R., Kurup, P.M. (Eds.), Quality Requirements of Spices for Export Spices Board, Cochin, Kerala State, India.

Spices Board 2011. Quality standards Agmark grade specifications for spices 2005. <http://www.indianspices.com/> (accessed 10.05.2011).

Sreekumar, M.M., Sankarikutty, B., Nirmala Menon, A., Padmakumari, K.P., Sumathikutty, M.A., Arumughan, C., 2002. Fresh flavoured spice oil and oleoresin. Spice India 15 (4), 15–19.

Takahashi, M., Osawa, K., Sato, T., Ueda, J., 1982. Components of amino acids of *Zingiber officinale* Roscoe. Ann. Rep. Tohuku. Coll. Pharm (Tohuku, Japan). 29, 75–76.

Varier, N.S., 1945. A note on the essential oil from ginger scrapings. Curr. Sci. 14, 322.

24 Production, Marketing, and Economics of Ginger

Ginger is an important commercial crop grown for its aromatic rhizomes, which are used both as a spice and as a medicine. India accounts for about 30% of world production, followed by China at 20%. The world production is approximately 0.75–0.8 million tons annually from an area of around 0.3 million hectares. Table 24.1 gives an account of this. During the same period, the export was around 20% of the total world production valued at US$ 105.73 million. Even though India is the largest producer of ginger in the world, the country occupied only the seventh position in export during 1999–2000, after China at the number one position, followed by Thailand, Brazil, Taiwan, Nigeria, and Indonesia. The major importing countries are the United Kingdom, the United States, Japan, and Saudi Arabia.

In India, the most amount of ginger-producing states are Kerala, Meghalaya, Odisha, West Bengal, Andhra Pradesh, Karnataka, Sikkim, and Himachal Pradesh. Official statistics on area, production, and productivity, although conflicting and often confusing, are available through FAOSTAT (statistical data of FAO) and SPICESSTAT (Spices Statistical database of the Integrated National Agricultural Resources Information System, INARIS of the Indian Council of Agricultural Research, India, located at the Indian Institute of Spices Research, Calicut, Kerala State, India). However, trade-related data available are relatively complete and make a distinction between dried and fresh ginger. A multitude of processed ginger products entering the world market are not taken into account separately. Despite certain limitations in the availability, this chapter will make use of the time-series data on production, export, and import. The objective of this effort is to obtain a broad indication on the possible changes which have taken place in the economy of ginger production during the last three to four decades starting from 1970 to 1971, and examine further prospects, both national and global, for the crop. From 1975 to 1980, India was the major producer of ginger with a total share of 30–35% of the total production, followed by China at 15%. China increased its share of world production up to 24% in the recent past, but India was on the decline and only produced 28% during this period. Indonesia totaled about 15% of world production, and together these three countries contribute more than one-third of world's production of ginger. The decline in ginger production in India in the 1970s has a parallel in cardamom production, coinciding with the same period, and while in the case of cardamom, Guatemala surpassed India, in the case of ginger, China started approaching India. Table 24.2 gives a global picture with regard to production by major ginger-producing countries.

The supply of ginger on a country-wise basis is computed taking into consideration area, production, and exports. The analysis brings out inconsistencies in yield and expansion of area. In order to make a meaningful analysis, ginger-producing

The Agronomy and Economy of Turmeric and Ginger. DOI: http://dx.doi.org/10.1016/B978-0-12-394801-4.00024-7

Table 24.1 Ginger Production: A World Scenario

Country	Area	Percentage of World Acreage	Total Production	Percentage of Total Production
Traditional	14,5344	45.84	650,330	84.37
Bangladesh	6,879	2.17	38,000	4.93
China	13,200	4.16	160,000	20.76
Dominica	45	0.01	100	0.01
Dominican Republic	400	0.13	1,500	0.19
Fiji Islands	65	0.02	2,500	0.32
India	83,220	26.25	281,160	36.48
Jamaica	180	0.06	620	0.08
Korea	4,255	1.34	7,950	1.03
Malaysia	1,000	0.32	2,500	0.32
Nigeria	17,400	5.49	90,000	11.68
Philippines	4,700	1.48	28,000	3.63
Sri Lanka	2,000	0.63	8,000	1.04
Thailand	12,000	3.78	30,000	3.89
Newcomers	15,261	4.81	12,4948	16.21
Australia (1990)	150	0.05	4,500	0.58
Bhutan (1980)	350	0.11	3,100	0.40
Cameroon	1,370	0.43	7,500	0.97
Costa Rica	1,600	0.50	21,000	2.72
Ethiopia (1993)	150	0.05	400	0.05
Ghana	0.00	60	0.01	–
Indonesia (1981)	10,000	3.15	77,500	10.05
Kenya (1989)	55	0.02	150	0.02
Madagascar (1992)	8	0.00	30	0.00
Mauritius (1985)	70	0.02	200	0.03
Nepal (1985)	12,00	0.38	3,200	0.42
Pakistan (1994)	78	0.02	28	0.00
Reunion (1985)	30	0.01	500	0.06
Saint Lucia (1985)	25	0.01	60	0.01
Uganda (1990)	50	0.02	120	0.02
United States (1985)	125	0.04	6,500	0.84
Zambia	0.000	100	0.01	–
World Total	31,7055	100.00	770,778	100.00

Figures in parentheses indicate the earliest year of initiating ginger production in the country mentioned.
Source: FAO (2003).

countries are grouped into two major categories: (i) traditional producers; and (ii) newcomers. Data in Table 24.1 give a bird's-eye view of the entire situation. The groupings suggest that up to 1980 there were about 15 countries engaged in the production of ginger. Since ginger cultivation and processing are labor intensive, most of the African countries have neglected this crop, and consequently, they are not very active now in the world market, though there is tremendous potential for the production of this crop on the African continent. But many other countries have entered the

Table 24.2 Ginger Production in Major Ginger-Producing Countries:
A World View (1975–2002)

Period	Percentage Share of the Total				World Production (mt)
	India	China	Indonesia	Others	
1975	30.67	11.68	–	57.65	147,213
1980	33.47	20.75	–	45.78	246,316
1985	35.37	12.89	12.56	39.18	390,259
1990	31.35	11.05	16.27	41.33	491,153
1995	30.11	20.00	11.34	38.55	728,376
2000	28.58	23.70	15.52	32.20	962,060
2002	27.83	23.98	15.18	33.01	988,182

Source: FAO (2003).

ginger market, as the data in Table 24.1 show. The number has almost doubled to date. The average share of newcomers in total global ginger production during the recent past (1998–2001) is 16.21%, which is on the increase. Among the newcomers, Indonesia, which although it entered the field of ginger production quite late, has made remarkable progress, considering the fact that it entered the field of ginger production only in 1981 (Table 24.1). It accounts for more than 10% of total world production now. It is quite possible that many more countries like Indonesia have entered this lucrative market, but reliable statistical data is difficult to find.

Area Expansion

An analysis of the world scenario on ginger production in terms of the acreage covered reveals the following:

1. China recorded the highest growth in acreage during 1991–2002 (10.969%) among all the ginger-growing countries. Indonesia and India, the other major producers, showed moderate growth at 5.6% and 3.006%, respectively, during the same period.
2. Other countries showing considerable growth in ginger acreage during the above-mentioned period are Sri Lanka, (0.26%), the United States (5.92%), Costa Rica (7.57%), Mauritius (1.31%), Bangladesh (1.34%), and Nigeria (5.8%). On the other hand, Uganda showed negative growth (−20.35%), as did Fiji (−8.29%), Pakistan (−13%), and Jamaica (−10.41%). These countries are newcomers.
3. The Philippines, Nepal, and Thailand showed relatively lower decline in acreage at −3.35%, −3.35%, and −4.69%, respectively.
4. An interesting phenomenon observed in terms of fluctuations in growth in acreage is that a high growth rate in a specific region (country) during a particular period is generally followed or preceded by a period of low or negative growth.
5. There is no striking difference between performances of traditional growers and newcomers. In terms of growth in acreage, some newcomers have fared well, whereas some others have failed miserably. The same argument holds true in respect of traditional ginger-producing countries, as well.

The world scenario inasmuch as ginger production is concerned highlights the recent trends:

1. China recorded the highest growth (11.39%) during 1991–1997, followed by Mauritius (11.15%), and Kenya (9.95%). Next in order are Nigeria (8.56%), Malaysia and Sri Lanka (both 6.78%), Madagascar (5.96%), and South Korea (4.36%).
2. A number of countries have recorded a high rate of negative growth, Uganda with the highest negative growth (−21.67%), followed by Fiji (−17.24%).
3. Between the above two extremes lie the other countries, some showing moderate positive growth, while others moderate negative growth.

Regarding growth, the cyclical nature of the growth pattern was observed over the decades for both area and production. With the exception of Fiji, the nature of fluctuations in acreage and production was almost identical (in terms of both peak and trough) for other countries. Again, on average, the growth pattern in production is also not group specific.

Yield

In terms of productivity performance, the world scenario shows the following:

1. Except in Fiji, South Korea, the Philippines, and Nigeria, the traditional growers of ginger, the cyclical growth fluctuations are not that sharp in other countries.
2. The fluctuations are highly erratic in Fiji, recording a high negative growth during 1971–1980 and a very high positive growth in the following decade, only to come down to around 3% during 1991–1997.

In order to analyze the salient features of major ginger-producing and ginger-consuming countries individually, an effort is made to present country-wise details separately.

India

Ginger is grown in almost all the states in India. However, the major ginger-producing states are Kerala, Odisha, Meghalaya, West Bengal, Karnataka, Sikkim, Andhra Pradesh, and Himachal Pradesh. Kerala accounts for the major share of production (19%) and acreage (19%). This figure has remained more or less unchanged during the last more than three decades. Odisha comes after Kerala, followed by Meghalaya. These three traditional ginger-growing states in India account for approximately 40% of India's production.

In South India, although ginger cultivation was confined only to Karnataka State in the earlier years, during the past decade it has been making inroads into the paddy fields of Kerala and Tamil Nadu, including Karnataka. In Karnataka State, ginger is cultivated on a commercial scale in the Coorg and Chickmagalur districts, with a reported area of approximately 5000 ha (Korikanthimath and Govardhan, 2001a). Enterprising farmers from the adjoining Waynad district of Kerala State lease paddy fields for ginger cultivation. Fresh ginger harvested during the months of January–March has buyers coming from as far as Nagpur district and Mumbai in Maharashtra State and also from Bengaluru in Karnataka State. A sizeable quantity of fresh ginger goes to the traditional ginger-growing districts of Ernakulam and Kottyam in Kerala State for processing into dry ginger. In Kerala State, Waynad and Idukki districts contribute the most for the export of quality ginger. Incidentally, these two districts have the most intensive ginger cultivation in India.

Farmers from Karnataka have a unique practice of putting back a certain portion of harvested ginger in the ground and preserving it as "old ginger" for the succeeding year. This is because of the low ginger price during harvest season. During the following year, this ginger is used for sowing, and more rhizomes develop, and the farmers hope to get more money for the fresh produce and the old produce. The major cultivars used are Himachal, Maran, and Rio de Janeiro (Spices Board, 1988).

Production Economics

Examination of the time-series data indicates that the coefficient of variation for the farm price of ginger was higher than the production cost incurred over a period, indicating the huge fluctuation in the market price of ginger in India. This had a greater impact on the production economics of the farming community. The problem can be better understood by the following facts. Farmers buy seed rhizomes for prices as high as Rs 50/kg (approximately US$ 1) at times, but the harvest price could fetch them only one-fifth of this price paid. In order to preempt price-related risks, the farmer cultivates ginger as an intercrop in the main crop, such as coffee. In the major ginger-growing state, Kerala, approximately one-fourth of the cultivated area is in the uplands as pure crop, whereas in the major area (45%) is the garden lands category and the rest is under a mixed cropping system. A study on economics showed that banana + ginger + vegetable (cowpea) intercropping system fetched close to an equivalent of US$ 2000/ha. The benefit–cost ratio was also highest in the banana + ginger mixture (2.28), whereas the lowest benefit–cost ratio (1.56) was recorded in the banana + turmeric system (Regeena and Kandaswamy, 1987). It takes about approximately 10 cents to produce a kilogram of ginger in Kerala.

An investigation conducted in Maharashtra to work out the economics of ginger production revealed that the average production cost per quintal (100 kg) is approximately US$ 40 and the estimated cost–benefit ratio is 1.38. The cost of the seed rhizome accounts for approximately 40% of the total cost of production (Gaikwad et al., 1998).

Korikanthimath and Govardhan (2001b) conducted an investigation to compare the economics of ginger cultivation in the uplands and paddy fields of Karnataka, which indicated that the cost–benefit ratio is more favorable in paddy fields (1.7) as compared to upland cultivation (1.11). This higher profitability is mainly due to higher productivity (23.5 t/ha) achieved in the paddy fields, when compared to the yield level of 13.5 t/ha in the uplands.

Trends in Area, Production, and Productivity

The time-series data on area, production, and productivity of ginger along with the growth index worked out for the last three decades indicates the following.

Area

The area under ginger cultivation has shown remarkable increase during the last three decades, with occasional fluctuations attributed to the ups and downs in price structure. Low profit in a specific year due to unfavorable price structure leads to reduction in the area cultivated, which reflects on the production levels in the subsequent years.

Production

India's production of ginger has been increasing steadily from 29.59 thousand tons/ha during 1970–1971 to 263.17 thousand tons/ha by 1999–2000. This is almost a 100-fold increase. Precisely, this works out to 789% increase, which resulted from a combined improvement in both area cultivated and the concomitant increase in productivity. Both Kerala and Meghalaya put together accounted for more than 65% of the total production in India. If one makes a region-wise grouping, the southern region comprising Kerala, Tamil Nadu, Karnataka, and Andhra Pradesh accounts for 52.4% of production with a corresponding total area of 42.4% in the period from 1990–1991 to 2000–2001. District-wise, ginger production area shows that ginger cultivation is mainly confined to Kerala and Meghalaya.

The state-wise area, production, and productivity of ginger for three periods, 1982–1983, 1992–1993, and 1998–1999 show that as against the national average yield of around 3371 kg/ha achieved during 1992–1993, states such as Meghalaya, Andhra Pradesh, Sikkim, and Tamil Nadu have consistently recorded a higher yield. Tamil Nadu achieved the highest yield of 19,450 kg/ha during the period and has attained a record productivity of 31,683 kg/ha during the 1998–1999 crop season. The insignificant change in area in Tamil Nadu is taken care of by a significant growth in yield in the state, thereby helping it to register a healthy growth in production. Nagaland, Mizoram, Arunachal Pradesh, and Meghalaya are the other states in order to achieve higher productivity (more than 5500 kg/ha) during the same period. Andhra Pradesh registered 7164 kg/ha, Meghalaya 5137 kg/ha, Mizoram 5000 kg/ha, while Odisha registered the lowest figures (1990 kg/ha) (DASD, 2002).

Productivity

Further analysis of the time-series data between 1970–1971 and 1997–1998 indicated that the yield level of ginger in India increased over the years from 1371 kg/ha during 1970–1971 to 3391 kg/ha during 1997–1998. The yield level which was approximately 1371 kg/ha during 1970–1971, did not show much improvement until the end of 1980, except for occasional fluctuations toward the higher side (up to 1991 kg/ha) during 1977–1978. It seems the yield increase during this period did not contribute

Table 24.3 Change in Ginger Production, Area, and the Relative Contribution of Changes in Area and Yield on the Changes in Production During Different Periods in India

Details	1980–1981/ 1984–1985 to 1985– 1986/1989–1990	1985–1986/ 1989–1990 to 1990– 1991/1994–1995	1990–1991/ 1994–1995 to 1995– 1996/1999–2000
Change in:			
Production	40.22	26.84	30.47
Area	18.28	10.10	21.47
Productivity	19.54	15.01	8.16
Change in production due to change in:			
Area	49.66	40.48	73.11
Productivity	52.81	58.83	29.49

Analysis based on the method followed by Librero et al. (1988).

much to the overall increase in production. The increase in production during that period was largely due to an increase in cropped area. However, the productivity level improved from 1980 to 1981 onward and reached an average of 3188 kg/ha from 1990–1991 to 1998–1999. Productivity registered during 2000–2001 was more than twofold the productivity of 1970–1971. The estimated growth index for the year 1998–1999 in production was 254% over the base year 1970–1971.

To ascertain the impact of area expansion and productivity on the overall production during different periods, period-wise data were analyzed using a simple technique followed by Librero et al. (1988). Results show that there is a positive sign in all three parameters, indicating the steady improvement in production due to both area expansion and productivity increase. Results are shown in Table 24.3. However, the detailed component analysis revealed that the change in productivity had a more positive role in the first two periods, whereas in the last, area expansion played a major role in the expansion of production.

Growth Estimates

In order to obtain the long-term trends in area, production, and productivity in the major ginger-growing states in India, semilogarithmic growth equations were estimated, which indicated that the overall trend in the area under ginger cultivation registered an average annual growth rate of 4.3% for the period 1990–1999. Growth in production was at the rate of 6.11% during the same period, indicating a slight improvement in productivity, which was approximately 1.82% for the period.

Production Constraints

A status paper prepared by the Spices Board (1990) on the ginger crop highlights the fact that mostly small and marginal growers cultivate ginger in India. They face enormous problems and constraints which hamper the ginger productivity. Major production constraints in ginger cultivation listed by various investigators, including the Spices Board of India (Kithu, 2003; Selvan and Thomas, 2003) are:

1. Low productivity (3391 kg/ha) compared to an achieved average productivity of more than 1 lakh kg/ha elsewhere in the world.
2. Prevalence of innumerable traditional cultivars, which are mostly poor yielders. Absence of an adequate supply of planting materials of improved cultivars.
3. Being a predominantly rainfed crop, failure of rains, and increased labor costs are some of the factors responsible for the higher cost of cultivation of ginger in India.
4. Nonadoption of integrated plant protection measures to control pests and diseases, such as rhizome rot, which cause heavy production and postharvest losses in the crop in many parts of the country.
5. Lack of suitable postharvest processing facilities and poor marketing facilities, especially in the northeastern states of India, which result in poor returns to the farmers.
6. Lack of remunerative prices for the produce in subsequent years, which leads to diminished enthusiasm of the farmers to cultivate ginger, which eventually leads to the neglect of the crop in the country, adversely affecting overall production and growth rate of the crop.

Taking the above points into consideration, there is an imminent need to develop cropping systems with ginger as a component. Although the crop is being cultivated as an intercrop in coconut and arecanut plantations, the researchers have yet to develop ideal cropping systems focusing primarily on the cost–benefit ratio for the famer, and other associated environmental considerations, such as soil degradation.

China

In China, ginger is grown extensively in all the central and southern provinces. It is cultivated as either an annual or perennial crop. China emerged as the second largest producer during the year 2002 (23.98% of the total world production), next to India. During 1990, China's production was 54,284 t, accounting for 11.05% of total world production. Within the next decade, there was a more than fourfold increase in production and the production level reached one-fourth of the total world production. This achievement is primarily due to high average productivity of 115,104 kg/ha, an unheard of production level anywhere in the world, and the highest level was 120,641 kg/ha in 1996.

Since Chinese ginger contains less fiber, the rhizomes being bigger in size, and the end product competitive in price, Chinese ginger commands the first place in the world production. In 1994, China exported 52.05% of total production and this continued until 2000, which accounted for 61.59% of total world exports. Of the annual export of 91,000 t, Chinese export accounts for more than 61%. Many importing countries prefer Chinese ginger for its price competitiveness and quality of produce.

Ginger from China is also exported in crystallized form in earthenware jugs and as syrup in wooden kegs. Harvesting of ginger in China commences in April and extends up to June. Harvested ginger is transported to processing plants in Chiang Rai for export, mostly to Japan. Young ginger is preserved in vinegar bottles and consumed as pickles.

Table 24.4 Australian Ginger Yield

Harvest	Time of Harvest	Yield (t/ha)
Early	Late February–Early March	12–50
Early-Late	April–August	20–50
Late	Mid June–Early October	38–75

Australia

Commercial cultivation of ginger in Australia was first started in Buderim in Southeast Queensland as early as the 1940s mainly for the domestic ginger market. Ginger is now grown in the *Caboolture*, *Nambour*, and *Gympie* areas for processing at *Yandina*. Twenty-four growers currently represent the Australian ginger industry with approximately 150 ha under cultivation. The bulk of production is processed, with smaller volumes sold in the domestic and export markets. Buderim Ginger Ltd. is the only ginger-processing facility in Australia. This factory, through production quotas and a differential pricing system, controls the quality and quantity of ginger production for processing. Most growers derive the majority of their income from processed ginger. A few also supply the domestic fresh ginger market, and only two to three growers export fresh product. In 1987, Royal Pacific Foods began exporting Buderim ginger to the United States. Now the Australian products under the brand name "The Ginger People" are freely available on the market shelves of many well-known food chain stores the world over. The Australian ginger farmer has achieved a reasonably higher productivity against the world average as shown in Table 24.4.

Thailand

Thailand produces about 32,000 t of ginger annually. The crop is cultivated extensively in the northern part of the country, especially in the mountains. Ninety percent of the production comes from the hills. Thailand has had a slow increase in ginger production. Without much improvement in the recorded productivity of 25,000 kg/ha, improvement in the overall production was achieved through area expansion. The estimated normal growth rate for the period 1990–2002 was 2.7%, 2.81%, and 0.10%, respectively, for area, production, and productivity. Ginger from Thailand is noticeably distinguishable compared to others because of its plumpness, roundness, and short internodes. The popular dried "Golden" variety from Thailand is packed and exported.

Marketing

Products of Commerce

Three primary products of ginger rhizome traded nationally and globally are (i) fresh ginger; (ii) preserved ginger in syrup or brine; and (iii) dried ginger.

Preserved ginger is made from the immature rhizome, whereas the pungent and aromatic dried spice is prepared from harvesting and drying the mature rhizome. Fresh ginger, consumed as a vegetable, is harvested both when immature and mature. The preserved and dried products are the major forms in which ginger is internationally traded. Fresh ginger is of less importance in international trade, but this is the major form in which ginger is consumed in the producing countries. Dried ginger is used directly as a spice and also for the preparation of extractives, ginger oleoresin, and ginger oil (ITC, 1995).

Commercial ginger in India is graded according to the region where it is produced, number of fingers contained in the rhizome, its size, color, and fiber content. Among the Indian States, it is only in the State of Madhya Pradesh, where grading of ginger is done. The first grade, popularly known as "Gola" in the local market, comprises very bold and round bits of dry ginger, which have maximum dry matter and low fiber content. The second grade, known as "Gatti," includes bits of bold, round to oblong pieces, which are smaller than the "Gola" variety. The third and fourth grades are smaller bits with low dry matter and high fiber content (Jaiswal, 1980). For export purposes, *Calicut* and *Cochin* ginger are graded into special, good, and nonspecial grades, depending on the size of the rhizomes and the percentage of the presence of extraneous material.

Dried ginger has been traditionally traded internationally in the whole or split forms and is ground in the consuming centers. Export of the ground spice from the dried ginger-producing countries is on an extremely small scale. The major use of ground dried ginger on a worldwide basis is for domestic culinary purposes, whereas in the industrialized Western countries it also finds extensive use in the flavoring of processed foods. Ground dried ginger is employed in a wide range of food stuffs, especially in bakery products and desserts (Anonymous, 1996).

Ginger oleoresin, an important value-added product, is obtained by solvent extraction of dried ginger and is prepared both in certain industrialized Western countries as well as in some of the spice-producing countries, most notably India and Australia. This product possesses the full organoleptic properties of the spice—aroma, flavor, and pungency—and finds similar applications as in the case of ground spice in the flavoring of processed foods. The oleoresin is also used in certain beverages and to a limited extent in pharmaceutical preparations. The new process developed by the Regional Research Laboratory in Thiruvananthapuram, Kerala State, India, for extracting oil and oleoresin from fresh ginger, will lead to a higher recovery of the oil superior organoleptic qualities, and will drastically reduce spoilage of fresh ginger during harvesting season. This technology, which is highly suitable for the northeastern states of India, can utilize the cheap raw material available during the harvesting season to convert it into high-priced value-added products. The operating cost of the fresh ginger-processing facility is much lower than that of conventional plants. Further, drying, peeling, and so forth are dispensed with, and because the processing is done during the ginger harvesting season, the raw material inventory can be reduced drastically. It is expected that adoption of this new technology can boost the country's prospects in adding value to the export of Indian ginger.

Ginger oil is distilled from the dried spice mainly in the major spice-importing countries of western Europe and North America, as well as in some of the spice-producing countries such as India. This product possesses the aroma and flavor of the spice but lacks the pungency. It finds its main application in the flavoring of beverages and also is used in confectionery and perfumery. Preserved ginger is prepared mainly in China, Hong Kong, Australia, and India, but smaller quantities of fresh ginger are processed in some importing countries as well. It is used both for domestic culinary purposes and in the manufacture of processed foods, such as jams, marmalades, cakes, and confectionery (Sreekumar and Arumugham, 2003).

Market Structure

Regarding the market structure, there are a number of firms and individuals actively participating in the ginger trade, especially in the case of dried ginger. A large number of brokers, dealers, and various other intermediaries between the dealer and the consumer or even within dealers, exist both in exporting and in importing countries. Singapore, London, New York, Hamburg, and Rotterdam are major trading centers. In the case of preserved ginger, Hong Kong is the major trading center. Fresh ginger is marketed through the fruits and vegetables trade network. The market framework indicated that in terms of the ratio between farm harvest price and retail price, it was observed that the ratio was higher in 1989 as compared to 1995. Moreover, fluctuations in the ratio were also less in 1989. The ratio between the farm harvest prices and wholesale prices has also gone down in recent years, indicating that the producer is able to obtain a better price for the produce, and there is less exploitation by middlemen in the ginger business.

Factors Controlling Demand/Export

A major factor that contributes to the export/demand potential of a commodity is quality. In ginger, quality parameters are fiber, volatile oil, and nonvolatile ether extract contents. Ginger grown in various parts of the country varies considerably in its intrinsic properties and its suitability for processing. This is perhaps more important with regard to preparing dried ginger than preserved ginger. The rhizome size is relevant in particular with regard to processing of dried ginger, and medium-sized rhizomes are generally the most suitable. Some areas grow ginger types yielding very large rhizomes, which are marketed as fresh ginger, but are unsuitable to convert to the dried spice, owing to their high moisture content. This causes difficulties in drying; frequently a heavy wrinkled product is obtained, and the volatile oil content is often low and below standard requirements. From the above point of view, ginger produced in certain pockets of Kerala is in more demand and has more export potential in the global market.

Indian Dried Ginger

Two types of Indian ginger entering the international market are (i) *Cochin* and (ii) *Calicut*, named after two production centers and major ports of the Malabar Coast in Kerala State, India. The bulk of Indian exports is rough-shaped, whole rhizomes.

In addition to this, some bleached or limed ginger is also produced, but this is mainly exported to the Middle East, as it is not favored in European and North American markets. *Cochin* and *Calicut* ginger have volatile oil contents in the range of 1.9–2.2%. They are characterized by a lemon-like aroma and flavor, which is more pronounced with the *Calicut* spice. They are starchier but are almost as pungent as Jamaican ginger. Their nonvolatile ether extract content is about 4.3%. They are widely used for blending purposes, and ginger beer manufacturers prefer these types (Spices Board, 1992).

Economics of Dry Ginger Production

In India, production of the dried ginger of commerce is confined exclusively to Kerala State and the product is of two types—*Cochin* and *Calicut*. *Cochin* type, preferred over the *Calicut* type, is grown in central Kerala, mainly concentrated in the Ernakulam and Idukki districts, while the *Calicut* type is confined to the Malabar region, including the Waynad district in northern Kerala State.

There is no other recognized commercial variety of dried ginger produced in other parts of India other than the *Cochin* and *Calicut* types. Kerala ginger is considered to be superior because of its low fiber content, boldness, and characteristic aroma and pungency. Ginger produced in other parts of India have more fiber, which are largely used for domestic consumption in the form of green ginger. Kerala State accounts for over 60% of the total dried ginger production and about 90% of India's ginger export trade.

In contrast to Jamaican ginger, which is clean peeled, Indian dried ginger is usually rough peeled or scraped. The rhizomes are peeled or scraped only on the flat sides of the hands; much of the skin between the "fingers" remains intact. The dry ginger so produced is known as the rough or unbleached ginger of commerce, and the bulk of the dried ginger produced in central Kerala consists of only this quality. Sometimes Indian ginger is exported unpeeled. For the foreign market, both *Cochin* and *Calicut* gingers are graded according to the number of "fingers" in the rhizomes. They are classified as follows: "B" for three fingers, "C" for two fingers, and "D" for pieces. In addition to these well-known types of Indian ginger, another type, Kolkata ginger, is occasionally seen in the market as well (Pruthi, 1989).

World Scenario

As ginger is mainly used as a spice and condiment, its per capita consumption is not high enough to sustain its world-level production with the growing number of new ginger-producing countries taking recourse to the international trade in ginger. However, the market information indicates that there is a "hot trend" in the US market, which means an escalating demand for many other spices like black pepper, chilies, and ginger. This is a clear reflection of the changing food habits of the Americans, which is veering toward spicy food, as compared to the earlier bland types. This could also emanate from the new ethnic mix in the population. Chinese food and Indian curry, in which ginger is a principal ingredient, are increasingly being favored by the Americans. There is also a growing demand for ginger products worldwide. A recent development noted in ginger trade is the increasing use of ginger oils, oleoresins, and powdered and processed ginger in major ginger-importing countries, especially the United States and Europe.

Main Suppliers

Major ginger exporters are India and China. Others which also export are Indonesia, Brazil, Sierra Leone, Australia, Fiji, Nigeria, and Jamaica. Indonesia, Taiwan, China, and Thailand are major exporters of fresh ginger. Important suppliers of preserved ginger are Hong Kong, which reexports refined fresh ginger, and Australia (ITC, 1995).

In the world trade, there are two categories: (i) producer exporters (countries engaged in ginger cultivation which usually export after meeting the domestic consumption); occasionally these countries may also import to meet domestic needs; and (ii) countries which reexport.

World Trade

Distribution Channels

Specialized importers still play an important role in the global ginger trade. A list of the importers can be obtained from the International Trade Center (ITC).

Dry Ginger

The traditional distribution system for dry ginger has declined as a result of the increase in purchase by dealers and processors directly from the source of production. There has also been an increase in trade in some countries among certain ethnic communities. Asians, in particular, have developed their own system of distribution based on direct trading with the producing countries and a network of small retail outlets.

Fresh and Preserved Ginger

The marketing structure for fresh and preserved ginger is similar to that of vegetables in India. The recent rise of supermarkets in India has eroded the position and clout of wholesale dealers as some importers sell directly to supermarkets. In some importing countries, however, ginger in its fresh form is seen almost exclusively in shops catering to ethnic communities.

Export

During 1994, China contributed 52.05% of total ginger export worldwide, followed by Thailand at 16.77%, Indonesia 9.73%, Brazil 6.24%, Taiwan 3%,

Costa Rica 2.23%, India 1.98%, Nigeria 1.61%, Vietnam 1.37%, Malaysia 1.36%, and the United States 0.93%. Clearly, despite India's preeminent production level in the world, it only contributed a small percentage of world export in 1995. China and Thailand maintained top export positions until 2000, in fact, showing a surge by China at 61.59%, followed by Thailand at 23%.

The ginger export of Jamaica and Sierra Leone is considered to be of high quality on account of their superior flavor and clean appearance. However, the price of Jamaican ginger is quite high, which has led importers to search for cheaper alternatives. Today, the ginger from Australia is regarded as being of high quality due to its standardized and clean appearance and steady price. Grinders have favored Chinese ginger, but the use of bleaching agents and sulfur dioxide has adversely influenced Chinese exports to Europe and North America, as they are highly conscious of additives in food stuff.

In order to examine the trend in returns from trade earned by the exporting countries, Datta et al. (2003) have used a simple index (VADD) defined as follows:

$$VADD = Unit\ Value\ of\ Exports - Unit\ Value\ of\ Imports$$

where

Unit Value of Exports = Total Value of Exports/Total Quantity Exported
Unit Value of Imports = Total Value of Imports/Total Quantity Imported

Datta et al. (2003) have ranked all the countries in terms of VADD in decreasing order and reported that:

1. Out of the top 15 exporting countries, only 3 belong to the producer–exporter group. The rest are all from the reexporter group.
2. Only two countries are the traditional producers.
3. Of the major producers, India ranked 40th with a VADD of 0.38, followed by China at 44th place with a VADD of 0.21. In the case of Indonesia, the VADD estimate turned out to be negative at −0.13, meaning that Indonesia imported ginger at a higher unit value than it exported.
4. Thus, reexporters have, in general, succeeded in achieving a greater value addition to their export of ginger into the world market.

A further analysis of the import–export reexport trade showed that the unit price (US$ 2.18/kg) earned by the European Union (EU) countries (reexporters) from export is much higher than the average unit price (US$ 1.53/kg) earned by other producer exporters to EU countries. The Netherlands, Germany, and the United Kingdom are the major reexporters of ginger in Europe. Obviously, value addition to the imported material turned out to be the biggest money spinner to the EU countries. This is also a good and clear lesson to be learned from the world ginger trade, and that is unless the producer countries have a clear road map to the value-addition process, merely producing ginger in large quantities is of no avail for reaping great pecuniary benefits from the ginger trade. This is a fundamental lesson all the developing countries have to learn in world trade. Extending this logic further, many examples can be cited in world agricultural trade, where the reexporter and not the producing countries benefit from world commodities trade. For instance, the best cocoa is grown in Africa, either in Ghana or in the Republic of Cameroon. But the pecuniary benefits in large measure go to Paris, not to the toiling African farmer sweating it out under the scorching sun. It is here that technology plays a crucial role. Many examples can be cited to prove this point.

Datta et al. (2003) have analyzed the export performance of Indian ginger economy between 1961 and 1996, which shows the following:

1. The physical volume of exports has increased approximately 2.96% annually, whereas the annual growth in value terms works out to be approximately 10%. The annual growth in the unit price realization over this period works out to be approximately 6.9%.
2. At a decadal disaggregated level, however, the performance of ginger export from India does not appear encouraging. There is a steady decline in unit value realization from ginger exports. During the 1960s, the unit value realization grew at an annual rate of more than 19%, despite the fact that there was a negative growth in the physical volume of exports. The growth in the physical volume of exports picked up considerably during the 1970s, although at the cost of a decline in the growth in unit value realization. The 1980s witnessed a fall in the growth rate of both of these attributes. During the first half of the 1990s, however, a spurt in the growth of physical exports is observed, accompanied by an almost stagnant unit value realization, despite considerable devaluation of the Indian rupee (Indian currency, which is now trading (at the time of writing this book, 2012) at Rs 55 per US$, a depreciation of almost 20% during the past few months).

Export Instability

In order to estimate instability in ginger exports in terms of quantity, value, and price, an instability analysis was done using the time-series data, and the results are presented in Table 24.5. It can be observed from Table 24.5 that there was instability in the case of volume of ginger exported, value, and unit value of ginger export. The instability was relatively higher in the case of the volume of ginger exported (72.91%) compared to the value of the export (57.41%) and the unit value of the export (29.15%). This instability index was a close approximation of the average year-to-year percentage variation in the three parameters studied, which was adjusted for the trends.

Composition of Indian Exports

As far as itemwise export of ginger from India is concerned, there has been a marked improvement in recent years. More than half of the total export value is earned by

Table 24.5 Instability Indices (Coppock's Instability Index) of Ginger Exports (1970–2000)

Particulars	1970–1971 to 1979–1980	1980–1981 to 1989–1990	1990–1991 to 1999–2000	1970–1971 to 1999–2000
Volume of export	47.95	60.37	84.71	72.91
Value of export	51.60	62.35	63.50	57.41
Unit value of export	49.44	68.81	34.62	29.15

Table 24.6 Contents of Indian Export Basket (1990–1991 to 1999–2000 Average)

Details	Percentage Share of the Total		
	Quantity (Mt)	Value (Approx. US$)	Unit Price (Approx. US$) (Per kg)
Dry ginger	4,587.80	405,000	90 Cents
Fresh ginger	10,138.07	200,000	10 Cents
Ginger powder	418.32	20,000	90 Cents
Ginger oil	7.63	18,000	40.00 US$
Ginger oleoresin	59.63	12,000	22.00 US$
Total	15, 211.45	8,000	60 Cents

Source: Spices Board (2008).

dry ginger, which accounts for 30.16% in terms of quantity (Table 24.6). Fresh ginger, though, accounts for 66.65% of the total quantity exported; in terms of value, the percentage share is 24.67. Ginger oil and oleoresin are the other products exported that have fetched a high value. As in the case of reexporting countries, especially EU countries, India has the potential to strengthen the processing industry to add more value-added products onto its export basket. India exports a sizeable quantity of fresh ginger through land custom stations in the northeastern states to Bangladesh. Although this export channel provides an opportunity to market the exportable surplus across the border at a reasonable price, whenever the price goes up, Bangladesh turns to a cheap supply from China and Indonesia. The same is the case with the other neighboring country, Pakistan (John, 2003).

Direction of Indian Exports

Until the end of 1980s, more than 30% of Indian exports was to Arabian countries, in general, which has shown a declining trend ever since then, and India is finding a new market in Pakistan and Bangladesh. However, these countries turn to other sources, wherever the prices are competitive. During 1991, India exported more than 49% to Pakistan. However, during 1999–2000, Pakistan's share of Indian import declined to 33%. A similar situation was witnessed in Bangladesh as well. Other main markets for Indian ginger are Saudi Arabia, the United Arab Emirates (UAE), Morocco, the United States, Republic of Yemen, the United Kingdom, and the Netherlands.

To analyze the concentration of ginger exports to various countries, both in terms of quantity and in value of export markets, the Hirschman index was estimated, which is presented in Table 24.7. Generally, the index number above 40% is considered to be a high concentration. Here, the estimated index for quantity is more than 40 during all the three periods indicating higher concentration. In the case of value as well, the index was more than 40% in the first period, and it was nearer to 40% in the remaining two periods, as well. This indicates that the country has a set of markets, which prefer Indian ginger.

Table 24.7 Hirschman Index for Export of Ginger

Period	Particulars	Quantity	Value
1981–1982	Volume of ginger export	44.35	45.47
1991–1992	Value of ginger export	48.52	38.63
1999–2000	Unit value of ginger export	48.42	36.87

Export Promotion Program

The Spices Board, Government of India, implements a number of programs to promote ginger export. These are:

1. Assistance in establishing improved cleaning and processing facilities.
2. Support for setting up high technology-assisted processing facilities.
3. Assistance in establishing and strengthening in-house quality laboratories to test various quality parameters.
4. Assistance for new product/end use development.
5. Assistance for improved packaging.
6. Assistance for undertaking promotional sale tours and participation in international spice fairs.
7. Support for promoting branded consumer-packed ginger in identified markets overseas.
8. Support for organic certification to process ginger derivatives.

Imports

The major importers globally are the United States, Japan, and the United Kingdom, where an increase in terms of volume and value has been observed over recent years. However, these countries import ginger mainly from China and Thailand and not from India.

Japan accounted for the major share (58.74%) of imported ginger during 1995, followed by the United States (9.93%), Hong Kong (7.62%), Singapore (5.16%), Saudi Arabia (4.37%), the United Kingdom (4.36%), Canada (2.39%), the Netherlands (1.93%), Germany (1.25%), and Malaysia (0.78%), and the rest by others, including India. Japan, a major importer, imports more than 50% of its requirement from China. Thailand exports mostly preferred fresh ginger in large quantities to Japan. Japan's import from India constitutes mainly dry ginger. Of late, the Spices Board is exploring the Japanese market to increase India's export to that country.

Indian Import of Ginger

A sizeable quantity of ginger imported into India is of the green form. The major imports in fresh form are from Nepal, where dried ginger is imported from China and Nigeria. The Indian import scenario is given in Table 24.8.

Table 24.8 Itemwise Import of Ginger into India During 1995–2000

Item	1995–1996		1996–1997		1997–1998		1998–1999		1999–2000	
	Q	V	Q	V	Q	V	Q	V	Q	V
Dry ginger	782.6	218.5	133.9	64.7	247.4	106.2	542.3	291.4	4695.0	1198.5
Fresh ginger	6682.2	429.0	9277.7	580.7	1185.4	703.1	9727.2	614.8	7614.2	688.7
Ginger powder	–	–	–	–	–	–	Negligible	0.03	13.0	6.4

Q = Quantity in metric tons, V = in lakhs (1 lakh Indian rupees = US$ 1995 Approx.)

Market Opportunities

According to an ITC market development paper (1995), consumption of spices is likely to increase due to an augmented production of highly flavored food by the food industry. In addition, an increasing interest in health food, consequently, "natural" instead of "artificially" flavored food, will also increase the consumption of natural, unadulterated spices.

Dry Ginger

A development noted in the ginger trade has been the increasing use of oils, oleoresins, and powdered and processed ginger in major importing countries, especially in the United States and EU countries. Ginger exports for the manufacture of powdered ginger must be fiber-free, whereas the products exported for the manufacture of ginger oil and oleoresins should have a high oil content. Export efforts should be based on increased productivity and improved postharvest technology.

Fresh Ginger

There may be some prospects for a modest increase in international trade in fresh ginger, mainly to cater to the ethnic market, especially of the Asian communities settled in the United States and European countries.

Preserved Ginger

Japan will continue to have the world's largest market in preserved ginger. This is because preserved ginger finds a place in an array of Japanese food. There is also the prospect of a modest market for preserved ginger in the United States and western European countries, especially to cater the ethnic taste. The Chinese segment is, in particular, important in this context. However, the growth in this segment can only be modest, as of now, and also in the future.

Table 24.9 Unit–Price Ratio for Various Exporting Countries From 1994 to 1998

Country	1994	1995	1996	1997	1998	1994–1998
China	0.90	0.83	1.17	1.03	0.94	0.96
Thailand	0.87	1.06	0.91	0.82	0.73	0.86
Indonesia	0.54	0.51	0.43	0.62	0.48	0.53
Brazil	1.61	1.64	1.11	1.26	1.36	1.35
Taiwan	3.18	3.83	2.00	2.43	3.21	2.76
Costa Rica	1.19	1.33	0.89	1.01	1.10	1.10
India	1.23	2.07	1.06	1.20	1.80	1.40
Nigeria	0.74	0.97	0.85	1.16	1.36	1.06
Vietnam	0.50	0.67	0.57	0.89	0.56	0.66
Malaysia	0.35	0.43	0.32	0.37	0.38	0.36
United States	1.99	1.96	0.86	1.04	1.45	1.29
Others	1.77	1.53	0.96	1.12	1.65	1.33

Competitiveness of Indian Ginger Industry

In order to understand better the position and competitiveness of individual exporters in world trade of ginger, market shares and unit value ratios were calculated and are presented in Table 24.9. In the absence of time-series data on prices for individual products from various countries, the unit price was worked out from the value of the export and quantity exported. While calculating unit price, individual items of export were not taken into account. Hence, there is bound to be slight variation depending on the share of value-added products in the export basket of individual countries. However, the estimated unit–value ratios help in comparing the prices of each exporting country with another and with the average of total exports. The ratio is computed by dividing the price received for a country's export by the world average price. When the unit–price ratio is less than 1, then it is considered that the country possesses competitiveness in the export market for its product. Accordingly, as can be observed from the data in Table 24.9 that Indonesia, China, Thailand, Vietnam, and Malaysia with a unit–price ratio less than 1 are highly competitive, whereas India, with an average unit–price ratio of 1.40, is less competitive in its price structure in the world market.

Any country's competitive power in exporting a commodity depends crucially on its relative price and the quality of that commodity over the competing countries. India has a weak competitive position in the international market for ginger, which is mainly because of a very low productivity of 3357 kg/ha as compared to 55,636 kg/ha of the United States, and an average world productivity of 10,179 kg/ha (FAO, 2003). Moreover, the increased cost of production due to lower productivity of Indian ginger compared to that of other producing countries makes it imperative for India to increase its ginger productivity, which also can reduce the cost of production. The country has enough potential to increase its ginger productivity. To be successful in the changing environment, it is essential to be innovative and proactive. India, being the major producer of ginger in the world, stands seventh in ranking when its performance is compared with that of other countries.

The gross margin is a good measure for comparing the economic and productive efficiency of similar-sized farms. More importantly, it represents the bare minimum that a farm must generate in order to stay in business. The cost–benefit ratio worked out for ginger production in the United States was 1.34. Productivity achieved on the ginger farms of Hawaii ranged from 50,000 lbs/acre to a low of 27,500 lbs/acre. The reported average returns for the farm with a productivity of 46,200 pounds depends not only on the yield but also price.

Risks and Uncertainty

Risk is inherent in all of agriculture, but the ginger industry appears to be more exposed to risk than many other agricultural endeavors (Fleming and Sato, 1998). A review by the Hawaii Agricultural Statistics Services (HASS) reveals considerable volatility in ginger price and yield, with relatively little correlation between the two variables. In addition to abruptly fluctuating prices, ginger is relatively susceptible to serious disease problems, providing an ever-present possibility for a disease problem to reduce yields sharply (Nishina et al., 1992). A sustainable ginger economy is possible only when these risks are minimized.

Along with price risk, cash-flow implications are the perceived crop risk for a crop such as ginger. This is related to age to first bearing and longevity of the crop. Production and marketing risks are greater the longer the crop takes to bear and the greater the life of the crop. The length of the harvest period also has its risks: the longer the harvest period, the greater the risk of failure. Vinning (1990), in the Australian Centre for International Agricultural Research (AICIAR) technical report for marketing perspectives on a "Potential Pacific Spice Industry," has given crop failure ratings for different spice crops, based on the above points. It was found that ginger topped the list as a "high risk" commodity, followed by vanilla.

The ginger industry is facing risk and uncertainty in different forms. Each country has to face considerable competition from other ginger-producing countries because of many new countries entering into the ginger industry in recent years. Over the years, India has lost its market to China and Indonesia, as is the case in the case of cardamom, where India lost the market to Guatemala. This is primarily because of the competitive price edge in the case of Chinese and Indonesian ginger as compared to India's. From the point of view of Indian farmers, the prices have been generally good during the past decade, although there was a drastic reduction in 1996 and 1997 and 2001 and 2002. During 1999–2000, ginger farmers received an all-time high price, which was more than double the price in the preceding year. The price was always above the break-even point, with the average of US 10 cents/kg for fresh ginger in the northeastern states, where fresh ginger is marketed, thus leading to profitable ginger farming.

The price for dry ginger was well below the break-even point in the 1980s and in the early 1990s, as well. From 1982 to 1984 and 1993 to 1995, the price almost doubled. In addition to this abrupt fluctuation in price, the ginger crop is also highly susceptible to serious disease problems, leading to a reduction in rhizome yield, and an unmarketable production. At times, the farmer might lose up to 80% of the crop toward harvest time. Thus, the ginger crop industry is influenced by the risk factors associated with yield and seasonal price fluctuations, though these factors seem unrelated on a long-term basis. However, the ANOVA indicates that the price variability of ginger is greater than the yield variability.

Some of the policy measures appropriate to the ginger industry are the following:

1. Healthy seed production through the "Seed Village Concept" by regular field monitoring for diseases and pests, and enforcing seed certification measures.
2. Impose quarantine measures to restrict free and uncontrolled movement from one place to another, especially where bacterial wilt is endemic.
3. An integral approach to control the most serious problem in ginger cultivation—the prevalence of rhizome rot. This could be further complicated by bacterial, fungal, and insect attack on the ginger plant.
4. High fiber content and high cost of production are the deterrents which make Indian ginger uncompetitive, both qualitywise and pricewise. Ginger breeding must focus especially on the former, while ginger agronomy should focus on the latter.
5. Develop cropping systems most suited to ginger as an intercrop, especially in coconut and arecanut gardens. In developing such cropping systems, adequate attention must be given to the cost–benefit ratio.
6. Value addition is a key component in making the ginger market attractive and profitable. The global scenario, where reexport from EU countries show that India has to learn how best to exploit this avenue.
7. Since the importing countries, especially Japan, the United States, and EU countries are highly quality conscious, there is an imminent need to focus more on postharvest measures, where the end produce is free of all extraneous materials, including pesticide residue. Good and attractive packing makes the ginger market more consumer friendly.
8. Indian farmers need to be urgently educated on various aspects of ginger production, right from sowing to harvest and proper postharvest technology. As of now, many are not properly oriented, neither is there a concerted effort to bring them up to international standards.
9. High-value organic farming is an emerging market. There is a great urgency to popularize the concept among as many ginger farmers as possible to reap the full benefit of global ginger marketing from this segment.
10. There should be no mistaking that value addition confines only to the manufacture and sale of ginger derivatives, such as oleoresins and volatile oils. One EU document reveals that in the export market, "Buyers are looking for clean, well flavored, artificially dried product with high hygiene levels, in contrast to the bulk of the materials which has been sundried on the ground" (Commonwealth Secretariat, 1996, p. 45).

For the successful implementation of the above-mentioned policy-related suggestions, there is an imminent need to develop a special database regarding all aspects of ginger-based crop production, which include marketing, employment potential, production techniques, cost of cultivation, and value addition. This in turn, will help in creating decision support systems to benefit the stakeholders, both producer and consumer.

References

Anonymous, 1996. Guidelines for Exporters of Spices to the European Market Commonwealth Secretariat, London UK.

DASD, 2002. Arecanut and Spcices Database. Directorate of Arecanut and Spices Development (DASD), Calicut, Kerala State.

Datta, S.K., Singh, G., Chakrabarti, M., 2003. Ginger and its products: management of marketing and exports with special reference to the eastern Himalayan region. National Consultative Meeting for Improvement in Productivity and Utilization of Ginger, 12–13 May 2003, Aizwal, Mizoram, India.

FAO, 2003. FAOSTAT, FAO, Rome <www.Faostat.org/>.

Fleming, K., Sato, K., 1998. Economics of ginger root production in Hawaii. Agribusiness December 1998 AB-12 College of Tropical Agriculture and Human Resources, University of Hawaii.

Gaikwad, S.H., Thorve, P.V., Bhole, B.D., 1998. Economics of ginger (*Zingiber officinale* Rosc.) production in Amravati District (Maharashtra, India). J. Spices Aromat. Crops 7 (1), 7–11.

ITC, 1995. Market Development: Market brief on Ginger. International Trade Centre, UNCTAD/WTO, Rome, Italy.

Jaiswal, P.C., 1980. Handbook of Agriculture. ICAR, New Delhi, pp. 1179–1184.

John, K., 2003. Organic agriculture and possibilities for organic production of ginger in Mizoram. National Consultative Meeting for Improvement in Productivity and Utilization of Ginger, 12–13 May 2003, Aizwal, Mizoram, India.

Kithu, C.H., 2003. Marketing and export prospects of ginger. National Consultative Meeting for Improvement in Productivity and Utilization of Ginger, 12–13 May 2003, Aizwal, Mizoram, India.

Korikanthimath, V.S., Govardhan, R., 2001a. Resource productivity in ginger (*Zingiber officinale*) cultivation in paddy fields and upland situations in Karnataka. Indian J. Agron. 46 (2), 368–371.

Korikanthimath, V.S., Govardhan, R., 2001b. Comparative economics of ginger (*Zingiber officinale* Rosc.) cultivation in paddy fields and uplands (open, vacant areas). Mysore J. Agric. Sci. 34, 368–371.

Librero, A.R., Nineveth, E.E., Myrna, B.O., 1988. Estimating Returns to Research Investment in Mango in the Philippines. Phillippino Council of Agriculture, Forestry and Natural Resources Research and Development, Los Banos, Laguna, Philippines.

Nishina, M.S., Sato, D.M., Nishijima, W.T., Mau, R.F.I., 1992. Ginger root production in Hawaii Commodity fact sheet GIN-3(A). College of Tropical Agriculture and Human Resources, University of Hawaii.

Pruthi, J.S., 1989. Post-harvest technology of spices and condiments—pretreatments, curing, cleaning, grading and packing. Report of the Second Meeting of International Spice Group, 6–11 March 1989, Singapore.

Regeena, S., Kandaswamy, A., 1987. Economics of ginger cultivation in Kerala. S. Indian Hortic. 40 (1), 53–56.

Selvan, T. Thomas, K.G., 2003. Development of ginger in India. National Consultative meeting for Improvement in Productivity and Utilization of Ginger, 12–13 May 2003, Aizwal, Mizoram, India.

Spices Board, 1988. Report on the Domestic Survey of Spices (Part 1). Spices Board, Ministry of Commerce, Government of India, Cochin, Kerala.

Spices Board, 1990. Production and Export of Ginger Status. Spices Board, Cochin, Kerala State.

Spices Board, 1992. Report of the Forum for Increasing Export of Spices. Spices Board, Ministry of Commerce, Government of India, Cochin, Kerala State.

Spices Board, 2008. Report on the Domestic Survey of Spices (Part I). Spices Board, Ministry of Commerce, Government of India, Cochin, Kerala State.

Sreekumar, M.M. Arumugham, C., 2003. Technology of fresh ginger processing for ginger oil and oleoresin. National Consultative Meeting of Improvement in Productivity and Utilization of Ginger, 12–13 May 2003, Aizwal, Mizoram, India.

Vinning, G., 1990. Marketing Perspectives on a Potential Pacific Spice Industry. Australian Centre for International Agricultural Research, Canberra, Australia, (ACIAR Technical Reports No 15. p. 60).

25 Pharmacology and Nutraceutical Uses of Ginger

Ginger rhizomes have been widely used as a cooking spice and herbal remedy to treat a variety of human conditions, diseases, or otherwise. Fresh and dried ginger is used for different clinical purposes in traditional Chinese medicine (*Kampo*). Fresh ginger (*Zingiberis Recens Rhizoma*; *Sheng Jiang* in Chinese, *Shoga* in Japanese) is used as an antiemetic, antitussive, or expectorant, and is used to induce perspiration and dispel cold, whereas dried ginger (*Zingiberis Rhizoma*; *Gan Jiang* in Chinese) is used for stomachache, vomiting, and diarrhea accompanied by cold extremities and faint pulse (Benskey and Gamble, 1986). Dried ginger, either simply dried in the shade (*Gan Sheng Jiang*, or simply *Gan Jiang* in Chinese, *Shokyo* in Japanese) or processed ginger that is heated in pans with hot sand (*Rhizoma Zingiberis Preparata*, *Pao Jiang* in Chinese) is often used in China. The simply dried ginger and the processed ginger are not clearly differentiated in clinical use. On the other hand, different types of dried gingers have been used in traditional Japanese medicine, such as dried (*Shokyo* in Japanese, as cited above) and steamed and dried ginger (*Zingiberis Siccatum Rhizoma*, *Kankyo* in Japanese). Steamed and dried ginger are rarely used in traditional Chinese medicine. Here we describe the "simply dried ginger" as "dried ginger" (*Shokyo*) and the "steamed and dried ginger" as steamed ginger (*Kankyo*). Gingerols and shogaols are identified as the main components of dried ginger (*Shokyo*) and steamed ginger (*Kankyo*), respectively. However, before this study was conducted, only little was known about the scientific reasons why *Shokyo* and *Kankyo* were used for different clinical purposes.

The juice from freshly squeezed ginger (containing gingerols) has been reported to be hypoglycemic in diabetic rats. The diabetic state alters the microvascular function and affects the synthesis of prostacyclin, thromboxane, and leukotrienes. Similarly, the gingerols have been reported to inhibit both cyclooxygenase and lipoxygenase and to diminish the production of prostaglandins and leukotrienes. The chemical structures of gingerols are similar in part to those of prostaglandins.

The therapeutic application of gingerol in the diabetic state (i.e., gingerol lowering the blood glucose level) is an area of much interest. A lot of investigations have been carried out in this connection, using mice as test animals. However, data from investigations in the case of humans is rare. Recent research at the University of Sydney found that ginger has the power to control blood glucose by using muscle cells. Professor Basil Roufogalis, specializing in pharmaceutical chemistry, who led the research, said, "Ginger extracts obtained from Buderim Ginger were able to increase the uptake of glucose into muscle cells independently of insulin. This assists in the management of high levels of blood sugar that create complications for long-term diabetic patients, and may allow cells to operate independently of insulin." He

The Agronomy and Economy of Turmeric and Ginger. DOI: http://dx.doi.org/10.1016/B978-0-12-394801-4.00025-9

also added, "The components responsible for the increase in glucose were gingerols, the major phenolic components of the ginger rhizome. Under normal conditions, blood glucose level is strictly maintained within a narrow range, and skeletal muscle is a major site of glucose clearance in the body." This investigation was published recently in the reputable publication *Planta Medica* (Business Line, August 8, 2012).

Ethanolic extract of ginger fed orally produced significant antihyperglycemic effect in diabetic rats, lowered serum total cholesterol, triglycerides, and increased the HDL-cholesterol levels. Ginger extract treatment lowered the liver and pancreas thiobarbituric acid reactive substances (TBARS) values compared to pathogenic diabetic rats. The results of the test drug were comparable to gliclazide, a standard antihyperglycemic agent, indicating that the ethanolic extract of ginger can protect the tissues from lipid peroxidation. The extract also exhibited significant lipid-lowering activity in diabetic rats (Bhandari et al., 2005).

Treatment with extracts of ginger increased the lowered levels of RBC and WBC counts and PCV in diabetic rats, and it also decreased the elevated levels of platelets and glucose concentration of diabetic rats. Thus, oral ginger administration might decrease the diabetes-induced disturbances of some hematologic parameters in alloxan-induced diabetic rats (Olayaki et al., 2007).

Antimicrobial Property of Ginger

The essential oil of ginger showed antimicrobial activity against all Gram-positive and Gram-negative bacteria tested, as well as against yeasts and filamentous fungi, using the agar diffusion method (Martins et al., 2001). Ginger oil extracts, especially fresh extracts, exhibited significant activity against *Salmonella typhii* (Chaisawadi et al., 2005).

Antifungal Activity of Ginger

There are but very few reports on antifungal proteins from rhizomes, and there is none from the family Zingiberaceae. An antifungal protein with a novel N-terminal sequence was isolated from ginger rhizomes. It exhibited an apparent molecular mass of 32 kDa and exerted antifungal activity toward various fungi including *Botrytis cinerea, F. oxysporum, Mycosphaerella arachidicola*, and *Physalospora piricola* (Wang and Ng, 2005). Endo et al. (1990) elucidated the structures of antifungal diarylheptenones, gingerenones A, B, C, and isogingerenone B, which were isolated from the rhizomes. The essential oils of ginger exhibited high activity against the yeasts *Saccharomyces cerevisiae, Cryptococcus neoformans, Candida albicans, C. tropicalis*, and *Torulopsis glabrata* (Jantan et al., 2003).

Gingerols, the pungent constituents of ginger, were assessed as agonists of the capsaicin-activated vanilloid receptor (VR1). [6]-Gingerol and [8]-gingerol evoked capsaicin-like intracellular Ca^{2+} transients and ion currents in cultured dorsal root ganglion neurons. These effects of gingerols were blocked by capsazepine, the VR1 receptor antagonist. The potency of gingerols increased with increasing size of the side chain and with the overall hydrophobicity in the series. It is concluded that gingerols represent a novel class of naturally occurring VRI receptor agonists that may contribute to the medicinal properties of ginger, which have been known for centuries. The gingerol structure may be used as a template for the development of drugs acting as moderately potent activators of the VRI receptor (Dedov et al., 2002).

Larvicidal Activity of Ginger

Essential oil of ginger served as a potential larvicidal and repellant agent against the filarial vector *Culex quinquefasciatus* (Pushpanathan et al., 2008).

Anthelmintic Activity of Ginger

Crude powder and crude aqueous extract of dried ginger exhibited a dose- and a time-dependent anthelmintic effect in sheep, naturally infected with mixed species of gastrointestinal nematodes, thus justifying the age-old traditional use of this plant in helminth infestation (Iqbal et al., 2006).

Insecticidal Property of Ginger

Vacuum-distilled ginger extracts are repellents toward the adult maize weevil, *Sitophilus zeamais*, in both the absence and the presence of maize grains (*Zea mays*). Fractions containing oxygenated compounds accounted for the repellent activity; three major compounds identified in the behaviorally active functions to be 1,8-cineole, neral, and geranial, in a ratio of 5.48:1:2.13, respectively. The report by Ukeh et al. (2009) provides the scientific basis for the observed repellent properties of ginger and demonstrates the potential for their use in stored-product protection at the small-scale farmer level.

Pharmacological Studies on Ginger Extracts and Active Components

In vitro studies have demonstrated that an aqueous extract of fresh ginger inhibits the activities of cyclooxygenase, as a result of which, it inhibits arachidonic acid metabolism and platelet aggregation (Srivastava, 1984). Oral administration of the acetone extract of dried ginger promotes gastrointestinal motility in rats (Mustafa and Srivastava, 1990). In a Danish study, blood thromboxane B_2 levels were lowered after consumption of fresh ginger, an effect which must be due to inhibition of cyclooxygenase by the active components of fresh ginger.

Medicinal Uses of Ginger

The past decade has witnessed a considerable increase in interest in the use of various traditional herbs and plant extracts in primary health care and conventional medicine. They form part of traditional systems of medicine, and a vast body of anecdotal evidence exists supporting their use and efficiency. Some of these traditional medicines (especially in the Chinese system of medicine) have stood the test of time and have been substantiated by modern clinical investigations and interpretations, whereas the so-called miracle cures of others have been either disproved or not substantiated. There is, evidently, a lack of scientific data from well-planned clinical trials, and the situation is further complicated by the fact that the herbs are almost always used as complex polyherbal mixtures. Among the herbal drugs, one raw drug that has undergone considerable study is ginger. Ginger is perhaps most widely used in the Indian system of medicine known as *Ayurveda*. Ginger is also a very important drug in both the Chinese and the Japanese systems of medicine.

Ginger in Indian System of Medicine

In the *Ayurvedic* system of medicine, both fresh and dry ginger are used. Ginger has been widely used as a common household remedy for various illnesses since ancient times. The properties and uses of ginger in *Ayurvedic* medicine are available from authentic ancient treatises like *Charaka Samhitha* and *Susrutha Samhitha*, which are the basics for this system. Descriptions of ginger are available from similar documents of Chinese and Sanskrit literature written in the subsequent centuries. Dry ginger seems to be an essential ingredient in several *Ayurvedic* preparations, and hence, ginger is called *Mahaoushadha*, the great cure. This emphasizes the extensive usage of ginger in *Ayurveda*.

Fresh and dry ginger in *Ayurveda* are used in the following ways:

1. As a single medicine for internal use
2. As an ingredient in compound medicines
3. For external use
4. As an adjuvant
5. As an antidote
6. For the purification of some mineral drugs.

In Sanskrit literature, ginger has several synonymous words, which are indicative of its properties, and in *Ayurveda* different terms are used to denote the usage of ginger in different contexts. The following synonyms are quoted by Aiyer and Kolammal (1966):

1. *Ardrika, Ardraka*: that which waters the tongue and also shows the relation to the star Ardra
2. *Sringivera, Sringa, Sringika*: that which resembles the shape of the horns of animals
3. *Chatra, Rahuchatra*: that which dispels diseases
4. *Sunti, Kaphahari, Soshana*: that which overcomes diseases due to kapha (phlegm)
5. *Mahaushadha*: the great cure

6. *Viswa, Viswabeshaja*: universal remedy
7. *Nagara*: that which is commonly found in towns
8. *Katubhadra*: drug that has a pungent taste, which is capable of bringing goodness
9. *Katootkata, Katuka, Katu, Ushna, Katigranthi*: drug that has a pungent taste
10. *Gulma moola*: rhizome (root) which is generally spongy in nature
11. *Saikatesta*: that which grows generally in sandy places
12. *Anoopaja*: the plant that requires plenty of water for its growth.

Properties

Fresh and dry ginger are similar in their properties. The only difference is that fresh ginger is not so easily digested, while the dried one is (Aiyer and Kolammal, 1966). All the *Ayurvedic* classics like *Charaka Samhita, Susrutha Samhitha, Ashtanga Hridaya*, and *Nighantus* give the same properties for ginger as shown below:

Rasa (taste)—*Katu* (pungent)
Guna (property)—*Laghu, Snigdha* (light and unctuous)
Veerya (potency)—*Ushna* (hot)
Vipaka (metabolic property)—*Madhura* (sweet).

Ginger rhizome has a pungent taste and is considered to be converted to sweet products after metabolic changes. Being hot and light, ginger is easily digestible. It has an unctuous quality. In *Bhavaprakasa* (an important and ancient text in *Ayurveda*, authored by Bhavamisra), fresh ginger is called *rooksha*, meaning dry. It acts as an appetizer, carminative, and stomachic. Ginger is acrid, anodyne, antirheumatic, antiphlegmatic, diuretic, aphrodisiac, and cordial. It has anti-inflammatory and antiedematous action according to *Dhanwantary nighantu*, yet another ancient text on *Ayurveda*, written by *Rishi* (meaning *Sage* in the ancient Indian language, Sanskrit) *Dhanwantari*. It cleanses the throat, is good for the voice (corrective of larynx affections), reduces vomiting, relieves flatulence and constipation, and relieves neck pain (*Saligrama nighantu*, yet another ancient *Ayurvedic* text). Due to its hot property, ginger is capable of causing dryness and is thus antidiarrheal in effect. *Bhavaprakasa* specifically emphasizes the antiarthritic and antifilarial effects of dry ginger. It is also good for asthma, bronchitis, piles, eructation, and ascites.

Kittikar and Basu (1935) mentioned a remedy for cough in which fresh ginger is made into pills along with honey and ghee, and taken in a dose of four pills a day. It is applied externally to boils and enlarged glands, and internally as a tonic in Cambodia (Kittikar and Basu, 1935). The outer skin of ginger is used as a carminative and is said to be a remedy for opacity of the cornea. In acute ascites with dropsy arising from liver cirrhosis, complete subsidence by the use of fresh ginger juice is reported. The juice also acts as a strong diuretic (Kittikar and Basu, 1935).

Ginger strengthens memory, according to Nadkarni (1976), and removes obstructions in the blood vessels, incontinence of urine, and nervous diseases. Dry ginger paste with water is effective in remedying the adverse effects of fainting and helps the patient recover fast. This paste can also be applied to the eye lids, and the ginger powder can also be used as a snuff.

Bhaishajya ratnavali (another ancient *Ayurvedic* text) gives an important combination of dry ginger, rock salt, long pepper, and black pepper, powdered and mixed with fresh ginger juice, to be gargled after warming, as a specific drug for phlegmatic affections of the heart, head, neck, and thoracic region. It is very effective for all types of severe fevers and their associated symptoms. Ginger is made use of in veterinary science as a stimulant and carminative in indigestion in horses and cattle, in spasmodic colic of horses, and to prevent gripping by purgatives (Pruthi, 1979). The ginger sprouts and shoots do not have any conspicuous taste and are said to aggravate*Vata* and *Kapha* (*Saligrama nighantu*)[1] (Aiyer and Kolammal, 1966).

Indications

Apart from its extensive use as a spice, ginger plays an important and unavoidable role in traditional medicine, with a wide range of indications. On account of its carminative, stimulant, and digestive properties, ginger wet or dry is commonly used in fever, anorexia, cough, dyspnea, vomiting, cardiac complications, constipation, flatulence, colic, swelling, elephantiasis, dysuria, diarrhea, cholera, dyspepsia, diabetes, tympanitis, neurological disorders, rheumatism, arthritis, and inflammation of liver. It is also indicated in all the phlegmatic conditions and respiratory problems, such as asthma and cough.

Contraindications

In diseases such as leprosy, anemia, leukoderma, painful micturition, hematemesis, ulcers, and fevers of *Pitha* predominance and in hot seasons, wet or dry ginger is not indicated (*Bhavaprakasa* and *Saligrama nighantu*).

Experimental and Clinical Investigations

Effect on Digestive System

Goso et al. (1996) investigated the effect of ginger on gastric mucin against ethanol-induced gastric injury in rats and found that the oral administration of ginger significantly prevented gastric mucosal damage. Patel (1996) and Patel and Srinivasan (2000) investigated the influence of dietary spice on digestive enzymes experimentally.

---

[1] According to the Indian *Ayurvedic system* of medicine, all physiological functions of the human body are governed by three basic biological parameters: *Thridoshas* or the three basic qualities, namely, *vata, pitta,* and *kapha (kafa)*. *Vata* is responsible for all voluntary and involuntary movements in the human body; *pitta* is responsible for all digestive and metabolic activities; and *kapha* provides the static energy (strength) for holding tissues together, and also provides lubrications at various joints of friction. When these three qualities (*doshas*) are in a normal state of equilibrium, the human body is healthy and sound, but when they lose the equilibrium and become vitiated, by varying internal and external factors, they produce varied diseases. *Ayurvedic* treatment of any disease is aimed at restoring the basic equilibrium of the three *doshas* or qualities.

Dietary ginger prominently enhanced the secretion of saliva and intestinal lipase activity by chymotrypsin and pancreatic amylase as well as the disaccharides sucrose and maltose. The positive influence of this terminal enzyme of the digestive process could be an additional feature of this spice to stimulate digestion. Ginger contains proteolytic enzymes which promote the digestive process and also enhance the action of the gall bladder and protects the liver against toxins (Yamahara et al., 1990).

Yoshikawa et al. (1994) analyzed ginger for its stomachic principles. An antiulcer constituent (6-gingesulfonic acid), three monoacyl digalactosyl glycerols (gingeroglycolipids A, B, C), and (+)-angelicoidenol (2-O-β-D-glucopyranoside) were isolated.

Cardiovascular and Related Actions

Suekawa et al. (1984) reported that gingerol and shogaol present in ginger juice cause vagal stimulation leading to a decrease in both the blood pressure and heart rate. Janssen et al. (1996) studied the effect of the dietary consumption of ginger on platelet thromboxane production in humans. The result of the clinical assay of the raw and cooked ginger does not support the hypothesis on the antithrombolic activity of ginger in humans.

Lamb (1994) investigated the effect of dried ginger on human platelet function, thrombogen, and hemostasis. The use of ginger as an antiemetic in the preoperative period has been criticized because of its effect on thromboxane synthetase activity and platelet aggregation. When administered to the elderly volunteer, ginger had no dose-dependent effect on thromboxane synthetase activity or such an effect only occurs in the fresh state of the product, that is, the ginger. However, in a previous investigation involving 10 male healthy volunteers, it was shown that 5 g of ginger taken with a high-fat meal for seven days was able to inhibit significantly the enhanced tendency to platelet aggregation normally seen after a high-fat food intake (Verma et al., 1993). It has also been reported earlier that ginger, in addition to inhibiting platelet aggregation, also reduces platelet thromboxane synthesis in both *in vivo* and *in vitro* conditions (Srivastava, 1989). However, this effect on *in vivo* conditions was seen after consumption of 5 g of ginger/day for seven days. It is unknown whether this positive effect would also be seen under normal patterns of food consumption. It is unlikely that 5 g of ginger/day would be part of a normal consumption pattern, and this amount is far in excess of what is currently available in ginger-containing food products.

Bordia et al. (1997) showed that the dose of 10 g powdered ginger administered to patients suffering from coronary artery disease produced a significant reduction in ADP- and epinephrine-induced platelet aggregation. An aqueous extract of ginger has strong anticlotting properties. Some components present in ginger have been shown to prevent blood clotting through physiological changes exerted on the arachidonic acid metabolism and cicosanoid metabolism (Srivastava, 1986).

Antiemetic and Antinauseant Properties

One of the best-known and best-studied areas of ginger use is for the treatment of various forms of nausea. Many animal and clinical trials have been conducted to investigate the use of ginger in preventing nausea of various types. Arfeen et al. (1995) carried out a double-blind randomized clinical trial to investigate the effect of ginger on the nausea and vomiting following gynecological laparoscopic surgery. Both 0.5 and 1.0 g rates of ginger were effective in reducing vomiting. Phillips et al. (1993) reported that ginger was as effective as metoclopromide in reducing postoperative nausea and vomiting.

Suekawa et al. (1984) reported that 6-gingerol and 6-shogaol suppressed gastric contraction, but increased gastrointestinal motility and spontaneous peristaltic activity in laboratory animals. However, these positive effects were observed only when ginger was administered orally or intravenously. This suggests that direct contact with the intestinal mucosa and not delivery by the blood is required for the action of ginger; possibly, the hepatic metabolism is involved in the action. Treatment of morning sickness, where ginger is used, is a widely researched topic. Fischer-Rasmussen et al. (1991), in a double-blind randomized crossover trial, found that 1 g/day of ginger was effective in reducing the symptoms of morning sickness and did not appear to have any adverse side effects on pregnancy.

Recently, Keating and Chez (2003) administered ginger syrup in water to study the ameliorating effect of ginger on nausea in early pregnancy. This double-blind study showed a positive improvement in 77% of the cases investigated. The authors concluded that 1 g ginger in syrup for in a divided dose daily is useful in some patients who experience nausea and vomiting during the first trimester of pregnancy. Fulder and Tenne (1996) reported that ginger is an over-the-counter medicine recommended to manage pregnancy-related nausea in many Western countries.

Ernst and Pittler (2000) carried out a systematic review of the evidence accumulated from randomized clinical trials on the effect of ginger in treating nausea and vomiting. They observed that only six trials satisfied all the experimental conditions. Among the three trials on postoperative nausea and vomiting, only two indicated that ginger was superior to placebo and equally effective as metoclopramide. The pooled information indicated only a statistically nonsignificant difference between ginger and placebo groups when 1 g ginger was taken prior to surgery. One study was found for each of the following conditions: seasickness, morning sickness, and chemotherapy-induced nausea. These studied collectively favored ginger than placebo.

Ginger is also reported to be an effective remedy for travel and morning sickness. Mowrey and Clayson (1982) have conducted the best-known experiments in this area. Among 39 men and women investigated, who reported a high susceptibility to motion sickness, the following observations were made: motion sickness was induced when subjected to a rotating chair, while blindfolded, under controlled conditions. It was found that ginger was significantly more effective in reducing motion sickness than antihistaminic dimenhydrinate and a placebo.

A Danish controlled trial on seasickness involved 80 naval cadets who were unaccustomed to sailing in rough seas. The subjects reported that ginger consumption reduced the tendency to vomiting and cold sweating better than the placebo.

The general hypothesis of the mode of action of ginger is that it ameliorates nausea associated with motion sickness by preventing the development of gastric dysrhythmias and the elevation of plasma vasopressin. Ginger also prolonged the latency before nausea onset and shortened the recovery time. Hence, ginger is recommended as a novel agent in the prevention and treatment of motion sickness (Holtmann et al., 1989).

Anti-Inflammatory Properties: Effect on Rheumatoid Arthritis and Musculoskeletal Disorders

In the Indian system of medicine (*Ayurveda*), ginger is used as an anti-inflammatory drug. It has been suggested that ginger may be useful as a remedy for arthritis, and a number of commercial preparations are available for this use. For instance, Bio-organics Arthri-Eze Forte (Bullivants, Australia) and Extralife Artri-care (Felton Grimwade & Brickford Ltd, Australia) are marketed as arthritis treatments and contain 500 mg dried, powdered ginger rhizome. Srivastava and Mustafa (1989) reported that more than 75% of patients receiving 3–7 g of powdered ginger a day for 56 days had a significant reduction in pain and swelling associated with either rheumatoid or osteoarthritis. Adverse effects have not been reported. These results prove the anti-inflammatory properties of ginger. Follow-up studies carried out (from three months to 2.5 years) in patients using 0.5–1 g/day exhibited significant reduction in pain and swelling to the extent of 75% and 100%, respectively, of arthritis (rheumatoid arthritis and osteoarthritis) and muscular discomfort. The World Health Organization (WHO) document 2000 reports that 5–10% ginger extract administration brought about full or partial pain relief, or recovery of joint function and a decrease in swelling in patients with chronic rheumatic pain and also lower back pain.

Kishore and Pande (1980) clinically evaluated the effects of a ginger and *Tinospora cordifolia* combination in rheumatoid arthritis. The combination showed better results compared to other traditional medicines. The antiarthritic effect of ginger and eugenol was studied by Sharma et al. (1997), who reported that ginger significantly suppressed the development of adjuvant arthritis.

Bliddal et al. (2000) carried out a randomized, placebo-controlled, crossover study on the effects of ginger extract and ibuprofen (a commonly prescribed pain-killer in patients with common arthritic/back pain problems) to study their effect on osteoarthritis. Ginger extract showed significant positive effect in pain reduction.

A common adverse side effect of treating inflammation with modern allopathic drugs is that it leads to ulcer formation in the intestines due to acidity formation. This can lead to ulcer formation in the digestive tract. Ginger not only prevents these symptoms of inflammation but also prevents ulcer formation in the digestive tract (Anonymous, 2003).

Chemoprotective Properties

There is considerable emphasis on identifying potential chemoprotective agents present in foods consumed by the human population. In prior *in vitro* studies, water or organic solvent extract of ginger was shown to possess antioxidative and anti-inflammatory properties. Sharma and Gupta (1998) investigated the effect of ginger in reversing the delay in gastric emptying caused by the anticancer drug Cisplatin.

Cisplatin causes nausea, vomiting, and inhibition of gastric emptying. Acetone and 50% ethanolic extracts of ginger (100, 200, and 500 mg/kg p.o.) and ginger juice (2 and 4 ml/kg body weight) were investigated against Cisplatin effects on gastric emptying in rats. Ginger administration significantly reversed Cisplatin-induced delay in gastric emptying. The ginger juice and acetone extracts were more effective than the 50% ethanolic extract. The reversal produced by the ginger acetone extract was similar to that caused by the 5-HT$_3$ receptor antagonist ondansetron; however, ginger produced better reversal than ondansetron. These investigators suggested that ginger, used as an antiemetic in cancer therapy, may also be useful in improving the gastrointestinal side effects of cancer chemotherapy.

Chih Peng et al. (1995) showed that the extract of dried ginger rhizome exhibited biphasic effects on the secretion of cytokines by human peripheral blood mononuclear cells *in vitro*. The stimulatory effect of the extract on cytokine secretion was shown to be time dependent; a significant increase in the secretion of cytokines was noted in the presence of low doses of ginger extract (10–30 mg/ml) 18–24 h after administration. At a higher concentration of the extract, cytokine production was suppressed.

Kim et al. (2002) found that the four types of shogaols from ginger protect IMR 32 neuroblastoma and normal human umbilical vein endothelial cells from β-amyloid at EC50 = 4.5–8.0 μM/l. The efficacy of cell protection from β-amyloid insult by the shogaols was shown to improve as the length of the side chain increases.

Hypolipidemic Effect

Sharma et al. (1996) studied hypolipidemic and antiatherosclerotic effects of *Z. officinale* in cholesterol-fed rabbits. The administration of GE increased the fecal excretion of cholesterol, thus suggesting a modulation of absorption; the treatment reduced total serum cholesterol and low-density cholesterol (LDC) levels. The atherogenic induct was reduced from 4.7 to 1.12.

Lamb (1994) investigated the effect of dried ginger on human platelet function, thrombogen, and hemostasis. The use of ginger as an antiemetic in the preoperative period has been criticized because of its effect on thromboxane synthetase activity and platelet aggregation. Bordia et al. (1997) studied the effect of ginger and fenugreek on blood lipids, blood sugar, and platelet aggregation in patients with coronary artery disease (CAD). A single dose of ginger powder (10 g) administered to CAD patients significantly reduced platelet aggregation induced by ADP or epinephrine and found no effect on blood lipid or blood sugar.

Antimicrobial and Insecticidal Properties

Hiserodt et al. (1998) successfully isolated 6-, 8-, and 10-gingerol and evaluated their activity to inhibit *Mycobacterium avium* and *M. tuberculoses*. 10-Gingerol was identified as an active inhibitor of these two bacteria *in vitro*.

The protective effect of a traditional Chinese medicine, *Shigyakuto* (containing 50% ginger), against infection of herpes T virus was investigated by Ikemoto et al. (1994).

The drug was found to be protective through the activation of $CD8^+$ T cells. No virucidal or virostatic activity was observed.

The bioactivity of ginger extract on adult *Schistosoma mansoni* worms and their egg production under *in vitro* and *in vivo* conditions in laboratory mice showed the following results. Ethyl acetate extract of ginger at a concentration of 200 mg/l of extract killed almost all worms within 24 h. Male worms are more susceptible than the females. The cumulative egg output of surviving worm pairs *in vitro* was considerably reduced when exposed to the extract. After four days of exposure to 50 mg/l, the cumulative egg output was only 0.38 eggs per worm pair compared to 36.35 for untreated worms. However, *in vivo* GE did not show any significant effect on worms or their egg-laying capacity. Extract of the rhizome of *Z. corallium* was shown to be effective in killing the larvae of *S. japonicum cercaria* (Shuxuan et al., 2001).

Anxiolytic-Like Effect

A combination of ginger and *Ginkgo biloba* was studied experimentally in elevated plus maze by Hasenohrl et al. (1996). The investigation evidently proved anxiolytic effect of ginger comparable to diazepam. But the action is biphasic; in high doses, it may also have an anxiogenic effect. The known antiserotonergic action of ginger and *G. biloba* is considered to be the responsible factor for the anxiolytic-like action.

Effect on Liver

Yamaoka et al. (1996) investigated a *kampo* medicine for its action in augmenting natural killer (NK) cell activity. This medicine is used in Japan to treat chronic hepatitis, distress, and dullness in the chest and in the ribs. Ginger is one of its seven ingredients, and studies showed that extracts of ginger and *Zizyphus jujuba* and the other three components augmented NK cell activity.

Sohni and Bhatt (1996) studied the activity of a formulation in hepatic amebiasis and in immunomodulation studies. The ingredients in the formulation were *Boerhavia diffusa*, *Tinospora cordifolia*, *Berberis aristata*, *Terminalia chebula*, and *Z. officinale*. The formulation showed a maximum cure rate of 73%. Humoral immunity was enhanced. The cell-mediated immune response was stimulated as observed in the leukocyte migration inhibition test.

Other Properties

Dedov et al. (2002) showed that gingerol functions as an agonist of the capsaicin-activated VRI. 6-Gingerol and 8-gingerol evoked capsaicin-like intracellular Ca^{2+} transients and ion currents in cultured dorsal root ganglion neurons. These effects of gingerols were blocked by capsazepine, the VRI receptor antagonist. The potency of gingerols increased with the increasing size of the side chain. The authors concluded that gingerol represents a novel class of naturally occurring VR1 receptor agonists that may contribute to the medicinal properties of ginger.

Ginger has also been shown to possess an antivertigo activity similar to Dramamine (Tyler, 1996). A significant decrease in induced vertigo indicated a possible inhibitory action of ginger on the vestibular impulses to the brain (Grontved and Hentzer, 1986).

Ginger has a strong antitussive effect and helps dispel bronchial congestion. A cup of hot ginger tea containing one-fourth tablespoon of ginger powder mixed with honey relieves bronchial congestion and has been reported to be more effective, when a cough mixture proved ineffective (McCaleb, 1996). Both gingerol and shogaol possess an antitussive property; shogaol is more potent (Miller and Murray, 1998).

Toxicity

Normally, ginger is a safe drug without any adverse reactions to the human body and has a wide range of utility. Paradoxically, it is included in the list of plants containing poisonous principles (Chopra et al., 1958) because of its oxalic acid content.

Studies on ginger oleoresin on adult Swiss mice showed the following results. The oleoresin exhibited a marked action on the CNS. A single dose up to 0.5 g/kg body weight resulted in vasodilation, activeness, and alertness in animals. A dose up to 3 g/kg body weight was nonlethal, whereas doses above that resulted in mortalities, an abnormal gait associated with abdominal cramps, and gastric irritation. The LD_{50} is 6.284 g/kg body weight. Death may be due to the direct action on CNS resulting in respiratory failure as well as circulatory arrest.

Purification

Fresh and dry ginger are used as such and are generally not subjected to any purification methods. Yet, there are some references in *Ayurvedic* preparations where purification is used. Methods for purification of dry ginger and fresh ginger juice are available from *Arogyakalpadruma* (an *Ayurvedic* text which concentrates on pediatrics). Purification of ginger may therefore be intended only for pediatric use, that is, to reduce the potency and pungency for infant use.

Purification of ginger involves immersing the rhizome in lime water for an hour to an hour and a half, washing with sour gruel or sour rice washings, and drying in bright sunlight. The expressed fresh juice is to be kept undisturbed until the particles settle down. The supernatant alone is then poured into a red-hot iron vessel, which then gets purified.

Ginger in Home Remedies (Primary Health Care)

1. Decoction of dry ginger with jaggery (a dark brown form of Indian sugar sold in cubes obtained from fresh crystallized sugarcane juice) relieves dropsy (excessive accumulation of watery fluid in any of the tissues or cavities of the human body).
2. Hot decoction of dry ginger is stomachic and digestive which relieves cough, asthma, colic, and angina pectoris.

3. Ginger juice with an equal quantity of milk is indicated in ascites (abnormal accumulation of fluid in the peritoneal cavity). The *ghee* (Indian molten butter, which is a common ingredient in many vegetarian dishes of the country) prepared with 10 times the ginger juice also has a similar effect.

4. Warm juice of ginger mixed with gingelly (sesamum) oil, honey, and rock salt is a good eardrop for otalgia (pain in the ear).

5. Paste of ginger made with *Ricinus communis* (castor) root decoctions is cooked over red-hot coals after covering with mud, and the juice is collected with this special method called *Pudapaka swarasa*. This juice, if taken along with honey, cures the symptoms of rheumatic fever.

6. Juice of ginger with old jaggery cures urticaria (nettle rash) and is a digestive.

7. Ghee prepared with ginger juice, ginger paste, and milk relieves edema, sneezing, ascites, and indigestion.

8. Ginger juice along with lemon juice and mixed with little rock salt powder is effective in controlling flatulence (presence of excessive gas in the stomach and intestine), indigestion, and anorexia (having no appetite for food).

9. Dry ginger is effective in all the symptoms due to the ingestion of jackfruit.

10. Ginger immersed in lime water (CaOH) and applied to the skin can remove warts.

11. Ginger juice and clear lime water mixed and applied cures corn (a small painful growth on the sole of the foot or the toes, akin to a callus).

12. Ginger juice and honey (from *Apis indica*) in equal proportions is hypotensive in action, and of course, is excellent in relieving cough.

13. Application of ginger juice around the umbilical region is a good cure for diarrhea.

14. Purified ginger juice, onion juice, and honey in equal parts is taken at bedtime is anthelmintic in action.

15. Dry ginger pounded in milk and then the expressed juice is used as a nasal drop, which relieves headache and associated symptoms.

16. Dry ginger powder, tied in a small piece of cloth, if massaged onto the scalp after heating will cure alopecia (hair loss, a condition in which the hair falls from one or more round or oval areas, leaving the skin smooth and white) and promotes hair regrowth.

17. Dry ginger paste, taken along with milk is indicated in jaundice, and when applied to the forehead relieves headache.

18. Dry ginger boiled in buttermilk is antipoisonous and is administered for internal use.

19. Dry ginger paste taken internally with hot water and applied over the whole body is an antidote for the toxic effects of *Gloriosa* (spider lily).

20. In snake poisoning, the external application with ginger over the snake bite wound and cold body parts and drinking of ginger decoction is believed to be an effective antidote for snake bite.

21. Ginger juice is an excellent adjuvant for the medicinal preparation *Vettumaran* (an *Ayurvedic* preparation) which is indicated in such conditions as fever, chicken pox, and mumps.

22. Ginger juice is used in the purification of cinnabar (HgS) before incinerating it to lessen its toxicity and to make it biologically acceptable.

Ginger forms a component of a large variety of *Ayurvedic* preparations. However, the following precautionary measures are indicated. Ginger has *ushna* (hot) and *tikshna* (intense, pungent) attributes, and hence is contraindicated in anemia, burning sensation, calculus (a concretion formed in any part of the body, usually by compounds of salts of organic or inorganic acids), hemorrhage of liver, leprosy, and blood-related diseases. Its consumption should be reduced or avoided in the hot summer season. Green ginger should not be used for medicinal purposes according to Nadkarni (1976). Ginger is also used in homeopathy and the *Unani* system of medicine (Indian). In the former, it is used to treat albuminemia (the presence of serum albumin and serum globulin in urine), bad breath, dropsy, and retention of urine in the bladder. In the latter, ginger is used for its anthelmintic, aphrodisiac, carminative, digestive, and sedative properties and also in treating headache, lumbago, nervous diseases, pain in the human body, rheumatism, and to strengthen memory (Nadkarni, 1976).

Ginger is also used in veterinary medicine in horses and cattle for rheumatic complaints, and as an antispasmodic, and as a carminative in atonic indigestion (Blumenthal, 1998).

Ginger in Chinese and Japanese Systems of Medicine

Ginger rhizome is an important drug in the Chinese and Japanese systems of medicine (known as *Sheng Jiang* in Chinese [Mandarin] and *Shokyo* in Japanese). In fact, in the Chinese medicine, fresh and dry ginger are used for different clinical purposes. Generally, fresh ginger (*Zingiberis Recens rhizome, Sheng Jiang*) is used as an antiemetic, antitussive, or expectorant, and is used to induce perspiration and to dispel cold. Dried ginger (*Zingiber Rhizoma, Gan Jiang* in China) is used for stomachache, vomiting, and diarrhea accompanied by cold extremities and faint pulse (Benskey and Gamble, 1986). In China, ginger dried in the sun as well as heated and dried in pans with or without hot sand is used. In Japanese medicine, ginger dried in the sun (*Shokyo* in Japanese) as well as steamed and dried (*Kankyo* in Japanese) are used differently.

In Chinese *Materia Medica* (Benskey and Gamble, 1986), ginger is indicated to have the following functions and clinical uses:

1. Releases the exterior and disperses cold: used for exterior cold patterns;
2. Warms the middle burner and alleviates vomiting: used for cold in the stomach, especially when there is vomiting;
3. Disperses cold and alleviates coughing: used for cough emanating from acute wind, cold cough patterns, and chronic lung disorders and phlegm;
4. Reduces the poisonous effects of toxic herbs: used to detoxify or treat overdoses of other herbs such as, *radix, aconity carmichaeli praeparata* (Fuzi), or *Rhizoma Pinelliae Ternate (Ban Xia)*;
5. Adjusts the nutritive and protective *Qi*: used for exterior deficient patients who sweat without an improvement in their condition.

In the *Divine Husbandman's Classic of the Materia Medica* of China, ginger rhizome is indicated to have the following functions and chemical uses: vomiting, diarrhea, light-headedness, blurred vision, and numbness in the mouth, and extremities. In advanced cases, there can be premature atrial contractions, dyspnea, tremors, incontinence, stupor, and a decrease in temperature and blood pressure.

Functions and Clinical Uses

1. Warms the middle body (stomach region) and expels cold: used to warm the spleen and stomach, especially in deficiency cold patterns with such manifestations as pallor, poor appetite, cold limbs, vomiting, diarrhea, cold painful abdomen and chest, a deep, slow pulse, and a pale tongue with a moist, white coating;
2. Rescues devastated *Yang* and expels interior cold: used in patterns of devastated or deficient *Yang* with such signs as a very weak pulse and cold limbs;
3. Warms the lungs and transforms phlegm: used in cold lung patterns with expectoration of thin, watery, or white sputum;
4. Warms the channels and stops bleeding: used for deficiency cold patterns that may present with hemorrhages of various types, especially uterine bleeding. Ginger is used in hemorrhage only if the bleeding is chronic and pale in color and is accompanied by cold limbs, ashen white face, and a soggy, thin pulse.

Major Combinations

With *Radix Glycyrrhizae Uralensis* (*Gan Cao*) for epigastric pain and vomiting due to cold-deficient stomach and spleen;

With *Rhizoma Alpiniae Officinari* (*Gao Liang Jiang*) for abdominal pain and vomiting due to cold stomach;

With *Rhizoma Pinelliae Ternate* (*Ban Xia*) for vomiting due to cold-induced congested fluids. Add radix ginseng (*Ren Shen*) for vomiting due to deficiency cold;

With *Rhizoma Coptidis* (*Huang Lian*) for epigastric pain and distension, dysentery-like disorders, and indeterminate gnawing hunger. The latter is a syndrome characterized by a feeling of hunger, vague abdominal pain, or discomfort sometimes accompanied by belching, distension, and nausea, which gradually culminates in pain;

With *Cortex Magnoliae Officinalis* (*Hou Po*) for epigastric distension and pain due to cold-induced congealed fluids;

With *Rhizoma Atractyloids Macrocephalae* (*Bai Zhu*) for deficient spleen and diarrhea. If both herbs are charred, they can be used for bloody stool and excessive uterine hemorrhage;

With *Fructus Schisandrae Chinensis* (*Wu Wei Zi*) for coughing and wheezing due to congested fluids preventing the normal descent of the lungs' *Qi*;

Compared to *Rhizoma Zingiberis Officinalis Recens* (*Sheng Jiang*), *Rhizoma Zingiberis Officinalis* (*Gan Jiang*) is more effective in warming the middle burner and expelling interior cold, whereas *Rhizoma Zingiberis Officinalis Recens* (*Sheng Jiang*) promotes sweating and disperses exterior cold.

Cautions and Contraindications

1. Contraindicated in deficient *Yin* patterns with heat signs;
2. Contraindicated in reckless marauding of hot blood;
3. Use cautiously during pregnancy.

Chinese Healing with Moxibustion: The Ginger Moxa

Moxibustion is a Chinese treatment practice used along with acupuncture for conditions ranging from bronchial asthma to arthritis with amazing success. In moxibustion, the leaves of the herb (*Artmesiae vulgaris*, Chinese mugwort) are dried, rolled into pencil-like sticks, and burned, and this burning stick is used for the treatment. The ginger moxa is one type of treatment that combines the therapeutic properties of moxibustion with those of ginger. A slice of ginger, 1–2 cm thick, is cut and pierced with tiny holes. Dried mugwort leaves are then rolled into a short cone. The ginger disk is placed on the umbilicus of a patient suffering from diarrhea or abdominal pain. The moxa cone is placed on the ginger disk and then carefully lit with a small flame. The burning nugget of moxa and ginger remain on the umbilicus until the patient perspires and the area turns red. New cones are added as the original cones burn down and this continues until four to five cones are consumed. Ginger moxa also has been proven to be beneficial in the treatment of painful joints (Balfour, 2003).

Ginger in Traditional Medical Care in Other Countries

Ginger is used in primary health care in almost all ginger-growing countries. The most important use is to cure indigestion and stomachache. The expressed juice of fresh ginger mixed with sugar or honey is used widely for these purposes.

References

Aiyer, K.N., Kolammal, 1966. Pharmacognosy of Ayurvedic Drugs of Kerala, vol. 9. Department of Pharmacognosy, University of Kerala, Trivandrum, Kerala State, India.
Anonymous, 2003. The medicinal properties of ginger Buderim Ginger Consumer promotion. <http://www.buderimginger.com/> (accessed 30.07.03.).
Arfeen, Z., Owen, H., Plummer, J.L., Isley, A.H., Sorby-Adams, R.A., Doecke, C.J., 1995. A double blind randomized controlled trial on ginger for the prevention of postoperative nausea and vomiting. Anaesth. Intensive Care 23, 440–452.
Balfour, T., 2003. Chinese healing with moxibustion: burrn your ailments away. <www.balfourhealing.com/mox> (accessed 30.07.03).
Benskey, D., Gamble, A. (Eds.), 1986. Chinese Herbal Medicine: *Materia Medica* Eastland Press, Seattle, WA.
Bhandari, U., Kanojia, R., Pillai, K.K., 2005. Effect of ethanolic extract of *Zingiber officinale* on dyslipidaemia in diabetic rats. J. Ethnopharmacol. 97 (2), 227–230.
Bliddal, H., Rosetzsky, A., Schlchting, P., Weidner, M.S., Anderson, L.A., Ibfelt, H.H., et al., 2000. A randomized, placebo-controlled, cross-over study of ginger extracts and ibuprofen in osteoarthritis. Osteoarthr. Cartil. 8, 9–12.
Blumenthal, M. (Ed.), 1998. The Complete German Commission E. Monographs Therapeutic Guide to Herbal Medicine American Botany Council, Austin, TX.
Bordia, A., Verma, S.K., Srivastava, K.C., 1997. Effect of ginger (*Zingiber officinale* Rosc.) and fenugreek (*Trigonella foenumgraecum* L.) on blood lipids, blood sugar and platelet aggregation in patients with coronary artery disease. Prostaglandins Leukot. Essent. Fatty Acids 56, 379–384.
Business Line, August 8, 2012.
Chaisawadi, S., Thongbute, D., Methawiriyasilp, W., Pitakworarat, N., Chaisawadi, A., Jaturonrasamee, K., et al., 2005. Preliminary study of antimicrobial activities on medicinal herbs of Thai food ingredients. Acta Hortic. 675, 111–114.

Chih Peng, C., Jan Yi, C., Fang, Y.W, Jan Gowth, C., 1995. The effect of Chinese medicinal herb *Zingiberis rhizoma* extract on cytokine secretion by human peripheral blood mononuclear cells. J. Ethnopharmacol. 48, 13–19.

Chopra, R.N., Chopra, I.C., Handa, K.L., Kapoor, L.D., 1958. Indigenous Drugs of India. Academic Publishers, Calcutta, West Bengal State, India.

Dedov, V.N., Tran, V.H., Duke, C.C., Connor, M., Christie, M.J., Mandadi, S., et al., 2002. Gingerols: a novel class of vanilloid receptor (VRI) agonists. Br. J. Pharmacol. 137 (6), 793–798.

Endo, K., Kanno, E., Oshima, Y., 1990. Structures of antifungal diarylheptenones, gingerenones A, B, C and isogingerenone B, isolated from the rhizomes of *Zingiber officinale*. Phytochemistry 29 (3), 797–799.

Ernst, E., Pittler, M.H., 2000. Efficiency of ginger for nausea and vomiting: a systematic review of randomized clinical trials. Br. J. Anaesth. 84, 367–377.

Fischer-Rasmussen, W., Kajaer, S.K., Dahl, C., Aspong, U., 1991. Ginger treatment of hyperemis gravidarum. Eur. J. Obstet. Gynecol. Reprod. Biol. 38, 19–24.

Fulder, S., Tenne, M., 1996. Ginger as an anti-nausea remedy in pregnancy—the issue of safety. Herbalgram 38, 47–50.

Goso, Y., Ogara, Y., Ishikara, K., Hotta, K., 1996. Effects of traditional herbal medicine on gastric mucin against ethanol-induced gastric injury in rats. Comp. Biochem. Physiol. Pharmacol. Toxicol. Endocrinol. 113, 17–21.

Grontved, A., Hentzer, E., 1986. Vertigo-reducing effect of ginger root. J. Otorhinolaryngol. Relat. Spec. 48, 282–286.

Hasenohrl, R.U., Nichau, C., Frisch, C., de Souza Silva, M.A., Huston, J.P., Mattern, C.M., et al., 1996. Anxiolytic-like effect of combined extracts of *Zingiber officinale* and *Gingko biloba* in the elevated plus-maze. Pharmacol. Biochem. Behav. 53 (2), 271–275.

Hiserodd, R.D., Franzbleau, S.G., Rosen, R.T., 1998. Isolation of 6-, 8-, and 10-gingerol from ginger rhizome by HPLC and preliminary evaluation of inhibition of *Mycobacterium avium and Mycobacterium tuberculosis*. J. Agric. Food Chem. 46, 504–508.

Holtmann, S., Clarke, A.H., Scherer, H., Hohm, M., 1989. The automotion sickness mechanism of ginger, a comparative study. Acta Otolaryngol. 3–4, 168–174.

Ikemoto, U.T., Utsunomiya, T., Ball, M.A., Kobayashi, M., Pollard, R.B., Suzuki, F., 1994. Protective effect of *shigyaku*-to, a traditional Chinese herbal medicine on the infection of herpes simplex virus type-1 (HSV-1) in mice. Experientia 50, 456–460.

Iqbal, Z., Lateef, M., Akhtar, M.S., Ghatyur, M.N., Gilani, A.H., 2006. *In vivo* anthelmintic activity of ginger against gastrointestinal nematodes of sheep. J. Ethnopharmacol. 106 (2), 285–287.

Janssen, P., Meyboom, S., Staveren Van, W.A., de Vegt, F., Katan, M.B., 1996. Consumption of ginger (*Zingiber officinale* Roscoe.) does not affect *ex vivo* platelet thromboxane production in humans. Eur. J. Clin. Nutr. 50 (11), 772–774.

Jantan, I.B., Yassin, M.S.M., Chin, C.B., Chen, L.L., Sim, N.L., 2003. Antifungal activity of the essential oils of nine Zingiberaceae species. Pharm. Biol. 41 (5), 392–397.

Keating, A., Chez, R.A., 2003. Ginger syrup as an antiemetic in early pregnancy. Altern. Ther. Health Med. 8 (5), 89–91.

Kim, D.S.H.L., Dongseon, K., Oppel, M.N., 2002. Shogaols from *Zingiber officinale* protect IMR 32 human neuroblastoma and normal human umbilical vein endothelial cells from β-amyloid insult. Plant. Med. 68, 375–376.

Kishore, P., Pande, P.N.R., 1980. Role of *Sunthi gudochi* in the treatment of *Amavata* (rheumatoid arthritis). J. Res. Ayur. Siddha. 1 (3), 417–428.

Kittikar, K.R., Basu, B.D., 1935. Indian Medicinal Plants, vol. 4. Bishen Singh Mahendrapal Singh, Dehradun, Himachal Pradesh, India.

Lamb, A.B., 1994. Effect of dried ginger on human platelet function. Thromb. Haemost. 71, 110–111.

Martins, A.P., Salgueiro, L., Goncalves, M.J., Cunha, A.P., Vila, R., Canigueral, S., et al., 2001. Essential oil composition and antimicrobial activity of three Zingiberaceae from S.Tome e Principe. Planta Med. 67 (6), 580–584.

McCaleb, R., 1996. Ginger: the anti-inflammatory pain relievers from nature. <http://sunsite. unc.edu/herbs/> (quoted from Pakrashi and Pakrashi, 2003).

Miller, L.G., Murray, W.J., 1998. Herbal Medicines—A Clinician's Guide. Pharmaceutical Press, New York, NY.

Mowrey, D.B., Clayson, D.E., 1982. Motion sickness, ginger and psychophysics. Lancet 1 (8273), 655–657.

Mustafa, T., Srivastava, K., 1990. Ginger (Zingiber officinale) in migraine headache. J. Ethnopharmacol. 29, 267–273.

Nadkarni, K.M., 1976., third ed. Indian Materia Medica, vol. 1 Popular Prakashan, Bombay, Maharashtra State, India.

Olayaki, L.A., Ajibade, K.S., Gesua, S.S., Soladoye, A.O., 2007. Effect of Zingiber officinale on some hematologic values in alloxan-induced diabetic rats. Pharm. Biol. 45 (7), 556–559.

Pakrashi, S.C., Pakrashi, A., 2003. Ginger. Vedams New Delhi.

Patel, K. 1996. Cited from Patel and Srinivasan, 2000.

Patel, K., Srinivasan, R., 2000. Influence of dietary spices and their active principles on pancreatic digestive enzymes in albino rats. Nabrung 44, 42–46.

Phillips, S., Ruggier, R., Hutchinson, S.E., 1993. Zingiber officinale (Ginger)—an antiemetic for day case surgery. Anaesthesia 48, 715–717.

Pruthi, J.S., 1979. Spices and Condiments. National Book Trust of India, New Delhi, India.

Pushpanathan, T., Jebanesan, A., Govindarajan, M., 2008. The essential oil of Zingiber officinale Linn. (Zingiberaceae) as a mosquito larvicidal and repellent agent against the filarial vector Culex quinquefasciatus Say (Diptera: Culicidae). Parasitol. Res. 102 (6), 1289–1291.

Sharma, I., Gosain, D., Dixit, V.P., 1996. Hypolipidaemia and antiatherosclerotic effects of Zingiber officinale in cholesterol fed rats. Phytober. Res. 10, 517–518.

Sharma, S.S., Gupta, Y.K., 1998. Reversal of cisplatin-induced delay in gastric emptying in rats by ginger (Zingiber officinale). J. Ethnopharmacol. 62, 49–55.

Sharma, J.N., Ishak, F.I., Yusuf, A.P.M., Srivastava, K.C., 1997. Effects of eugenol and ginger oil on adjuvant arthritis and the kallikreins in rats. Asia Pacific J. Pharmacol. 12, 9–14.

Shuxuan, J., Xiu Qin, H., Zheng, X., Li Mei, M., Ping, L., Juan, D., et al., 2001. Experimental study on Zingiber corallium Hance to prevent infection with Schistosoma japonicum cercaria. Chin. J. Schisto. Control 13, 170–172.

Sohni, Y.R., Bhatt, R.M., 1996. Activity of a crude extract formulation in experimental hepatic amoebiasis and in immunomodulation studies. J. Pharmacol. 54, 119–124.

Srivastava, K.C., 1984. Aqueous extracts of onion, garlic, and ginger inhibit platelet aggregation and alter arachidonic acid metabolism. Biomed. Biochem. Acta 43, 335–346.

Srivastava, K.C., 1986. Isolation and effects of some ginger components on platelet aggregation and cicosanoid biosynthesis. Prostaglandins Leukot Essent Fatty Acids 35, 183–185.

Srivastava, K.C., 1989. Effect of onion and ginger consumption on platelet thromboxane production in humans. Prostaglandins Leukot. Essent. Fatty Acids 35, 183–185.

Srivastava, K.C., Mustafa, T., 1989. Ginger (Zingiber officinale) and rheumatoid disorders. Med. Hypoth. 29, 25–28.

Suekawa, M., Ishige, A., Yuasa, K., Sudo, K., Aburada, M., Hosoya, E., 1984. Pharmacological studies on ginger I. Pharmacological actions of pungent constituents, (6)-gingerol and (6)-shogaol. J. Pharmacobiodyn. 7, 836–848.

The New Sunday Express Magazine, August 19, 2012. <www.drweil.com/>.

Tyler, V.E., 1996. What pharmacists should know about herbal remedies. J. Am. Pharm. Assoc. (Wash) 36, 29–37.

Ukeh, D.A., Birkett, M.A., Pickett, J.A., Bowman, A.S., Mordue, A.J., 2009. Repellent activity of alligator pepper, *Aframomum melegueta* and ginger, *Zingiber officinale*, against the maize weevil, *Sitophilus zea mais*. Phytochemistry 70 (6), 751–758.

Verma, S.K., Singh, J., Khamesra, R., Bordia, A., 1993. Effect of ginger on platelet aggregation in man Indian. J. Meat Res. 98, 240–242.

Wang, H., Ng, T.B., 2005. An antifungal protein from ginger rhizomes. Biochem. Biophys. Res. Commun. 336 (1), 100–104.

Yamahara, J., Huang, Q.R., Li, Y.H., Xu, L., Fujimura, H., 1990. Gastrointestinal motility enhancing effect of ginger and its active constituents. Chem. Pharm. Bull. 38, 430–431.

Yamaoka, Y., Kawakita, T., Kaneoko, M., Nomoto, K., 1996. A polysaccharide fraction of *Zizyphi fructus* in augmenting natural killer activity by oral administration. Biol. Pharm. Bull. 19, 936–939.

Yoshikawa, M., Yamaguchi, S., Kunimi, K., Matsuda, H., Okuno, Y., Yamahara, J., et al., 1994. Stomachic principles in ginger. Chem. Pharm. Bull. 42, 1226–1230.

26 Ginger as a Spice and Flavorant

Spices are added to contribute flavor to bulk foods, which are generally insipid, to increase their acceptability and intake. The term *flavor* is usually used to mean a combination of taste and aroma, but a comprehensive definition is the total effect provided in the mouth, when a prepared food is eaten. This includes, besides aroma and taste, other perceptions such as, pungency, astringency, warmth, and cold. It is essentially these sensations, which produce the physiological reactions leading to humeral and hormonal secretions, and in turn give the cues to their acceptance or rejection reactions. Apart from salt (sodium chloride), spices are considered to be the most important enhancers of taste and flavor. Spices are often used in association with the term *condiments*. Both terms are used indiscriminately and interchangeably. However, for the chef, food technologist, and connoisseur of food, spices and condiments mean different things. Spices are fragrant, aromatic, or pungent edible plant products, which contribute flavor and relish or piquancy to foods or beverages. Condiments, on the other hand, are prepared food compounds containing one or more spices or spice extractives, which when added to a food after it has been served, enhances the flavor of the food (Farrell, 1985). Hence, condiments are compound food additives and they are added after the food has been served. *Seasoning* is another term that is related to both spices and spice extractives, which when added to food, either during its manufacture or in its preparation, before it is served, enhances the natural flavor of the food and thereby increases its acceptance by the consumer (Farrell, 1985). Seasonings are added before or during the preparation of a food, whereas condiments are added after the preparation of the food or after it has been served.

Spices have various effects: they impart flavor, pungency, and color, and they also have antioxidant, antimicrobial, nutritional, and medicinal functions. In addition to these direct effects, spices have complex or secondary effects when used in cooking, such as salt reduction and improvement of texture of certain foods.

Forms of Ginger Used in Cooking

Ginger is, more or less, a universal spice, although its use is more predominant in certain countries such as China. Ginger is used in cooking, in various forms such as immature ginger, mature fresh ginger, dry ginger, ginger oil, ginger oleoresin, dry soluble ginger, ginger paste, and ginger emulsion. Essential oil, oleoresin, and other extractives are standardized by the manufacturers to yield the same aromatic and flavorant characteristics of the specific spice. Manufacturers usually determine the ground spice equivalence of the extract before marketing. Spice equivalence of

The Agronomy and Economy of Turmeric and Ginger. DOI: http://dx.doi.org/10.1016/B978-0-12-394801-4.00026-0

Table 26.1 Spice Extractive Equivalencies of Ginger

Extractives	Type	g/Type Equivalent to 1 oz Ground Spice	Percentage Extractives on Soluble Dry Edible Carrier	Minimum Percentage Volatile Oil in vol/wt	Federal Specifications	
					Extract on Dry Carrier Percentage	Volatile Oil Extract in vol/wt
Ginger (African)	OS	1.134	4	28	4	25
Ginger (*Cochin*)	SR	1.134	4	28	4	25
Ginger (Commercial)	SR	1.134	4	28	4	25

OR = Oleoresin, SR = superresin (a blend of oleoresin and essential oil).
Source: Farrell (1985).

extract is defined as the number of pounds of oleoresin required to equal 100 pounds of freshly ground spice in aromatic and flavorant characteristics (Farrell, 1985). The weight of oleoresin is added to sufficient salt, sugar, dextrose, or other edible dry material as a carrier to 100 pounds of dry, soluble spice. One pound of such dry, soluble spice is equivalent to one pound of the corresponding freshly ground spice. The spice extractive equivalencies of ginger are given in Table 26.1.

Spices have little value as nutrients, as they are used only in very small quantities. However, when ginger is consumed as ginger beer or ale, the intake also may be significant nutritionally. The nutritional composition of ginger is given in Table 26.2.

Ginger as a Flavorant

Spices are used in food for four basic purposes: (i) for flavoring; (ii) for masking or deodorizing; (iii) for imparting pungency; and (iv) for adding color. In addition, they also have ancillary properties such as antimicrobial and nutritional. However, when spices are used in cooking, complex secondary effects often result, leading to changes such as salt reduction and improved texture for certain foods. Each spice has one of the basic functions mentioned previously; in addition, it may perform one or more secondary functions (Hirasa and Takemasa, 1998).

When a spice is used in cooking, its overall quality has to be evaluated. Heating can lead to the loss of essential oil, and hence, it results in the reduction of overall quality of flavor. Because of this diverse effect of heat, the timing of adding a spice is a very important aspect. In addition, the taste of a food often changes in combination with another food or beverage. When one food component enhances the taste of another component, it results in a significant synergistic effect. On the other hand, certain tastes are decreased in intensity by combining with certain other food components resulting in a suppressive effect or "offset effect." The synergistic effect is said to involve a "taste illusion"; the best-known example is the taste enhancement caused by the addition of sodium glutamate. In fact, ginger has a remarkable synergistic effect when it is used in soft drinks as well. Ginger has a typical earthy

Table 26.2 Nutritional Composition of Ginger (Ground)

Composition	USDA Handbook 8-2[a]	ASTA[b]
Ground		
Water (g)	9.38	7.0
Food energy (kcal)	347.00	380.00
Protein (g)	9.12	8.5
Fat (g)	5.95	6.4
Carbohydrates (g)	70.79	72.4
Ash (g)	4.77	5.7
Calcium (g)	0.116	0.1
Phosphorus (mg)	148.00	150.00
Sodium (mg)	32.00	30.00
Potassium (mg)	1,342	1,400.00
Iron (mg)	11.52	11.3
Thiamine (mg)	0.046	0.050
Riboflavin (mg)	0.185	0.130
Niacin (mg)	5.155	1.90
Ascorbic acid (mg)	–	ND
Vitamin activity (RE)	15	15

[a]Composition of Foods: Spices and Herbs, USDA, Agricultural Handbook 8, 2 January 1977.
[b]The nutritional composition of spices. ASTA Research Committee, February 1977.
ND = Not detected.
Source: Tainter and Grenis (2001).

smell but has a refreshing flavor and imparts a "freshness stimulus." Such qualities enhance the freshness of some soft drinks when ginger is added to them (Hirasa and Takemasa, 1998).

The functionally significant components of ginger are primarily its aroma and secondarily its pungency. Many investigations have been published on the chemical component, which contribute to its functional qualities. Salzer (1975) has suggested the following determinants of ginger quality:

1. Citral and citronellyl acetate are important codeterminants of odor.
2. Zingiberene and β-sesquiphellandrene are the main components of the freshly prepared oil.
3. These components are converted to ar-curcumene with storage.
4. The ratio of zingiberene + β-sesquiphellandrene to ar-curcumene is indicative of the age of the oil.

However, the sensory evaluation studies conducted subsequently, mainly by Govindarajan and his group (1982a) have given us a much greater insight into the flavorant characteristics of ginger. They have developed the TLC aromagram technique to evaluate the flavor quality of individual components of ginger oil. Such studies have highlighted the importance of compounds such as borneol, α-terpineol, citral, and nerolidol to the total ginger aroma. The ginger aroma should have the proper blend of lemony, camphory, stale coconut (sweet rooty), and flavorant aromatic notes; the full flavor requires the impact of the pungency as well.

Ginger as a Deodorizing Agent

Spices perform a deodorizing function in food. According to the Weber–Fecher law, the strength of an odor perceived by the sense of the smell is proportional to the logarithm of the concentration of the smelled compounds. In other words, the sensational strength perceived with the five senses is proportional to the logarithm of the actual strength of these stimuli. Hence, even if 99% of the total smelled compound is eliminated chemically, the sensational strength perceived is reduced to 66% (Hirasa and Takemasa, 1998). In food items, spices are employed to mask or deodorize the remaining 1%. Spices differ much in their ability to mask odors. Ginger is, in fact, very weak in this property, having a deodorizing rate of only 4%, as shown in Table 26.3.

Frequency Patterning Analysis of Ginger

Food technologists have analyzed the use of spices in various countries and in various preparations and developed a frequency patterning analysis of each spice. This analysis gives information on natural trends and the suitability of a spice for a particular type of preparation. The information on ginger is summarized below (Hirasa and Takemasa, 1998).

1. Ginger is more suitable for dishes in Japan, China, India, Southeast Asia. But, it is less preferred in countries like United Kingdom.
2. Ginger is suitable for preparations of meat, seafood, milk, egg, grains, vegetables, fruit, beans, seeds, and beverages.

Table 26.3 Deodorizing Rate of Ginger in Comparison with Other Common Spices[a]

Spice	Deodorizing Rate (%)
Ginger	4
Turmeric	5
Cardamom	9
Black pepper	30
Star anise	39
Allspice	61
Cloves	79
Coriander	3
Fennel	0
Cumin	11
Anise	27
Celery	44
Mint	90
Rosemary	97
Thyme	99

[a]Deodorizing rate–percent of methyl mercaptan (500 mg) captured by methanol extract of each spice.
Source: Tohita et al. (1984).

3. Ginger is also suitable for boiled, baked, fried, deep fried, steamed, food processed with sauce, pickled, and fresh food; however, it is more suitable for fried and steamed dishes.
4. Ginger is used to impart pungency to food in Japan, China, Southeast Asia, India, and the United Kingdom; it is most commonly used in Chinese cooking.

Analysis of the ethnic foods in the United States has indicated that ginger is an important spice in the dishes of the following ethnic groups: Armenian, Cambodian, Chinese, Cuban, Danish, Egyptian, Finnish, French, German, Hungarian, Indonesian, Iranian, Korean, Lebanese, Norwegian, Polynesian, Puerto Rican, Russian, Spanish, Swedish, Syrian, Turkish, and Vietnamese (Farrell, 1985). However, compared to other major spices (e.g., black pepper and cinnamon) ginger is not usually found in seasoning (including spice blends). The spices usually used in Asian cooking formulations are ginger, cumin, coriander, basil, mint, and celery. Special seasoning masalas (Indian cooking medium with several spices) often create an almost magical effect on fish and meat dishes. In most of such blends, ginger is essential and in certain dishes ginger is a predominant component, as for instance, ginger chicken, ginger fish, and ginger vegetables.

Ginger is an ingredient in many curry powder formulations. The compositions of some typical curry powder blends are provided in Tables 26.4 and 26.5. The federal specifications of curry powder (Sp. No. EES-631) contain not less than 3% of ginger. Ginger also forms a part of a typical pickling spice combination, ranging from 0% to 5% in various brands. Ginger is a component of the popular pumpkin pie spice formulation (Table 26.6). A selection of curry flavor formulations in vogue in various countries is given in Table 26.7.

Table 26.4 Typical Curry Powder Formulations
Containing Ginger

Ingredients	Typical Range (%)
Coriander	10–50
Cumin	5–20
Turmeric	10–35
Fenugreek	5–20
Ginger	5–20
Celery	0–15
Black pepper	0–10
Red pepper	0–10
Cinnamon	0–15
Nutmeg	0–15
Cloves	0–15
Caraway	0–15
Fennel	0–15
Cardamom	0–15
Salt	0–10

Source: Tainter and Grenis (1993).

Table 26.5 Curry Powder Blends as per US Specifications

Freshly Ground Spice	US Standard		General-Purpose Curry Formula		
	Formula No. 1 (%)	Formula No. 2 (%)	Formula No. 3 (%)	Formula No. 4 (%)	Formula No. 5 (%)
Coriander	32	37	40	35	25
Turmeric	38	10	10	25	25
Fenugreek	10	0	0	7	5
Cinnamon	7	2	10	0	0
Cumin	5	2	0	15	25
Cardamom	2	4	5	0	5
Ginger (*Cochin*)	3	2	5	5	5
Pepper (white)	3	5	15	5	0
Poppy seed	0	35	0	0	0
Cloves	0	2	3	0	0
Cayenne pepper	0	1	1	5	0
Bay leaf	0	0	5	0	0
Chili peppers	0	0	0	0	5
Allspice	0	0	3	0	0
Mustard seed	0	0	0	3	5
Lemon peel (dried)	0	0	3	0	0

Formula 1 is the US Military specifications Mil-C-35042A. Formula 2 is considered a mild curry. Formula 3 is a sweet curry. Formula 4 is a hot curry type. Formula 5 is a very hot pungent Indian-style curry.
Source: Farrell (1985).

Table 26.6 Pumpkin Pie Spice Formulation

Ingredients	Typical Range (%)
Ground cinnamon	40–80
Ground nutmeg	10–20
Ground ginger	10–20
Ground cloves	10–20
Ground black pepper	0–5

Ginger as a Flavorant

In many other cases, ginger may not form a complement in the formulation itself but is added while cooking with, for example, fresh ginger, ginger paste, or ginger powder.

Flavor Properties of Ginger

The aromatic compounds present in ginger contribute to its flavorant properties. The pungency and hotness are the principal sensations, which make it more palatable.

Table 26.7 Curry Blends and Masala Mixes with Ginger as a Component

Spice	Ingredients
Japanese seven spice blend	Ginger, red pepper, orange/tangerine peel, dried nori/seaweed flakes, white and black pepper, sesame seed, and white poppy seeds.
Teriyaki blend and sauce	Soy sauce, sugar, vinegar, sweet wine, ginger, chives, sesame, and fish stock.
Indian curry blends	Basic curry blend consists of coriander, cumin, red pepper, and turmeric. Special blends (e.g., for fish or meat) contain, in addition to the above, ginger, cardamom, cloves, cinnamon, mustard, fenugreek, curry leaf, mint, coriander leaf, and celery seeds depending on the particular blend.
Chat masala	A specific north Indian curry masala blend consisting of green chilies, coriander leaf, coriander, cumin, ajwain, black pepper, ginger, asafetida, and dried mango powder (*amchoor*).
Tandoori masala	An aromatic, spicy masala blend used to marinate meat before baking in a tandoor (a type of clay oven). The blend consists of ginger, chilies, cumin, coriander, cloves, cinnamon, nutmeg, mace, cardamom, pepper, and bay leaf.
Pickling masala blends	Many different types of pickling blends are in vogue. The important ingredients are mango or lime pieces, chili, pepper, ginger, garlic, mustard oil, mustard seed paste, turmeric, sesame seeds, mint, and cilantro. Mango, lime, and mixed fruit pickles are the common ones.
Burmese curry blend	Onion, garlic, ginger, turmeric, fish sauce, chilies, and tamarind.
Malaysian curry blend	Lemongrass, star anise, ginger, galangal, pandan leaf, tamarind, mint, coriander, turmeric, and shallot.
Japanese curry blend	Hot and fiery curry blend consisting of ginger, galangal, black pepper, red pepper, and cassia cinnamon. Apart from these spices, meat and fish stocks and a variety of other ingredients are also used.
Mediterranean spice blend	Cardamom, ginger, cassia cinnamon, black pepper, cumin, fenugreek, lovage, mace, cubeb, long pepper, allspice, nutmeg, rose petals, lavender blossoms, orange blossoms, grains of paradise, chilies, nigella, onion, thyme, rosemary, and turmeric.
French spice blend	White/black pepper, ginger, nutmeg, cloves, mace, cinnamon, all spice, bay leaf, sage, marjoram, and rosemary.
Mozambique blend (*piri-piri*)	Hot and fiery curry blend consisting of bird's eye chilies, garlic, ginger, onion, hot paprika, black pepper, olive oil, and lemon juice.
West African spice blend	Black pepper, red chilies, pink pepper, ginger, cubebs, and grains of paradise.
Global spice blend (green)	Leafy spices (e.g., basil, cilantro, parsley, mint), green pepper, ginger, garlic, and lemon.

Source: Compiled from various sources worldwide.

Although, generally, volatile compounds contribute to flavor, in ginger, both volatile and nonvolatile constituents are important to impart the totality of flavor properties such as taste, odor, and pungency. The flavor quality depends on factors such as variety, geographical origin, processing methods, and storage conditions. African ginger has a harsh, strong flavor as compared to the mild, sweet flavor of Jamaican ginger. Peeling of green ginger for drying should be carried out carefully to avoid the loss of volatile oil due to damage of oil cells present below the epidermal layer. *Cochin* ginger has a softer and richer flavor than African ginger. Flavorant properties of ginger depend both on volatile oil and its nonvolatile fraction. Volatile oil is composed mainly by sesquiterpene hydrocarbons, of which α-zingiberene, β-bisabolene, and ar-curcumene are the major compounds. The major flavor constituents for the pungent principle of ginger have been reported to be 6-, 8-, or 10-shogaols. The type of ginger plays a prominent role in the flavor; the method of adoption for extraction influences the type of compounds and their quality. Bartley and Foley (1994) have reported neral, geranial, zingiberene, α-bisabolene, and β-sesquiphellandrene as major flavoring compounds of Australian ginger, and reported that 6-gingerol is the major contributor to the pungency (Bartley, 1995). Nishimura (1995) investigated the volatile compounds responsible for the aroma of fresh rhizomes of ginger, and the compounds with high dilution factor were linalool, geraniol, neral, isoborneol, borneol, 18-cineol, 2-pineol, geranyl acetate, 2-octenal, 2-decenal, and 2-dodecenal. The author also reported that linalool, 4-terpineol, borneol, and isoborneol contribute to the characteristic aromatic flavors of Japanese fresh ginger. The pungent principle of ginger, 6-gingerol, has been reported to be a potential antioxidant among 10 phenolic compounds separated by TLC.

Ginger as an Antioxidant

Ginger has a high content of antioxidants and has been grouped as one of the spices with good antioxidant activity, with a 1.8 index rating (Chipault et al., 1952). This makes it a free radical scavenger (Lee and Ahn, 1985). Sethi and Aggarwal (1957) reported that dried ginger has weak antioxidant properties. The antioxidant property of ginger in comparison with other common spices is given in Table 26.8.

Peroxidation Value

Studies on the antioxidant properties of the chemical compounds of many ginger spices revealed that the shogaol and zingiberene found in ginger exhibited strong antioxidant activities. The antioxidant activity of ginger is dependent on the side-chain structures and substitution patterns on the benzene ring. Twelve compounds showed higher activity than α-tocopherol. Mainly, the antioxidant activity is exerted by gingerol and hexahydrocurcumin (Tsushida et al., 1994). Ginger added at a level of 1–5% to soybean oil and cottonseed oil exerted antioxidant activity during storage and the activity was equivalent to BHT. In general, the increase in concentration of

Table 26.8 Antioxidant Activity of Ginger in Comparison with Other Common Spices Against Lard

Spice	Ground Spice POV (meEq/kg)	Petroleum Ether-Soluble Fractions POV (mEq/kg)	Petroleum Ether-Insoluble Fractions POV (mEq/kg)
Ginger	40.9	24.5	35.5
Turmeric	399.3	430.6	293.7
Black pepper	364.5	31.3	486.5
Chilies	108.3	369.1	46.2
Cardamom	423.8	711.8	458.6
Cinnamon	324.0	36.4	448.9
Cloves	22.6	33.8	12.8
Mace	13.7	29.0	11.3
Nutmeg	205.6	31.1	66.7
Rosemary	3.4	6.2	6.2
Sage	2.9	5.0	5.0

Concentration added was 0.02%.

crude gingerol increases the antioxidant activity. However, the thermal stability studies by heating the gingerol component at 165°C for 30 min indicated the retention of the antioxidant activity only to 10%. Shogaol has been shown to be a compound with high antioxidant activity when a methanolic extract of the spices was further fractionated by ethyl acetate and the activity potency was similar to tocopherol. In animal experiments (Ahmed et al., 2001), the diet containing ginger showed a highly protective effect against the malathion-induced oxidative damage exhibiting the antioxidative activity. Incorporation of salt and ginger extract to precooked lean beef retarded rancidity during storage, increased the tenderness, and extended shelf life (Kim and Lee, 1995).

Antimicrobial Activity

Although used in food preservation, ginger is not very effective in preventing spoilage of food due to microbial contamination and oxidative degradation. Ginger has only mild antimicrobial activity. The minimum inhibitory concentration (MIC) of ginger against *Clostridium botulinum* (the bacterium causing severe food poisoning) was shown to be about 2000 µg/ml. The essential oil in ginger was shown to inhibit both cholera and typhoid bacteria. The components of oil responsible for this antimicrobial effect were shown to be gingerone and gingerol (Hirasa and Takemasa, 1998). Other investigations reporting the antimicrobial properties of gingerols are in relation to *B. subtilis* and *E. coli* (Yamada et al., 1992) and *Mycobacterium* (Galal, 1996).

Ginger stimulates appetite, acts as an antioxidant, antimicrobial, and antiflatulent and hence has a tremendous use in processed food products. Ginger has occupied the

Table 26.9 Processed Foods Containing Ginger

Processed Food	Concentration	References
Plum appetizer	1.5% ginger extract	Barwal and Sharma (2001)
Cake	Ginger	Donovan (2001)
Chicken feet jokpyun	0.1% ginger	
Apple–ginger-based squash	10% ginger	Lal et al. (1999)
Baby food	Ginger	Theuer (2000)
Beverage	Ginger	Edjeme and Stapanka (1999)
Bread	Ginger	Ludewig et al. (1999)
Fragrant beef	Ginger	Feng and Cai (1998)
Wooung kimchi	1.3% ginger	
Chicken patties	Ginger	Nath et al. (1996)
Semidry fish sausage	0.1% ginger	Joshi and Rudrashetty (1994)
Meat pickle	10% ginger	Dhanapal et al. (1994)
Pickle	Ginger	Sachdev et al. (1994)

Table 26.10 Processed Foods Containing Ginger with Specific Actions

Food Products	Action	References
Meat and meat products	Tenderizing	Naveena and Mendirattta (2001a)
	Antioxidant	
	Antimicrobial	
Chicken meat	Antioxidant, antimicrobial	Naveena et al. (2001)
Chicken meat	Tenderizing	Naveena and Mendiratta (2001b)
Sheep meat	Tenderizing, antioxidant	Mendiratta et al. (2000)
Rape seed oil	Antioxidant	Takacsova et al. (1999)
Infant food	Reduce gastrointestinal reflux	Theuer (2000)
Beef patties	Antioxidant	Mansour and Khalil (2000)
Meat patties	Antioxidant	Abd-El-Alim et al. (1999)
Ready-to-serve tomato juice	Increased flavor and taste	Manimegalai et al. (1996)
Meat products	Superior sensory quality	Syed-Ziauddin et al. (1995)
Lean beef, cooked	Antioxidant	Kim and Lee, 1995
Korean cereal product	Antioxidant	Lee and Park (1995)
Buffalo meat	Antimicrobial	Syed-Ziauddin et al. (1995)
Soybean oil, cottonseed oil	Antioxidant	

pride of place in many food products, such as masala powders, curry mixes, ready-to-eat foods, and pastes. A list of processed foods, processed foods with specific actions, manufactured products, and a selection of dishes with ginger is provided in Tables 26.9–26.12.

Table 26.11 Manufactured Products Containing Ginger

Masalas	Meat, tandoori chicken, garam, channa, instant khara bath, kitchen king, chat, tea, gobi manchurian, pav bhaji, fish, kabab, chole, biryani/pulav, rajma, stuffed vegetable
Pickle and chutney	Ginger pickle, mango–ginger pickle, ginger thokku, spice-up tomato chutney, red chili chutney, onion chutney
Sauce	Chinese chili sauce, stromy sauce-chili tomato, manchurian, Schezuan hot
Powder and paste	Dry ginger powder, ginger–garlic paste, ginger paste
Mixes/ready-to-eat products	Butter chicken mix, tikka gravy, paneer makhanwala, vegetable jawa mix, shahi gravy, channa gravy, chicken kolhapuri mix, yellow curry paste, chicken mughlai mix, chole gravy mix, gobi manchurian mix, navaratna ready-to-eat, mixed vegetable curry (ready-to-eat), pav bhaji mix, vegetable pulav, instant kara bath
Others	Ginger biscuits, ginger and mint, ginger fresh (mouth freshener), oriental stir fry

Source: All of the food products listed are very popular Indian vegetarian and nonvegetarian dishes; compiled from various sources.

Table 26.12 Selection of Dishes Using Ginger as a Spice and Flavorant

Nonvegetarian dishes	Ginger chicken, ginger fried chicken, chicken pulao, mutanjan, taash kabab, mutton vindaloo, mutton chops with curd, mutton curry, mince pie, masala liver, minced fish, mallai curry, fish kebabs, fish roast
Vegetables	Spicy vegetable pie, end of the month pie, cauliflower in ginger manchurian, potatoes and aubergines with ginger, mushroom–ginger manchurian, sweet and sour ginger curry, spicy besan pakories, ginger pachadi, sindhi channa, ginger pakories, ginger dip
Ginger sweets and desserts	Ginger preserve, crystallized ginger, potato–ginger halwa, quick ginger pudding, ginger–custard pudding, gingerbread pudding, crisp fruit pudding, ginger refrigerated cake, oriental sundae, ginger snow, ginger snaps, brandy snaps, gingered pears, baked semolina pudding with ginger sauce, pumpkin pie, creamy custard flan, carrot fruit custard, ginger muffins, apple–ginger tarts, ginger crispies, ginger sugar puffs, ginger cake doughnuts
Ginger cakes, ginger breads, and ginger biscuits	American hot cheese cakes, ginger dundee cake, ginger pastry, ginger cake (eggless), ginger preserve cake, ginger sponge with treacle, date–ginger cake, fruit and nut bars, ginger dessert cake, ginger fruit cake, ginger nuts, simple ginger biscuits, frosted ginger cake, fruit gingerbread, lemon gingerbread, grantham gingerbread, grasmere gingerbread, spice leaf, sindhi masala bread, gingerbread nuts, ginger rock bun, Canadian gingerbread

(Continued)

Table 26.12 (Continued)

Ginger jams, ginger pickles, and ginger chutneys	Apple–ginger jam, tomato marmalade, beetroot jam with ginger, parvel preserve with ginger, hot chili pickle, ginger–prawn pickle, lime–ginger pickle, ginger–mango pickle, ginger–garlic pickle, ginger–onion pickle, ginger–apple pickle, mixed fruit pickle, Ceylonese (now Sri Lankan) mixed pickle, mustard mango pickle, mutton pickle, dry fruit pickle, plantain stem pickle, prawn belches, sour lime pickle, beetroot pickle in water, tomato chutney, ginger chutney, ginger–pineapple chutney, plum chutney, ripe tomato chutney, mango bud chutney, sweet pumpkin chutney, hot mango chutney, sweet mango chutney
Ginger soft drinks, beverages	Melon–ginger cocktail, ginger wine, rich ginger wine, lemon–ginger punch, orange–ginger punch, raspberry–ginger punch, cool ginger mint, hot lemon–ginger, ginger pop, ginger beer, gingerade, ginger tonic, ginger–orange milkshake, pineapple–ginger cocktail, ginger punch, ginger–vegetable soup, ginger–Chinese soup, ginger–mutton soup, ginger–chicken soup, ginger–seafood chowder soup, marwari tea, kothamalli (coriander) tea

Source: Compiled from various sources. All the indigenous names refer to Indian cuisine (ginger-based).

References

Abd-El-Alim, S.S.I., Lugasi, A., Hovari, J., Dworschak, E., 1999. Culinary herbs inhibit lipid oxidation in raw and cooked minced meat patties during storage. J. Sci. Food Agric. 79 (2), 277–285.

Ahmed, N., Katiyar, S.K., Mukhtar, H., 2001. Antioxidants in chemoprevention of skin cancer. Curr Probl Dermatol. 29, 128–139.

Bartley, J.P., 1995. A new method for the determination of pungent compounds in ginger. J. Sci. Food Agric. 68 (2), 215–222.

Bartley, J.P., Foley, P., 1994. Supercritical fluid extraction of Australian grown ginger. J. Sci. Food Agric. 66 (3), 365–371.

Barwal, V.S., Sharma, R., 2001. Standardization of recipe for the development of plum appetizer. J. Food Sci. Technol. 38 (3), 248–250.

Chipault, J.R., Mizuna, G.R., Hawkins, J.M., Landberg, W.O., 1952. The antioxidant properties of natural spices. Food Res. 17, 46–49.

Dhanapal, K., Ratnakumar, K., Indra Jasmine, G., Jayachandran, P., 1994. Processing chunk meat into pickles. Fish. Technol. 31 (2), 188–190.

Donovan, M.E., 2001. No fat, no cholesterol cake and process for making the same. US Patent 6235334B1.

Edjeme, G.A.J., Stapanka, S., 1999. Beverage based preparation based on ginger, oranges and lemons. French Patent 19980407.

Farrell, K.T., 1985. Spices, Condiments and Seasonings. The AVI Publ Co Inc, Westport, CN.

Feng, Z.Y., Cai, H.M., 1998. Techniques for producing soft pack fragrant beef. Meat Res. 2, 20–22.

Galal, A.M., 1996. Antimicrobial activity of 6-paradol and related compounds. Int. J. Pharmacog. 34, 64–66.

Govindarajan, V.S., 1982a. Flavour quality of ginger. In: Nair, M.K., Premkumar, T., Ravindran, P.N., Sarma, Y.R. (Eds.), Ginger and Turmeric. Proceedings of the National Seminar CPCRI, Kasaragod, Kerala State, pp. 147–166.

Hirasa, K., Takemasa, M., 1998. Spices—Science and Technology. Marcel Dekker, New York, NY.

Joshi, V.R., Rudrashetty, T.M., 1994. Effect of different levels of spice mixture, salt in the preparation of semidried fish sausages. Fish. Technol. 31 (1), 52–57.

Kim, K.j., Lee, Y.B., 1995. Effect of ginger rhizome extract on tenderness and shelf life of pre-cooked lean beef. Asian Aust. J. Anim. Sci. 8 (4), 343–346.

Lal, B.B., Joshi, U.K., Sharma, R.C., Sharma, R., 1999. Preparation and evaluation of apple and ginger based squash. J. Sci. Indian Res. 58 (7), 530–532.

Lee, I.K., Ahn, S.Y., 1985. The antioxidant activity of gingerol. Korean J. Food Sci. Technol. 17 (2), 55–59.

Lee, J.H., Park, K.M., 1995. Effect of ginger and soaking on the lipid oxidation of Yackwa. J. Korean Soc. Food Sci. 11 (2), 93–97.

Ludewig, H.G., Stoffels, I., Bruemmer, J.M., 1999. Investigation with the handmade production and shelf life of brown ginger bread. Getreide Mehl Und Brot 53 (2), 112–118.

Manimegalai, G., Premalatha, M.R., Vennila, P., 1996. Formulation of delicious tomato RTS. S. Indian Hortic. 44 (1–2), 52–54.

Mansour, E.H., Khalil, A.H., 2000. Evaluation of antioxidant activity of some plant extracts and their application to ground beef patties. Food Chem. 69, 135–141.

Mendiratta, S.K., Anjaneyulu, A.S.R., Lakshmanan, V., Naveena, B.M., Right, G.S., 2000. Tenderizing and antioxidant effect of ginger extract on sheep meat. J. Food Sci. Technol. 37 (6), 651–655.

Nath, R.I., Mahapatra, C.M., Kondaiah, N., Singh, J.N., 1996. Quality of chicken patties as influenced by microwave and conventional oven cooking. J. Food Sci. Technol. 33 (2), 162–164.

Naveena, B.M., Mendirattta, S.K., 2001a. Ginger as a tenderizing, antioxidant and antimicrobial agent for meat and meat products. Indian Food Ind. 20 (6), 47–49.

Naveena, B.M., Mendiratta, S.K., 2001b. Tenderization of spent hen meat using ginger extract. Br. Poultry Sci. 42 (3), 344–349.

Naveena, B.M., Mendiratta, S.K., Anjaneyalu, A.S.R., 2001. Quality of smoked hen meat treated with ginger extract. J. Food Sci. Technol. 38 (5), 522–524.

Nishimura, O., 1995. Identification of the characteristic odorants in fresh rhizomes of ginger. J. Agric. Food. Chem. 43 (11), 2941–2945.

Sachdev, A.K., Gopal, R., Verma, S.S., Kapoor, K.N., Kulshrestha, S.B., 1994. Quality of chicken gizzard pickle during processing and storage. J. Food Sci. Technol. 31 (1), 32–35.

Salzer, U.H., 1975. Analytical evaluation of seasoning extracts (oleoresins) and essential oils from seasoning. Int. Flavours Food Addit. 6, 206–210.

Sethi, S.C., Aggarwal, J.C., 1957. Stabilization of edible fats by spices. J. Sci. Indian Res. 16, 81–84.

Syed-Ziauddin, K., Rao, D.N., Amla, B.L., 1995. Effect of lactic acid, ginger extract and sodium chloride on quality and shelf life of refrigerated buffalo meat. J. Food Sci. Technol. 33 (2), 126–128.

Tainter D.R., Grenis A.T., 2001. The Spices and Seasonings, second ed. John Wiley and Sons Inc., New York, p. 97.

Takacsova, M., Kristianova, K., Nguyen, D.V., Dang, M.N., 1999. Influence of extracts from some herbs and spices on stability of rapeseed oil. Bull. Potravinarskebo-Vyskumu 38 (1), 17–24. Food Chem. 80 (2) 135–141.

Theuer, R.C., 2000. Ginger containing baby food preparation and methods therefore. US Patent 6051235.

Tohita et al., 1984. The Spices and Seasonings, second ed., John Wiley and Sons Inc., New York, p. 97.

Tsushida, T., Suzuki, M., Kurogi, M., 1994. Evaluation of antioxidant activity of vegetable extracts and determination of some active compounds. J. Jpn Soc. Food Sci. Technol. 41 (9), 611–618.

Yamada, Y., Kikuzaki, H., Nakatani, N., 1992. Identification of antimicrobial gingerols from ginger (*Zingiber officinale*). J. Antibact. Antifungal Agents 20, 309–311.

27 Additional Economically Important Ginger Species

The genus *Zingiber*, the type genus of the family Zingiberaceae, forms an important group of the order Zingiberales. The word "ginger" refers to the edible ginger of commerce, *Z. officinale*. Ginger is also the common term for the members of the family Zingiberaceae, which includes the many other species of *Zingiber* besides *Z. officinale*, worth growing as ornamentals, while some others are of great medicinal value. Many species are grown in the garden as ornamental crops. They bear showy, long-lasting inflorescences and often brightly colored bracts and floral parts; they are widely used as cut flowers in floral arrangements. The gradual changing of the inflorescence bracts from green to yellow to various shades of red and finally to deep red adds to the beauty of the inflorescence. Many wild species have great potential as ornamentals. Some of them are good foliage plants due to their arching form and shining leaves. Leaves exhibit shades of light green to dark green, variegated with yellow and white, or with deep purple undersurfaces. Many of the inflorescence bracts, when squeezed, release a thick juice with the form of mucilage or a shampoo-like substance. Hence, those gingers which have this mucilage in their bracts are called "shampoo gingers." The following is a brief description of *Zingiber* species which have economic importance as local medicine, as spice, or as ornamental plants.

General Features

The ginger plants are perennial in growth and are medium-sized herbs with stout rhizomes. Most of the species produce the inflorescence on a separate shoot directly from the rhizome at the tips of a short or long peduncle. In a few species, the inflorescence develops at the tips of the leafy shoots (*Zingiber capitatum*). The inflorescence possesses a number of closely overlapping bracts, each bearing a single flower. The flowers are peculiar in that the lateral staminodes are fused with the labellum, whereas in the other genera they are free, highly reduced, or absent. The anther is unique in having a curved beak or horn-like appendage. This genus resembles the other genera such as *Alpinia, Amomum, Hedychium*, and so on in the vegetative stage, but it can be distinguished from the others because it has a pulvinus at the base of the petiole. The rhizome and pseudostem of *Zingiber* spp. are fleshier when compared with *Alpinia, Amomum*, and so on. The duration of the flowers in the genus is very short and differs from one species to another, but it is constant for each species. In *Zingiber zerumbet*, the flowers open from morning until evening in the inflorescence.

The Agronomy and Economy of Turmeric and Ginger. DOI: http://dx.doi.org/10.1016/B978-0-12-394801-4.00027-2

The flowers are usually cross pollinated. The pollination in the species of *Zingiber* is rather simple because of the specially modified anther structure and the nature of the staminodes. An insect visiting a flower first lands on the labellum and moves to the throat of the corolla tube. When the insect's front portion pushes the base of the anther, the anther bends forward and dusts the pollen grains on the back of the insect. As it bends forward, the stigma protrudes and arches through the long anther crest and presses against the proboscis of the insect. Thus, pollen grains from other flowers deposited on the back of the insect stick to the stigma, and pollination is effected.

Zingiber mioga *Roscoe*

Z. mioga (*myoga* ginger or Japanese ginger) is a perennial woodland species, endemic to Japan, where it is most popular. It is grown for its edible flowers and young shoots, both of which are used extensively as vegetables. The flowers are mostly sterile, and the propagation of this species is through rhizome division, as in the case of true ginger (*Z. officinale*). The species is unique in having a pentaploid chromosome number of $2n = 55$. In Japan, it is widely grown as a seasonal crop, and the flower bud-producing season is summer. Forced production in glass houses and heated polyhouses occurs during the winter months, and the product attracts a premium price (Sterling et al., 2003). From Japan, *myoga* cultivation has spread to China, Vietnam, Taiwan, Thailand, Australia (Queensland), and New Zealand.

The *myoga* plant needs well-drained and fertile soil to grow well. In poor drainage, the growth of the plant is retarded, and the rhizome can rot. *Myoga* shoots emerge in the spring and produce dense foliage on robust stalks. The sterile flowers are produced at the ground level from the underground stem sprouts in spring and growth continues. The crop is propagated by planting 25 cm long rhizome pieces (sets), planted about 10 cm deep and 40 cm apart in rows. Harvesting the flower buds begins in the second year. An annual fertilizer application of 200–300 kg/ha of nitrogen–phosphorus–potassium fertilizer is suggested (Paghat, 2003).

Myoga flower buds are picked before they emerge above the soil surface from the underground shoots. To facilitate harvest, a 10–15 cm layer of saw dust is used to mulch the base of the plants. The buds are located in the saw dust and are harvested individually at the appropriate stage twice to thrice each week. Export-grade buds need to be more than 6 g and be plum or pink in color. The harvested flower buds can be kept in cold storage, and the production is around 8–13 t/ha in a second-year crop. Sterling et al. (2003) showed that flower bud production is influenced by photoperiod.

Myoga cultivation has become quite popular in Australia and New Zealand of late. In Australia, a superior type of *myoga* plant has been identified and multiplied through tissue culture on a large scale for distribution among farmers. The *myoga* industry in both Australia and New Zealand depends on this superior line, and cultural conditions have been standardized to grow this type of ginger (Clark and Warner, 2000).

A dwarf variegated variety, known as "Dancing Crane," is an ornamental plant, growing about 45 cm tall which produces yellow flowers. This variant has leaves with white stripes on a green background.

Myoga ginger is used in Japan as a spice and as a substitute for true ginger. Two compounds, namely galanal A and galanal B, were isolated from *myoga* rhizomes. These are known to contribute to the characteristic flavor of the *myoga* rhizomes. The pungent principle in *myoga* was identified as (E)-8β(17)β-epoxylated-12-ene-15, 16-diel, commonly known as myogadial. When isolated, this compound, and also galanal A and galanal B which are reduced, and myogadinol are tasteless. The pungency of myogadial depends on the presence of the $\alpha\beta$-unsaturated -1,4-dialdehyde group (Abe et al., 2002). In the Chinese pharmacopoeia, *myoga* ginger is used to treat fever and also as a vermifuge.

Zingiber montanum *(Koenig) Link ex Dietr. (*=Zingiber cassumunar *Roxb.)*

Z. montanum is a native of India and is present throughout the Malaya Peninsula, Sri Lanka, and Java. It is also cultivated in tropical Asia. Its rhizome is an ingredient in many traditional medicines. It is usually used together with the rhizomes of *Zingiber americanus, Zingiber aromaticum, Z. officinale,* and *Kaempferia galanga.* In the Philippines, the decoction from these rhizomes is used to relieve cough and asthma (Quisumbing, 1978). The rhizome is also used as an antidiarrheal medicine in its powdered form or it is made into a paste with rice water (Saxena et al., 1981). As a paste, it can also be given twice daily for three days as an antidote to snake bite. It has been shown that the oil of *Z. montanum* has antibacterial and antifungal properties. The rhizome is also considered a good tonic and appetizer. It is given with black pepper against cholera and is also used as a vermifuge (Barghava, 1981). *Z. montanum* is cultivated in the United States as garden ginger and is often called "Chocolate Pinecone Ginger." The stems are tall and thin, and the bracts are brown, thus the common name.

Z. cassumunar is propagated vegetatively through the division of suckers. The rhizomes on storage are easily affected by fungi, which causes rotting. Rhizome pieces with one or two emerging shoots are used for planting immediately after separation from the mother stock. Poosapaya and Kraisintu (2003) developed a tissue culture multiplication protocol. Shoot tips cultured in LS medium, supplemented with 4 mg/l BAP, produced an average of 13 shoots within eight weeks. The incorporation of antibiotics is essential to suppress microbial contamination. The rooting of shoots was achieved in a medium with a low concentration of NAA or with the addition of activated charcoal.

Many investigations have been made on the medicinal properties, especially of the anti-inflammatory effect, of *Z. casssumunar.* Ozaki et al. (1991) isolated three compounds identified as (E)-1-(3,4-dimethoxyphenyl)but-1-ene, (E)-1-(3,4-dimethoxyphenyl) butadiene, and zerumbone. The methanol extract was found to possess the antiinflammatory and analgesic activities, which comes from the first compound (E)-1-(3,4-dimethoxyphenyl)but-1-ene.

Masuda et al. (1995) isolated cassumunarins A, B, and C, three anti-inflammatory antioxidants, and determined their structures. Cassumunarins are complex curcuminoids. Their antioxidant efficiency was determined by the inhibition of linoleic acid's antioxidation in a buffer-ethanol system. The anti-inflammatory effect was measured by the inhibition of an edema formation on a mouse ear, induced by 12-o-tetrade-canoyl-phorbol-13-acetate. The cassumunarins showed greater activity than curcumin in both assays.

Masuda and Jitoe (1995) isolated four phenylbutanoid monomers from the fresh rhizomes of *Z. cassumunar* from Indonesia as follows:

1. (*E*)-4-(4-hydroxy-3-methoxyphenyl)but-2-en-1-ylacetate
2. (*E*)-4-(4-hydroxy-3-methoxyphenyl)but-2-en-1-ol
3. (*E*)-2-hydroxy-4-(3,4-dimethoxyphenyl)but-3-1-ol
4. (*E*)-2-methoxy-4-(3,4-dimethoxyphenyl)but-3-en-1-ol

In addition, Masuda and Jitoe (1995) also isolated three phenylbutanoid monomers that are already known.

Nugroho et al. (1996) screened the rhizomes of 18 species for insecticidal activity, and the rhizomes of *Kaempferia rotunda* and *Z. cassumunar* exhibited a marked insecticidal activity in chronic feeding bioassays at concentrations of 2500 and 1250 ppm, respectively. Bioassay-guided isolation led to two phenylbutanoids from the rhizomes of *Z. cassumunar* (an LC_{50} of 121 and 127 ppm). Both compounds were active in the residue-contact bioassay (LC_{50} of 0.5 and 0.36 µg/cm^2). The presence of oxygenated substitutes in the side chain nullified the insecticidal activity.

Panthong et al. (1997) assayed the anti-inflammatory activity of the compound D ((*E*)-4-(3'4'-dimethoxy-phenyl)but-3-en-2-ol) isolated from the hexane extract of *Z. cassumunar* rhizome using various inflammatory models in comparison with aspirin, indomethacin (indimetacin), and prednisolone. The results showed that the anti-inflammatory effect of compound D mediated prominently on the acute phase of inflammation. It exerted a marked inhibition of carrageenan-induced rat paw edema, exudate formation, leukocyte accumulation, and prostaglandin biosynthesis in carrageenan-induced rat pleurisy. Compound D possessed only a slight inhibition of both the primary and secondary lesions of adjuvant-induced arthritis and had no effect on cotton-pellet-induced granuloma in rats. Compound D elicited analgesic activity when tested on the acetic acid-induced writhing response in mice but had weak inhibitory activity on the tail flick responding to radiant heat. Compound D possessed marked anti-pyretic effect when tested on yeast-induced hyperthermia in rats.

Nagano et al. (1997) investigated the effect of cassumunarins A and B, isolated from *Z. cassumunar*, in dissociated rat thymocytes suffering from oxidative stress induced by 3 mM H_2O_2 and ethidium bromide by using flow cytometry. The effects were then compared with those of curcumin. The pretreatment of rat thymocytes with cassumunarins (100 and 3.0 µM) dose dependently prevented H_2O_2-induced decrease in cell viability. Cassumunarins were also more effective when administered before the start of the oxidative stress. The respective potencies of cassumunarins A and B in protecting the cells suffering from H_2O_2-induced oxidative stress were greater than that of curcumin.

Bordoloi et al. (1999) investigated the essential oil composition of *Z. cassumunar* from northeast India using GC-MS, analyzing oil hydrodistilled from rhizomes and leaves. The rhizome essential oil contained terpenen-4-ol (50.5%), *E*-1-(3,4-dimethoxyphenyl)buta-1,3-diene (19.9%), *E*-1-(3,4-dimethoxyphenyl)but-1-ene (6.0%), and β-sesquiphellandrene (5.9%) as major constituents out of the 21 compounds identified. In the leaf essential oil, 39 compounds were identified. The main components were 1(10),4-furanodien-6-one (27.3%), curzerenone (25.7%), and β-sesquiphellandrene (5.7%).

Zingiber zerumbet *(L.) Smith*

This is native of tropical Asia, occurring at an altitude of 1200 m above MSL. It is commonly known as "Shampoo Ginger;" it is also known as "Pinecone Ginger" in the southern United States. The inflorescence resembles a tight pinecone and releases a thick juice when squeezed. This juice is used to make Paul Mitchell and Freemans shampoos.

Pseudostems grow 0.6–2.0 m, the leaves are sessile or petiolate, and the rhizomes are light yellow inside. The inflorescence, produced at the tip of a long peduncle, is green when young. The flowers are white in color and three to four are produced at a time. The color of the inflorescence changes from green to red on aging and lasts many weeks. Hence, it is widely used as a cut flower for decorative purposes. Capsule is ellipsoid, seeds are black. Some of the variegated forms of this species are also grown in gardens. The variegated form, called *Darceyi*, is very popular in the United States. Another form, called "Twice as Nice," produces both basal inflorescences and occasional terminal spikes on a very compact plant.

Srivastava (2003) studied the pharmacognosy of this species on which the following discussion is based. The rhizome is 7–15 cm long, 1–2.5 cm broad, and is irregularly branched. In commerce, the rhizome is found in pieces 4–7 cm long, which are irregular, wrinkled, brown, and show a large central pith. The dried rhizome is hard, brittle, with a fragrant odor and an aromatic, spicy taste. A transverse section of the rhizome shows a single, layered epidermis, below which are 7–10 layers of thin-walled cork cells. The cortex consists of several layers of parenchymatous cells, with intercellular air spaces, and oil cells are present. The endodermis is made of a single layer of cells. The stele consists of a broad central zone of ordinary parenchymatous cells. Closed collateral vascular bundles are found in a circle just inside the endodermis. Throughout the remaining region of the stele, closed collateral bundles, partially covered by sclerenchymatous fibers are scattered. The tracheids are nonlignified and have reticulate, spiral, or scalariform thickening on the wall. The very tender rhizomes are eaten. The decoction of the rhizomes is used to cure various kinds of diseases.

Tewtrakui et al. (1997) analyzed the water-distilled volatile oil composition of *Z. zerumbet* by GC and MS. The main component of the volatile oil was found to be zerumbone (8-oxohumulene) (56.48%). Other components in significant amounts were 1,8-cineole (1.07%), *o*-caryophyllene (2.07%), α-humulene (25.70%), caryophyllene oxide (1.41%), humulene epoxide (3.62%), and humulene epoxide 11 (2.45%). Also, 3,4-*o*-diacetylafzelin and zerumbone epoxide were isolated from rhizomes (Rastogi and Mehrotra, 1993).

The most investigated chemical compound is the sesquiterpene zerumbone, which is reported to be a potent inhibitor of a tumor promoter, the

12-*o*-tetradecanoyl-phorbol-13-acetate (TPA)-induced Epstein–Barr virus activation. The IC_{50} value of zerumbone (0.14 μM) was noticeably lower than those of the antitumor promoters hitherto obtained (Murakami et al., 1999). Murakami et al. (2002) reported that zerumbone has potent anti-inflammatory and chemo-preventive qualities. They found that zerumbone effectively suppressed TPA-induced superoxide anion generation from both nicotinamide adenine dinucleotide (reduced) (NDAH)-oxidase in dimethyl sulfoxide-differentiated HL-60 human acute promyelocytic leukemia cells and xanthine oxidase in AS 52 Chinese hamster ovary)-2, together with the release of tumor necrosis factor (α) in RAW 264.7 mouse macrophages, were also markedly diminished. These suppressive effects were accompanied with a combined decrease in the medium concentrations of nitrite and prostaglandin E_2, whereas the expression level of Cox-1 was unchanged.

Zerumbone inhibited the proliferation of human colonic adenocarcinoma cell lines in a dose-dependent manner, whereas the growth of the normal human dermal (2FO-C25) and colon (CCD-1 8CO) fibroblasts was less affected. It also induced apostosis in COLO205 cells, as detected by the dysfunction of the mitochondria transmembrane, the Annexin N-detected translocation of phosphatidyl serine, and the chromation condensation. α-Humulene, a structural analog lacking only the carbonyl group in zerumbone, was virtually inactive in all the experiments, indicating that the α, β-unsaturated carbonyl group in zerumbone may play some pivotal role in the interactions with an unidentified target molecule. The results strongly support the claim that zerumbone is a food phytochemical (nutraceutical), which has a potential use in anti-inflammation, chemoprevention, and chemotherapy strategies. Dai et al. (1997) reported that zerumbone has a potent HIV-inhibitory action.

Zingiber americanus *BI.*

Z. americanus is a medicinally important spice species. The rhizome is small, yellow, hard, weakly fragrant, and bitter. The propagation is done through rhizome cuttings. It prefers shady, humid soils, rich in humus. It grows wild in teak forests of Southeast Asia, where it is used for several medicinal purposes. The old rhizomes are used as an ingredient in various traditional medicines. The pounded rhizome is usually used as a poultice for women after childbirth. It has gained importance as an attractive garden plant and is grown widely in the United States. The young rhizomes are eaten as a vegetable in Java (Prance and Sarket, 1997).

Zingiber aromaticum *Val.*

This plant grows up to a height of about 1.5 m. The yellow flowers come from a striking red cone produced from the base of the plant. The rhizome is strongly aromatic and fibrous, resembling *Z. americanus* in taste and aroma. The specific epithet is derived from the strong aroma of the rhizome. It is considered to be a native of tropical Asia and is called *Puyung in* Indonesia. *Z. aromaticum* is also widely cultivated in kitchen gardens and also as an ornamental plant. Its fresh and tender shoots and flowers are eaten and used to flavor foods. The rhizomes are used as an ingredient in folk medicines as well as a poultice (Prance and Sarket, 1997). From the rhizome of this species, zerumbone and 3″ c,4″-o-diacetylafzelin were isolated. Zerumbone exhibited HIV-inhibitory and cytotoxic activities (Dai et al., 1997).

Zingiber argenteum *J. Mood and I. Theilade*

This species is endemic to Sarawak, Malaysia. *Z. argenteum* is a small plant, the pseudostem reaching about 75 cm. Its leaves grow to 15–18 cm in length, with an upper surface that is silvery green with a dark green cloud along the midrib, while the lower surface is green. It produces cream-colored flowers and spikes 8–9 cm long which are broadly elliptic. Its bracts are bright orange colored, with the lower ones turning red. This species is related to *Z. coloratum* and *Z. lambi*. It is a very attractive plant and is valued much by ginger lovers. Under cultivation, the plant becomes highly floriferous (Theilade and Mood, 1997).

Zingiber bradleyanum *Craib.*

This plant is now cultivated in the United States and is mainly grown for its foliage; its leaves have a beautiful silvery stripe along the midrib. The plant has a natural dormancy and therefore may be winter-hardy in subtropical climates.

Zingiber chrysanthum *Rosc.*

This is a medium-sized *Zingiber* with stems about 1–1.5 m tall. It produces a typical basal inflorescence, but the individual flowers are long-lasting and very colorful, with a spotted lip. The seed capsules are also ornamental bright red in color and remain on the stalk until the stalks wither in winter.

Zingiber citriodorum *J. Mood and I. Theilade*

This species is from Thailand and has been under cultivation for several years in the United States as an ornamental plant under the name "Chiang Mai Princess." It was recently described as a new species, *Z. citriodorum*. It is a beautiful plant, with sharply pointed bracts starting out green and maturing to bright red color. The flowers are white. The pseudostems and foliage have a silvery gray color. This species can be difficult to grow in cool climates, as the rhizomes are subject to rotting during dormancy if kept too wet. This is a valuable ornamental plant used as a potted plant or as a cut flower for decorative purposes.

Zingiber clarkii *King ex Benth*

Z. clarkii is a native of the Sikkim Himalayas of India and has been adopted as a valuable ornamental plant in the subtropical and temperate countries. The plants are tall with a foot-long inflorescence which appears from the main stem rather than from the ground. The bracts form a tight, cone-shaped spike, and the individual flowers are dull yellow in color, with a dotted lip. Among the *Zingiber* species, this is unique because the spike is produced laterally and not radially as in the other species.

Zingiber collinsii *J. Mood and I.Thaleide*

A recently described beautiful *Zingiber* species, *Z. collinsii*, was discovered and introduced by Mark Collins. This one has silvery stripes across the leaves, some-what similar to *Alpinia pumila*, but much taller in height. This species has become a favorite of plant lovers in the United States and Europe.

Zingiber corallinum *Hance*

This *Zingiber* species is a medium-sized plant producing long, pointed, red-bracted inflorescences near the ground. The foliage is medium green in color. This species has been proven to be winter-hardy, down to 20°F. The rhizome is used in tradi-tional Chinese medicine. The effect of this species in preventing skin invasions by *Schistosma japonicum cercaria* was investigated in mice. A worm reduction rate of 91% was found at 5% concentration of the rhizome extract (Shuxuan et al., 2001). *Z. corallinum* is valued as an ornamental plant as well as a medicinal spice of immense value.

Zingiber eborium *J. Mood and I. Theilade*

This species is endemic to Borneo in Indonesia and is the unique white ginger, ivory ginger, or ivory spike ginger, in the parlance of nurserymen. It grows about 0.75–1 m in height, with dark green, ovate leaves, glabrous above, and with fine silky hairs below. It has a pale reddish scape sheath, with an inflorescence about 7–8 cm long. Its bracts are ivory white in color and flowers are orange colored. The white spike and the orange flower make this a very attractive garden plant and a favorite among the ginger lovers in the Western world. Under cultivation, the plant flowers profusely, making it a valuable potted plant (Theilade and Mood, 1997). It tolerates freezing.

Zingiber griffitbii *Baker*

This is a Malaysian species, having oblong, glabrous leaves, a short peduncled cylin-drical spike, and bright red, obovate, obtuse bracts. The tip of the bract is yellowish white and three lobed. It is a common ornamental plant. Zakaria and Ibrahim (1986) reported four phenolic compounds of the catechol and pyrogallol type, three flavo-noids, and terpenols in its volatile oil.

Zingiber graminuem *Noronha*

This plant is cultivated in US gardens under the common name "Palm Ginger." It is a thin-stemmed and narrow-leafed plant.

Zingiber lambi *J. Mood and I. Theilade*

This is a Malaysian ginger endemic to the eastern region of the country. It is a small plant with a leafy stem about 60 cm tall. Its leaves are glabrous, with a silvery green upper surface, ribbed side nerves, and a green lower surface. Its fusiform spike grows up to 10–12 cm in length. The bracts are orange colored, greenish toward the apex, and they turn pink with age. This species produces yellow flowers and a light yellow labellum. Z. *lambi* is unusual for its beautiful silvery green leaves, orange spikes, and yellow flowers. It has become a valuable garden plant within a short span of time. The species is related to Z. *argenteum*.

Zingiber longipedunculatum *Ridley*

Reportedly this plant has been in cultivation in Australia for many years and has been a valuable garden plant used for cut-flower purposes. The spikes are often used in floral arrangements.

Zingiber malaysianum *C.K. Lim*

This garden plant has become extremely popular and is widely sold in the United States under the name "Midnight Beauty." It has shiny dark brown, almost black leaves and produces bright red inflorescences at its base. It is an evergreen species and must be protected in winter from frost. It is a very attractive ornamental plant and is much in demand by plant lovers to be used as a potted plant or in garden beds. Its spikes are used as cut flowers and in floral arrangements.

Zingiber neglectum *Valet*

This is another popular plant in cultivation in the United States, one of the most sought-after plants by ginger plant collectors. It has very long inflorescences with beautiful cup-shaped bracts similar to *Zingiber spectabile*, with purple flowers.

Zingiber niveum *J. Mood and I. Theilade*

Zingiber niveum is an ornamental ginger plant sold commonly in the United States for several years under the name "Milky Way." It produces milky white, rounded spikes that look very unusual, with yellow flowers. The stems and leaves are silvery gray, and this attractive ornamental plant grows to a height of about only 3 ft.

Zingiber attensi *Valet*

This is a native of Southeast Asia. The stem is reddish and attractive, and hence is widely cultivated as an ornamental. The inflorescence is more or less similar to Z. *zerumbet*, but the color is red from the beginning and it persists for a long time. The peroxidase isozyme studies have proven that this is very close to Z. *zerumbet* and Z. *montanum*.

The rhizome is used as a poultice in postnatal treatment and also as an appetizer. Three sesquiterpenes—humulene, humulene epoxide, and zerumbone—and a diterpene, (E)-landa-8 (17), 12-diene-, 15,16-diel-, were isolated from dried rhizome (Sirat, 1994). It is a valuable ornamental plant, both as a potted one and also as a cut flower.

Zingiber pachysiphon *B.L. Burtt and R.M. Sm*

This is in cultivation in Australia. The plant has a beautiful purplish-colored inflorescence with white edges to the bracts. The species is rather rare and valued much by the ginger plant lovers as a very attractive potted plant.

Zingiber rubens *Roxb.*

This species is from the Indo-Malaysian region and has been introduced to the gardens in the United States as a potted plant, where it has become very popular. The plant is about 6–8 ft tall, leaves are 4–5 in. long, which are pubescent beneath the surface. Its spike is dense and globose, with a small peduncle, bright red bracts, red corolla segments, with an oblong lip that is much spotted and streaked with red color. The paste of the rhizome (10 g) is given after heating and is also applied to the forehead of those who suffer from giddiness (Saxena et al., 1981).

Zingiber spectabile *Griff.*

This is also known as "Beehive Ginger," owing to the peculiar shape of the spike. The plants are large, grow up to a height of 8 ft, and the inflorescence is large and very attractive. The spike turns from yellow to red on aging. As it has a shelf life of a few weeks, it is widely used as a cut-flower plant. It is also used as a flavoring agent and an ornamental plant (Holttum, 1950). This species is widely used in Malay traditional medicine (Burkill and Haniff, 1930) and is very popularly cultivated in the United States. It comes in two varieties, one with red bracts, and the other with golden yellow bracts. The latter is often sold as "Golden Shampoo Ginger." Both are very much valued as potted plants and for making ornamental cut flowers.

Ibrahim and Zakaria (1987) analyzed the rhizome using TLC and GC and reported 18 compounds, the major ones being *trans*-D-bergamontene, β-elemene, isobutyl benzoate, β-copaene, and β-terpenol.

Zingiber vinosum *J. Mood and I. Theilade*

This species of ginger is a native of Salah, East Malaysia, and its cultivation as an ornamental plant has spread rapidly in the United States. It is a short-statured plant with dark green upper foliage that is deep maroon on the underside. Its leafy shoot grows to about 1–1.25 m long and is dark burgundy at the base. Its spike can be 15–30 cm long, and the bracts are burgundy. The flowers are white, with a snow-white labellum. The attractive foliage and red inflorescence make this species a highly popular garden plant. The rhizome is moderately aromatic.

In addition to the aforementioned species, there are also others which can be groomed into attractive garden, potted, and cut-flower plants.

Wild Ginger

In the United States and Europe, the term "Wild Ginger" commonly denotes *Asarum canadense* (Aristolochiaceae), which is not related to ginger in any way. It is an inconspicuous, herbaceous perennial, about 30 cm long in height, found growing in rich soil on roadsides and in the woods. The plant is almost stemless, usually possessing two heart-shaped leaves and carrying a solitary bell-shaped flower between the two petioles at the base. The root stock is yellowish and creeping in structure and is sold in pieces 10–12 cm long. The plant has brownish ends that are wrinkled on the outside, whitish inside, fragrant, aromatic, spicy, and slightly bitter in taste (Grieve, 2003).

It is called "Wild Ginger" because it emanates a ginger-like smell. Native Americans have used the root to flavor foods, just like the true ginger. They have also used the root to treat digestive ailments, especially the problem of gas formation in the stomach, and it is also used as a poultice on skin sores. It is often used to promote sweating, which reduces fever, and also to counteract cough and sore throat. The plant extract has also been shown to have antimicrobial properties (Reed, 2003).

References

Abe, M., Ozawa, Y., Uda, Y., Yamada, Y., Morimitsu, Y., Nakamura, Y., et al., 2002. Labdane-type diterpenoid dialdehyde, pungent principle of myoga. *Zingiber mioga* Roscoe. Biosci. Biotech. Biochem. 66, 2698–2700.

Clark, R.J., Warner, R.A., 2000. Production and marketing of Japanese ginger (*Zingiber mioga*) in Australia RICRDC Pub. 00/117. Rural Ind. Res. & Dev. Corporation, Australia, <http://www.rirdc.gov.au/reports/AFO/00-117.sum.html> (accessed 05.10.03.).

Barghava, N., 1981. Plants in folk life and folkore in Andaman and Nicobar Islands. In: Jain, S.K. (Ed.), Glimpses of Indian Ethnobotany. Oxford and IBH Publishing Co, New Delhi, pp. 329–344.

Bordoloi, A.K., Sperkova, J., Leclercq, P.A., 1999. Essential oil of *Zingiber cassumunar* Roxb. from Northeast India. J. Ess. Oil Res. 11, 441–445.

Burkill, T.H., Haniff, M., 1930. The Malay village medicines. Gard. Bull. Sing. 6, 262–268.

Dai, J.R., Cardelina, J.H., McMahan, J.B., Boyd, H.R., 1997. Zerumbone, an HIV-inhibitory and cytotoxic sesquiterpene of *Zingiber aromaticum* and *Zingiber zerumbet*. Nat. Prod. Lett. 10, 115–118.

Grieve, M.B., 2003. A modern herbal. <http://www.botanical.com/botanical/mgmh/g/ginwil14.html> (accessed 10.05.03).

Holttum, R.E., 1950. Zingiberaceae of the Malay Peninsula. Gard. Bull. Sing. 13, 1–249.

Ibrahim, H., Zakaria, M.B., 1987. Essential oils from three Malaysian Zingiberaceae species. Malays. J. Sci. 9, 73–76.

Masuda, T., Jitoe, A., 1995. Pheylbutenoid monomers from the rhizomes of *Zingiber cassumunar*. Phytochemistry 39, 459–461.

Masuda, T., Jitoe, A., Mabry, M.J., 1995. Isolation and structure determination of cassumu-
    narins A, B and C: new anti-inflammatory antioxidants from a tropical ginger, *Zingiber
    cassumunar*. J. Am. Oil Chem. Soc. 72, 1053–1057.
Murakami, A., Takahashi, M., Jiwajinda, S., Koshimizu, K., Ohigashi, H., 1999. Identification of
    zerumbone in *Zingiber zerumbet* Smith as a potent inhibitor of 12-0-tetradecanoylphorbol-
    13-acetate-induced Epstein–Barr virus activation. Biosci. Biotech. Biochem. 63, 1811–1812.
Murakami, A., Takahashi, D., Kiroshita, T., Koshimizu, K., Kim, H.W., Yoshihiro, A., et al.,
    2002. Zerumbone, a Southeast Asian ginger sesquiterpene, markedly suppresses free
    radical generation, pro-inflammatory protein production, and cancer cell proliferation
    accompanied by apostasies: the α and β-unsaturated carbonyl group is a prerequisite.
    Carcinogenesis 23, 795–802.
Nagano, T., Oyama, Y., Kajita, N., Chikahisa, I., Nakata, M., Ozaki, E., et al., 1997. New cur-
    cuminoids isolated from *Zingiber cassumunar* protect cells suffering from oxidative stress: a
    flow-cytometric study using rat-thymocytes and H_2O_2. Jpn. J. Pharmacol. 75, 363–370.
Nugroho, E.W., Schwarz, B., Wray, B., Proksch, P., 1996. Insecticidal constituents from rhi-
    zomes of *Zingiber cassumunar* and *Kaempferia rotunda*. Phytochemistry 41, 129–132.
Ozaki, Y., Kawahara, N., Harada, M., 1991. Anti-inflammatory effect of *Zingiber cassumunar*
    Roxb. and its active principles. Chem. Pharm. Bull 39, 2353–2356.
Paghat, 2003. Paght's garden: *Zingiber mioga* "Dancing Crane" <http://www.paghat.com/
    japaneseginger.html> (accessed 15.10.03).
Panthong, A., Kanhanapothi, D., Niwatananant, W., Tuntiwachurittikul, P., Reutrakul, V.,
    1997. Anti-inflammatory activity of compound D (E)-4-(3′4 ′-dimethoxyphenyl)but-3-
    en-2-ol) isolated from *Zingiber cassumunar* Roxb. Phytomedicine 4, 207–212.
Poosapaya, P., Kraisintu, K., 2003. Micropropagation of *Zingiber cassumunar* Roxb. ISHS.,
    <http://www.Acta.Hort.org/books/344/344-64.htm> (accessed 10.05.03.).
Prance, M.S., Sarket, D., 1997. Root and Tuber Crops. SDE-40, Bogor, Indonesia.
Quisumbing, E., 1978. Medicinal Plants of Philippines. JMC Press Inc, Queen City,
    Philippines, pp. 186–202.
Rastogi, R.P., Mehrotra, B.N., 1993. Compendium of Indian Medicinal Plants, 1980–1984,
    vol. 3. NISCOM, New Delhi.
Reed, D., 2003. Wild ginger (*Asarum canadense*). Wild flowers of the Southern United States
    <http://2bnthewild.com/plants/H36.htm> (accessed 10.05.2003).
Saxena, H.O., Braamam, M., Dutta, P.K., 1981. Ethnobotanical studies in Orissa. In: Jain,
    S.K. (Ed.), Glimpses of Indian Ethnobotany Oxford and IBH Publishing Co, New Delhi,
    pp. 232–244.
Shuxuan, J., Xiu Qin, H., Zheng, X., Li Mei, M., Ping, L., Juan, D., et al., 2001. Experimental
    study on *Zingiber corallium* Hance to prevent infection with *Schistosoma japonicum*
    ceraria. Chin. J. Schisto. Control 13, 170–172.
Sirat, H.M., 1994. Study on the terpenoids of *Zingiber ottensii*. Planta Med. 60, 497.
Srivastava, A.K., 2003. Pharmacognostic studies on *Zingiber zerumbet* (L.) Sm. Aryavaidyan
    16, 206–211.
Sterling, K.J., Calrk, R.J., Brown, P.H., Wilson, S.J., 2003. Effect of photoperiod on flower
    bud initiation and development in myoga (*Zingiber mioga* Roscoe.). <http://agsci.eliz.
    tased.edu.au/jgp/jgp7.htm> (accessed 10.05.03).
Tewtrakui, S., Sardsangjun, C., Itchaypruk, J., Chaitongruk, P., 1997. Studies on volatile oil
    components in *Zingiber zerumbet* rhizomes. Songklanakarin J. Sci. Technol. 19, 197–202.
Theilade, I., Mood, J., 1997. Two new species of *Zingiber* (Zingiberaceae) from Sabah,
    Borneo. Sandakania 9, 21–26.
Zakaria, M.B., Ibrahim, H., 1986. Phytochemical screening of some Malaysian species of
    Zingiberaceae. Malays. J. Sci. 8, 125–128.

Printed in the United States
By Bookmasters